Doppler Radar and
Weather Observations

DOPPLER RADAR
AND WEATHER
OBSERVATIONS

Second Edition

Richard J. Doviak

Dušan S. Zrnić

National Severe Storms Laboratory
National Oceanic and Atmospheric Administration
Norman, Oklahoma

and

Departments of Electrical Engineering and Meteorology
University of Oklahoma
Norman, Oklahoma

ACADEMIC PRESS, INC.
Harcourt Brace Jovanovich, Publishers
San Diego   New York   Boston
London   Sydney   Tokyo   Toronto

Academic Press, Inc.
1250 Sixth Avenue, San Diego, California 92101-4311

United Kingdom Edition published by
Academic Press Limited
24–28 Oval Road, London NW1 7DX

Library of Congress Cataloging-in-Publication Data

Doviak, R. J.
    Doppler radar and weather observations / Richard J. Doviak and
Dusan S. Zrnic. – 2nd ed.
        p.   cm.
    Includes bibliographical references and index.
    ISBN 0-12-221422-6 ✓
    1. Radar meteorology–Observations.   2. Doppler radar.   I. Zrnic
Dusan S.   II. Title.
QC973.5.D68   1992
551.6'353–dc20

                                                            92-23494
                                                            CIP

PRINTED IN THE UNITED STATES OF AMERICA

93 94 95 96 97 98 QW   9 8 7 6 5 4 3 2 1

Contents

Preface   xiii

Preface to the First Edition   xv

List of Symbols   xvii

1. Introduction   1
  1.1   Historical Background   1
  1.2   The Plan of the Book   7

2. Electromagnetic Waves and Propagation   10
  2.1   Waves   10
  2.2   Propagation Paths   14
    2.2.1   Refractive Index of Air   14
    2.2.2   Refractivity N   16
    2.2.3   Spherically Stratified Atmospheres   18
    Problems   28

3. Radar and Its Environment   30
  3.1   The Doppler Radar (Transmitting Aspects)   30
    3.1.1   The Electromagnetic Beam   32
    3.1.2   Antenna Gain   34
  3.2   The Scattering Cross Section   35
  3.3   Attenuation   38
    3.3.1   Attenuation in Rain   39
    3.3.2   Attenuation in Clouds   43
    3.3.3   Attenuation in Snow   44
    3.3.4   Attenuation in Gases   44

3.4   The Doppler Radar (Receiving Aspects)   45
       3.4.1   The Radar Equation   46
       3.4.2   The Incoherent Receiver   48
       3.4.3   The Coherent Receiver (In-Phase and Quadrature
                Components)   50
3.5   Practical Considerations   54
       3.5.1   System Noise Temperature   54
       3.5.2   Bandwidth   57
       3.5.3   Filtered Waveform   58
       3.5.4   Signal-to-Noise Ratio, Matched Filters   60
3.6   Ambiguities   60
       Problems   62

4. Weather Signals   64
4.1   Weather Signal Samples   64
4.2   The Power Sample   67
4.3   Signal Statistics   69
4.4   The Weather Radar Equation   72
       4.4.1   Receiver Calibration   74
       4.4.2   The Range-Weighting Function   75
       4.4.3   Finite Bandwidth Power Loss   79
       4.4.4   The Resolution Volume   80
       4.4.5   Reflectivity Factors   82
4.5   Signal-to-Noise Ratio for Distributed Scatterers   83
4.6   Correlation of Samples along Range Time   84
       Problems   85

5. Doppler Spectra of Weather Signals   87
5.1   Spectral Analysis of Weather Signals   87
       5.1.1   Discrete Fourier Transform   87
       5.1.2   Convolution and Correlation   92
       5.1.3   Power Spectrum of Random Sequences   95
       5.1.4   Bias, Variance, and the Window Effect   98
       5.1.5   Expressing Spectral Estimates in Terms of the True
                Spectrum   100
       5.1.6   Variance of the Periodogram   105
5.2   Weather Signal Spectrum and Its Relation to Reflectivity and Radial
       Velocity Fields   106

5.2.1   Power Spectrum for Uniform Shear and Reflectivity   110

5.2.2   Contributions of Independent Meteorological Mechanisms to the Power Spectrum   112

5.2.3   Probability Distribution of Turbulent Velocities Related to the Power Spectrum   115

5.3   Velocity Spectrum Width   116

Problems   118

6. Weather Signal Processing   122

6.1   Spectral Moments   122

6.2   Weather Signals in a Receiver   123

6.3   Signal Power Estimation   125

6.3.1   Sample Time Averaging   125

6.3.2   Range Time Averaging   129

6.4   Mean Frequency Estimators   130

6.4.1   Autocovariance Processing: The Pulse Pair Processor   131

6.4.2   Spectral Processing   135

6.5   Estimators of the Spectrum Width   136

6.5.1   Autocovariance Processing   136

6.5.2   Spectral Processing   140

6.6   Minimum Variance Bounds   141

6.7   Performance on Data   143

6.8   Signal Processing for Coherent Polarimetric Radar   145

6.8.1   Reflectivity and Differential Reflectivity   146

6.8.2   Mean Radial Velocity and Differential Propagation Phase   147

6.8.3   Specific Differential Phase   153

6.8.4   Spectrum Width   155

6.8.5   Correlation Coefficient   155

6.8.6   A Signal Processing Scheme for Echoes with Alternating Polarization   157

6.9   Concluding Remarks   157

Problems   158

7. Considerations in the Observation of Weather   160

7.1   Range Ambiguities   160

7.2   Velocity Ambiguities   164

7.3   Signal Coherency   165

7.4   Techniques to Mitigate the Effects of Ambiguities   167
    7.4.1   Phase Diversity   167
    7.4.2   Spaced Pairs with Polarization Coding   170
    7.4.3   Staggering the PRT to Increase the Unambiguous
           Velocity   171
    7.4.4   Interlaced Sampling   175
    7.4.5   Correcting Aliased Velocities   177
7.5   Methods to Decrease the Acquisition Time   179
    7.5.1   Frequency Diversity   180
    7.5.2   Random Signal Transmission   181
7.6   Pulse Compression   184
7.7   Artifacts   187
    7.7.1   Quantization and Saturation Noises   188
    7.7.2   Amplitude and Phase Imbalances   190
    7.7.3   Phase Jitter   192
7.8   Effective Pattern of a Scanning Radar   193
7.9   Antenna Sidelobes   197
7.10   Clutter   199
    7.10.1   Ground Clutter and Its Suppression   200
    7.10.2   Other Clutter   206
    Problems   207

8. Precipitation Measurements   209
8.1   Drop Size Distributions   210
    8.1.1   Cloud Drop Size Distributions   210
    8.1.2   Raindrop Size Distributions   212
    8.1.3   Hailstone Size Distributions   215
8.2   Terminal Velocities   216
8.3   Rainfall Rate, Reflectivity, and Liquid Water Content   218
    8.3.1   Liquid Water Content   218
    8.3.2   Reflectivity Factor Z   219
    8.3.3   Rainfall Rate   222
8.4   Single-Parameter Measurement of Precipitation   223
    8.4.1   Reflectivity Factor Method   223
    8.4.2   Attenuation Method   231
    8.4.3   Differential Phase Method   234
8.5   Multiple-Parameter Measurements of Precipitation   235

       8.5.1   *Dual Wavelength*   236

       8.5.2   *Polarization Diversity*   239

       8.5.3   *Application of Dual Polarization*   252

       8.5.4   *Rain Gauge and Radar*   271

  8.6   Distribution of Hydrometeors from Doppler Spectra   274

      Problems   278

9. Observations of Winds, Storms, and Related Phenomena   280

  9.1   Thunderstorm Structure   281

  9.2   Wind Measurement with Two Doppler Radars   288

       9.2.1   *Wind Field Synthesis*   289

       9.2.2   *Supercell Thunderstorms*   293

       9.2.3   *Ordinary Thunderstorms*   301

  9.3   Wind Measurement with One Doppler Radar   304

       9.3.1   *Linear Wind Fields*   306

       9.3.2   *Uniform Wind along a Circular Arc*   310

       9.3.3   *Linear Wind over a Circle (VAD)*   311

       9.3.4   *Prestorm Application*   317

       9.3.5   *Large-Scale Horizontal Wind*   323

       9.3.6   *Large-Scale Vertical Wind*   325

       9.3.7   *Flow Models*   328

  9.4   Severe Storms   328

  9.5   Mesocyclones and Tornadoes   335

       9.5.1   *Smoothing of a Vortex Signature*   338

       9.5.2   *High-Resolution Images of a Tornado*   340

       9.5.3   *Doppler Spectra of Tornadoes*   344

  9.6   Downdrafts and Outflows   351

       9.6.1   *Density Currents*   352

       9.6.2   *Convergence Bands*   357

       9.6.3   *Downbursts and Microbursts*   359

  9.7   Buoyancy Waves   363

       9.7.1   *Observation in a VAD*   364

       9.7.2   *Large-Amplitude Buoyancy Waves*   365

  9.8   Large Weather Systems   371

       9.8.1   *Mesoscale Convective Systems*   371

       9.8.2   *Hurricanes and Typhoons*   376

    9.8.3   Cold Fronts and Dry Lines   378

9.9   Lightning   380

    9.9.1   Physical Characteristics Determined by Radar   380

    9.9.2   Lightning and Storm Structure   382

    Problems   383

10. Measurements of Turbulence   386

10.1   Statistical Theory of Turbulence   386

    10.1.1   Turbulence Spectra and the Correlation Function   386

    10.1.2   Structure Functions, Locally Homogeneous Fields   391

    10.1.3   Structure and Spectral Functions, Locally Isotropic Fields   393

    10.1.4   Chandrasekhar's Theory   395

10.2   Spatial Spectra of Point and Average Velocities   398

    10.2.1   Filtering by the Weighting Function   398

    10.2.2   Variance of Point and Average Velocities   403

    10.2.3   Turbulence Parameters from a Single Radar   404

    10.2.4   Turbulence Parameters from Two Doppler Radars   407

10.3   Doppler Spectrum Width and Eddy Dissipation Rate   408

10.4   Doppler Spectrum Width in Severe Thunderstorms   410

    Problems   422

11. Observations of Fair Weather   424

11.1   Reflection, Refraction, and Scatter: Coherence   424

11.2   Formulation of the Wave Equation for Inhomogeneous and Turbulent Media   426

11.3   Solution for Fields Scattered by Irregularities   430

11.4   Small Volume Scatter   435

    11.4.1   Bragg Scatter   437

    11.4.2   Radio Acoustic Sounding System (RASS)   440

    11.4.3   Expected Scattered Power Density   443

11.5   Common Volume Scatter   452

    11.5.1   Correlation Length Shorter Than the Fresnel Length   453

    11.5.2   Correlation Length Comparable to or Larger Than the Fresnel Length   455

    11.5.3   The Spectral Representation   460

    11.5.4   Scattering from Anisotropic Irregularities, an Example   463

11.6   Characteristics of Refractive Index Irregularities   464

   11.6.1   Dependence of the Structure Parameter on the Height   469

   11.6.2   Inertial Subrange   475

   11.6.3   Criteria for Detection of Refractive Index Irregularities   477

11.7   Observations of Clear-Air Reflectivity   479

11.8   Observations of Wind, Waves, and Turbulence in Clear Air   486

   11.8.1   Wind Profiling   487

   11.8.2   Kinematic Structure of the Convective Boundary Layer   496

11.9   Other Fair-Weather Observations   503

   Problems   505

APPENDIX A Geometric Relations for Rays in the Troposphere   507

   A.1   Integral Solution for Ray Path in a Spherically Stratified Medium   507

   A.2   Relating a Scatterer's Apparent Range and Elevation Angle to Its True Height and Great Circle Distance   508

APPENDIX B Correlation between Signal Samples as a Function of Sample Time   510

APPENDIX C Correlation of Echoes from Spaced Resolution Volumes   513

   C.1   Signal Sample Correlation versus Range Difference $c\,\delta\tau_s/2$ 513

   C.2   Correlation of Signals from Azimuthally Spaced Resolution Volumes   515

APPENDIX D Geometric Optics Approximation to the Wave Equation   518

APPENDIX E Derivation of Green's Function   520

References   523

Index   547

Preface

Much has happened in operational radar meteorology since the first edition of this book was published. The National Weather Service, the Federal Aviation Administration, and the Air Force Weather Service have begun deploying a network of Doppler weather radars (WSR-88D) to replace the aging noncoherent systems throughout the country. Simultaneously the FAA is acquiring Terminal Doppler Weather Radars (TDWR) to monitor hazards and better manage routing of aircraft into and out of terminals at nearly 50 major airports. The National Oceanic and Atmospheric Administration has just installed a demonstration network of about 30 wind-profiling radars that might lead to improved understanding and short-term forecasts of mesoscale weather phenomena. We have also witnessed new technological developments in airborne and spaceborne radars and lidars, and the emergence of a Radio Acoustic Sounding System (RASS) that measures vertical profiles of temperature. Our revised book is meant to be a reference for users and developers of these systems. The addition of a problem set at the end of each chapter should make this edition better suited for graduate courses on radar meteorology.

As with the previous edition, we have benefited from teaching the course on radar meteorology at the University of Oklahoma. Furthermore, we were privileged to lecture in short courses organized by George Washington University, and to present the material to members of MIT's Lincoln Laboratory and NOAA's Program for Regional Observing and Forecasting Services. These interactions and constructive criticisms by our colleagues led to crucial revisions and numerous clarifications throughout this edition.

The new material, in addition to sets of problems, consists of an expanded Chapter 1, which now contains a short history of radar, sections on polarimetric measurements and data processing, an updated section on RASS, and a section on wind profilers. Furthermore, Chapters 9–11 have been expanded and updated to include new figures of phenomena observed with the WSR-88D. These figures were obtained from NOAA's Operational Support Facility managed by Dr. R. L. Alberty. Dr. V. Mazur provided the material for inclusion in the updated treatment of lightning, and Dr. J. M. Schneider and Mei Xu contributed to the section in Chapter 10 on measurement of turbulence in the planetary boundary layer.

Preface to the First Edition

To be able to observe remotely the internal motions of a tornadic thunderstorm that presents a hazardous threat to human communities is an impressive experience. We were fortunate to have entered the field of radar meteorology at a time when the use of Doppler radar was rapidly growing. Such advances were made possible by the general availability of inexpensive digital hardware, facilitating the implementation of theory developed in the early years of radar. The Doppler weather radar techniques developed by radar engineers and meteorologists may soon find applications in the National Oceanic and Atmospheric Administration's (NOAA) NEXRAD (next generation radar) program. A network of Doppler weather radars is planned to replace the present aging radar system used by the National Weather Service (NWS). Improvements in the techniques to provide warnings of tornadoes and other hazardous phenomena continue to be made. Doppler weather radar has already found a home in several television stations that broadcast early warnings of storm hazards.

To a large extent this book is based on lectures given in a course on radar meteorology taught by both authors at the University of Oklahoma. A considerable portion of Chapter 11 is derived from a graduate course in wave propagation through random media given earlier by R. J. Doviak at the University of Pennsylvania. Material from this book has also been used in a one-week course on radar meteorology offered nationally by the Technology Service Corporation. The opportunities available to us at the National Severe Storms Laboratory were vital for the pursuit of research in the diverse disciplines required to develop and apply Doppler radar in remote sensing of severe thunderstorms. Such research fostered comprehensive and detailed treatment of the theory and practice of radar design, digital signal processing techniques, and interpretation of weather observations.

In this book we have aimed to enhance radar theory with observations and measurements not available in other texts so that students can develop an understanding of Doppler radar principles, and to provide practicing engineers and meteorologists with a discussion of timely topics. Thus we present Doppler radar observations of tornado vortices, hurricanes, and lightning channels. In order to better relate radar observations to weather events commonly observed by eye, radar data fields are correlated with photographs of the physical phenomena such as gust fronts, downbursts, and tornadoes.

While our focus is on meteorology, the theory and techniques developed and discussed here have applications to other geophysical disciplines. Propagation in and scatter from media having a random distribution of discrete targets or from media described by continuous temporal and spatial random variations in their refractive properties also occur in the nonstormy atmosphere and in the ocean, where waves or turbulence, or both, can be generated by intrinsic physical phenomena, by vehicles (such as aircraft, ships, and rockets) traversing the media, or by perturbations purposely created to examine media characteristics. Radar specialists, who often are more interested in the detection and tracking of vehicles and who treat storms as annoying clutter, should find the observations of weather phenomena and the characterization of their properties described herein useful in design studies aimed to maximize target echoes and to minimize interference by weather.

List of Symbols

The following symbols are those used most frequently.

a_e	Effective earth radius
A_e	Effective aperture area of the antenna
B_n	Noise bandwidth
B_6	Receiver–filter bandwidth, 6-dB width
c	Speed of light, 3×10^8 m s^{-1}
C_n^2	Structure parameter of refractive index
D_a	Diameter of the antenna system
D_e	Diameter of an equivalent volume spherical raindrop
D_0	Median volume diameter
E	Electric field intensity
f	Frequency
f_d	Doppler shift
f_N	Nyquist frequency
$f^2(\theta, \phi)$	Normalized one-way power gain of radiation pattern
g	Gravitational constant (9.81 m s^{-2})
g_t, g_r	Power gain of transmitting and receiving antenna
$I(\mathbf{r}, \mathbf{r}_1)$	Weighting function
$I(t)$	In-phase component of the complex signal
k	Specific attenuation (m^{-1})
k	Boltzmann constant [1.38×10^{-23} (W s K^{-1})]
k	Electromagnetic wave number ($2\pi/\lambda$)
k_c	Specific attenuation due to clouds (m^{-1})
k_g	Specific attenuation due to air (m^{-1})
k_r	Wind shear in the r direction
k_θ	Wind shear in the θ direction
k_ϕ	Wind shear in the ϕ direction
K	Specific attenuation (dB km^{-1})
K	Wave number of an atmospheric structure ($2\pi/\Lambda$)
K_{DP}	Specific differential phase (deg km^{-1})
K_r	Specific attenuation due to rain (dB km^{-1})
K_s	Specific attenuation due to snow (dB km^{-1})

l	One-way propagation loss due to scatter and absorption
ln	Natural logarithm
log	Logarithm to base 10
l_r	Finite bandwidth receiver loss factor
L_r	$10 \log l_r$ (dB)
m	Complex refractive index of water
M	Number of signal samples (or sample pairs) along sample time axis
M	Liquid water content
M_I	Number of independent samples
n	Atmosphere's refractive index
N	Refractivity $= (n-1) \times 10^6$
N	White noise power
N_s	Surface refractivity ($N_s = 313$)
$N(D)$	Drop size distribution
P_r	Received signal power
P_t	Peak transmitted power
P_w	Partial pressure of water vapor
$P(\tau_s)$	Instantaneous weather signal power
Q_w	Total water content
$Q(t)$	Quadrature phase component of the complex signal
r	Range to scatterer
r_a	Unambiguous range
r_t	Vortex radius
r_6	6-dB range width of resolution volume
$\mathbf{r}_0$	Vector range to the resolution volume V_6 center
R	Rainfall rate
R	Gas constant for dry air ($287.04 \text{ m}^2 \text{ s}^{-2} \text{ K}^{-1}$)
$R(T_s)$	Autocorrelation of $V(nT_s)$
S	Signal power
$S_n(f)$	Normalized power spectral density
SNR	Signal-to-noise ratio
T	Absolute temperature (K)
T_s	Pulse repetition time (PRT) or sample time interval
T_v	Virtual temperature
v	Velocity
v_a	Unambiguous velocity
v_r	Radial component of velocity (Doppler velocity)
V_6	Resolution volume
$V(kT_s)$	kth complex signal sample
w	Vertical velocity
w_t	Terminal velocity
$W(r)$	Range weighting function
Z	Reflectivity factor
Z_e	Equivalent reflectivity factor

Z_{DR}	Differential reflectivity
α	Antenna rotation rate
δ	Wind direction
ε	Eddy dissipation rate
η	Reflectivity (cross section per unit volume)
η_0	377-Ω space impedance
θ	Angular distance from the beam axis; also, potential energy
θ_e	Elevation angle
θ_1	One-way beamwidth between half-power points
λ	Electromagnetic wavelength
Λ	Structure wavelength of an atmospheric quantity (turbulence)
ρ	Mass density of air
$\boldsymbol{\rho}$	Distance in lag space
$\rho_{hv}(0)$	Correlation coefficient between horizontally and vertically polarized return signals
ρ_w	Density of water
σ_a	Absorption cross section
σ_b	Backscattering cross section
σ_d	Spectrum width due to different fall speeds of hydrometeors
σ_e	Extinction or attenuation cross section
σ_o	Spectrum width due to change in orientation and/or vibration of hydrometeors
σ_s	Spectrum width due to shear
σ_t	Spectrum width due to turbulence
σ_v	Doppler velocity spectrum width
σ_α	Spectrum width due to antenna rotation
σ_θ^2	Second central moment of the two-way radiation pattern
τ	Pulse width
τ_s	Range time delay
ϕ	Azimuth
ϕ_a	Effective radiation pattern width
ϕ_{DP}	Differential phase
ω	Angular frequency
ω_d	Doppler shift (rad s^{-1})

Introduction

The capability of microwaves to penetrate cloud and rain has placed the weather radar in an unchallenged position for remotely surveying the atmosphere. Although visible and infrared cameras on satellites can detect and track storms, the radiation sensed by these cameras cannot probe inside the storm's shield of clouds to reveal, as microwave radar does, the storm's internal structure and the hazardous phenomena that might be harbored therein. The Doppler radar is the only remote sensing instrument that can detect tracers of wind and measure their radial velocities, both in the clear air and inside heavy rainfall regions veiled by clouds—clouds that disable lidars (i.e., radars that use radiation at optical or near optical wavelengths) because optical radiation can be completely extinguished in several meters of propagation distance. This unique capability supports the Doppler radar as an instrument of choice to survey the wind and water fields of storms and the environment in which they form. Pulsed-Doppler radar techniques have been applied with remarkable success to map wind and rain within severe storms showing in real time the development of incipient tornado cyclones, microbursts, and other storm hazards. Such observations should enable weather forecasters to provide better warnings and researchers to understand the life cycle and dynamics of storms.

1.1 Historical Background

The term *radar* was suggested by S. M. Taylor and F. R. Furth of the U.S. Navy and became in November 1940 the official acronym of equipment built for *ra*dio *d*etecting *a*nd *r*anging of objects. The acronym was by agreement adopted in 1943 by the Allied powers of World War II and thereafter received general international acceptance. The term *radio* is a generic term applied to all electromagnetic radiation at wavelengths ranging from about 20 km (i.e., a frequency of 15,000 Hz—Hz or hertz is a unit of frequency in cycles per second that commemorates the pioneering work of Heinrich Hertz, who in 1886–1889 experimentally proved James Clerk Maxwell's thesis that electrical waves are identical except in length to optical waves) to fractions of a millimeter.

Perhaps the earliest documented mention of the radar concept was made by Nikola Tesla in 1900 when he wrote in *Century Magazine* (June 1900, LX,

1

p. 208): "When we raise the voice and hear an echo in reply, we know that the sound of the voice must have reached a distant wall, or boundary, and must have been reflected from the same. Exactly as the sound, so an electrical wave is reflected ... we may determine the relative position or course of a moving object such as a vessel at sea, the distance traveled by the same, or its speed ...."

The first recorded demonstration of the detection of objects by radio is in a patent issued in both Germany and England to Christian Hulsmeyer for a method to detect distant metallic objects by means of electromagnetic waves. The first public demonstration of his apparatus took place on 18 May 1904 at the Hohenzollern Bridge, Cologne, Germany, where river boats were detected when in the beam of generated radio waves (not pulsed) of wavelength about 40 to 50 cm (Swords, 1986).

Although objects were *detected* by radio waves as early as 1904, *ranging* by pulse techniques was not possible until the development of pulsed transmitters and wideband receivers. The essential criteria for the design of transmitters and receivers for pulsed oscillations were known in the early 1900s (e.g., pulsed techniques for the acoustical detection of submarines were vigorously developed during World War I), but the implementation of these principles into the design of practical radio equipment first required considerable effort in the generation of short waves.

The first successful demonstration of radio detection and ranging was accomplished using continuous waves (cw). On 11 December 1924, E. V. Appleton of King's College, London, and M. A. F. Barnett of Cambridge University in England used frequency modulation (FM) of a radio transmitter to observe the beat frequency due to interference of waves returned from the ionosphere (i.e., a region in the upper atmosphere that has large densities of free electrons that interact with radio waves) and those propagated along the ground to the distant receiver. The frequency of the beat gives a direct measure of the difference of distance traveled along the two paths and thus the height or range of the reflecting layer. This technique is based on exactly the same principles used in the FM-cw radars that are comprehensively described in Section 7.10.3 of the first edition (1984) of this book.

Pulse techniques are commonly associated with radar and in July 1925 G. Breit and M. A. Tuve (1926) in their laboratory of the Department of Terrestrial Magnetism of the Carnegie Institution obtained the first ranging with pulsed radio waves. They cooperated with radio engineers of the United States Naval Research Laboratory (NRL) and pulsed a 71.3-m wavelength NRL transmitter (Station NKF, Bellevue, Anacostia, D. C.) located about 10 km southeast of their laboratory, and detected echoes from a reflecting layer about 150 km above the earth. The equipment of Appleton and Barnett can be considered perhaps the first FM-cw radar, and that of Breit and Tuve the first pulsed radar. On the other hand, because the height of ionospheric reflection is a function of the radio wavelength, these radio systems might not be considered radars because they did not locate an object well defined in space as an aircraft (Watson-Watt, 1957, Chap. 21). Nevertheless, these radar-like systems were assembled for atmospheric studies and not for the location of aircraft, which was the impetus for the explosive growth of radars in the late 1930s and early 1940s.

It is likely that the first attempt to use pulsed radar principles to measure ionospheric heights came from a British physicist, W. F. G. Swann, who during the years 1918 to 1923 joined the University of Minnesota in Minneapolis where Breit was an assistant professor (1923–1924) and Tuve was a research fellow (1921–1923) (Hill, 1990). It was at the University of Minnesota that Swann and J. G. Frayne made unsuccessful attempts to measure the height of the ionosphere using radar techniques. Although many have contributed to the development of radar as we know it today, the earliest radars were developed by men interested in research of the upper atmosphere and methods to study it (Guerlac, 1987, p. 53).[1]

The role of atmospheric scientists in the early development of radar is also evident from the British experience. It was in January 1935 that the Committee for the Scientific Survey of Air Defense (CSSAD) approached Robert A. Watson-Watt to inquire about the use of radio waves in the defense against enemy aircraft. Sir Watson-Watt graduated as an electrical engineer, and in 1915 joined the Meteorological Office to work on a system to provide timely thunderstorm warnings to World War I aviators. After this wartime effort, it was realized that meteorological science was an essential part of aviation. He therefore was able to continue his research on direction finding of storms using radio emissions generated by lightning.

The CSSAD inquiry triggered Watson-Watt and his colleague A. F. Wilkins to propose, in a memo dated 27 February 1935, a radar system to detect and locate aircraft in three dimensions. The feasibility of their proposal was based on their calculation of echo power scattered by an aircraft, and was supported by earlier published reports by British Post Office engineers who detected aircraft that flew into the beam of Postal radio transmitters (Swords, 1986, p. 175). It was in July 1935, less than five months after their proposal, that Watson-Watt and his colleagues successfully demonstrated the radio detection and ranging of aircraft. This radar system, after considerable modifications and improvements, led to the Chain Home radar network that provided British aviators with early warning of approaching German aircraft.

---

1. Dr. William Blair, who studied the properties of microwaves for his Ph.D. at the University of Chicago and was involved in the development of an atmospheric sounding system known as the radiosonde, was a scientist in the U.S. Army Signal Corps Laboratories at Fort Monmouth, N.J., when he made a proposal in 1926 to the U.S. Army for a "Radio Position Finding" project. However, he was unsuccessful in obtaining support. Nevertheless, he actively pursued the development of radar theory and its practical realization. For this, he was granted the U.S. patent for radar on 24 August 1957. It is interesting to note that Dr. Blair's pursuits were without official authorization, which came several years later.

Although many have contributed to the development of radar, Watson-Watt credits many of his remarkable achievements to the earlier nonmilitary work of atmospheric scientists. To quote Watson-Watt (1957, p. 92): "... without Breit and Tuve and that bloodstream of the living organism of international science, open literature, I might not have been privileged to become ... the Father of Radar."

Throughout the 1930s, independent parallel efforts in radar development took place in the United States, Germany, Italy, Japan, France, Holland, and Hungary. The almost simultaneous and similar radar developments in all these countries should not be surprising because the ideas basic to radar principles had been repeatedly presented for many years preceding its development. It was during this period that the threat of faster and more lethal military aircraft, and the looming of global conflict, gave tremendous impetus to the development of equipment for the early detection and location of aircraft. On 28 April 1936, scientists at NRL obtained the first definitive detection and ranging of aircraft, and on 14 December the U.S. Army's Signal Corps, in an independent work, succeeded in locating an airplane by the pulse method. For a detailed description of these efforts and those in other countries, the reader is referred to the excellent and comprehensive books by Swords (1986) and Guerlac (1987).

The use of microwaves in radars for longrange detection did not become practical until early in 1940, when a powerful and efficient transmitting tube, the multiresonant cavity magnetron, was developed. The magnetron, as we know it today, evolved in many stages from a primitive device used initially as a switch and a high-frequency oscillator (Hull, 1921). In 1924, an important modification led to the split anode design, which allowed the generation of useful ultrahigh frequency waves first described by Erich Habann in his dissertation (Habann, 1924). This early work led, in 1924, to the discovery by August Žáček that the split anode magnetron was able to produce appreciable microwave power at wavelengths as short as 29 cm (Žáček, 1924). Apparently, Japanese investigators developed the split anode magnetron independently and in 1927 reported intense microwave power at wavelengths of about 40 cm (Okabe, 1928). However, the breakthrough in the production of truly powerful microwaves came when J. T. Randall and H. A. Booth at Birmingham University in England combined the resonant cavity feature of the klystron with the high current capacity of the magnetron cathodes to conceive a multiresonant cavity structure that is the basic design of today's magnetrons. The first magnetron built by Randall and Booth produced on 21 February 1940 an impressive 400 watts of continuous microwave power at wavelengths near 10 cm (Guerlac, 1987). This robust design is commonly used to this day in microwave ovens in many homes.

It is difficult to trace the origin of the first radar detection of precipitation, no doubt because of wartime secrecy. But beginning in July 1940 a 10-cm radar system was operated by the General Electric Corporation Research Laboratory in Wembley, England, a place where Dr. J. W. Ryde was working. There is no documented evidence that this radar (or another like it, which was at about this time located also in England) detected echoes from precipitation in 1940; but the work of Ryde (1946) to estimate the attenuation and echoing properties of clouds and rain is strong evidence that this study was undertaken because precipitation echoes were observed, and because there was concern for effects this might have on detection of aircraft (Probert-Jones, 1990). Thus it seems likely that radar first detected precipitation in the latter half of 1940.

The origins of radar meteorology are hence traced to this early work of Ryde. Although weather radar is commonly associated with detection of precipitation and storms, the earliest of what we now call meteorological or weather radars detected echoes from the nonprecipitating troposphere. However, only in the last few decades has this capability been exploited to explore the structure of the troposphere; more recently it has led to measurements of winds and temperature in all weather conditions. These particular radars are now known as *Profilers*.

Detection of echoes from the clear troposphere can be traced to the 1935 observations of Colwell and Friend (1936) in the United States and Watson-Watt and others (1936) in England. These researchers used vertically pointed radio beams to detect echoes from layers at heights as low as 5 km. These echoes at first were thought to originate from ionized layers, but Englund and his associates (Englund *et al.*, 1938) at Bell Laboratories clearly showed both experimentally and theoretically that short waves (wavelengths of about 5 m) are reflected from the dielectric boundaries of different air masses. In 1939 Friend (1939) was able to perfectly correlate his observations of tropospheric reflection heights with air mass boundaries located with *in situ* measurements made aboard an aircraft. After the war, Friend completed his Ph.D. studies and initiated experiments to locate air mass boundaries using a 300-MHz (one-meter wavelength) radar. This effort was continued by Peter Harbury, who constructed a vertically pointed 50-m diameter antenna to resolve returns from the troposphere. Tragically, Harbury was electrocuted while working on the radar modulator and this experiment was shortly thereafter discontinued (Swingle, 1990). If this work had continued it seems likely that Harbury and Friend would have found echoes from throughout the troposphere, and the development of Profilers might have commenced much earlier.

Pulsed-Doppler radar was developed during World War II to better detect aircraft and other moving objects in the presence of echoes from sea and land that are inevitably illuminated by microwave emissions through sidelobes (i.e., radiation in directions outside the beam or mainlobe) of the antenna's radiation pattern. Although pulsed-Doppler radar was developed in the early 1940s, Doppler effects were observed in radio receivers when echoes from moving objects were received simultaneously with direct radiation from the transmitter or scattered from fixed objects. Actually these observations preceded the development of radar, and in fact provided the incentive for radar because it was shown in the early 1920s that moving objects such as ships and aircraft were detectable.

The earliest pulsed-Doppler radars were called MTI (moving target indication) radars in which a coherent continuous-wave oscillator, phase-locked to the random phase of the sinusoid in each transmitted pulse, is mixed (i.e., beated) with the echoes associated with that pulse. The mixing of the two signals produces a beat or fluctuation of the echo intensity at a frequency equal to the Doppler shift (Doviak and Zrnic, 1988). Although these early MTI radars were used to suppress the display of echoes from fixed targets so that only moving target echoes were displayed, they are based on exactly the same physical principles used in pulsed-Doppler radars. The only significant difference is that MTI radars detect moving targets but do not measure their velocities, whereas pulsed-Doppler radars do both. The rapid development of pulsed-Doppler radar was impeded by the formidable amount of signal processing that is required to extract quantitative estimates of the Doppler shift at each of the thousand or more range locations that a radar can survey. It was only in the late 1960s and early 1970s that solid-state devices made practical the implementation of Doppler measurements at all resolvable ranges.

The first application of pulsed-Doppler radar principles to meteorological measurements was made by Ian C. Browne and Peter Barratt of the Cavendish Laboratories at Cambridge University in England in the spring of 1953 (Barratt and Browne, 1953). They used an incoherent version of the MTI radar in which the reference phase of the coherent oscillator is replaced by a signal reflected from ground objects at the same range as the meteorological targets of interest. The beam of the radar was pointed vertically into a rain shower while part of a magnetron's output was directed horizontally to ground objects. Barratt and Browne showed that the shape of the Doppler spectrum agreed with the spectrum expected from raindrops of different sizes falling with different speeds, but that the measured spectrum was displaced by an amount consistent with a downdraft of about 2 m s^{-1} (Rogers, 1990).

Those readers interested in a comprehensive presentation of the evolution of radar meteorology since 1940 are encouraged to examine "Radar in Meteorology" (Atlas, 1990), which has 18 chapters that contain the history of radar meteorology in various countries and principal organizations.

By far the most comprehensive treatment of radar techniques is found in the collected works compiled by M. I. Skolnik in his "Radar Handbook" (1970). Battan's text (1973) on weather radar applications is probably the most widely used by meteorologists, and Atlas (1964) also gives a concise and informative review of many weather radar topics. Both of these works emphasize the electromagnetic scatter and absorption by hydrometeors. A book by Nathanson (1969) emphasizes *the total radar environment* as well as radar design principles. The radar environment as defined by Nathanson is also the source of unwanted reflection (clutter) from the sea and land areas. (Precipitation is said to produce clutter when aircraft are the targets of interest.) Thus, precipitation echoes are comprehensively treated. The anomalous propagation of radar signals enhances ground clutter. A good general reference on the propagation of electromagnetic waves through the stratified atmosphere is the book by Bean and Dutton (1966). Sauvageot (1982) has distilled the essence of over 500 references in his book "Radarmétéorologie" that includes much of the radar meteorological work accomplished during the 1970s. The book "Radar Observation of Clear Air and Clouds" by Gossard and Strauch (1983) emphasizes the potential of radars for studying storms in their early evolutionary stage and for studying clear-air structure and wind profiles. "Applications of Weather Radar Systems" by Collier (1989) is a guide to uses of radar data in meteorology and hydrology that contains a fair amount of system concepts. A comprehensive treatment of weather radars on board satellites is contained in "Spaceborne Weather Radar" by Meneghini and Kozu (1990); this is a timely topic with obvious significance for global monitoring of precipitation. Rinehart's (1991) book "Radar for Meteorologists" is an up-to-date text for undergraduates and professionals, with color figures of radar displays of meteorological and nonmeteorological phenomena.

1.2   The Plan of the Book

The book "Doppler Radar and Weather Observations" by Doviak and Zrnic (1984a) emphasizes the application of Doppler radar for the observations of stormy and clear weather. The 1984 edition was intended to be a reference book on radar theory and techniques applied to meteorology. To have an updated text that is also useful to students, meteorologists, and atmosphere scientists not familiar with Doppler radar, we have revised the 1984 text to provide additional explanatory material. To stimulate further investigation and understanding we have included problems at the end of each chapter. This present edition also discusses the fundamental principles underlying recent developments such as polarimetric Doppler radar and radio acoustical soundings systems. As in the earlier text, this edition lightly touches on subjects comprehensively treated elsewhere (e.g., the scattering properties of hydrometeors), but presents a comprehensive treatment of the techniques used in extracting meteorological information from weather echoes, and relates radar and signal characteristics to meteorological parameters. Chapter 2 introduces the essential properties of radio waves needed to understand radar principles and describes the effect that the atmosphere has on the path of the radar pulse and its echo. In Chapters 3 to 5 we develop weather radar theory starting from fundamental principles, most of which are covered in undergraduate physics and mathematics. In Chapter 3 we trace the path of the transmitted pulse, through the antenna, along the beam to a single hydrometeor, and its return as an echo to the receiver, highlighting along the way the important aspects of the signal properties. We immediately consider the coherent or Doppler radar, but equations derived can directly be applied to the incoherent weather radar commonly used for over 40 years.

In Chapter 4 we extend radar principles, developed in Chapter 3 for single hydrometeors, to the more complex precipitating weather systems which are a conglomerate of hydrometeors that produce a continuous stream of echoes with random fluctuations of amplitude and phase. We show the origin of these fluctuations and develop the weather radar equation for the echo power in terms of radar and meteorological parameters. We show the limitations on the detection and spatial resolution of weather systems.

We treat the discrete Fourier transform in Chapter 5 and apply it to weather signals so as to make a connection between the Doppler spectrum and shear and turbulence of the flow. In Chapter 6 we analyze the weather Doppler spectrum and outline the signal processing methods used to derive the principal moments of the Doppler spectrum, emphasizing results rather than processor details.

The very important topic of range and Doppler velocity ambiguities as they pertain to distributed scatterers, as well as other considerations in observing weather, are presented in Chapter 7. The limitations imposed by antenna sidelobes, ground clutter, signal decorrelation, and power are discussed, together with techniques to mitigate these limitations; a comprehensive treatment of pulse compression is also given. We develop the theory needed to explain commonly encountered artifacts in the signal and show that antenna rotation coupled with signal averaging produces an apparent broadened beam of radiation.

The physics behind a variety of methods of rainfall estimation is discussed in Chapter 8. These methods are divided into single- and multiple-parameter techniques, depending on the number of independent measurements. Considerable space is devoted to polarization diversity and its utility for quantitative measurements of precipitation and discrimination between hydrometeor types.

A brief introduction to storm structure, in Chapter 9, is followed by examples of wind fields, obtained from the analysis of Doppler radar data on storms. Photographs are provided of several significant phenomena associated with storms. The important research subject of data analysis from more than one Doppler radar is briefly discussed. Multiple Doppler data synthesis to map the wind field with high resolution confirms the interpretation of single Doppler signatures of severe weather events.

Although much of the discussion of thunderstorms is focused on their hazards, one should not be led to believe that thunderstorms bring only misery. Each storm can release on the order of 10^{10} kg of beneficial rain water and, at the cost of about 31 cents per ton, the water is worth nearly a million dollars if properly stored and distributed. We need to learn methods by which losses due to storms can be lessened while their benefits continue. Proper warning and the protection of life and property are the first defense. The modification of storms to reduce their hazards without a loss of rain appears to be a long-term effort. Each storm releases energy at the rate of about 10^7 MW of latent heat (Sikdar et al., 1974). This prodigious amount of energy spread over large volumes is indeed difficult to control.

In Chapter 10 the theory of turbulence is reviewed, with emphasis on topics applicable to the radar measurements. Spatial spectra of velocity fields filtered by the resolution volume and examples of eddy dissipation rate fields are presented. The contributions of turbulence and shear to the Doppler spectrum width in a severe storm are examined.

A theory based on Fourier spectral representations is developed in Chapter 11 to explain radar echoes from clear-air refractive index irregularities. Existing theories are extended to develop a formulation for the Fresnel scatter from horizontally extended irregularities. These theories are amply illustrated with specific examples and are used to explain actual observations. Waves and turbulence in the earth's convective boundary layer are revealed by the use of the Doppler weather radar to observe echoes from refractive index irregularities. Radar reflectivity is related to the dynamics of atmospheric flow, and the potential of weather radars for mapping the kinematic structure of the atmosphere is discussed. Implications are made concerning vertical profiling of winds with specialized radars, and measurement of temperature with the radio acoustic sounding system is explained.

2

Electromagnetic Waves
and Propagation

To understand the remote sensing of weather by radar requires knowledge of a few basic properties of electromagnetic waves and the effects that the atmosphere has on these waves as they propagate between the radar and the hydrometeors. This chapter reviews fundamental wave concepts and presents elementary theories that describe wave propagation. Useful formulas are derived that quantify some of the important effects that the environment has on the radar's capability to assign a location from which scatterers principally contribute to a sample of the weather signal.

2.1 Waves

Electromagnetic or radio waves are electric $\mathbf{E}$ and magnetic $\mathbf{H}$ force fields that propagate through space at the speed of light and interact with matter along their paths. These interactions cause the scattering, diffraction, and refraction also common to visible electromagnetic radiation (light). These waves, focused into beams by the antenna system, have sinusoidal spatial and temporal variations; the distance or time between successive wave peaks of the electric (magnetic) force defines the wavelength λ or wave period (i.e., the reciprocal of the frequency f, in hertz). These two important electromagnetic field parameters are related to the speed of light c.

$$c = \lambda f = 3 \times 10^8 \text{ m s}^{-1}. \tag{2.1}$$

Microwaves are electromagnetic forces having spatial wavelengths between 10^{-3} and 10^{-1} m, whereas visible radiation has a wavelength of about 6×10^{-7} m (Fig. 2.1). The upper end (0.01–0.1 m) of the microwave band is used by weather and aircraft surveillance radars.

The electric field wave far from the transmitting antenna has time t and range r dependence generally given by

$$\mathbf{E}(r, \theta, \phi, t) = \frac{\mathbf{A}(\theta, \phi)}{r} \cos\left[2\pi f\left(t - \frac{r}{c}\right) + \psi \right] \text{ V m}^{-1}, \tag{2.2a}$$

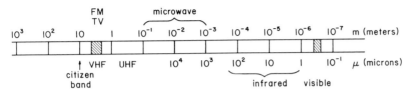

Fig. 2.1 The electromagnetic spectrum, showing the locations of the very high frequency (VHF), ultrahigh frequency (UHF), microwave, infrared, and optical bands.

where **A** depends on θ, ϕ (the direction of r from the radiation source), and ψ is usually an unknown but constant transmitter phase angle. The dependence (2.2a) of **E** and **H** on r, t, θ, and ϕ is characteristic of all electromagnetic waves propagating in space devoid of matter, be they radio waves or light. Equation (2.2a) approximates well, at weather radar frequencies, the properties of waves propagating through the earth's atmosphere. Because a force has direction, **E** is a vector quantity. The waves propagate in the direction of **r**; that is, an observer's range r must increase at a rate c to stay on a wave crest ($t - r/c =$ const). The vectors **E**, **H** are perpendicular to each other and lie in the plane of polarization that is perpendicular to **r** if r is large (Section 3.1.2) compared to the antenna dimensions.

The magnitude and direction of **E**, or **H**, will be known if the magnitude and phase angle of two orthogonal components of **E** are known (e.g., the magnitudes and phase angles, E_x and ψ_x, E_y and ψ_y, of the electric field in the x, y directions for propagation in the z direction). If the phase angle difference, $\psi_x - \psi_y$, between the orthogonal components is zero or an integer multiple of π, the wave is said to be linearly polarized. If **E** has only a horizontal component, the wave is said to be horizontally polarized. If **E** lies totally in the vertical plane, the wave is said to be vertically polarized. If both horizontal and vertical components of the wave are simultaneously present, the wave is, in general, elliptically polarized. If the phase angle difference is $\pi/2$, and the amplitudes A_x, A_y of the two Cartesian components are equal, the wave is right-hand circularly polarized (that is, the fist of right-hand fingers indicates the direction of electric field rotation when the thumb points in the direction of propagation in a right-handed coordinate system); if the phase angle difference is $-\pi/2$, it is left-hand circularly polarized and the electric vector rotates in the counterclockwise direction when viewed in the direction of propagation (Fig. 2.2).

Because the principal factors characterizing a periodic electric field are the amplitude $\mathbf{A}(\theta, \phi)/r$ and phase $2\pi f(t - r/c) + \psi$, it is convenient and instructive to use complex-number or phasor notation to describe these parameters. The electric field (2.2a) is then expressed as

$$\mathbf{E} = \frac{\mathbf{A}(\theta, \phi)}{r} \exp\left[j2\pi f\left(t - \frac{r}{c}\right) + j\psi \right], \qquad (2.2b)$$

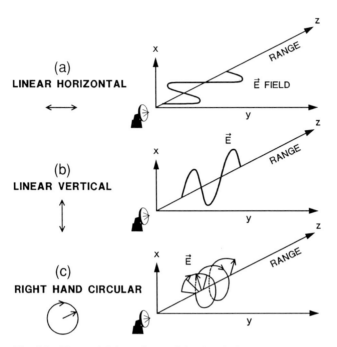

(a)
LINEAR HORIZONTAL

(b)
LINEAR VERTICAL

(c)
RIGHT HAND CIRCULAR

Fig. 2.2 The spatial dependence of the electric field vector for (a) horizontally, (b) vertically, and (c) circularly (right-handed) polarized transmissions.

which, according to Euler's formula, can be represented by a two-dimensional phasor diagram (Fig. 2.3). This diagram clearly describes the time and space dependence of amplitude and phase β to within an integral multiple of 2π. The time rate of change of β is frequency, and it can be seen from Fig. 2.3 that frequency is composed of two terms: $\omega = 2\pi f$ is the transmitted or carrier frequency (radians s^{-1}), and $(2\pi f/c)\, dr/dt$ is a Doppler shift that would be experienced by an observer at r moving at velocity dr/dt. It is understood that

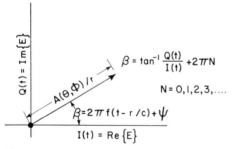

Fig. 2.3 Phasor diagram. Re{E} is a real part of the complex electric field **E** and Im{E} is the imaginary part.

we need to take the real or imaginary part of Eq. (2.2b) to obtain the time and space dependence in terms of real numbers. The time dependence of the phase is of paramount importance in understanding the principles of Doppler radar.

Another important electromagnetic field quantity is the time-averaged power density $S(r, \theta, \phi)$.

$$S(r, \theta, \phi) = \frac{1}{2} \frac{\mathbf{E} \cdot \mathbf{E}^*}{\eta_0} = \frac{A^2(\theta, \phi)}{2\eta_0 r^2} \quad \mathrm{W\,m}^{-2}, \tag{2.3}$$

where * denotes complex conjugation, the dot indicates a scalar product, and $A = \sqrt{A_x^2 + A_y^2}$ is the magnitude of $\mathbf{A}$. S is the magnitude of the complex Poynting vector $\mathbf{S} = \mathbf{E} \times \mathbf{H}^*/2$ which gives the direction of energy flux. It can be shown that $\frac{1}{2}\mathbf{E} \cdot \mathbf{E}^*$ is the time average of the square of Eq. (2.2a). The factor η_0 is the wave impedance (in space, or in earth's atmosphere at radar wavelengths, η_0 is a constant equal to 377 ohms) and is the ratio of the electric to the magnetic field amplitude. Time averages represented by Eq. (2.3) are averages of power over a cycle or period f^{-1} of the wave; but if power is *pulsed* (i.e., transmitted in bursts of energy), then A and S are functions of time and, moreover, the average of power over a cycle of f can change during the pulse. The significance of $S(r, \theta, \phi, t)$ is that it represents the power density flowing outward from a source, either continuously (i.e., independent of time) or in bursts, and products of S with areas, later to be specified, represent the power that is received, absorbed, scattered, and so on.

Most remote probing of our physical environment is performed at the short wavelength electromagnetic radiation visible to the human eye. The angular resolution of scatterers (i.e., the discrimination between two adjacent similar objects at the same range) is dependent on the wavelength and antenna size. The angular resolution or diameter of the first null in a circular diffraction pattern is well approximated by

$$\Delta\theta \approx 104\lambda/D \text{ deg} \tag{2.4}$$

(Born and Wolf, 1964, p. 415), where D is the diameter of the antenna system. Thus, long radio waves (i.e., $\lambda = 10^2$–10^3 m) require huge antenna installations to achieve an angular resolution of a few degrees—rather poor for optical antenna systems. For example, the human eye has an angular resolution of about $0.02°$. It is evident that remote radio sensing even at microwave frequencies is characterized by poor resolution compared to optical standards.

The essential distinguishing feature favoring microwaves for weather radars is their property to penetrate rain and cloud and thus to provide a view inside showers and thunderstorms, day or night. Rain and cloud do attenuate microwave signals, but only slightly (for $\lambda \geq 0.05$ m) compared to the almost complete extinction of optical signals. Raindrops scatter electromagnetic energy, and the portion scattered constitutes the signal whose characteristics are diagnosed to determine storm properties. Scattered signal strength measures rain intensity, and the time rate of phase change (Doppler shift) measures the raindrop speed in the radial (**r**) direction.

2.2 Propagation Paths

In free space, waves propagate in straight lines because everywhere the dielectric permittivity ε_0 and magnetic permeability μ_0 are constants related to speed of propagation $c = (\mu_0 \varepsilon_0)^{-1/2}$. However, the atmosphere's permittivity ε is larger than ε_0 and is vertically stratified; therefore microwaves propagate at speeds $v < c$ along curved lines, and sometimes the beam is refracted (bent) back to the surface (anomalous propagation), causing distant ground objects normally not seen to appear on displays. It is common practice today to make simple corrections for refractive effects, but although these work well most of the time, there exist atmospheric conditions that require sophisticated methods to determine the path of the radar signal.

The path of radar signals depends principally on the change in height of the atmosphere's refractive index $n = c/v$, or relative permittivity $\varepsilon_r = \varepsilon/\varepsilon_0 = n^2$ (because relative permeability of air μ_r is unity). We shall show how the refractive index is related to temperature, pressure, and water vapor contet. Then, given a vertical profile of these meteorological variables, one can determine the path to and strength of the radar signal at a scatterer. Because refractive index is related to temperature, pressure, and water vapor content. expressions developed herein useful in Chapter 11, where we discuss radar echoes from clear air.

2.2.1 Refractive Index of Air

The refractive index is proportional to the density of molecules and their polarization. Molecules that produce their own electric field without external forces are called *polar*. Polar molecules possess a permanent displacement of opposing charges within their internal structure, thus causing a dipole electric field that reaches far beyond the molecule's interatomic spacing. The water vapor molecule is polar. Although dry air molecules do not possess a permanent dipole moment, they become polarized when an external electric force (e.g., radar signal) is impressed on them. Without external forces, polar molecules have their dipole moments (with direction along the axis connecting the centers of opposing charge) randomly oriented due to thermal agitation. External forces can align these molecules so that their dipole fields add constructively to enhance the net electric force acting on each molecule. Thus the electric force acting on any molecule is the sum of the external electric field plus that produced by the polarized molecules.

The permittivity of a gas depends only on the density N_v of molecules and a factor α_T proportional to the molecule's level of polarization. It can be shown that

$$(\varepsilon_r - 1)/(\varepsilon_r + 2) = N_v\alpha_T/3. \tag{2.5}$$

By a remarkable coincidence, this relation was found independently by two scientists of almost identical names, L. V. Lorenz and H. A. Lorentz; accordingly, Eq. (2.5) is called the Lorenz–Lorentz formula. It applies to gases at all but extreme pressures. For air, ε_r is very near unity (i.e., $\varepsilon_r \approx 1.000300$), so

$$\varepsilon_r = n^2 \approx 1 + N_v\alpha_T. \tag{2.6}$$

Laboratory measurements of ε_r can be used to deduce α_T. For a gas that contains a mixture of molecules of types (1), (2), $\ldots$,

$$n^2 = 1 + N_v^{(1)}\alpha_T^{(1)} + N_v^{(2)}\alpha_T^{(2)} + \cdots. \tag{2.7}$$

Now, the number density of molecules of any type is proportional to the mass density ρ of the gas constituent.

$$N_v = \rho/M, \tag{2.8}$$

where M is the mass of the molecule under consideration. The mass density, pressure, and temperature obey the equation of state.

$$\rho = \rho_0(273/T)(P/1013), \tag{2.9}$$

where ρ_0 is the mass density at standard temperature and pressure (0°C and 760 mm Hg), P is the pressure in millibars, and T is the temperature in kelvins. The number of molecules per unit volume of any gas at fixed temperature and pressure is independent of the gas type (Avogadro's law). At standard temperature and pressure we have a number equal to

$$N_{v0} = 2.6873 \times 10^{25} \text{ m}^{-3} \tag{2.10}$$

and

$$N_v = \frac{273}{1013}(N_{v0}P/T). \tag{2.11}$$

Thus Eq. (2.7) can be written

$$n^2 = 1 + \frac{273}{1013}(N_{v0}/T)(P_1\alpha_T^{(1)} + P_2\alpha_T^{(2)} + \cdots), \tag{2.12}$$

where T is outside the parentheses because we assume thermal equilibrium, and P_1 is the partial pressure of gas 1. For the troposphere we need to consider only the contribution from dry air (a nonpolar gas) and water vapor (a polar gas). For dry air

$$n_d^2 = 1 + C_d \frac{P_d}{T}, \tag{2.13}$$

where C_d is a constant. For a combination of dry air and water vapor, Eq. (2.7) has the form

$$n^2 = 1 + C_d P_d/T + C_{w1} P_w/T + C_{w2} P_w/T^2, \tag{2.14}$$

where the last term is the contribution to the refractive index from the permanent dipole moment of the water vapor molecule; P_d and P_w are the partial pressures of dry air and water vapor. The constant parameters (e.g., N_{vo}) are contained in the constants C.

2.2.2   Refractivity N

Because the relative permittivity ε_r and refractive index n of the atmosphere are so near unity at microwave frequencies, it becomes convenient to introduce a different measure of the refractive properties of air. The refractivity N is defined as (Bean and Dutton, 1966, p. 357)

$$N \equiv (n - 1) \times 10^6. \tag{2.15}$$

From Eq. (2.6) one can express n as

$$n = [1 + (\varepsilon_r - 1)]^{1/2}. \tag{2.16a}$$

Now, $\varepsilon_r - 1$ is small compared to 1, and expansion of Eq. (2.16a) gives the linear relation

$$N = \tfrac{1}{2}(\varepsilon_r - 1) \times 10^6 \tag{2.16b}$$

between N and ε_r. Using Eq. (2.14) and the preceding equations, N can be written in the form

$$N = C_d' P_d/T + C_{w1}' P_w/T + C_{w2}' P_w/T^2. \tag{2.17}$$

Bean and Dutton (1966) give a survey of the various measurements and estimates of the constants C'. From their work we have

$$C_d' \simeq 77.6 \ \text{K mbar}^{-1}, \tag{2.18a}$$

$$C_{w1}' = 71.6 \ \text{K mbar}^{-1}, \tag{2.18b}$$

$$C_{w2}' = 3.7 \times 10^5 \ \text{K}^2 \ \text{mbar}^{-1}. \tag{2.18c}$$

Equation (2.17) can be approximated to an accuracy of about 0.1 by the simplified form

$$N = (77.6/T)(P + 4810 P_w/T), \tag{2.19}$$

where P is the total pressure in millibars and T is in kelvins.

For example, given a relative humidity of 60% and $T = 17°C$ at sea level,

$$P_w \simeq 10 \text{ mbar}, \qquad P \simeq 1000 \text{ mbar}, \qquad T \simeq 300 \text{ K}.$$

Thus

$$N \simeq 0.26 \times (10^3 + 1.6 \times 10^2) \simeq 300$$

and

$$n = 1 + N \times 10^{-6} = 1.000300.$$

It is apparent that the refractive index of the atmosphere differs very little from that of free space. Nevertheless, a change in n in the fifth and sixth significant digits is sufficient to have a measureable effect on electromagnetic wave propagation and scatter.

Both pressure and temperature usually decrease with height from sea level up to about 10 km, at which altitude the temperature becomes relatively constant for several kilometers (Fig. 2.4). In the troposphere the fractional decrease in pressure is larger than that for temperature, so N normally decreases with altitude. When the rate of decrease in N exceeds a certain value (i.e., $dN/dh \leq -157 \text{ km}^{-1}$) electromagnetic beams are bent toward the surface of the earth (i.e., trapped), as we shall demonstrate next. This condition is usually brought about by inversion layers, that is, layers of atmosphere in which the temperature departs from its usual decrease with height, thus causing the slope dN/dh to be more negative. In addition to systematic smooth variations in atmospheric properties, there are small-scale fluctuations in temperature, pressure, water vapor content, etc., that cause N to have small-scale variations. Electromagnetic wave scattering occurs from these refractive index irregularities. (The properties of the scattered waves are given in Chapter 11.) We shall ignore for now these

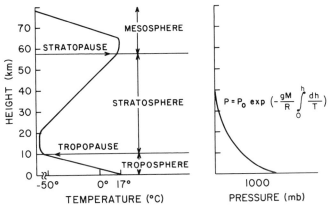

Fig. 2.4 Dependence of the temperature and pressure on height. (R is the universal gas constant and M is the molecular weight in atomic mass units.)

irregularities and consider the effects of a spherically stratified refractive index (i.e., N that is dependent only on height) on electromagnetic propagation through the lower altitudes of the troposphere.

2.2.3 Spherically Stratified Atmospheres

For most applications we can assume temperature and humidity to be horizontally homogeneous so that the refractive index is a function only of height above ground. At microwave frequencies it is permissible to assume that wave fronts are perpendicular to and propagate along rays, analogous to optical propagation. In Appendix A we show that the ray path in a spherically stratified atmosphere is given by the integral

$$s(h) = \int_0^h \frac{aC\,dh}{R[R^2 n^2(h) - C^2]^{1/2}}, \tag{2.20a}$$

$$C = an(0)\cos\theta_e, \tag{2.20b}$$

where $s(h)$ is the great circle distance (along the earth's surface) to a point directly below the ray at height h above the surface, a is the earth's radius, and R is the radial distance from the center of the earth (Fig. 2.5). The refractive index $n(h)$ is assumed to be smoothly changing (i.e., within a wavelength the relative changes in N are small) so that ray theory is applicable. The elevation angle θ_e is that of the ray at the transmitter, and $n(0)$ is the refractive index at that location. It can also be shown that Eq. (2.20) is a solution to the exact

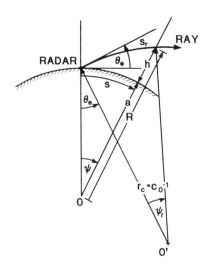

Fig. 2.5 Circular path of a ray in an atmosphere in which n is linearly dependent on height.

differential equation

$$\frac{d^2h}{ds^2} - \left(\frac{2}{R} + \frac{1}{n}\frac{dn}{dh}\right)\left(\frac{dh}{ds}\right)^2 - \left(\frac{R}{a}\right)^2\left(\frac{1}{R} + \frac{1}{n}\frac{dn}{dh}\right) = 0, \qquad (2.21)$$

which describes the ray path (Hartree *et al.*, 1946).

2.2.3.1 Equivalent Earth Model

The curvature C_0 of any line (e.g., the ray path) in a plane is given by

$$C_0 \equiv \frac{\left[R^2 + 2\left(\frac{dR}{d\psi}\right)^2 - R\frac{d^2R}{d\psi^2}\right]}{\left[R^2 + \left(\frac{dR}{d\psi}\right)^2\right]^{3/2}}, \qquad (2.22a)$$

where ψ is the polar angle and $R = a + h$. Because our interest lies in the atmosphere below 20 km, where $h \ll a$, R can be replaced by a, and dR by dh. Hence Eq. (2.22a) is simplified to

$$C_0 \simeq \frac{1 + 2\left(\frac{dh}{ds}\right)^2 - a\frac{d^2h}{ds^2}}{a\left[1 + \left(\frac{dh}{ds}\right)^2\right]^{3/2}} \qquad (2.22b)$$

in which $s = a\psi$ has been substituted. Under these conditions, but noting that $n \simeq 1$, Eq. (2.21) can be reduced to

$$\frac{d^2h}{ds^2} - \frac{2}{a}\left(\frac{dh}{ds}\right)^2 - \frac{1}{a}\frac{dn}{dh}\left[\left(\frac{dh}{ds}\right)^2 + 1\right] \simeq 0. \qquad (2.23)$$

Substituting d^2h/ds^2 from Eq. (2.23) into Eq. (2.22b), the curvature formula reduces to

$$C_0 \simeq \frac{-\left(\frac{dn}{dh}\right)}{\left[1 + \left(\frac{dh}{ds}\right)^2\right]^{1/2}} \qquad (2.24a)$$

Because rays of importance in weather radar studies are usually those launched at small elevation angles that remain at heights $h \ll a$, $(dh/ds) \ll 1$ and

$$C_0 \simeq -\frac{dn}{dh}. \qquad (2.24b)$$

If n is well approximated by a linear function in the height interval containing the rays, the ray paths will have constant curvature. That is, all rays near earth lie on circles having the same radius of curvature r_c, but have centers displaced by θ_e from the line passing through the radar and earth's center (Fig. 2.5).

Using accepted approximations we now develop simple means to determine the height of the curved ray as a function of arc distance from the radar. Consider an equivalent homogeneous atmosphere in which the ray path is straight, and compute the radius a_e of an equivalent earth such that the height of the straight ray above it is the same as the actual height h for the *same arc distance s of ray travel*. For the sake of mathematical simplicity we solve the problem for a ray launched at $\theta_e = 0$. Because all rays launched at low elevation angles have the same curvature (in an atmosphere having n that is linearly dependent on h), the ray heights can be found easily once the solution for $\theta_e = 0$ is found.

By applying the law of sines to the sides b and c of triangle ABC (Fig. 2.6a) we find that

$$\frac{b}{\cos \psi_r} = \frac{c}{\cos \psi},$$

where

$$c = r_c \left[\left(\frac{1}{\cos \psi_r} \right) - 1 \right]; \qquad b = h_2 - h; \qquad h_2 = a \left[\left(\frac{1}{\cos \psi} \right) - 1 \right].$$

Solving for h we obtain

$$h = a \left[\left(\frac{1}{\cos \psi} \right) - 1 \right] - r_c \frac{(1 - \cos \psi_r)}{\cos \psi}. \tag{2.25}$$

Now applying the law of sines to the triangle OAO′, a second expression

$$h = \frac{r_c \sin \psi_r}{\sin \psi} - a$$

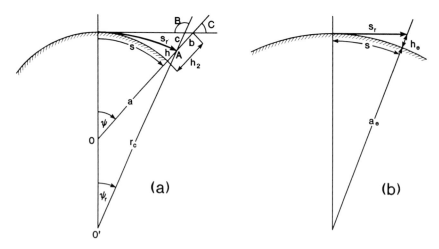

Fig. 2.6 (a) Circular path of a ray launched at an elevation angle $\theta_e = 0$. (b) Straight path of the same ray above a modified earth of radius a_e.

for height is obtained. By equating these two, h is eliminated and an expression for ψ_r is

$$\frac{\sin(\psi_r - \psi)}{\sin \psi} = \frac{a - r_c}{r_c}. \tag{2.26}$$

It is obvious that if $-(dn/dh)^{-1} = r_c = a$, then $\psi_r = \psi$ and hence $h = 0$. Therefore a ray (e.g., the beam axis) at $\theta_e = 0$ remains at the earth's surface and is said to be trapped. Rays launched at higher elevation angles require larger gradients of n in order to be trapped.

For our purposes the solution $\psi_r = a\psi/r_c$ of Eq. (2.26) to first order in ψ is adequate. Substituting this into Eq. (2.25) and retaining terms to second order in $\psi = s/a$ and ψ_r we find that

$$h = \frac{s^2}{2a} - \frac{s^2}{2r_c}.$$

Now referring to the equivalent earth of radius a_e (Fig. 2.6b), the height of the straight ray launched at $\theta_e = 0$ and traveling the same arc distance s is

$$h_e = \frac{s^2}{2a_e}.$$

Equating h to h_e, using the relation $r_c = -(dn/dh)^{-1}$, and solving for a_e, we obtain

$$a_e = \frac{a}{1 + a\left(\dfrac{dn}{dh}\right)} = k_e a. \tag{2.27}$$

Thus whenever the refractive index gradient is well approximated by a constant in the height interval of interest, we can use the equivalent earth of radius a_e to determine the height

$$h = k_e a \left[\frac{\cos \theta_e}{\cos(\theta_e + s/k_e a)} - 1 \right] \tag{2.28a}$$

of a ray leaving the radar at an elevation angle θ_e.

The following two equations relate h and s to radar-measurable parameters, the range r and θ_e.

$$h = [r^2 + (k_e a)^2 + 2rk_e a \sin \theta_e]^{1/2} - k_e a, \tag{2.28b}$$

$$s = k_e a \sin^{-1}\left(\frac{r \cos \theta_e}{k_e a + h} \right). \tag{2.28c}$$

Researchers have found that the gradient of n in the first kilometer or two is typically $-1/4a$, so the effective radius of the earth is

$$a_e = \tfrac{4}{3}a. \tag{2.28d}$$

Although the effective earth's radius model conveniently determines beam height as a function of range or arc length, two limitations need to be discussed.

1. n is linearly dependent on h.
2. The development of Eq. (2.27) assumed $dh/ds \ll 1$, which imposes a limit on the use of an effective earth's radius.

The gradient of the refractive index is not always a constant, and we have particularly severe departures from linearity when there are strong temperature inversions or large moisture gradients. Furthermore, the refractive index cannot decrease linearly without bound, because at large heights it must asymptotically approach unity. The unrealistic profile of n assumed by the model with an earth's radius of $4a/3$ is contrasted in Fig. 2.7 with a realistic dependence of n on h, as given by a reference atmosphere model that agrees closely with measured N data. Both models assume surface refractivity $N_s = 313$.

It is obvious that for $h \geq 2$ km there is a considerable difference between the N values. We may well wonder whether the effective earth radius model would be useful for ray paths above 2 km and, because our derivation assumed $dh/ds \ll 1$, for elevation angles larger than a few degrees.

For weather radar applications it can be shown that the model having 4/3 earth's radius can be used for all θ_e, if h is restricted to the first 10–20 km and if n

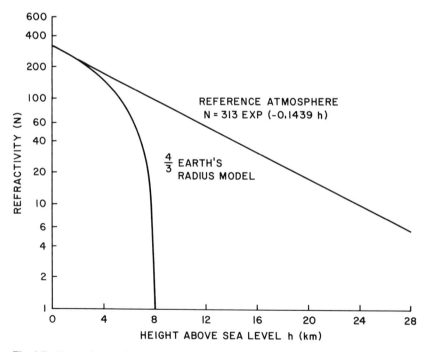

Fig. 2.7 Dependence of refractivity on height for a reference atmosphere contrasted with that implied by the effective earth's radius $a_e = 4a/3$.

has a gradient of about $-1/4a$ in the first kilometer of the atmosphere. Figure 2.8 shows a comparison of ray paths for $a_e = 4a/3$ and an exponential reference atmosphere where

$$n(h) = n_s e^{-0.1439h}. \tag{2.29}$$

We can see from Fig. 2.8 that, although the difference in n is large for $h \geq 2$ km, the difference in the ray paths is less than 1 km at a range of 250 km for $\theta_e = 0$, and the difference in the ray path height decreases rapidly as θ_e is increased from 0. Furthermore, most weather radar beamwidths are of the order of 1°, and errors in height are small compared to the beamwidth at these ranges. We therefore conclude that if the refractive index is well represented by Eq. (2.29), use of effective radius $a_e = 4a/3$ predicts beam height with sufficient accuracy for weather radar applications.

What are the effects of large temperature inversions? The following section discusses the ray paths in the lower atmosphere when there is an anomalously large gradient of refractivity.

2.2.3.2 Ground-Based Ducts, Reflection Height

We shall show how ray paths can be determined for refractive index profiles that depart considerably from those associated with the exponential reference

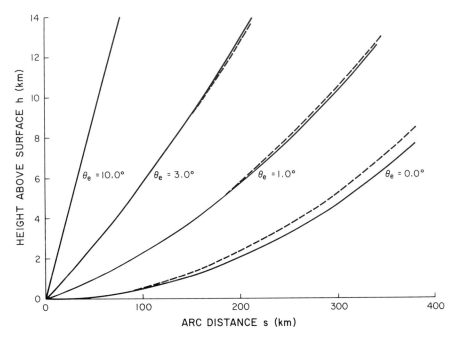

Fig. 2.8 Comparison of the ray paths for an $a_e = 4a/3$ model and an atmosphere with an exponentially dependent refractive index; ——, exponential atmosphere; - - -, 4/3 earth's radius.

atmosphere. We assume that the refractive index profile can be approximated by a piecewise linear model of N versus h. Consider the refractive index profile depicted in Fig. 2.9. This model shows a large gradient $(dN/dh = -300 \text{ km}^{-1})$ of refractivity for the first 100 m and thereafter a gradient associated with an effective earth's radius of $(4/3)a$.

When $h \ll a$, the integrand of Eq. (2.20) can be linearized with respect to h and thus integrated to produce the following relation for $0 \text{ km} \leq h \leq 0.1 \text{ km}$.

$$s(h) = [(\cos \theta_e)/(1 + \beta_0 a)]\{[a^2 \sin^2 \theta_e + 2a(1 + \beta_0 a)h]^{1/2} - a \sin \theta_e\}, \quad (2.30)$$

where $\beta_0 = -3 \times 10^{-4} \text{ km}^{-1}$ is the gradient of n at the surface, and we have substituted 1 for $n(0)$. There is a like expression for $0.1 \text{ km} \leq h$, where the gradient of n is β_1.

$$s'(h') = [(\cos \theta'_e)/(1 + \beta_1 a')]$$
$$\times \{[a'^2 \sin^2 \theta'_e + 2a'(1 + \beta_1 a')h']^{1/2} - a' \sin \theta'_e\}, \quad (2.31)$$

where $s'(h')$ is the arc distance from the point of emergence of the ray from the layer, h' is the height of the ray above h_1, $a' = a + h_1$, and θ'_e is the angle made by the ray at $h_1 = 0.1 \text{ km}$. This angle is

$$\theta'_e = \tan^{-1}(dh/ds) \qquad \text{at} \qquad h_1 = 0.1 \text{ km}. \quad (2.32)$$

Differentiating Eq. (2.30) and substituting into Eq. (2.32), we obtain

$$\theta'_e \simeq \tan^{-1}\{[a^2 \sin^2 \theta_e + 2ah_1(1 + \beta_0 a)]^{1/2}/a \cos \theta_e\}. \quad (2.33)$$

Because $\beta_0 a < -1$, the angles $\theta_e \leq \theta_p$ cause the radical to be imaginary, where

$$\theta_p = \sin^{-1}[-2h_1(1 + \beta_0 a)/a]^{1/2} \quad (2.34)$$

is the penetration angle. All rays having $\theta_e \leq \theta_p$ are reflected within the layer $0 \text{ km} \leq h \leq 0.1 \text{ km}$. The height h_r of the ray at the point of reflection is obtained

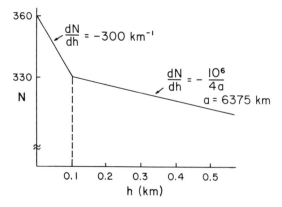

Fig. 2.9 Refractivity N profile for a model atmosphere in which there is a strong ground-based temperature inversion (i.e., the temperature increases with height) up to $h_1 = 0.1 \text{ km}$.

by solving for h_r in the differential equation

$$\frac{dh}{ds} = \frac{[a^2 \sin^2 \theta_e + 2ah_r(1 + \beta_0 a)]^{1/2}}{a \cos \theta_e} = 0, \tag{2.35}$$

obtained by differentiating Eq. (2.30). The solution of Eq. (2.35) is

$$h_r = (-a \sin^2 \theta_e)/2(\beta_0 a + 1). \tag{2.36}$$

In this example $\theta_p = 0.31°$, and rays having an elevation angle less than $0.31°$ are trapped in the layer. A ray having $\theta_e = 0.2°$ has a height of reflection $h_r = 43$ m, and reflection occurs at an arc distance obtained by solving Eq. (2.30).

$$s(h_r) = (-a \cos \theta_e \sin \theta_e)/(\beta_0 a + 1), \tag{2.37}$$

which in this example is 24.4 km. The ray returns to earth at an arc distance $2s(h_r)$. Thus an object at a distance of 49 km, which would not be visible to radar under normal propagation conditions, becomes visible if the beam elevation angle is $0.2°$ for the given refractivity profile. A few sample ray paths for this case are shown in Fig. 2.10. The ray path for $h \geq 0.1$ km is obtained using an effective earth's radius of $(4/3)a$ and θ' from Eq. (2.33) as the initial elevation angle of the ray emerging above the layer h_1.

Also apparent from Fig. 2.10 is the effect of the temperature inversion on spatial resolution. For example, suppose we observe scatterers at a distance of

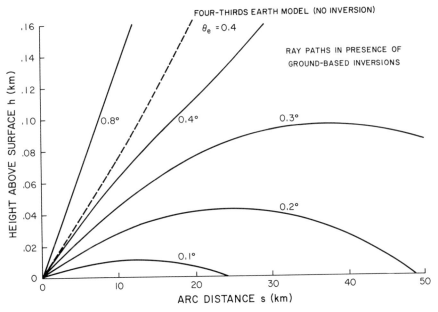

Fig. 2.10 Ray paths in an atmosphere modeled as shown in Fig. 2.9. A surface-based inversion exists in the first 100 m of height.

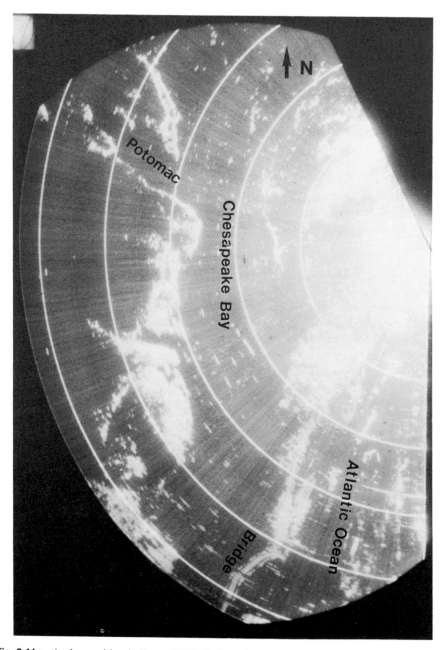

Fig. 2.11a A plan position indicator (PPI) display of ground objects made visible because of a strong ground-based temperature inversion. Circular arcs are range marks spaced 10 nautical miles apart.

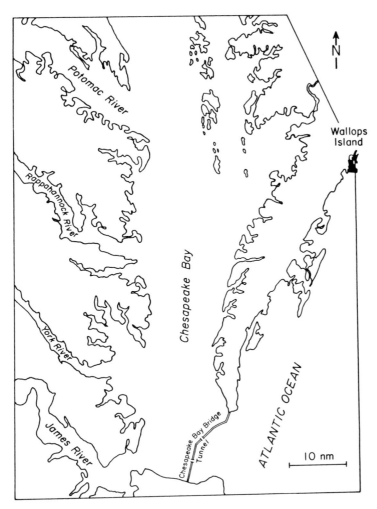

Fig. 2.11b Map of the area scanned by the radar beam.

50 km using a radar with 0.2° angular resolution and pointed at an elevation angle of 0.3°. The vertical beamwidth in the absence of the inversion would have been 175 m, but in the presence of inversion the beamwidth broadens to 270 m. Such broadening not only leads to deterioration of the resolution but can also result in erroneous measurements of an object's cross section, because the power density at the object is reduced.

A striking example of echoes in the presence of a strong ground-based inversion is shown in Fig. 2.11a. These anomalous propagation data were obtained with a 10-cm radar at Wallops Island, Virginia. The radar beamwidth was about 0.4°, and the elevation angle was 0.5°. Judging from the large range over which echoes are detected, we conclude that parts of the beam must have been grazing the ground. Clearly visible on the figure are islands and waterways in the Chesapeake Bay as well as the bridge–tunnel complex. For comparison, the chart of the area and the radar location are drawn in Fig. 2.11b.

Problems

2.1   An antenna is radiating isotropically in free space. The transmitter emits power in short bursts (pulses). (a) What is the average power density in each pulse at 105 km from the antenna if each pulse has one megawatt (MW) of average power? (b) Find the **E** field magnitude and phase (relative to the phase at the transmitter) at 52.5 and 105 km for a 10-cm radar. What is the phase difference of the transmitted signal, at two raindrops spaced one centimeter apart? Repeat the calculation for a 5-cm radar. (c) If the 1-MW pulses are 5 μs long, what is the minimum time separation (i.e., the pulse repetition time, PRT) between pulses and maximum PRF in order not to exceed the transmitter specifications that average power (i.e., average over a pulse-to-pulse interval) must be less than one kilowatt (kW)? (d) Same as (c) but pulse length is 1 μs.

2.2   Compute error in range to a scatterer at $r = 300$ km if one assumes free space propagation velocity is $c = (\mu_0 \varepsilon_0)^{-1/2}$ instead of $v = c/n$ where $n = 1.000300$. What would be the difference in echo phase for these two different speeds of propagation?

2.3   (a) Use an early morning atmospheric sounding (any will do) to determine the height dependence of refractivity N computed with and without water vapor. Compare and discuss. Explain the origin of marked departures of N from the exponential reference profile if there are any. (b) Using the computed profile of N, determine a mean gradient of n in the lower atmosphere (i.e., first few hundred meters) and compute the effective earth's radius.

2.4   Using Eq. (2.33) find the gradient of N that causes surface-based ducting of electromagnetic waves.

2.5   Assume that Eq. (2.30) accurately describes the height $h(s)$ of the centroid of a packet of radiation as it propagates in the atmosphere along ray paths. Show that the effective earth's radius model is sufficiently accurate for computing ray paths for most weather radar applications (hint: all elevation angles, heights to 20 km, angular resolution worse than 0.5°, and ranges to 300 km) provided that the refractive index gradient is about $-1/4a$ [e.g., compare the $h(s)$ from Eq. (2.30) to the h obtained using the effective earth radius model].

2.6   In the late evening you are observing a storm at 60-km range with a 10-cm radar having a beam width of 1.0°. The beam axis is pointed at an elevation of 0.5°. Rawinsonde data show a strong surface-based inversion layer in which both dry bulb and dew point temperatures vary nearly linearly between the two data levels shown in the table.

Rawinsonde data (6/19/80): KOUN: 2215 CST						
	P (mb)	h (m) MSL	T (°C)	T_D	P_W (mb)	N
surface	970	450	26	20		
	900	1000	26	11		

(a) Complete the table entries and determine the height of the beam axis above the earth at the storm and compare its height assuming (1) a 4/3 earth radius model, or (2) no correction for refractive index effects.

(b) Using the equation for ray height versus great circle distances, determine the vertical width of the beam at the storm. Compare with the beamwidth (in km) assuming (1) or (2) in (a).

2.7 Assume a refractive index height dependence

$$n = n_s - \beta h.$$

If $h \ll a$, show that the height of a ray above the earth versus great circle distance s is

$$h(s) \approx \frac{\left[\dfrac{s(1 - \beta a)}{(\cos \theta_e)} + a \cdot \sin \theta_e \right]^2 - a^2 \sin^2 \theta_e}{2a(1 - \beta a)}.$$

2.8 Sketch the height dependence of temperature and partial pressure of water vapor that might lead to anomalous reception of echoes from ground objects below the radar's normal horizon.

2.9 Show that the integral

$$s(h) = \int_0^h \frac{aC \, dh}{R\sqrt{R^2 n^2(h) - C^2}},$$

where $C = a \cdot n(0) \cos \theta_e$, is a solution to the differential equation (2.21) which describes the ray path.

3

Radar and Its Environment

We now describe the important elements of a pulsed-Doppler radar, with particular emphasis on its application to observation of weather. In this chapter, Doppler radar principles for detection and ranging of a single point scatterer are summarized; in the following chapter they are extended to the conglomerate of scatterers (e.g., aerosols, hydrometeors such as raindrops, snow, etc.). We sequentially describe aspects of the radar (starting with the transmitter) and examine the effects of attenuation as the radar pulse propagates out to and is scattered by a hydrometeor and returns to the receiver where it is transformed from a microwave pulse to one that can be displayed on video equipment.

3.1 The Doppler Radar (Transmitting Aspects)

Figure 3.1 is a block diagram of the principal components of a simplified pulsed-Doppler radar. This is a schematic of a homodyne system in which there are no intermediate frequency (e.g., 60-MHz) circuits found in most radars to improve performance. Nevertheless, this simplified homodyne radar illustrates all the basic principles of a Doppler radar. The stabilized local oscillator (STALO) generates a continuous-wave (cw) signal of nearly perfect sinusoid form (i.e., an extremely coherent signal) which is modulated (e.g., pulsed on and off) and amplified by a klystron to produce intense microwave power. The combination of an oscillator and power amplifier (MOPA: *m*aster oscillator ad *p*ower *a*mplifier) is usually used as a transmitter because it produces a high-power microwave pulse of fine spectral purity (i.e., the absence of power at frequencies other than the intended ones).

Development of high-gain klystron amplifiers in the 1950s made practical the generation of high-power microwaves that are phase coherent from pulse to pulse, a requirement for pulsed-Doppler radars if the velocity of objects is to be measured. Radar pulses are phase coherent from pulse to pulse if the phase angle ψ_t (Section 2.1; a subscript is added to distinguish the transmitter phase angle from other phases introduced in later sections of this chapter) for each pulse is fixed (e.g., by a STALO in a MOPA transmitter) or measured. This measurement is necessary when phase incoherent oscillators such as magnetrons

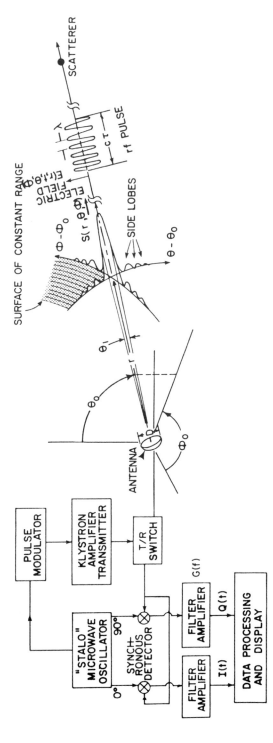

Fig. 3.1 Simplified Doppler radar block diagram.

are used as transmitters. But, the spectral purity of a magnetron oscillator is not as clean as that achieved with a MOPA transmitter. Spectral purity of the transmitted pulse is of practical importance for achieving good ground clutter cancellation (i.e., suppressing echoes from stationary objects on the ground), and in using Doppler spectral analysis to detect, for example, weakly reflecting tornadoes in presence of strong echoes from other scatterers outside the tornado (Chapter 9).

The pulse modulator generates a train of microwave pulses that are spaced at the pulse repetition time (PRT) T_s interval; each pulse has duration τ of about 1 μs. An idealized transmitted pulse of power density can be represented as $S(r, \theta, \phi)U(t - r/c)$, where

$$U(t - r/c) = \begin{cases} 1, & r/c \leq t \leq (r/c + \tau), \\ 0 & \text{otherwise.} \end{cases} \tag{3.1}$$

Equation (3.1) only approximates the actual pulse shape usually found in most radars, so in practice the pulse width τ is defined as the time between instances when the power is one-quarter of the peak (Taylor and Mattern, 1970). $S(r, \theta, \phi)$ illuminates hydrometeors as the pulse propagates within a narrow beam, and a tiny fraction of this radiation is scattered toward a receiver located, in most cases, at the transmitter site. Furthermore, for economic reasons, the same antenna is shared by the transmitter and receiver.

The transmit/receive (T/R) switch connects the transmitter to the antenna during the time τ, whereas the receiver (i.e., the synchronous detector and amplifiers) is connected during the time interval $T_s - \tau$, the "listening period." The T/R switching is not performed instantaneously, and there is a period of time (usually a few tens of microseconds) wherein the receiver does not have full sensitivity for detection.

The Doppler-shifted (if the scatterer has a component of velocity toward or away from the radar) echo pulse and the cw output of the STALO are applied to a pair of synchronous detectors (Fig. 3.1). The receiver is then said to be coherent (Section 3.4.3). If the STALO is not connected to the detector, the receiver is said to be incoherent (Section 3.4.2).

3.1.1 The Electromagnetic Beam

The microwave pulse leaves the antenna in an essentially collimated beam of diameter D_a equal to that of the antenna–reflector (Fig. 3.2). But, because of diffraction, the electromagnetic beam begins to spread at a range $r = D_a^2/\lambda$ into a conical one having an angular width given by Eq. (2.4). The beamwidth θ_1 (Fig. 3.1) is commonly specified as the angle (i.e., the 3-dB beamwidth) within which the microwave radiation is at least one-half its peak intensity.

The radiation pattern $S(\theta, \phi)$ describes the angular distribution of power density that emanates from the antenna. It is impossible to confine all the energy into a narrow conical beam, and some of it inevitably falls outside the mainlobe

Fig. 3.2 The National Severe Storms Laboratory's weather radar antenna–reflector, shown inside its protective geodesic radome (radar cover dome). The reflector, a paraboloid of revolution, has a diameter of 9.14 m. The radiation source is the horn at the end of the curved (black) waveguide. The tubes extending to the right support the source and waveguide.

into sidelobes (Fig. 3.1). Usually the power density in any sidelobe is less that 1/100th of the peak density in the mainlobe. Furthermore, the sum total of power in sidelobes can often be held to just a few percent of that transmitted within the mainlobe (Sherman, 1970). This integrated sidelobe power level is an important consideration in the design of weather radars because scatterers are distributed over vast regions.

The antenna–reflector is usually a paraboloid of revolution and is illuminated by a source located at the focal point (Fig. 3.2). The illumination is made to be nonuniform across the reflector in order to reduce sidelobe levels, and often its intensity versus distance ρ from the axis has the dependence

$[1 - 4(\rho/D_a)^2]^2$. In this case the normalized power density pattern $f^2(\theta)$ has the following angular dependence symmetrical about the reflector axis (Sherman, 1970):

$$f^2(\theta) = \frac{S(\theta)}{S(0)} = \left\{ \frac{8J_2[(\pi D \sin \theta)/\lambda]}{[(\pi D \sin \theta)/\lambda]^2} \right\}^2, \qquad (3.2a)$$

where θ is the angular distance from the beam axis and J_2 is the Bessel function of second order. This formula describes quite accurately the radiation pattern in the angular region containing the first few sidelobes, but outside this region actual radiation patterns often have larger contributions from power scattered by imperfections in the reflector and structures supporting the source. When the beamwidth is small compared to 1 rad, Eq. (3.2a) shows that the 3-dB beamwidth θ_1 is

$$\theta_1 = 1.27\lambda/D \qquad \text{rad}, \qquad (3.2b)$$

and the width between the first nulls becomes $3.27\lambda/D_a$ (rad).

The transmitted microwave energy is within a $c\tau$-thick spherical shell that expands (i.e., propagates) at a speed c. Thus, the power density $S_i(\theta, \phi)$ incident on scatterers decreases inversely with r^2, even though the pulse power P'_t transmitted through any enclosing sphere is constant. This is the reason for the $1/r$ dependence of $\mathbf{E}$ in Eq. (2.2). Transmitted power is not constant during the pulse, so pulse power is defined as the average, over the duration of the pulse, of the power averaged over the *period of the microwave signal* (Section 2.1). Because of losses in the antenna, its transmission lines, and its protective radome, the pulse power P_t delivered to the antenna's input port is larger than P'_t.

3.1.2   Antenna Gain

If P'_t were radiated equally in all directions, S would be equal to $P'_t/4\pi r^2$. However, the antenna focuses radiation into a narrow angular region (i.e., the beamwidth) wherein the peak radiation intensity S_p is many times stronger than $P'_t/4\pi r^2$. The ratio

$$S_p/(P'_t/4\pi r^2) \equiv g'_t \qquad (3.3)$$

defines the *maximum directional* gain of the antenna. Measurement of P'_t is difficult, and the antenna engineer therefore measures the power P_t delivered to the antenna's input port and S_p at distances far (i.e., $r > 2D_a^2/\lambda$) from the antenna. In this case the computed gain g_t accounts for losses of energy associated with the antenna system (e.g., radome and waveguide). Then, in the absence of attenuation (Section 3.3), the incident radiation power density at range r is given by

$$S_i(\theta, \phi) = P_t g_t f^2(\theta, \phi)/4\pi r^2, \qquad (3.4)$$

where $f^2(\theta, \phi)$ is the normalized [i.e., $f(\theta, \phi) = 1$ at θ_0, ϕ_0; Fig. 3.1] power gain pattern, and g_t is simply the antenna *power* gain along the beam axis.

3.2 The Scattering Cross Section

The *cross section* σ of a scatterer (e.g., hydrometeor) is an apparent area that intercepts a power σS_i which, if radiated (i.e., scattered) isotropically, produces at the receiver a power density.

$$S_r = S_i \sigma(\theta', \phi')/4\pi r^2 \qquad (3.5)$$

equal to that scattered by the actual hydrometeor. The definition (3.5) indicates that the scatterers do not radiate power isotropically, and hence the scatterer's cross section $\sigma(\theta', \phi')$ can depend on the relative location of the transmitter and receiver. The polar angles θ', ϕ' to a receiver are referenced to a polar axis drawn from the antenna to the scatterer.

It is easy to deduce that the scattering cross section may have no resemblance to the scatterer's physical cross section. In fact, thin metallized fibers (chaff) of real cross section small compared to λ^2 can have scattering cross sections many times larger than their physical area. For example, consider a chaff element (needle) with a 0.1-mm diameter that has length l parallel to the electric field vector and that is resonant (e.g., $l = \lambda/2$) at the wavelength λ of an incident wave. It has a maximum backscattering cross section $\sigma_{bm} = 0.857\lambda^2$ (Nathanson, 1969, p. 223) equal to 8.6×10^{-3} m^2 at $\lambda = 10$ cm, whereas the physical cross section is only 5×10^{-6} m^2. For a needle oriented so that **E** is not parallel to its axes, $\sigma_b = \sigma_{bm}(1 - \sin^2 \psi/\sin^2 \theta) \cos^2[(\pi/2) \cos \theta]$. The angle ψ is the angle between the chaff axis and the plane containing the Poynting vector $\mathbf{S}_i$ and **E**, and θ is the angle of $\mathbf{S}_i$, relative to the axis.

On the other hand, a large plain sheet of metal can have an extremely small backscattering cross section when it is oriented so its normal is not along the line to the radar. The scattering cross section of the metal sheet can be many orders of magnitude smaller than its physical area projected on a plane perpendicular to $\mathbf{S}_i$.

The backscattering cross section σ_b of a spherical water drop of diameter D small compared to λ (i.e., $D \leq \lambda/16$; large drops are oblate and their cross sections depend on polarization of the incident field, Section 8.5.3) is well approximated by

$$\sigma_b \approx (\pi^5/\lambda^4)|K_m|^2 D^6 \qquad (3.6)$$

where $K_m = (m^2 - 1)/(m^2 + 2)$ and $m = n - jn\kappa$ is the complex refractive index of water. The refractive index is n, and κ is the attenuation index (Born and Wolf, 1964, p. 613). Some authors (Battan, 1973) define an absorption coefficient $k = n\kappa$ and list its value as a function of wavelength and temperature. Raindrop diameters can be as large as 8 mm (see, for example, Fig. 8.1).

Therefore, only for wavelengths about 10 cm and longer can the simple formula (3.6) be applied to rain scatter. At wavelengths around 10 cm, values of n for water are about 9 and are relatively independent of temperature, whereas k has values ranging from 0.63 to 1.47 for temperatures from 20 to 0°C. On the other hand, the refractive index of ice is 1.78, also independent of temperature; but k is relatively small, ranging from 2.4×10^{-3} to 5.5×10^{-4} for temperature from 0 to −20°C. Values of m at other wavelengths are given by Battan (1973). $|K_m|^2$ for water varies between 0.91 and 0.93 for wavelengths between 0.01 and 0.10 m and is practically independent of temperature. Ice spheres have $|K_m|^2$ of about 0.18 (for a density 0.917 g cm^{-3}), a value independent of temperature as well as wavelength in the microwave region.

The relation (3.6) is called the Rayleigh approximation and it has the same wavelength dependence as the cross section of atmospheric molecules, which scatter light and whose diameters are small compared to optical wavelengths. Equation (3.6) shows that waves at shorter λ are more strongly scattered—a fact that Rayleigh used to explain why the sky is blue.

Because the refractive index of liquid or ice water is nearly independent of frequency within the weather radar band, Atlas and Ludlum (1961) were able to estimate the backscattering cross sections of water spheres at various wavelengths from Aden's (1951) calculations at the single wavelength of 16.23 cm. Herman and Battan (1961) made calculations at 3.21 cm using then-available values for the dielectric constants of ice and water. Plots in Fig. 3.3 were obtained using up-to-date values of the dielectric constants (Ray, 1972). These cross sections have more ripples for ice spheres at 3.21 cm than determined by Atlas and Ludlum, but are in agreement with Herman and Battan (1961). A satisfactory explanation for these fluctuations did not emerge until Probert-Jones (1984) reported that internal resonance was responsible.

Although water drops larger than 6 mm are unlikely, water-coated hailstones can have backscatter cross sections nearly equivalent to water spheres of the same diameter. For example, if one-twentieth of the diameter of a 4-mm ice sphere melts to form a uniform shell of water around the stone, the backscattering cross section is nearly the value for a water drop of the same diameter (Battan, 1973).

In examining Fig. 3.3, it is surprising that there are ranges of diameter for which the backscattering cross section of ice is significantly larger than for water spheres of the same diameter. Although this result was reported by Ryde (1946), who made the first calculations on the backscatter and attenuation of microwaves incident on hydrometeors, early radar meteorologists expressed doubt that ice could scatter so much more than water (Atlas and Donaldson, 1954). This doubt arose because ice has a significantly smaller scattering cross section than water of the same diameter for diameters extremely small (i.e., the Rayleigh approximation holds) or extremely large (i.e., geometrical optics solution can be applied). In the limit of infinitely large diameters the normalized backscatter cross section $4\sigma_b / \pi D^2$ is equal to the power reflection coefficient of a

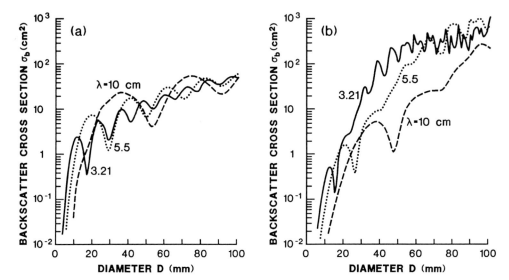

Fig. 3.3 Backscattering cross section σ_b for spheres: (a) Liquid water; the complex relative permittivity m^2 (at 0°C) is $80.255 + j24.313$, $65.476 + j37.026$, and $44.593 + j41.449$ at wavelengths of 10 cm, 5.5 cm and 3.21 cm (Ray, 1972). (b) Ice (m^2 is $3.16835 + j0.02492$, $3.16835 + j0.01068$, and $3.16835 + j0.0089$) at the indicated wavelengths (computed by N. Balakrishnan, Indian Institute of Science, Bangalore, India).

plane surface and is given by the formula (Born and Wolf, 1964, p. 41),

$$\frac{4\sigma_b}{\pi D^2} = \frac{|m-1|^2}{|m+1|^2}. \tag{3.7}$$

Applying Eqs. (3.6) and (3.7) it can be shown that the backscatter cross section of ice is about 0.2 times that of water spheres if $D \ll \lambda$ and about 0.12 times if $D \to \infty$. A set of carefully conducted experiments and independent numerical calculations, however, confirmed that ice spheres can have considerably larger backscatter cross section than water spheres of the same diameter (Atlas et al., 1960). Atlas et al. (1960) gave a physical explanation for the enhanced backscatter in terms of a Luneberg lens mechanism.

Calculation shows that ice spheres have total scattering cross sections (defined as the integral of scattered power density in all directions, Section 3.3) not too much different than water spheres (Battan, 1973, Fig. 6.5) even though the backscatter for ice is an order of magnitude larger. Thus larger backscattering cross section for ice implies a scattering cross section dependence on θ', ϕ' in which the radiation pattern of scattered energy is more directive in backscatter for an ice sphere than for a water sphere.

If the particle is small compared to wavelength, the scattered energy is radiated nearly isotropically. Large objects have a much more directive scatter radiation pattern, and often more radiation flows in directions other than back to the transmitter. For example, scattered power density in the direction of **r** (forward scatter) can be 100–1000 times larger than that returned in the direction of the source (the Mie effect; Born and Wolf, 1964).

There is an abundance of experimental and theoretical work that relates a particle's cross sections to its shape, temperature, size, and mixture of phases (e.g., water-coated ice spheres). These works are reviewed by Battan (1973) and Atlas (1964).

3.3 Attenuation

Were it not for electromagnetic energy absorption by water or ice drops, radars with shorter wavelength radiation would be much more in use because of their superior angular resolution [Eq. (3.2b)]. But short-wavelength (e.g., $\lambda = 3$ cm) radars suffer echo power loss that can be 100 times larger than that of radars operated with $\lambda \geq 10$ cm.

Each drop absorbs an amount of power P_L, that can be expressed simply

$$P_L = \sigma_a S_i, \tag{3.8}$$

where σ_a is the absorption cross section, an apparent area that intercepts from the incident radiation a power equal to the power dissipated as heat in the drop. There is no simple formula for σ_a applicable to drop diameters found in moderate-to-heavy rain, even if diameters satisfy the condition $D \leq \lambda/16$ (Section 3.3.1). Thus one needs to resort to numerical evaluations of complicated expressions or measurements; furthermore for nonspherical hydrometeors σ_a depends on the polarization of the incident electric field.

A wave suffers power loss both from energy absorption and scatter. Analogous to the backscattering cross section, there is a total scatter cross section σ_s that accounts for the total power scattered by a particle. Total scatter cross section is an area that, when multiplied by the incident power density, gives a power equal to that scattered by the particle; σ_s is proportional to the integral of scattered power density $S_r(\theta', \phi')$ over a sphere enclosing the particle. For small spheres $\sigma_s = 2\sigma_b/3$ (Battan, 1973). Thus the total power extracted from a wave is proportional to the sum $\sigma_a + \sigma_s$, defined as the attenuation or extinction cross section σ_e.

If we assume that drops within an elemental volume $\Delta V(r)$ do not significantly alter the incident power density S_i within this volume (i.e., we use the Born approximation, which neglects the scattering of the scattered field), then the power density change ΔS_i in a wave propagating a short distance Δr through the volume is

$$\Delta S_i = -\frac{\Delta r}{\Delta V} \sum_{n=1}^{N} (\sigma_{an} + \sigma_{sn}) S_i, \tag{3.9}$$

where the negative sign signifies loss, the summation extends over all N drops within ΔV, and σ_{an}, σ_{sn} are the absorption and scattering cross section, respectively, of the nth particle. In the limit $\Delta r \rightarrow 0$, S_i can be considered a constant within ΔV and hence can be placed outside the summation. The rate of change in power density is then

$$\lim_{\Delta r \rightarrow 0}\left(\frac{\Delta S_i}{\Delta r}\right) = \frac{dS_i}{dr} = -kS_i \tag{3.10}$$

and the power density at any range r is the integral solution of Eq. (3.10),

$$S_i(r_2) = S_i(r_1)\exp\left(-\int_{r_1}^{r_2} k\, dr\right), \tag{3.11a}$$

where

$$k \equiv \lim_{\Delta r \rightarrow 0} \sum \frac{\sigma_{an} + \sigma_{sn}}{\Delta V}$$

$$= \int_0^{\infty} N(D, \mathbf{r})\sigma_e(D)\, dD \tag{3.11b}$$

is the specific attenuation, or (as denoted by some authors) the attenuation coefficient. $N(D, \mathbf{r})$, named the drop size distribution, is the expected number density of hydrometeors per unit diameter. The product $N(D, \mathbf{r})\, dD$ gives the number of hydrometeors per unit volume having diameters in the interval dD about D. The specific attenuation expressed in decibels per kilometer is

$$K \equiv \frac{d}{dr_2}\left[10 \log \frac{S(r_1)}{S(r_2)}\right] = 4.34 \times 10^3 k \quad \text{dB km}^{-1} \tag{3.12}$$

when k has units of inverse meters.

Combining Eqs. (3.4), (3.5), and (3.11) we deduce that the echo power density S_r at the radar antenna is

$$S_r(r, \theta, \phi) = P_t g_t f^2(\theta, \phi)\sigma_b/(4\pi r^2)^2 l^2 \quad \text{W m}^{-2}, \tag{3.13a}$$

$$\ell = \exp\left(\int_0^{r} (k_g + k)\, dr\right), \tag{3.13b}$$

where l is the one-way transmission loss due to hydrometeors (i.e., k) as well as that caused by atmospheric gases (i.e., k_g), which we discuss in the next few pages.

3.3.1 Attenuation in Rain

For linearly polarized radiation, the rate at which incident energy is reduced by scatter and absorption is proportional to the amplitude component of the forward-scattered wave having the same polarization as the incident one (Born and

Wolf, 1964, p. 658). Thus, one can simply compute the forward scattering cross section using the Mie (1908) series to obtain the composite effects of absorption and total scatter. For relatively small drop-diameter wavelength ratios, the leading terms in the series solution formulated by Mie for the sphere scattering problem give the following approximation for σ_a and σ_s:

$$\sigma_a \approx (\pi^2 D^3/\lambda)\text{Im}(-K_m) \tag{3.14a}$$

and

$$\sigma_s \approx (2\pi^5 D^6/3\lambda^4)|K_m|^2. \tag{3.14b}$$

Comparison of these terms shows that $\sigma_s < \sigma_a$. Because of this, one might be tempted to use σ_a for attenuation estimation at wavelengths $\lambda \geq 10$ cm, for which all raindrops satisfy the Rayleigh condition $D \leq \lambda/16$. We must be cautious, however, because the complete solution to the scattering problem shows that there is a significant contribution to absorption from the remaining terms (Gunn and East, 1954, also reproduced in Doviak and Zrnic, 1984a, Fig. 8.10) of the series (even at $\lambda = 10$ cm although $D \leq \lambda/16$ for all raindrops). In rain at low rates and for wavelengths less than 1 cm, σ_s can be larger than σ_a (Setzer, 1970).

To determine the attenuation in rain at microwave wavelengths, we must retain higher-order terms of the Mie solution for σ_e. Examples of $\sigma_e(D)$ for four wavelengths are given in Fig. 3.4. The range of drop diameters for each wavelength is shown at the top of the figure. The solutions are for spherical water drops, and because drops noticeably flatten when $D \gtrsim 0.3$ mm (Fig. 8.14), the solutions are approximate.

We note that as the drop diameter becomes large compared to the wavelength, the extinction coefficient approaches an asymptotic value of 2.0. This result appears somewhat paradoxical, because with objects large compared to wavelength one would have expected the geometrical optics approximation to apply so that the sum of scatter and absorption cross section would be equal to $\pi D^2/4$. The explanation of this apparent contradiction is that, no matter how large the scatterer, there is always a narrow region (the neighborhood of the edge of the geometric shadow) where the geometric optics approximation does not hold. For further discussion, the reader is referred to the article by Sinclair (1947).

Given the extinction cross section $\sigma_e(D)$, the specific attenuation is determined if $N(D)$ is known. The combination of Eqs. (3.11b) and (3.12) yields

$$K = 4.34 \times 10^3 \int_0^\infty N(D)\sigma_e(D)\,dD \quad \text{dB km}^{-1}$$

$$= K(\text{absorption}) + K(\text{scatter}). \tag{3.15}$$

At centimeter wavelengths, absorption loss dominates for all rain rates; but when wavelength is less than 1 cm, scatter can dominate absorption at low rain rates (Fig. 8.12, Doviak and Zrnic, 1984a). There is no easy solution to Eq. (3.15), so investigators have resorted to numerical methods using specified

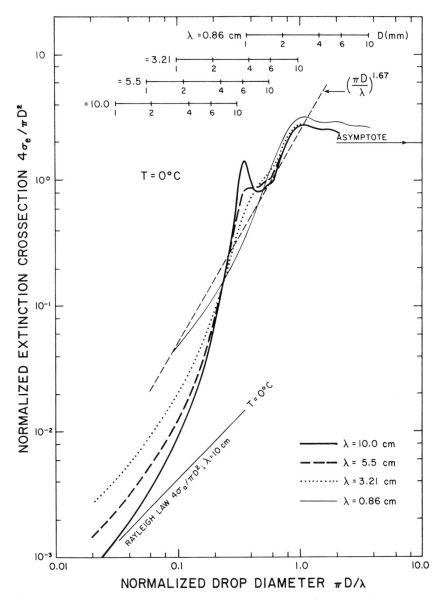

Fig. 3.4   The normalized extinction cross section versus normalized drop diameter for spherical water drops at a temperature of 0°C and at four wavelengths. (Data from Herman *et al.*, 1961.) The solid straight line is the asymptotic absorption cross section in the Rayleigh limit. The dashed straight line shows the power law that fits the data at $\lambda = 0.86$ cm (Section 8.4.2).

drop size distributions. Burrows and Attwood (1949) have used the Laws and Parsons (1943) drop size data (Fig. 8.5) to compute specific attenuation at various wavelengths and temperature. A least squares fit applied to the logarithms of the Burrows and Attwood data yields the following one-way, specific attenuations at a temperature of 18°C (Fig. 3.5):

$$K_r = \begin{cases} 0.000343R^{0.97} \ (\text{dB km}^{-1}), & \lambda = 10 \ \text{cm}, & (3.16a) \\ 0.0018R^{1.05}, & \lambda = 5 \ \text{cm}, & (3.16b) \\ 0.01R^{1.21}, & \lambda = 3.2 \ \text{cm}, & (3.16c) \end{cases}$$

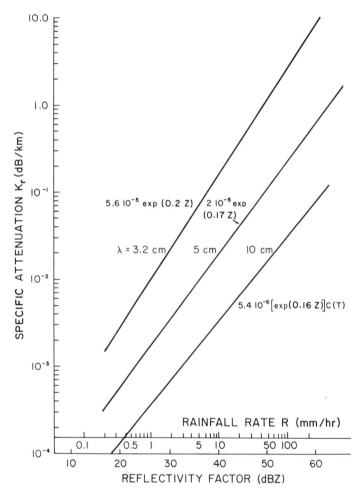

Fig. 3.5 Specific attenuation versus rainfall rate ($T = 18°C$). The Laws and Parson (1943) drop size distribution is assumed. The relation between the rainfall rate and drop size distribution is developed in Chapter 8, and the reflectivity factor is defined in Section 4.4.5. Temperature adjustment factor $C(T) = 2 \exp(-0.035T)$. $Z = 10 \log(400R^{1.4})$. (Z is in units of dBZ.)

where R is the rainfall rate in millimeters per hour. At other temperatures a multiplicative correction to Eqs. (3.16) must be applied. The correction factor is independent of the rainfall rate for a 10-cm wavelength, and an empirical formula valid for temperatures between $0°$ and $40°C$ is presented in the caption of Fig. 3.5. At shorter wavelengths the dependence on rainfall rate and temperature is tabulated by Burrows and Attwood (1949).

Remove measurements of short wavelength attenuation is difficult. A usual approach is to use nonattenuating wavelengths to measure the reflectivity and relate it to short wavelength attenuation; but there is considerable scatter between these two variables due to variations in the drop size distributions. On the other hand specific differential phase (Section 8.4.3) is much less affected by drop size distributions and hence is more tightly related to attenuation (Bringi *et al.*, 1990).

Note that rainwater on radomes significantly attenuates weather signals (Collier, 1989).

3.3.2 Attenuation in Clouds

Cloud and fog particles (liquid or ice) generally are not detected with weather radars but nevertheless can cause measurable attenuation of radar signals. Because the total scattering cross section σ_s [Eq. (3.14b)] is proportional to the sixth power of the drop diameter, it is usually small compared to σ_a [Eq. (3.14a)] for cloud droplets if $D \ll \lambda/16$. At wavelengths usually used for meteorological radars (i.e., ≥ 3 cm), $\sigma_s \ll \sigma_a$ and therefore specific attenuation for cloud is well approximated by

$$k_c \approx \frac{\pi^2}{\lambda} \mathrm{Im}(-K_m) \int_0^\infty N(D,\mathbf{r})D^3 \, dD. \tag{3.17a}$$

Now liquid water density M is

$$M = \frac{\rho_w \pi}{6} \int_0^\infty N(D,\mathbf{r})D^3 \, dD, \tag{3.17b}$$

where the water density $\rho_w = 10^3$ kg m^{-3}. Thus the specific attenuation by cloud water can be expressed in terms of liquid water density as

$$k_c \approx \frac{6\pi M}{\rho_w \lambda} \mathrm{Im}(-K_m). \tag{3.17c}$$

For weather radar wavelengths between 3.2 and 10 cm, and clouds at $0°C$ (Battan, 1973, p. 39),

$$\mathrm{Im}(-K_m) \simeq \frac{1.1 \times 10^{-3}}{\lambda} \tag{3.17d}$$

with an accuracy of about 3%, where λ is in meters. For example, at $\lambda = 10$ cm, a two-way attenuation of 0.9 dB is computed for propagation through 50 km of 1.0 g m^{-3} liquid-water cloud. At lower temperatures (i.e., supercooled cloud water), attenuation is larger. Thus, even though the radar might not detect the presence of clouds (e.g., σ_b, for a cloud droplet of 10 μm diameter is twelve orders of magnitude less than σ_b for a precipitation particle of 1 mm diameter), its attenuation could be significant. Tables of specific attenuations for other wavelengths and temperatures are given by Battan (1973, p. 70).

3.3.3 Attenuation in Snow

Battan (1973) gives the following formula for specific attenuation in dry snow at 0°C:

$$K_s = 3.5 \times 10^{-2}(R^2/\lambda^4) + 2.2 \times 10^{-3}(R/\lambda) \quad \text{dB km}^{-1}, \tag{3.18}$$

where the first term is due to scattering and the second to absorption, R is in millimeters per hour of melted snow, and λ is measured in centimeters. At a wavelength of 10 cm, K_s can be safely neglected; but at shorter wavelengths it might be appreciable, especially if the rate R is large. Wet snow and water-coated ice attenuate more than dry ice, but because of the irregular shapes of snowflakes there are no simple formulas. Graphs of specific attenuation by dry and water-coated ice spheres are given by Battan (1973).

3.3.4 Attenuation in Gases

Besides attenuation due to rain and cloud droplets, there is attenuation due to absorption of energy by the atmosphere's molecular constituents, mainly water vapor and oxygen. Because the total scattering cross section σ_s for an air molecule is much smaller than its absorption cross section, the attenuation due to atmospheric gases is well approximated by the absorption loss. This specific attenuation k_g is appreciable when storms are far away and beam elevation is low, even at $\lambda = 10$ cm.

The gaseous attenuation depends not only on the propagation path length but also on the depth of the troposphere penetrated. This dependence for the standard atmosphere is illustrated in Fig. 3.6, which has been plotted for the case $\lambda = 10$ cm from theoretical curves published by Blake (1970), who also gives the absorption loss or wavelengths from 3 m to 3 cm. At $\lambda = 3$ cm the two-way absorption loss can be as large as 5.6 dB for distant scatterers near the horizon. An empirical formula that approximates the two-way attenuation $2L_g$ for elevation angles $\theta_e < 10°$ and slant ranges $r < 200$ km is

$$2L_g \equiv 20 \log l_g = [0.4 + 3.45 \exp(-\theta_e/1.8)]$$
$$\times (1 - \exp\{-r/[27.8 + 154 \exp(-\theta_e/2.2)]\}) \quad \text{dB.} \tag{3.19}$$

This formula approximates the theoretical loss curves (Fig. 3.6) to within

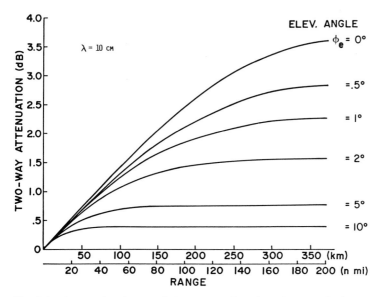

Fig. 3.6 Attenuation (two-way) for propagation through a standard atmosphere ($\lambda = 10$ cm). Central Radio Propagation Laboratory's (CRPL) exponential reference atmosphere (Fig. 2.7) is used to compute ray paths, and the pressure–temperature profile is based on the International Civil Aviation Organization (ICAO) standard atmosphere.

0.2 dB. The two-way gaseous attenuation at two other wavelengths often used by radar meteorologists (i.e., $\lambda = 5$ and 3.3 cm) can be obtained to within the same accuracy (i.e., ~0.2 dB) by multiplying $2L_g$ with the coefficient 1.2 (for $\lambda \simeq 5$ cm) and 1.5 (for $\lambda \simeq 3.3$ cm).

3.4 The Doppler Radar (Receiving Aspects)

The echo power P_r (in watts) collected by the antenna system from a wave scattered by a hydrometeor at r, θ, ϕ, is

$$P_r = S_r(r, \theta, \phi)A_e(\theta, \phi), \tag{3.20}$$

where A_e is the effective collection area of the antenna for radiation from direction θ, ϕ. We shall prove that

$$A_e = (g_r\lambda^2/4\pi)f^2(\theta, \phi), \tag{3.21}$$

where g_r is the gain of the receiving antenna, equal to g_t if the transmitting antenna is used for echo reception and P_r is measured at the same location in the antenna system as P_t. Consider two antennas 1 and 2 at a distance r from each other, and assume that antenna 1 is radiating while 2 is receiving the power:

$$P_{r2} = [P_{t1}g_1f_1^2(\theta, \phi)/4\pi r^2]A_{e2}. \tag{3.22a}$$

Interchanging the roles of the two antennas produces

$$P_{r1} = [P_{t2}g_2f_2^2(\theta, \phi)/4\pi r^2]A_{e1}. \tag{3.22b}$$

If the transmitted powers are made equal, $P_{t2} = P_{t1}$, then by reciprocity the receiving powers must be equal; i.e., $P_{r1} = P_{r2}$. From this we find

$$A_{e1}/g_1f_1^2(\theta, \phi) = A_{e2}/g_2f_2^2(\theta, \phi) = \text{const.} \tag{3.23}$$

Because the constant in Eq. (3.23) must be independent of antenna type, one can calculate it for the simplest possible antenna [e.g., the dipole for which one obtains $\lambda^2/4\pi$ (Jordan and Balmain, 1968, p. 377)]. Thus, Eq. (3.21) follows. For aperture antennas and scatterers along the beam, A_e is smaller than the physical area of the antenna reflector. For example, the weather radar antenna shown in Fig. 3.2 has $A_e = 42 \text{ m}^2$, whereas its physical area is 65.7 m^2. The significantly smaller area A_e is caused principally by a taper in the illumination of the reflector, which in the reception mode is equivalent to weighting the power density reflected from the parabolic surface. Radome and waveguide losses are other contributors to the reduction of A_e.

3.4.1 The Radar Equation

Combining Eqs. (3.13a), (3.20), and (3.21), we arrive at the radar equation for a point scatterer having backscatter cross section σ_b.

$$P_r = P_t g^2 \lambda^2 \sigma_b f^4(\theta, \phi)/(4\pi)^3 r^4 l^2, \tag{3.24}$$

where we have substituted $g_t = g_r = g$ because the same antenna is used for transmitting and receiving. This equation relates echo power P_r to the radar parameters and scatterer location; it assumes P_r is measured at the same point in the radar as P_t, and that g includes transmission line losses (e.g., in waveguides) from the antenna to the point where P_t and P_r are measured. Often P_r is expressed in decibel units dBm, where $\text{dBm} \equiv 10 \log_{10} P_r$, when P_r is in units of milliwatts.

Although Eq. (3.24) stipulates echo power given radar parameters, it doesn't specify limits on the detection of weak echoes. Among signal processing and radar parameters [some of which are contained in Eq. (3.24)] that determine this limit, system noise power is of singular importance. This unwanted microwave power at the input to the receiver arises from emissions in the radar's electronic circuits as well as emissions from the air, ground, and space that are intercepted by the antenna (Section 3.5.1). Echo signal as weak as receiver noise power is readily detected in the output of the radar's signal processor or on its displays. Furthermore, Doppler velocity measurements at this level of echo power often have acceptable accuracies (Section 6.4) for meteorological purposes (Section 11.8).

As an example of the sensitivity that a weather radar has to detect particles of extremely small cross section, consider the parameters of the new WSR-88D radar operated by the National Weather Service (NWS) (Table 3.1). Substitut-

Table 3.1
WSR-88D Specifications

Antenna subsystem

Radome

Type	Fiber glass skin foam sandwich
Diameter	11.89 m
rf loss—two way	0.3 dB at 2995 MHz

Pedestal

Type	Elevation Over Azimuth	
	Azimuth	Elevation
Scanning rate—maximum	$30° \ s^{-1}$	$30° \ s^{-1}$
Acceleration	$15° \ s^{-2}$	$15° \ s^{-2}$
Mechanical limits	$-1°$ to $+60°$	

Reflector[a]

Type	Paraboloid of revolution
Polarization	Linear[b]
Reflector diameter	8.54 m
Gain[c]	44.5 dB
Beam width	1°
First sidelobe level	−26 dB (with radome)

Transmitter and receiver subsystem

Transmitter

Type	Master oscillator power amplifier
Frequency	2700 MHz to 3000 MHz
Pulse power[c]	475 kW
Pulse width	1.57 μs and 4.57 μs
rf duty cycle	0.002 maximum
PRFs	
short pulse	
(eight selectable)	320 Hz to 1300 Hz
long pulse:	320 Hz and 450 Hz

Receiver

Type	Linear
Dynamic range	93 dB
Intermediate frequency	57.6 MHz
System noise power[c]	−113 dBm

Filter

Short pulse:	Analog filter; bandwidth (3 dB): 0.63 MHz
	bandwidth (6 dB): 0.80 MHz
Long pulse:	Additional digital filtering; 3 samples (spaced 0.25 km) of I and Q are averaged. Output samples are spaced at 0.5 km intervals.

System performance
A reflectivity factor of -7.5 dBZ at 50 km must produce an SNR > 0 dB.

[a] Antenna specifications include the effects of the radome.
[b] Initially the first radars will have circular polarization which will be changed.
[c] Transmitted power, antenna gain, and receiver noise power are measured at the antenna port.

ing the appropriate parameters from this table into Eq. (3.24) and assuming that an echo power equal to receiver noise results in a detectable signal, a particle at a range of 20 km with a backscatter cross section as small as σ_b(minimum) $= 2 \times 10^{-7}$ m^2 can be detected. That is, a single water drop with a diameter of 6.3 mm could be detected at 20 km. It would not take, therefore, a very large number of drops to provide a detectable signal at larger range (e.g., 200 km); moreover the transmitted power and receiver performance of this radar is only moderate in relation to advanced systems in use today (such as deep-space probes). Scatterers such as birds and insects have sufficiently large backscattering cross section and therefore are easily visible to the radar. Table 3.2 is a list of backscattering cross sections of common objects. Some have large variations that depend on the object's aspect.

3.4.2 The Incoherent Receiver

If the synchronous detector is replaced with a video detector and the STALO output is not added to the incoming echoes, the radar receiver is said to be incoherent. Doppler measurements require a coherent receiver (Section 3.4.3).

Because of the large span of echo amplitudes, a logarithmic amplifier usually precedes the video detector; Fig. 3.7 shows the logarithm of echo intensity A^2 (amplitude squared, which is proportional to echo power). In this way, both strong echoes from nearby objects on the ground and weak echoes from atmospheric scatterers in clear air can be displayed simultaneously. The echoes displayed in Fig. 3.7 were obtained during a period when the sky was clear. This figure shows the signal at the output from the amplifier-filter (Fig. 3.1 without the STALO connected to the detector) when a multitude of scatterers (both airborne and grounded) are being illuminated by the radar. The decibel range, $10 \log[A^2]$, of echo intensity displayed in Fig. 3.7 is about 90 dB; that is, the strongest echo is about 10^9 times more intense than the receiver noise power ($\approx 10^{-14}$ W or -110 dBm), which is actually larger than the power of atmo-

Table 3.2
Backscattering Cross Sections (at $\lambda = 10$ cm)

Object	σ_b (m^2)
C-54 aircraft	10 to 1000
Man	0.14 to 1.05
Weather balloon, sea gull	10^{-2}
Small birds	10^{-3}
Wingless hawkmoth	10^{-5}
Bee, dragonfly	3×10^{-6} to 10^{-7}
Water sphere ($D = 2$ mm)	1.8×10^{-10}
Free electron	8×10^{-30}

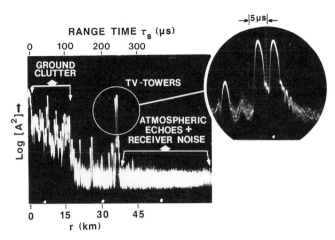

Fig. 3.7 Range time delay τ_s and strength of echoes from point scatterers (e.g., TV towers at about 40 km) and distributed scatterers. Ground clutter is a conglomerate of indistinguishable echoes from objects on the ground. The vertical scale is proportional to the logarithm of the echo power. The transmitted pulsewidth τ is 1.2 μs. Inset shows an expanded view of the tower echoes.

spheric echoes at ranges beyond 60 km. Instantaneous receiver noise power fluctuates as do weather echoes, but it has an average value (i.e., an average of power samples at a fixed range time delay τ_s) that is independent of range.

If no more than one scatterer lies in the range interval $c\tau/2$, and if the scatterer's dimensions are small compared to the smaller of $c\tau/2$ or $r_i\theta_1$, where r_i is the range to the ith scatterer, echoes are said to be returned from a "point" scatterer. Each echo is then a replica of the transmitted pulse, but in passing through the receiver, these replicated pulses are distorted by the receiver, which also adds electronic noise. These echoes can be displayed on video equipment (Fig. 3.7 highlights two from a pair of adjacent TV towers) to show their relative strengths and delay (i.e., range time delay) referenced to the time of the transmitted pulse. Weather echoes from a very large number of hydrometeors are composites of such echoes (Section 4.2), each of which is similar, but of much smaller amplitude, to those seen from TV towers.

The quasicontinuous distribution of echoes (as in Fig. 3.7) is called clutter; both atmospheric scatterers and objects on the ground contribute to it. For radars designed to detect objects such as aircraft, weather clutter is a nuisance that often obscures observation. On the other hand, weather clutter contains the needed meteorological information, and for weather radars, ground clutter and echoes from aircraft, birds, et cetera, obscure meteorological observation; consequently we reserve the term clutter for such interferring echoes.

The stretch of fluctuating amplitudes decreasing beyond about 40 km in Fig. 3.7 shows mainly echoes from airborne insects or atmospheric scatterers (e.g., refractive index irregularities) superimposed on receiver noise; whereas for ranges inside 40 km, echoes from ground objects are added to the mixture. If the density of atmospheric scatterers is spatially homogeneous and large (i.e., many insects or irregularities in refractive index within the range extent $c\tau$ of the transmitted pulse), and scatterers move relative to each other, the fluctuations of echo amplitude are exactly the same as receiver noise; therefore, it is difficult to distinguish (as in Fig. 3.7) noise from atmospheric echoes. However, Doppler measurements (not shown here) yield radial velocities, at ranges beyond 40 km, equal to those from aerosols or refractive index irregularities carried by wind.

Because of the logarithmic scale, the fluctuating signals of atmospheric echoes and receiver noise superimposed on the steady echo from ground objects appear to be less than that seen in the absence of ground clutter. Note that the receiver noise and atmospheric echoes are not evident on the expanded view of echoes from the TV tower. Nevertheless, the amplitude of atmospheric echoes is not diminished in the presence of ground clutter, and it becomes the purpose of the Doppler processor to sort out the strong steady signals due to ground objects, from sometimes weaker atmospheric echoes.

3.4.3   The Coherent Receiver (In-Phase and Quadrature Components)

The receiver in Fig. 3.1 is coherent because the output of the STALO is connected to synchronous detectors. If there is a point scatterer at range r and the bandwidth of low-noise microwave amplifiers (these are not shown in Fig. 3.1, but are usually inserted between the T/R switch and the synchronous detectors) is sufficiently large (Section 3.5.2), the echo voltage $V(t)$ replicates the transmitted waveform of the electric field E[Eq. (2.2)] and is proportional to it. That is,

$$V(t, r) = A\{\exp[j2\pi f(t - 2r/c) + j\psi]\}U(t - 2r/c), \qquad (3.25)$$

where $2r$ is the total path traversed by the incident and scattered wave, A is now the complex amplitude, $|A|\exp(j\psi_s)$ (i.e., it contains the phase shift ψ_s produced by the scatterer), at the input to the synchronous detector (Fig. 3.8), and, as before, $U = 1$ when its argument is between 0 and τ, zero otherwise. Range time, $\tau_s = 2r/c$, specifies, in units of time after the start of the transmitter pulse, the scatterer's location. The echo phase ψ_e is defined from Eq. (3.25) as

$$\psi_e \equiv -\frac{4\pi r}{\lambda} + \psi_t + \psi_s \qquad (3.26)$$

in which we have included the phase shift ψ_s upon scattering and the transmitter phase ψ_t, but have omitted the phase ωt. The echo phase is dependent on time because r and ψ_s are time dependent.

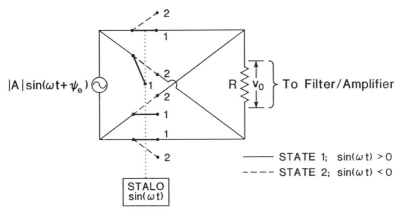

Fig. 3.8 Simplified schematic of a synchronous detector.

The Doppler radar usually has two synchronous detectors (without which the direction of the scatterer motion, toward or away, cannot be determined). The synchronous detector is often a bridge of diodes that function as high-speed switches that are opened and closed by the STALO voltage. The simplified schematic in Fig. 3.8 provides the essential elements to explain the operation of the synchronous detector. For example, when the STALO sinusoid has a positive value the diodes are in state 1, wherein the echo voltage is applied to the output in a "normal configuration." On the other hand, when the STALO voltage is negative the diodes are switched to state 2 and then the echo voltage is "inverted" (i.e., the sign of the echo voltage is changed across the output resistor R). Thus the sign of the echo voltage applied to the filter-amplifiers (Fig. 3.1) is switched at twice the frequency of the STALO. If the echo phase ψ_e were an integer of 2π, the output would be a series of positive half-sinusoids and the average output voltage would be positive maximum. If ψ_e were an integer multiple of 2π plus $\pi/2$ the average of the output voltage would be zero.

The filter-amplifier circuits remove the harmonic signals that are present and what remains is the average value which is proportional to

$$|A| \cos \psi_e \sin(\omega t) U(t - 2r/c). \tag{3.27}$$

Likewise, the average output of the quadrature-phase synchronous detector is proportional to

$$|A| \sin \psi_e\, U(t - 2r/c). \tag{3.28}$$

These components are proportional to the echo amplitude and are sinusoidal functions of echo phase. Thus we can express the filtered signals as

$$I(t) = (|A|/\sqrt{2})U(t - 2r/c)\cos(4\pi r/\lambda - \psi_t - \psi_s), \qquad (3.29\text{a})$$

$$Q(t) = (-|A|/\sqrt{2})U(t - 2r/c)\sin(4\pi r/\lambda - \psi_t - \psi_s), \qquad (3.29\text{b})$$

which are in-phase $I(t, r)$ and quadrature $Q(t, r)$ components, respectively, of the echo signal $V(t, r)$. For convenience we ignored losses and used a factor of $1/\sqrt{2}$ in Eq. (3.29) so that the sum of the powers in the I and Q channels equals the input power $|A|^2/2$ averaged over a cycle of the microwave signal. (In practice precise calibration is required to relate output voltage to echo amplitude at the receiver input.)

Echoes from stationary scatterers have signals in which the phase $\psi_e = -(4\pi r/\lambda) + \psi_t + \psi_s$ is time independent. If r changes in time, but ψ_t and ψ_s are time independent, the phase also changes and the time rate of phase change,

$$\frac{d\psi_e}{dt} = -\frac{4\pi}{\lambda}\frac{dr}{dt} = -\frac{4\pi}{\lambda}v_r = \omega_d. \qquad (3.30)$$

is the Doppler shift (in radians per second). It is to be noted that because water drops can vibrate as they fall in air, ψ_s is not in general time independent, and changes in ψ_s in the period between transmitted pulses (i.e., the pulse repetition time, PRT) will cause fluctuations in the Doppler shift that will broaden the Doppler spectrum (Section 5.2). A physical explanation for the Doppler shift $f_d = -2v_r/\lambda$ is as follows: A pulse of radiation impinging on a hydrometeor forces molecular vibrations in synchronism with the time-changing electric and magnetic fields. If the hydrometeor is moving toward the transmitter at velocity v_r, its vibrational frequency is higher by v_r/λ because the scatterer molecules experience more rapid fluctuations of electric and magnetic force. The vibrating molecules themselves generate electromagnetic fields, which in turn radiate outward from the scatterer. We see that the factor of 2 is the result of a two-step increase in the frequency. First the hydrometeor's electric vibrational frequency is increased by v_r/λ; second, the frequency of its radiation field in the direction of the receiver is increased by v_r/λ. Similar reasoning shows that there is no Doppler shift in the scattered electromagnetic field along the beam but beyond the hydrometeor (i.e., the forward-scatter path).

It is relatively easy to see from Eq. (3.29) that for usual radar conditions (i.e., $\tau \approx 10^{-6}$ s) and scatterer velocities of the order of tens of meters per second, the change in the trigonometric functions is extremely small during the time $U(t - 2r/c)$ is nonzero. Thus we measure the echo phase shift over the longer PRT $\equiv T_s \approx 10^{-3}$ s rather than during a pulse duration τ. Because of this the pulsed-Doppler radar behaves as a phase as well as an amplitude-sampling device; samples are at $t = \tau_s + (n-1)T_s$, where range time τ_s is the time delay between any transmitted pulse and its echo.

It is convenient to introduce another time scale, the sample time; that is, time is incremented in discrete steps of length T_s, the sample time, after $t = \tau_s$ (see Fig. 4.1 for a display of both time axes). It is important to realize that the echo phase and amplitude changes are examined in sample-time space at the discrete instants $(n-1)$ T_s for an echo at range time τ_s.

Samples of $I(\tau_s, T_s)$ and $Q(\tau_s, T_s)$ taken from actual observations of stationary and moving scatterers are shown in Fig. 3.9. Traces 1, 2, and 3 in this figure are echo samples at three successive sampling intervals. Because $I(\tau_s, T_s)$ is near zero and decreasing, whereas $Q(\tau_s, T_s)$ has a maximum amplitude and is also decreasing, the Doppler shift of the moving scatterer is positive, corresponding to a negative radial velocity or scatterer motion toward the radar. The transmitted pulsewidth τ is 1.2 μs, and because the moving scatterer produces an echo pulse that nearly replicates the transmitted one, we can assume that there is only one moving scatterer observed in the range interval (1500 m) displayed in Fig. 3.9. On the other hand, the echoes associated with the stationary scatterers are not a replica of the transmitted pulse because more than one scatterer lies in the range interval $c\tau/2$. Nevertheless, the I and Q samples do not change amplitude as a function of sample time mT_s (i.e., traces 1, 2, and 3 are identical) for echoes from the stationary scatterers.

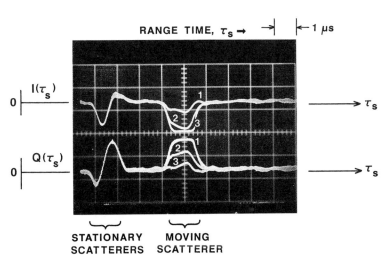

Fig. 3.9 Samples of in-phase $I(\tau_s)$ and quadrature phase $Q(\tau_s)$ components of echoes for moving and stationary scatterers. Echo traces versus τ_s are from three successive sampling intervals T_s, but have been superimposed to show the relative change of amplitude.

3.5 Practical Considerations

The preceding discussion treats an ideal radar to facilitate understanding of radar principles, but it omits some important limitations in radar measurements.

3.5.1 System Noise Temperature

An ideal noiseless receiver, in a noiseless environment, has no lower limit to detect the weakest signals. Generally all receivers detect echo power only above some limit imposed by the random voltage from the thermal agitation of electrons in the receiver (in Fig. 3.1 it is the synchronous detector). Some high-performance radar receivers have parametric amplifiers or other low-noise amplifiers prior to the synchronous detector so that microwave emission from the earth's surface and atmosphere can exceed the noise generated by the receiver.

Noise considerations are often unimportant for weather radars that observe storms, because the signal power is much larger than noise. But, when Doppler weather radars are used to measure wind in clear air, as discussed in Chapter 11, the noise contributions from components (such as transmission lines, radomes, or the T/R switch) between the receiver and scatterers have paramount importance because the echo power can be smaller than the noise power. Even when a signal is weaker than the noise, the processing techniques described in Chapter 6 can extract Doppler velocity information.

The noise power at some convenient reference point in the radar system (e.g., the receiver input) has contributions from several sources. Radiation from space (cosmic noise) and from the oxygen and water vapor molecules in the atmosphere is intercepted by the antenna and, after being attenuated by the radome and transmission lines (between the antenna and receiver), is present at the receiver input. Transmission lines and other components in the path between the antenna and receiver input not only attenuate external radiation but, because they are at a temperature well above absolute zero, also generate noise that passes to the receiver input. The noise power generated by an attenuating component is given by (Dicke et al., 1946).

$$P_n = kTB_n(1 - l^{-1}) \tag{3.31}$$

where l, the loss factor ($l > 1$), is the ratio of the power in to the power out of the lossy device; T is the temperature of the component in kelvins; $k = 1.38 \times 10^{-23}$ W s K^{-1} is the Boltzmann constant; and B_n is the noise bandwidth of the component. Usually power loss is expressed in decibels, and then $l = \log^{-1}(L/10)$. The power loss is caused by the thermal agitation of electrons in the walls of the device that confine or guide the electromagnetic signals. Transmission lines (i.e., waveguides) attenuate noise from sources outside the antenna and may reduce the noise at the input to the receiver but they also attenuate the signal. Although the noise power at the receiver might be less than the noise power without the attenuating waveguide, what really matters is the echo signal-to-noise ratio (SNR); any device that adds noise will decrease the SNR at the receiver's input. It is therefore important to use transmission lines that have small attenuation.

Because the thermal noise power at the receiver input is spread over band-widths that are large compared to the radar receiver's bandwidth, much of the noise power can be filtered. It is accepted practice to express the radar noise level at the reference point in terms of a system temperature

$$T_{sy} \equiv N/kB_n, \qquad (3.32)$$

where B_n is now the noise bandwidth of the receiver and N is the effective noise power at the reference point, which is usually at the input to the receiver.

Not only do the components in front of the receiver contribute noise, but the receiver, which is composed of a low-noise amplifier (LNA), detector, etc. (Figs. 3.1 and 3.10), adds noise to the signals that it amplifies. A figure of merit for an LNA is its noise temperature T_R, which is referenced to the LNA's input when the LNA is considered a noiseless device whose output is connected to noiseless components. Thus, although the noise power added by the LNA is not really present at its input terminals, we add this noise to that contributed by components before the LNA. Thus T_{sy} includes T_R even though N is not entirely present at the reference point. In practice the LNA is not connected to noiseless devices, so the LNA's output supplies power into subsequent com-ponents that add noise. But, if the gain of an LNA is sufficiently high to amplify the equivalent noise power at its input terminal to levels far above the additional noise due to subsequent detectors or amplifiers, then the noise contri-buted by them can be ignored.

Figure 3.10 shows components of the transmission path between the scat-terer and the reference point (input to LNA) that need to be considered in

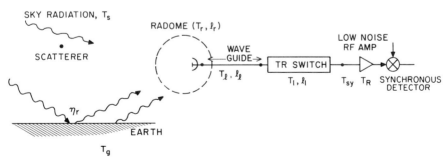

Fig. 3.10 Schematic of the absorbing elements that need to be considered when estimating the system noise temperature T_{sy} at the receiver input. The waveguide path may also contain other components, such as rotary joints, that add additional losses to the transmission path. Furthermore, rain and clouds add to the sky noise temperature.

computing T_{sy}. Based on Eqs. (3.31) and (3.32) we arrive at the following expression for T_{sy}:

$$T_{sy} = T_s(l_r l_l l_1)^{-1} + T_r(l_l l_1)^{-1}(1 - l_r^{-1})$$
$$+ T_l l_1^{-1}(1 - l_l^{-1}) + T_1(1 - l_1^{-1}) + T_R, \qquad (3.33)$$

where T_s is the sky noise temperature due to cosmic and atmospheric radiation and T_r, T_l, and T_1 are the temperatures of the radome, transmission line, and T/R switch, respectively. These latter temperatures can usually be set equal to 290 K, which approximates the temperature of the environment. The loss factor associated with each element is shown in Fig. 3.10.

The contribution to receiver noise from the sky temperature is a function of the direction in which the antenna points, because cosmic radiation is nonuniformly distributed over angular space (it is maximum along the galactic plane). Furthermore, when the antenna's beam is vertically directed, the absorption in the thin blanket of the earth's atmosphere is small and so is the atmosphere radiation intercepted by the antenna. As the beam points toward the horizon, however, more of the absorbing layer is within the field of view of the antenna. Thus both sources of sky noise have an angular dependence, which is summarized in Fig. 3.11. Figure 3.11 gives the sky noise temperature T_s for an idealized antenna having the sun in a unity-gain sidelobe but without earth directed sidelobes. Above 50 MHz but below 1000 MHz, cosmic or galactic noise dominates other noise sources. Above about 1000 MHz, cosmic noise becomes negligible and there remains only a contribution from oxygen and water vapor absorption and reradiation, which occur at their respective resonance peaks of 60 and 22 GHz. Although the oxygen resonant peak is farther from the weather radar frequencies, most atmospheric noise comes from the oxygen molecules.

Although the mainlobe may point at a relatively cool sky, a real antenna has sidelobes that are directed at a relatively warm and reflecting earth. Thus cosmic, atmospheric, and earth radiation contribute to an effective sky noise temperature

$$T_s' = T_s(1 - \chi) + \chi(1 - \eta_r)T_g + T_s\chi\eta_r, \qquad (3.34)$$

where χ is the fraction of the antenna's power pattern subtending the ground at temperature T_g, and η_r is the fraction of incident noise power reflected from the earth. A suggested conventional value for χ is 0.125, which results if an isothermal earth is viewed over a π-steradian solid angle by sidelobes averaging -3 dB gain (Blake, 1970). For example, consider an antenna pointed at a relatively high angle, and suppose that the main beam and first sidelobe, which contain about 90% of the radiated power, "see" an average temperature of 20 K due to oxygen absorption. The other 10% of sidelobe reception is from a warm earth at 300 K that is assumed to reflect 50% of the incident radiation. Then the effective sky temperature is

$$T_s' = 0.9 \times 20 + 0.5 \times 0.1 \times 300 + 0.5 \times 0.1 \times 20 = 25 \text{ K}.$$

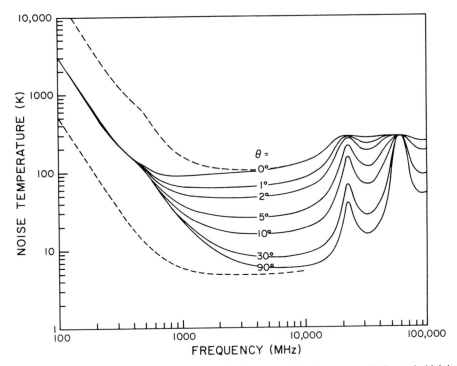

Fig. 3.11 Sky noise temperature of an idealized antenna (lossless, no earth-directed sidelobes) located on the earth's surface. Solid curves are for geometric-mean galactic temperature, sun noise ten times quiet level, sun in unity-gain sidelobe, cool temperate-zone troposphere, and 2.7 K cosmic blackbody radiation. The upper dashed curve is for maximum galactic noise (center of galaxy, narrow-beam antenna), sun noise 100 times quiet level, zero elevation angle, and other factors the same as for the solid curves. The lower dashed curve is for minimum galactic noise, zero sun noise, and elevation angle of 90°. The maxima at 22.2 and 60 GHz are due to water-vapor and oxygen absorption resonances. From "Radar Handbook," edited by M. I. Skolnik. Copyright © 1970 McGraw-Hill Book Company. Used with the permission of McGraw-Hill Book Company.

Thus considerable radiation could be intercepted by ground-directed sidelobes, and so T_s' should be used in place of T_s in Eq. (3.33).

Radome losses deserve further comment. These losses are due to absorption by the protective membrane and its supporting structure, as well as to scatter from the support frame. The scatter loss will result in increased sidelobe levels. Thus radome scatter loss contributes to system noise temperature only through the antenna radiation pattern and cannot be considered part of l_r in Eq. (3.33).

3.5.2 Bandwidth

The echo pulse, as any other signal, can be decomposed into Fourier spectral components. The receiver amplifies only a band of these spectral or frequency components; that is, the receiver has a finite bandwidth. The filter's (Fig. 3.1) amplitude transfer function $G(f)$ is usually a monotonically decreasing function of frequency; its bandwidth B_6 is specified as the band of frequencies where the power gain, $|G(f)|^2$, is within 6 dB of its highest level. Noise bandwidth B_n is not so simply specified because it depends on $G(f)$ as well as on the frequency dependence of noise power $N(f)$, but in practice it is nearly equivalent to the 3-dB width if noise is white [i.e., noise power is independent of frequency (Kraus, 1966, p. 265)].

The larger B_6 is, the better is the fidelity of the echo pulse shape; but noise power increases in proportion to B_n. Usually it is not important to detect the return echo with nearly perfect fidelity; nor is it imperative to excessively reduce noise. Rather, a compromise between the two conflicting effects is reached. We shall see later that both filter bandwidth and transmitter pulse width determine the range-dependent weight given to the scatterers' cross section (Section 4.4.2).

An important measurement made with radar is the range to scatterers. The radar often needs to resolve point scatterers (objects) when they are closely spaced and have largely different backscattering cross sections. A noiseless receiver of infinite bandwidth resolves objects if their spacing is wider than one-half the spatial pulse width (i.e., $c\tau/2$); objects so spaced return echoes that arrive at separate and distinct time intervals. Finite-bandwidth receivers distort the echo pulse by reverberating echo power within the receiver in amounts decreasing with time after the peak echo signal; an example is television tower echoes (Fig. 3.7). This causes the receiver output pulse to have widths larger than τ, and weak echoes could then be masked by strong ones.

3.5.3 Filtered Waveform

It is demonstrable that the filter output echo $I(t)$ or $Q(t)$ has a time waveform approximated by

$$I(t) = \tfrac{1}{2}I_0\{\text{erf}[aB_6(t + \tau/2)] - \text{erf}[aB_6(t - \tau/2)]\}, \qquad (3.35)$$

where I_0 is the prefilter amplitude $A/\sqrt{2}$ [Eq. (3.29)] of a rectangular echo pulse having a width τ (Fig. 3.12); $a = \pi/(2\sqrt{\ln 2})$; erf[ ] is the error function, given by

$$\text{erf}(x) = \frac{2}{\sqrt{\pi}} \int_0^x e^{-t^2}\, dt; \qquad (3.36)$$

and $t = 0$ is the time at which the output echo is maximum. This solution is for an assumed filter frequency response $G(f)$ described by the Gaussian function

$$G(f) = e^{-(4\ln 2)f^2/B_6^2}. \qquad (3.37)$$

Although Eq. (3.35) approximates the pulse shape well if the filter response is well approximated by Eq. (3.37), it has the defect of producing signal for all t: It

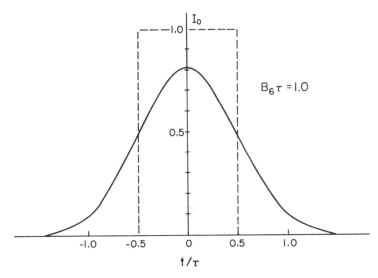

Fig. 3.12 Output (solid line) of Gaussian response filter for a rectangular pulse input (dashed lines). The delay in the output relative to input is not shown. The filter width B_6 is matched to pulse width τ (i.e., $B_6\tau = 1.0$).

does not show a starting time. This is because the Gaussian function is practically approached only in the limit of an infinite pole filter (i.e., an unbounded number of cascaded elementary electrical circuits), which would cause an infinite propagation delay through the filter. Nevertheless, Eq. (3.35) is instructive because it shows the interrelation between B_6, τ, and the range time τ_s (we relate τ_s to t in Section 4.1). Furthermore, one can measure accurately the time delay to the peak response, and (3.35) describes the actual response quite well about this peak. It is readily seen that if B_6 is much larger than τ^{-1}, the filtered echo pulse is nearly a scaled replica of the transmitted pulse.

The filter's frequency response specifies a range-dependent function that weights the cross sections that contribute to the sample of echoes from distributed scatterers such as rain (Chapter 4). In incoherent radar transmitters B_6 is often chosen significantly larger than would be required to produce an acceptable echo response in order to allow for frequency instabilities. A filter can also reject interference from extraneous sources, and smaller B_6 makes the receiver more selective. A filter with rectangularly shaped frequency response is much more selective than the Gaussian filter, but its time response to echo pulses may be objectionable because of larger reverberating signals. (These mask nearby weak echoes and, in rain showers, smear gradients of rain intensity.) A Gaussian filter is often a good compromise between frequency selectivity and a time resolution because its output power quickly decays after the peak signal time.

3.5.4   Signal-to-Noise Ratio, Matched Filters

An important parameter is the ratio of peak echo power, $S(0) \equiv I^2(0) + Q^2(0)$, to noise power. This signal-to-noise ratio, SNR, readily obtained from Eqs. (3.32) and (3.35) is

$$\text{SNR} = [I^2(0) + Q^2(0)] \ \text{erf}^2[aB_6\tau/2]/kT_{sy}B_n. \tag{3.38}$$

Because for a Gaussian filter $B_n = 1.06B_6/\sqrt{2}$, SNR is maximized if

$$B_6 = 1.04/\tau. \tag{3.39}$$

It can be shown that the SNR is maximized in general if the filter response is matched to the prefilter echo spectrum (Nathanson, 1969, p. 277). To simplify the filter hardware and to achieve better filtering of extraneous signals, a matched filter is often only approximated in practice. Taylor and Mattern (1970, pp. 5–25) demonstrate that for a wide variety of filter responses Eq. (3.39) optimizes the SNR, and condition (3.39) is therefore considered to match the echo spectrum.

Often radar performance is characterized by a minimum detectable signal (MDS); this minimum is a function of radar parameters and strongly depends on signal processing (Chapter 6). Nevertheless it is accepted practice to define MDS as that signal which equals noise power.

3.6   Ambiguities

Range and velocity ambiguities are inherent in Doppler radars. Often Doppler radars are operated with uniform PRT (i.e., T_s) so that when scatterers have range r larger than $cT_s/2$, their echoes for the nth transmitted pulse are received after the $(n + 1)$th pulse is transmitted (Fig. 3.13). Therefore, these echoes are received during the same time interval that scatterers at $r < cT_s/2$ return echoes from the $(n + 1)$th pulse. Thus, the range r to the distant scatterer may appear to have a value $r' = r - (N_t - 1)r_a$, which is ambiguous. N_t designates the number of $cT_s/2$ intervals (trips) to the scatterer, and

$$r_a = cT_s/2 \tag{3.40a}$$

is the unambiguous range within which all scatterers must lie in order to have their ranges directly measured. We emphasize that r_a does not necessarily limit the range to which the pulsed-Doppler radar can achieve useful measurement. If its STALO (Fig. 3.1) is phase coherent over many T_s intervals, the radar can accurately measure velocities of scatterers beyond r_a.

The second ambiguity relates to measurement of the scatterer's velocity. As discussed in Section 3.4.3, the echo phase ψ_e is sampled at intervals T_s and its change $\Delta\psi_e$ over the interval T_s is a measure of the Doppler frequency

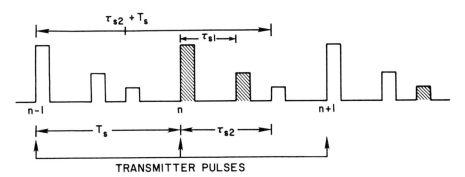

Fig. 3.13 Range-ambiguous echoes. The nth transmitted pulse and its echoes are crosshatched. This example assumes that the larger echo at delay τ_{s1} is unambiguous in range but the smaller echo, at delay τ_{s2}, is ambiguous. This second-trip echo, which has a true range delay $T_s + \tau_{s2}$, is due to the $(n-1)$th transmitted pulse.

$f_d = \omega_d/2\pi$. Unfortunately, given a set of sampled phases (computed from I and Q samples), we cannot relate them to one unique Doppler frequency. As Fig. 3.14 shows, it is not possible to determine whether $V(t)$ rotated clockwise or counter clockwise and how many times it circled the origin in the time T_s. Therefore any of the frequencies $\Delta\gamma/2\pi T_s + n/T_s$ (where $-\pi < \Delta\gamma \le \pi$ and n is a positive or negative integer) could be a correct Doppler shift. All such Doppler shifts are called *aliases*, and $f_N = (2T_s)^{-1}$ is the Nyquist (or folding) frequency. All Doppler frequencies between $\pm f_N$ are the principal aliases, and frequencies higher than f_N are ambiguous with those between $\pm f_N$. Thus, particle radial velocities must lie within the unambiguous limits

$$v_a = \pm\lambda/4T_s, \tag{3.40b}$$

to avoid ambiguity. But the transmission of pulses at two interlaced pulse repetition frequencies (PRFs) and signal processing can extend the unambiguous velocity (Chapter 7).

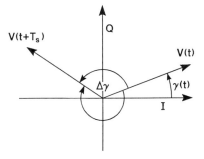

Fig. 3.14 Depiction of frequency aliasing with a phasor diagram.

Problems

3.1   If the antenna gain is 10,000 at a wavelength of 10 cm, find its effective area for the reception of power. Given a scattering cross section of 1 m^2, an elevation angle $0°$, propagation in air, peak transmitter power of 1 kW. If the radar receiver has a capability to detect an echo power of 10^{-12} W, what is the range to the scatterer beyond which detection will be lost?

3.2   (a) A 10-cm Doppler radar has a pulse repetition time $T_s = 1$ ms (i.e., 10^{-3} s). Find its unambiguous range and velocity. (b) The radar beam is directed along a squall line. The range of the line extends from 20 to 200 km. Will there be range ambiguities and, if so, which parts of the squall are range ambiguous and what is the apparent range of these parts? (c) Hydrometeors are moving at 40 m s^{-1} toward the radar. What Doppler shift (Hz) do they generate and what apparent shift (m s^{-1} and Hz) does the radar measure?

3.3   A reflecting balloon at 10 km is travelling straight toward the radar with a velocity of 10 m s^{-1}. The radar transmits 1-μs pulses at a rate of 1000 Hz. Plot the I and Q components over a 6-ms time interval if the radar wavelength is 5 cm. Plot the phasors on a phasor diagram. How fast must the balloon move to have the phasor change by π radians between successive echoes?

3.4   Show that the sum of instantaneous powers of the I and Q components is constant if $I(t) = \cos \omega_d t$ and $Q(t) = \sin \omega_d t$.

3.5   A continuous wave (cw) with a wavelength of 10 cm is transmitted in a Gaussian antenna pattern with a one-way half-power beamwidth of $1°$. The antenna rotates at 10 deg s^{-1}. The antenna scans past a single raindrop moving toward the radar at 5 m s^{-1}. (a) Sketch the in-phase (I) and quadrature (Q) components at the output of the filter amplifiers as the antenna beam scans past the raindrop. (b) Sketch the received power ($I^2 + Q^2$) at the output of the filter amplifiers as the antenna beam scans past the raindrop. (c) Sketch the path of the phasor; $I + jQ$, as a function of time.

3.6   (a) Compute one-way attenuation of waves at $\lambda = 3$ and 10 cm, if propagation is through a cloud containing a liquid water content of 1 g m^{-3}. For $\lambda = 3$ cm, Im$(-K_w)$ is 3.67×10^{-5} and at $\lambda = 10$ cm it is 1.1×10^{-5}. Assume the cloud is 5 km thick. (b) Let all the drops be of equal size. Choose a reasonable drop diameter and compute drop density.

3.7   The profile of rain rate R is

$$R = \begin{cases} -(r-100)^2/8 + 50 & \text{(mm h}^{-1}) \qquad \text{for} \qquad 80 \le r \le 120 \quad \text{km} \\ 0 & \text{otherwise} \end{cases}$$

Between the antenna and the rain, there are no other clouds and the elevation angle is $\theta_e = 1°$. Plot the specific attenuation versus range, and the total (cumulative) one-way attenuation for $\lambda = 10$ cm to 120 km in range. Don't ignore gaseous attenuation.

3.8   (a) Determine specific attenuation in dB km^{-1} per g m^{-3} of liquid water for $\lambda = 3.2$ cm and 10 cm at a temperature $T = 20°$C. Compare your results to that listed in Table 6.1 of Battan (1973). (b) Show that attenuation is independent of the drop size distribution and only depends on the cloud liquid-water density M.

3.9   Assume a receiver is perfectly matched (i.e., it has an amplitude transfer function $G(f)$ equal to the spectrum of the transmitted pulse) to a rectangular pulse of duration τ. (a) Determine the time response (i.e., shape) of the echo after it passes through the receiver. (Note: the spectrum of the echo at the receiver's output is the product of the echo's spectrum at the receiver's input and the frequency response of the receiver.) (b) Plot some expected waveforms of I (or Q) for the cases when two echoes are received from two drops spaced less the $c\tau/2$ apart. (c) Determine and sketch the shape of the echoes' instantaneous power versus time at the input and output of the receiver.

3.10 A Gaussian filter has an amplitude transfer $G(f) = \exp[-4(\ln 2)f^2/B_6^2]$. If $B_6 = 1$ MHz and the system noise temperature is 300 K, find the total noise power at the filter's output.

3.11 A radar beam passes through a large thunderstorm whose rainfall rate is 50 mm h^{-1} over a path of 20 km. Using Fig. 3.5, compute the one-way and two-way attenuation (in dB) for a 3-cm radiation passing through the storm (assume $T = 18°C$ and neglect atmospheric attenuation). Compute the loss factor l.

3.12 Two hailstones are within the radar beam. One is at 40 km, centered on the beam axis, and has a backscattering cross section of 0.5 cm^2. The other is off beam center at a 3-dB point of the one-way pattern; its distance from the radar is 20 km, and its backscattering cross section is 1 cm^2. Find the ratio of the echo powers at the receiver, from these two hailstones.

3.13 A one-way antenna power pattern $f^2(\theta)$ satisfies the equation

$$10 \log[f^2(\theta)] = -3\theta^2.$$

Find the 3-dB beamwidth of this pattern.

3.14 A radar technician has inadvertently switched the I and Q cables (i.e., what was the Q channel is now I and vice versa). Does this affect radar performance and if so, how?

Weather Signals

In radar literature dealing with the detection of discrete objects such as air-craft, weather signals are called weather clutter. Because weather clutter is a source of information useful to radar meteorologists, we define the composite of echoes from individual hydrometeors, or from refractive index irregularities in clear air, as weather signals. In this chapter we determine the statistical properties of the weather signal voltage $V(\tau_s)$ and instantaneous power $P(\tau_s)$, relate the average of successive $P(\tau_s)$ samples to the composite backscattering cross section of hydrometeors within the radar's resolution volume, and develop a form of the weather radar equation that accounts for range weighting. Here we show the correlation of $V(\tau_s)$ versus lag $\delta\tau_s$, whereas in Chapter 5 we discuss the correlation of echo samples versus T_s.

4.1 Weather Signal Samples

A weather signal is a composite of echoes from a very large number of hydro-meteors. After a delay (the round trip propagation time between the radar and the near boundary of the volume of hydrometeors), echoes are continuously received over a time interval equal to twice the time it takes the microwave pulse to propagate across the volume containing the scatterers. Because one cannot resolve the individual echoes, we resort to sampling the weather signal at discrete range time delays τ_s (Fig. 4.1). τ_s is the time [i.e., t in Eq. (3.29)] delay (after the start of a timing pulse) of a sampling gate, but τ_s also defines the approximate range, $r \approx c\tau_s/2$, of those scatterers that contribute most to the sample of weather signal. The timing pulse initiates, at periodic intervals of T_s, the formation of the transmitter pulse and the chain of sampling gates.

Radars use gating circuits to sample, nearly instantaneously (i.e., the gate width is much narrower then the transmitted pulse), the I and Q signals and convert their analog value of voltage or current to a digital number. For each gate or sample point there is a resolution volume in space within which hydrometeors contribute significantly to the sample. Although the resolution volume's location is determined by the time delay τ_s, its dimensions and the contribution of each hydrometeor to the weather signal sample are specified by a

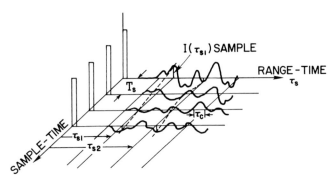

Fig. 4.1   Idealized traces for the in-phase component $I(\tau_s)$ of weather signals from distributed scatterers. Each trace represents a composite of echoes for a single transmitted microwave pulse. Instantaneous samples are taken at τ_{s1}, τ_{s2}, etc. The dashed line indicates a probable time dependence of the sample at τ_s. τ_c depicts the signal correlation time along τ_s and is related to $B_6\tau$. Samples at fixed τ_s taken at T_s intervals are used to construct the Doppler spectrum for scatterers located about the range $c\tau_s/2$.

weighting function that has an angular dependence related to the antenna's radiation pattern $f^2(\theta, \phi)$ and a range dependence related to the filter's amplitude transfer function $G(f)$ (Fig. 3.1) and the shape of the transmitted pulse (to be discussed in Sections 4.4.2 and 4.4.3). Gates (i.e., sample locations) usually have a time spacing equal to or less than the width τ of the transmitted pulse. It should be noted that gate spacings do not determine the resolution volume size, but only the range separation of these volumes.

Each scatterer in the resolution volume returns an echo that has a shape at the receiver's output similar to that sketched in Fig. 4.2. Depending on the precise position (i.e., within a wavelength) of the hydrometeor, the detected echo pulse at the output can have any value between maximum positive and negative excursions. These echoes constructively or destructively (depending on

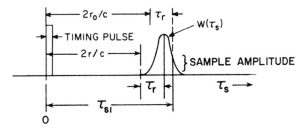

Fig. 4.2   A receiver's filtered output (I or Q) to an input echo from a point scatterer located at r. τ_r is the radar delay and r_0 is the range where a continuum of scatterers receive maximum weight from $W(r)$ when the sample gate is at τ_{s1}.

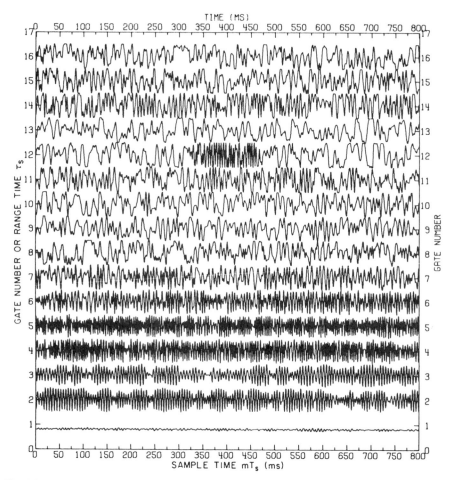

Fig. 4.3 The I or Q samples of weather signals at 16 range time locations $\tau_{s1} \ldots \tau_{s16}$ versus sample time mT_s. Sample time interval T_s is 0.75 ms and the samples themselves have a duration less than 10^{-3} ms. Although the separation between sample points is several hundreds times the sample duration, the points in this figure give a continuous trace because the resolution of the trace is coarse (i.e., ≈ 2 ms along mT_s).

their sign) interfere with each other to produce a composite phasor sample $V(\tau_{s1})$, often called a complex voltage sample $V(t) = I(t) + jQ(t)$.

The random size and location of scatterers cause the amplitude and phase of $V(\tau_{s1})$ to be random variables. The weather signal sample at another range gate at delay τ_{s2} is a composite signal from scatterers in a different resolution volume, and hence we expect $V(\tau_{s1})$ to differ from $V(\tau_{s2})$. Thus, $V(\tau_s)$ fluctuates as τ_s increases (Fig. 4.1) even when the scatterer density is spatially uniform. The correlation between $V(\tau_s)$ samples taken at different τ_s is related to the range extent of the resolution volume and the spacing $\delta\tau_s$ between samples (Section 4.6).

Figure 4.3 is an actual recording of the I or Q weather signal samples. In this figure range time τ_s is along the vertical and sample time mT_s is along the abscissa. The sample time dependence of the weather signal at 16 range times τ_s (i.e., range gate locations) is shown in this figure, where zero amplitude for each trace is a horizontal line (for clarity, not shown in Fig. 4.3) at the gate number location. The radar beam is vertical and the first resolution volume is 600 m aboveground. (Ground clutter received through the antenna sidelobes is the cause of the relatively constant signal level.) Each successive trace is the composite of echoes from hydrometeors in resolution volumes of about 170-m depth spaced 600 m apart corresponding to a gate separation of 4×10^{-6} s.

In the first four or five range gate locations of Fig. 4.3 we see relatively high Doppler shifts produced by some fast moving hydrometeors. The transient of high-frequency Doppler shift seen in gate number 12 at about $mT_s = 350$ to 500 ms is caused by echoes from lightning channels, which move with the velocity of the air and hence produce a Doppler shift different than that of precipitation, which falls through air. The mean squared value of $V(\tau_s)$ over sample time mT_s in each of the traces is about the same even though weather signal amplitude might vary considerably from gate to gate; the receiver in the radar that produced these traces had an automatic gain control circuit that adjusts the output to have a nearly constant mean squared value at each range time.

4.2 The Power Sample

As mentioned previously, the weather signal sample $V(\tau_s)$ is a composite of echoes,

$$V(\tau_s) = \frac{1}{\sqrt{2}} \sum_i A_i W_i e^{-j4\pi r_i/\lambda}, \qquad (4.1)$$

where $|A_i|/\sqrt{2} = (I_i^2 + Q_i^2)^{1/2}$ is the prefilter echo amplitude [Eq. (3.29)] of the ith scatterer located at r_i, ϕ_i, θ_i. W_i is a complex number (or a phasor—it contains amplitude and phase information) that is a range-dependent weight. We shall not yet be specific about the functional form of W_i. For now it suffices to say that W_i has a range dependence such that it principally weights those hydrometeors residing near a range r determined by τ_s (i.e., $r = c\tau_s/2$). Furthermore, the antenna weighting factor $f^2(\theta, \phi)$ is assumed to be part of A_i. We can ignore the constant phase ψ_t given in Eq. (3.29). We also assume a hydrometeor velocity sufficiently small that W_i is independent of it. This implies that the Doppler shift is small compared to B_6; otherwise, fast-moving hydrometeors would return signals at frequencies that fall outside the bandwidth of the receiver. Hydrometeors never move at speeds to shift echoes outside the receiver bandwidth.

The sample, at range time τ_s, of echo power $P(\tau_s)$ averaged over a cycle of the transmitted frequency f is proportional to

$$VV^* = \frac{1}{2} \sum_{i,k}^{N_s} A_i A_k^* W_i W_k^* \exp[j4\pi(r_k - r_i)/\lambda]$$

$$= \frac{1}{2} \sum_i^{N_s} |A_i|^2 |W_i|^2 + \frac{1}{2} \sum_{i \neq k}^{N_s} A_i A_k^* W_i W_k^* \exp[j4\pi(r_k - r_i)/\lambda]. \tag{4.2}$$

[See Eq. (2.3) and associated text. We have summed the power in the I and Q channels, and therefore the factor of $1/2$ in front of VV^* required for the power in either the I or Q channels of real signals is not present in Eq. (4.2)]. The instantaneous echo power $P(\tau_s)$ is for one transmission, and N_s is the number of scatterers. An instantaneous power is considered to be the power averaged over any one cycle of the transmitted frequency. Because scatterers move relative to one another, the weather signal sample at range time τ_s differs for each transmitted pulse, and the amount of change depends on the sampling time interval T_s (the interval between transmitted pulses) and the relative velocity of the scatterers. The equation defining $P(\tau_s)$ for successive transmission will have the same form as Eq. (4.2), but the r_i and r_k will have changed owing to scatterer motion. If scatterers within a sample volume move randomly a significant fraction of a wavelength (e.g., $\lambda/4$) between successive transmissions, each successive echo sample $V(\tau_s)$ (spaced T_s apart) will be uncorrelated. To make coherent Doppler measurements of the scatterer's mean radial speed, the time T_s between successive samples must be small enough that successive samples, at fixed delay τ_s, are correlated.

Whereas the second sum in Eq. (4.2) represents a rapidly fluctuating contribution to the instantaneous power $P(\tau_s)$, the first sum is relatively constant if the scatterer's displacement is small compared to the distances over which the weighting functions change appreciably. Hydrometeor displacements of the order of a wavelength can cause large changes in the second sum, but they need to be displaced a hundred meters or more (for typical weather radar operating parameters) before significant change occurs in the first sum.

Fluctuations are caused by hydrometeor displacement, which changes the phase of each elemental echo. Although the second sum can be significantly larger than the first [it has $N_s(N_s - 1)$ contributions, compared to N_s for the first term] for some weather signal samples, its average value over many successive samples (i.e., its sample time average) approaches zero as the number of samples increases without limit; this is because the average of the complex exponential term tends to zero. The first sum is thus the sample time mean power $\bar{P}(\tau_s)$.

The sample time mean power $\bar{P}(\tau_s)$ does not change with time if the scatterer's displacement on the average is small compared to the distances over which the weighting functions change significantly, or from another point of view if, during the sample time averaging interval, scatterers displaced far from the region where $f^2(\theta_i, \phi_i)W_i(r)$ has significant weight are replaced by others having the same statistical properties (i.e., if we assume statistical properties of the scattering medium are stationary). In practice, sample time averaging intervals are so small that hydrometeor displacement is usually less than a few tens of meters during the period of average power estimation. We stress, however, that $\bar{P}(\tau_s)$ estimates can only be made because hydrometeors are moving relative to one another so that the average of the second sum in Eq. (4.2) is small compared to the first. An accurate estimate of $\bar{P}(\tau_s)$ is important because it relates to the estimates of liquid water in the resolution volume.

4.3 Signal Statistics

The I and Q components of the weather signal sample are random variables if the scatterers' positions are unpredictable. Consider two consecutive echoes spaced T_s seconds apart. The first is given by Eq. (4.1), and the second one can be written

$$V(\tau_s, T_s) = \frac{1}{\sqrt{2}} \sum_i |A_i W_i| \cos \gamma_i - j \frac{1}{\sqrt{2}} \sum_i |A_i W_i| \sin \gamma_i, \qquad (4.3a)$$

where

$$\gamma_i = 4\pi r_i/\lambda + 4\pi v_i T_s/\lambda - \psi_{si} - \beta_i. \qquad (4.3b)$$

$v_i = \Delta r_i/T_s$ is the average radial velocity needed to move the ith scatterer by Δr_i, ψ_{si} is the phase shift upon scattering, and β_i is the phase of W_i. Because the range extent of the resolution volume is much larger than the wavelength ($c\tau/2 \gg \lambda$), and there are many scatterers, it is natural to expect that the values of $4\pi r_i/\lambda$ would be uniformly distributed between $-\pi$ and π. Even though the distribution (i.e., the density function) of $4\pi r_i/\lambda$ need not be uniform, its width usually spans many intervals of 2π so that multiple aliasing (of phases into the unambiguous 2π interval) causes the distribution across 2π to be, for all practical purposes, uniform. Therefore, regardless of the v_i or ψ_{si} density function, the phase γ_i is also uniformly distributed. This follows because the density function of the sum of random variables is obtained after convolving (on a circle from $-\pi$ to π) the individual densities (Papoulis, 1965, p. 189). Because one of them is uniform, the density function of the sum will always be uniform.

Now we are in a position to apply the central limit theorem to the real and imaginary parts of Eq. (4.3a). The theorem states that a sum of independent random variables tends to have a Gaussian density function if their number is large and none of the variables is dominant (i.e., much larger than the rest). Both conditions are certainly true for echoes from hydrometeors, and thus the $I(\tau_s, T_s)$ and $Q(\tau_s, T_s)$ have a Gaussian distribution with zero mean.

Figure 4.4a illustrates the distribution of I, Q values at a fixed range time. Because the sample $I_m = I(\tau_s, mT_s)$ is uncorrelated with the sample

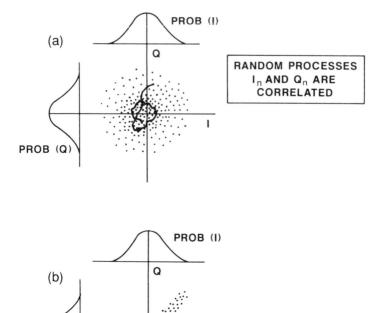

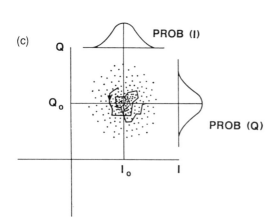

Fig. 4.4 Weather echo statistics. (a) The distribution of I and Q for weather signal samples that are uncorrelated and have a Gaussian distribution. (b) Same as in (a), but I and Q are correlated. (c) Distribution of I and Q for weather signal samples superimposed on ground clutter sample I_o, Q_o. Each increment of voltage change along the path in (a) and (c) occurs over a PRT interval.

$Q_m = Q(\tau_s, mT_s)$ (Appendix B), the distribution of I_m and Q_m is spherically symmetric, as shown in Fig. 4.4a. If I_m were correlated with Q_m the distribution of I_m and Q_m could still be Gaussian, but the values of I_m and Q_m would scatter about some line and have a distribution like that sketched in Fig. 4.4b.

Although the probability of the phase γ_i of $V(\tau_s, T_s)$ is uniform in the interval $\pm \pi$, the change in phase from sample to sample is not necessarily uncorrelated because the phasor can follow the well-ordered path sketched in Fig. 4.4a. That is, the phase γ_i can have similar changes in time, but its starting position has equal probability of being anywhere in the $\pm \pi$ interval. Although the mth I sample I_m is uncorrelated with Q_m, the random processes controlling the changes in I and Q are correlated. I and Q samples will follow a somewhat ordered path (as shown in Fig. 4.4a) only if sample spacing T_s is sufficiently short so that each sample is correlated with the previous one. If the path encircles the origin, there is a nonzero mean Doppler shift. If stationary ground clutter is mixed with weather echoes, the distribution can be like that sketched in Fig. 4.4c, where the path may encircle a point I_o, Q_o which is the phasor of the ground echo.

It is worth noting that

$$I(\tau_s, T_s) = \sum_i |A_i W_i| \cos \gamma_i = |V(\tau_s, T_s)| \cos[\theta(\tau_s, T_s)], \tag{4.4a}$$

and

$$Q(\tau_s, T_s) = -\sum_i |A_i W_i| \sin \gamma_i = |V(\tau_s, T_s)| \sin[\theta(\tau_s, T_s)]. \tag{4.4b}$$

The equalities in Eqs. (4.4) follow because a sum of sinusoids can always be expressed as a sinusoid with a phase $\theta(\tau_s, T_s)$ and an amplitude factor (envelope) $|V(\tau_s, T_s)|$. Nevertheless, this does not mean that I and Q have pure sinusoidal variation with time, but rather have changes like those shown in Fig. 4.3.

Because the in-phase and quadrature components are independent random variables, the joint probability distribution of I and Q is the product of the individual probabilities:

$$p(I, Q) = (1/2\pi\sigma^2) \exp(-I^2/2\sigma^2 - Q^2/2\sigma^2), \tag{4.5}$$

where σ^2 is the mean square value of I (equal to that for Q). Using well-established procedures (Papoulis, 1965, p. 418), the probability distributions of $|V|$, θ and the power $P(\tau_s)$ can be obtained from Eqs. (4.4) and (4.5). We can show that the phase θ is independent of the amplitude $|V|$ and is uniformly distributed, while the amplitude $|V| = (I^2 + Q^2)^{1/2}$ has a Rayleigh probability density

$$p(|V|) = (|V|/2\sigma^2) \exp(-|V|^2/2\sigma^2). \tag{4.6}$$

Because the power $P(\tau_s)$ is proportional to $I^2 + Q^2$, it follows from Eq. (4.5) that $P(\tau_s)$ is exponentially distributed with density

$$p(P) = (1/2\sigma^2)\,\exp(-P/2\sigma^2) \tag{4.7}$$

and a mean value $\bar{P}(\tau_s) = 2\sigma^2$. Constants of proportionality, and impedance factors that make the transition from Eq. (4.5) to Eq. (4.7) dimensionally correct, have been ignored. Radar receivers have gains and losses, and therefore the receiver needs to be calibrated with a known input power to relate accurately output $I^2 + Q^2$ to the weather signal power. The relationship between weather signal power and radar parameters is developed in Section 4.4.

In summary, although I and Q are independent random variables, the random processes controlling $I(\tau_s, mT_s)$ and $Q(\tau_s, mT_s)$ are not independent. This means that, in general, the expected value $E[I(\tau_s, mT_s)\,Q(\tau_s, kT_s)] \neq 0$ for $k \neq m$ (Appendix B, Eq. B.13). Furthermore, the correlation between two successive samples of the complex signal will be appreciably different from zero, only if the distribution of $4\pi v_i T_s/\lambda$ in Eq. (4.3b) is narrow compared to 2π, which is equivalent to saying that the distribution of the v_i is narrow compared to $\lambda/2T_s$ (i.e., the radar's unambiguous velocity interval).

4.4 The Weather Radar Equation

We now relate the mean weather signal power $\bar{P}(\tau_s)$ at the receiver's output to the radar parameters and backscattering cross section σ_b. The expected power contribution $E[P_i]$[1] from *each* scatterer to the mean echo power is, from Eq. (4.2),

$$E[P_i] = \alpha \tfrac{1}{2}|W_i|^2 E[|A_i|^2], \tag{4.8}$$

where α is a proportionality constant, and $\alpha E[|A_i|^2]/2$ is the ith scatterer's expected echo power (at the receiver's input), which can be directly expressed in terms of radar parameters and σ_b by using Eq. (3.24). Because a hydrometeor undergoes continuous change in shape or orientation as it falls through the air, its backscattering cross section fluctuates about a mean value. Therefore the expected value of σ_b (i.e., the mean over the ensemble of the scatterer's cross sections) is needed to determine $E[P_i]$.

Now consider an elemental volume dV containing hydrometeors. The expected echo power $E[dP]$ from dV is

$$E[dP] = \frac{1}{2}\alpha \sum_i |W_i|^2 E[|A_i|^2]. \tag{4.9a}$$

Strictly, echo voltages, not powers, are summed linearly. Nevertheless, because ensemble averaging (to obtain expected values) and summing (integration) of

1. The overbar and $E[\ ]$ have the same meaning.

elemental power P_i are permutable operations, and because the summation is taken over volumes whose range extent is many wavelengths, we can simply use Eq. (4.9a), rather than carry the second sum given in Eq. (4.2), a sum whose expected value is zero.

Now using Eq. (3.24), $E[dP]$ can be expressed as

$$E[dP] = I(\mathbf{r}_0, \mathbf{r}) \, dV \int_0^\infty \sigma_b(D) N(D, \mathbf{r}) \, dD, \tag{4.9b}$$

where

$$I(\mathbf{r}_0, \mathbf{r}) = \frac{C f^4(\theta - \theta_0, \phi - \phi_0) |W_s(r_0, r)|^2}{l^2(\mathbf{r}) r^4}$$

$$C = P_t g^2 \lambda^2 / (4\pi)^3 \tag{4.9c}$$

is a composite weighting function, the value of which at $\mathbf{r}$ depends on the beam direction θ_0, ϕ_0, as well as the sampling gate delay τ_s that determines $\mathbf{r}_0$, the range to a resolution volume (to be discussed in more detail in Section 4.4.2). A subscript "s" (stands for system) is appended to $W(r_0, r)$ to differentiate it from an adjusted $W(r_0, r)$ introduced in the next section. $\sigma_b(D)$ is the expected backscattering cross section for a hydrometeor of diameter D. $N(D, \mathbf{r})$, the particle size distribution, determines the expected number density of hydrometeors having equivolume diameters between D and $D + dD$. The equivolume diameter D is the diameter of a water sphere having the same mass as the actual hydrometeor. Thus, $\sigma_b(D) N(D) \, dD$ is the expected backscattering cross section per unit volume associated with all the hydrometeors within dV of diameters between D and $D + dD$.

The integral in Eqs. (4.9), which depends on $\mathbf{r}$, defines the reflectivity $\eta(\mathbf{r})$;

$$\eta(\mathbf{r}) = \int_0^\infty \sigma_b(D) N(D, \mathbf{r}) \, dD \tag{4.10}$$

or the expected backscattering cross section per unit volume. Substituting Eq. (4.10) into Eqs. (4.9) and integrating over all space, we obtain the following equation:

$$\bar{P}(\mathbf{r}_0) = \int_0^{r_2} \int_0^\pi \int_0^{2\pi} \eta(\mathbf{r}) I(\mathbf{r}_0, \mathbf{r}) \, dV, \tag{4.11}$$

where

$$dV = r^2 \, dr \sin\theta \, d\theta \, d\phi.$$

The upper limit for the r integration does not extend to infinity because scatterers beyond some range r_2 cannot return echoes soon enough to contribute to the echoes sampled at delay τ_s. In the next section we shall be more specific about the limits for r and the functional form of $W_s(r)$, but for now we need only recognize that the r^4 and f^4 dependence in Eq. (4.9c) are not valid for $r < 2D_a^2/\lambda$ (i.e., those ranges within which transmitted power density is relatively independent of r; D_a is the diameter of the antenna). Thus, the integral (4.11) is an approximation valid for $r > 2D_a^2/\lambda$.

In general, η and l are functions of $\mathbf{r}$, but let us assume that over the volume where $f^4(\theta, \phi)|W_s(r)|^2$ has a significant value the reflectivity and attenuation can be considered constant. Furthermore, we assume that the range r_0 to this volume is large compared to its range extent. Thus, we can approximate Eq. (4.11) by

$$\bar{P}(\mathbf{r}_0) \approx \frac{P_t g^2 \lambda^2 \eta(\mathbf{r}_0)}{(4\pi)^3 r_0^2 l^2(\mathbf{r}_0)} \int_0^{r_2} |W_s(\mathbf{r})|^2 \, dr \int_0^{2\pi} \int_0^{\pi} f^4(\theta, \phi) \sin \theta \, d\theta \, d\phi. \quad (4.12)$$

If the antenna radiation pattern is circularly symmetric, and if it can be approximated by a Gaussian shape, we can show that

$$\int_0^{\pi} \int_0^{2\pi} f^4(\theta, \phi) \sin \theta \, d\theta \, d\phi = \pi \theta_1^2 / 8 \ln 2, \quad (4.13)$$

where θ_1 is the 3-dB width (in radians) of the one-way pattern.

4.4.1 Receiver Calibration

The weather signal power given by Eq. (4.12) is at the output of the receiver (i.e., the filter-amplifier in Fig. 3.1). However, radar meteorologists usually refer to echo power at the antenna port where antenna gain, receiver noise temperature, and transmitted power are measured (Section 3.1.2). But at this place the filter-amplifiers cannot alter the shape and amplitude of the echo.

In the hypothetical case in which there are no losses or gains in a receiver having a filter bandwidth much larger than τ^{-1} (e.g., if $G(f)$ in Fig. 3.1 has a value of one), the integral of $|W_s(r)|^2$ equals $c\tau/2$, and using Eqs. (4.12) and (4.13) we compute the weather signal mean power at the antenna to be given by

$$\bar{P}(\mathbf{r}_0) = \frac{P_t g^2 \eta c \tau \pi \theta_1^2 \lambda^2}{(4\pi)^3 r_0^2 l^2 16 \ln 2}. \quad (4.14)$$

This is the usual form of the weather radar equation.

Power measuremments are always made at the receiver output and we must relate these to reflectivity. This requires a calibration of the receiver to determine the losses (usually gain) in echo power as it passes through the receiver. Calibration can be done (as in the NEXRAD system) with a signal generator, of known power, stable amplitude, and continuous wave (cw), to determine this loss (or gain). Calibration corrections are necessary to account for weather echo amplitude fluctuations if nonlinear receivers (i.e., the output power is non-linearly related to the input power as for logarithmic amplifiers) are employed (Doviak and Zrnic, 1984a, Section 6.3.1.1).

The use of a cw signal, set at the transmitted frequency, for calibration cannot account for losses in the receiver due to the finite bandwidth of $G(f)$. Because echo power is distributed over a relatively large spectral band (typically 1 MHz, determined by the transmitted pulse width and shape), $G(f)$ *will attenuate echo power even if the cw calibrating source indicates no losses (i.e., $G(0) > 1$) in the receiver*. Although attenuation of different spectral components by the receiver is not the same, we can conveniently account for this loss by expressing the range-weighting integral in Eq. (4.12) in terms of a receiver finite bandwidth loss factor l_r and the system power gain $g_s = G^2(0)$. In practice g_s is the net gain of the T/R switch, synchronous detector, and filter-amplifier (Fig. 3.1) when determined by a *cw signal generator*. The adjusted range-weighting function (to remove the system gain), $|W(r)|^2$, is therefore defined as $|W_s(r)|^2/g_s$.

If a rectangular pulse of width τ is transmitted and the receiver bandwidth is infinitely large (for practical purposes it needs to be an order of magnitude larger than τ^{-1}), only scatterers within a range interval $c\tau/2$ will contribute to the power of a weather signal sample. Since the integral of $|W(r)|^2$ equals $c\tau/2$ for infinite bandwidth, the finite bandwidth loss factor l_r can be defined as

$$l_r \equiv \frac{c\tau}{2} \bigg/ \int_0^\infty |W(r)|^2 \, dr. \qquad (4.15)$$

The loss factor l_r gives the weather signal power loss caused by the finite bandwidth of the receiver. This loss is a function of both the shape of the transmitted pulse and the receiver's frequency response $G(f)$, as will be shown in Section 4.4.2. For radar receivers having $G(f)$ perfectly matched to the spectrum of the transmitted pulse [e.g., $G(f)$ is of the form $(\sin x)/x$ when the pulse shape is rectangular], l_r is significant (i.e., $l_r = 1.5$ or in decibel units, 1.8 dB for the example cited) and needs to be taken into account in calibration of weather radars. Such radar receivers are called matched filter receivers; radar performance is improved when matched filters are used. In view of these considerations the weather radar equation (4.12) can be written as

$$\bar{P}(r_0) = \frac{P_t g^2 g_s \lambda^2 \eta c\tau \pi \theta_1^2}{(4\pi)^3 r_0^2 l^2 l_r 16 \ln 2}, \qquad (4.16)$$

where now $\bar{P}(r_0)$ is the expected weather signal power at the receiver's output.

4.4.2 The Range-Weighting Function

We now need to determine the weighting function $W(r)$. As before, assume that scatterers have random distribution in space, and let's sum the echo voltages from elemental volumes over θ and ϕ so as to produce, at the output of the synchronous detectors, an increment of complex voltage δV_r per unit range. That is, a shell of thickness dr at range r returns an echo voltage

$$\delta V_r(t, r) = M_r e^{-j4\pi r/\lambda} e_t(t - 2r/c) \equiv A_r e_t(t - 2r/c) \qquad (4.17)$$

that has an incremental magnitude M_r and phase $-4\pi r/\lambda$, and a waveform e_t that is the envelope of the transmitted pulse (e.g., see Eq. 3.25). Particles in a vanishingly thin spherical shell will generate, at the output of the filter-amplifier, a voltage having the same form as the envelope of the transmitted pulse.

The output of the filter-amplifier is obtained by convolving the input waveform with the impulse response $h(t)$ of the filter. (This is exactly equivalent to multiplying the filter's frequency response $G(f)$ by the spectrum of the envelope of the transmitted wave and then taking the inverse Fourier transform to obtain the time response at the receiver output. If the envelop of the transmitted pulse is rectangular and $G(f)$ has a Gaussian shape with a half-power width equal to τ^{-1}, the output of the receiver would roughly be that sketched in Fig. 4.2.) Thus the output of the receiver is

$$\delta V_0(\tau_s, r) = \int_{-\infty}^{\infty} A_r h(\tau_s - t) e_t(t - 2r/c) \, dt, \qquad (4.18a)$$

where $\delta V_0(\tau_s, r)$ is the output voltage per unit dr. The factor $2r/c$ shows that no voltage is present at the output until after the round trip delay $2r/c$ of the transmitted pulse to the elemental volume. Furthermore, because of causality, $h(t)$ has finite values only when its argument is positive (i.e., there cannot be any output until after the input is applied). By recognizing the boundaries of the regions within which contributions to the integral are obtained (Fig. 4.5), and substituting $t' = t - 2r/c$, the above integral can be written as

$$\delta V_0(\tau_s, r) = A_r \int_0^{\tau_s - 2r/c} h(\tau_s - 2r/c - t') e_t(t') \, dt'$$

$$\equiv A_r f(\tau_s, r). \qquad (4.18b)$$

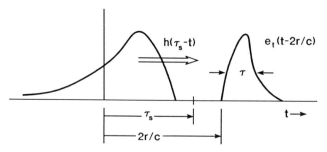

Fig. 4.5 The convolution of the impulse response $h(t)$ and the transmitted waveform e_t. The arrows indicate sliding of the impulse response with τ_s whereas the integration of the product $h e_t$ is over all t for fixed τ_s.

This equation gives the output voltage for the shell at r, but because the filter can have finite response after the cessation of the input voltage (Section 3.5.3), all elemental volumes at ranges less than $c\tau_s/2$ can contribute to the voltage sample at τ_s. Thus the total voltage at time τ_s is

$$V_0(\tau_s) = \int_0^{c\tau_s/2} A_r f(\tau_s, r)\, dr. \tag{4.19}$$

It can be shown that the range integral (4.19) can be expressed as the convolution of $f(\tau_s, r)$ with A_r; in the limit of infinite bandwidth this is simply the convolution of the transmitter envelope $e_t(t)$ with the range dependent A_r. Because scatterers are in random motion, the output voltage will be random and its mean square value

$$E[|V_0(\tau_s)|^2] = \iint\limits_0^{c\tau_s/2} E[M_{r1}, M_{r2}]\, \delta(r_2 - r_1) f(\tau_s, r_1) f^*(\tau_s, r_2)\, dr_1\, dr_2 \tag{4.20a}$$

is the quantity of fundamental importance; it is proportional to the mean weather signal power $\bar{P}(\tau_s)$. To arrive at Eq. (4.20a), use was made of the identity

$$E[M_r(r_1)e^{-j4\pi r_1/\lambda}M_r(r_2)e^{j4\pi r_2/\lambda}] = E[M_{r1}M_{r2}]\, \delta(r_2 - r_1). \tag{4.20b}$$

With Eq. (4.18b) and one integration, Eq. (4.20a) is reduced to

$$\bar{P}(\tau_s) = \int_0^{c\tau_s/2} \alpha(r) E[M_r^2] \left| \int_0^{c\tau_s/2-r} h(c\tau_s/2 - r - r')e_t(r')\, dr' \right|^2 dr \tag{4.21}$$

in which we have substituted $r' = ct'/2$ and where $\alpha(r)$ is a proportionality factor that can be obtained from Eqs. (4.9). From discussion leading to Eq. (4.10) we recognize that $E[M_r^2]$ is proportional to the reflectivity η and that the integral

$$|W(c\tau_s/2 - r)|^2 = \left| \int_0^{c\tau_s/2-r} h(c\tau_s/2 - r - r')e_t(r')\, dr' \right|^2 \equiv |W(r)|^2, \tag{4.22}$$

weights in range the reflectivity field (hence the name *range-weighting function*). For further discussion and illustrative examples refer to Zrnic' and Doviak (1978).

Figure 4.6a shows an actual receiver response to a rectangular microwave pulse (only the envelope of the input microwave pulse is displayed) for the filter bandwidth much larger that τ^{-1} (i.e., about 10 times). In Fig. 4.6b are the corresponding waveforms for a practically matched filter (i.e., $B_6 \approx \tau^{-1}$) receiver; the receiver-filter has a Gaussian frequency response $G(f)$ and thus is not perfectly matched to the spectrum of the transmitted pulse envelope. The ordinate scale is logarithmic to better display the "tail" of weak signals that

(a)   (b)

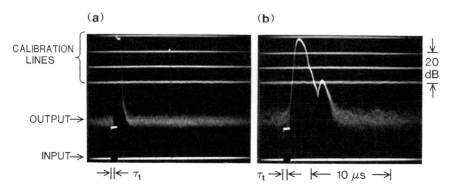

CALIBRATION
LINES

                                                              ↓
                                                              20
                                                              dB
                                                              ↑

OUTPUT→

INPUT→

→‖← τ_t τ_t →‖← ‖← 10 μs →‖

Fig. 4.6 Receiver inputs and outputs. The envelope (lower trace) of the input microwave pulse and output video pulse (upper trace) for (a) a wide bandwidth (i.e., $B_6\tau \simeq 10$) receiver and (b) matched filter (i.e., $B_6\tau \simeq 1$) receiver.

reverberate within the receiver well after the input pulse has ceased. Also note the delay τ_t in the initial rise of the output pulse. This delay is caused by propagation through the transmission line from the antenna to the receiver, as well as propagation through the receiver. Thus only if $\tau_t = 0$, and if transmitter delays are zero, would the filter output voltage sample, at time τ_s, receive contributions from all the scatterers between range zero and $c\tau_s/2$, as can be seen in Fig. 4.7. Furthermore, the scatterers nearest to the radar contribute a vanishingly small voltage increment to samples at increasingly large τ_s.

The receiver output peaks after the leading edge of the echo pulse (from a shell of particles located at $r = c\tau_s/2$) arrives at the antenna (Fig. 4.2). Radar delay τ_r is a sum of the delay from the timing pulse to the arrival of the transmitted pulse at the antenna (i.e., transmitter delay) and the delay between the echo arrival (at the antenna) and the peak at the receiver's output. τ_r can be larger than the transmitted pulse width (e.g., Fig. 4.6) and therefore it can increase significantly the apparent range, $c\tau_s/2$, from the actual position of those scatterers that contribute most to the weather signal. Radar delay τ_r must be subtracted from the time delay τ_s to the weather signal sample to estimate true range. Thus a voltage sample at τ_s receives maximum weight from

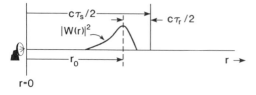

Fig. 4.7 The range-dependent part of the composite weighting function.

scatterers at range

$$r_0 = c(\tau_s - \tau_r)/2 \tag{4.23}$$

Conversely, we can state that, for the voltage increments contributing to the sample at τ_s, the receiver's filter effectively gives less weight to those scatterer's displaced in range relative to r_0. Thus a scatterer receives a weight that has the functional range dependence

$$W(r_0 - r + c\tau_r/2) \tag{4.24}$$

with maximum at $r = r_0$ (Fig. 4.7).

From Fig. 4.2 and Eq. (4.22) it becomes clear how the delays can be measured, and even how the range-weighting function can be determined. It suffices to insert a replica of the transmitter pulse into the receiver and to observe the shape of the response. This response is the mirror image of the weighting function.

4.4.3 Finite Bandwidth Power Loss

Consider the receiver to have a frequency transfer described by a Gaussian function and the transmitted pulse to be rectangular. Consequently, from Eq. (3.35) we conclude that a scatterer at r has a range-dependent weight

$$W(r) = \tfrac{1}{2}\{\mathrm{erf}[(2aB_6/c)(r_0 - r + c\tau/4)] - \mathrm{erf}[(2aB_6/c)(r_0 - r - c\tau/4)]\} \tag{4.25}$$

for the voltage sample taken at delay $\tau_s = 2r_0/c + \tau_r$ (Fig. 4.2). In Section 3.5.2 we noted that a perfect Gaussian filter has infinite delay (i.e., $\tau_r = \infty$) so, strictly speaking, we cannot reference r to a starting time $\tau = 0$ and sample time τ_s. However, Eq. (4.25) should describe accurately the weighting function about any range $r = r_0$ where it is maximum.

Note that Eq. (4.25) shows, in the limit of infinite bandwidth and no propagation delays in the transmitter and receiver, radar delay is one-half the transmitted pulse width (i.e., $\tau_r \to \tau/2$). Thus we have an equivalent radar delay $\tau/2$ even though the signal experiences no delays in the radar. This apparent contradiction arises because we have assigned a range to a scatterer corresponding to the time its echo contributes most to the signal sampled at delay τ_s. Although the location of the peak of Eq. (4.25) may not be the best means of assigning a range to an isolated point scatterer (because we do not know if the echo amplitude is maximum at the sample time τ_s), *it is a proper formula for locating the range to those distributed scatterers that contribute most to the signal sampled at τ_s.*

For a Gaussian transfer function it can be shown that

$$|W(x)| = [\mathrm{erf}(x + b) - \mathrm{erf}(x - b)]/2, \tag{4.26}$$

where

$$b = B_6 \tau \pi/4\sqrt{\ln 2}, \qquad a = \pi/2\sqrt{\ln 2}, \tag{4.27a}$$

$$x = (2aB_6/c)(r_0 - r). \tag{4.27b}$$

Note that the peak of $|W(x)|$ is less than one and decreases with a reduction in bandwidth. We can obtain an analytic solution for the finite bandwidth loss factor l_r [Eq. (4.15)] by approximating $\mathrm{erf}(y)$ with $\tanh(y)$. We find

$$l_r = [\coth(aB_6\tau) - 1/aB_6\tau]^{-1}. \qquad (4.28)$$

The finite bandwidth loss $L_r = 10 \log l_r$ is plotted in Fig. 4.8 along with an exact numerical solution. From Fig. 4.8 we see that if $B_6\tau \geq 1$, then Eq. (4.28) has an error less than 0.6 dB. However for $B_6\tau < 1$ we need the exact solution for accurate calibration. For $B_6\tau \ll 1$ the loss of signal power due to finite receiver bandwidth is

$$L_r \rightarrow 10 \log(aB_6\tau/3). \qquad (4.29)$$

Thus the weather signal sample power decreases linearly with B_6 for $B_6\tau \ll 1$. This may seem disastrous, but what really counts is the signal-to-noise ratio (Section 4.5).

Nathanson and Smith (1972) examined the perfectly matched (to a rectangular pulse) filter receiver and deduced L_r to be 1.8 dB. A practical matched filter (i.e., Gaussian frequency response and $B_6\tau = 1$), has $L_r \simeq 2.3$ dB, or about 0.5 dB more than the value obtained with a perfectly matched filter receiver.

4.4.4 The Resolution Volume

It is useful to define the resolution volume V_6 to be that circumscribed by the 6-dB contour of $|W^2(r)| f^4(\theta, \phi)$. The 6-dB width of $f^4(\theta, \phi)$ (the two-way antenna pattern function) is often taken to be the angular width of V_6. In

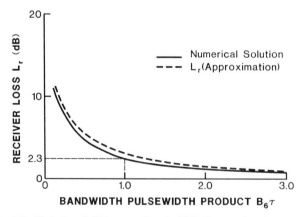

Fig. 4.8 Finite bandwidth power loss L_r (dB). The receiver frequency transfer is Gaussian, and the echo pulse is rectangular. The solid curve is a numerical solution of the exact response function, and the dashed curve is obtained from an analytical approximation [Eq. (4.28)].

an analogous manner we define the 6-dB width of $|W(r)|^2$ to be the range width r_6 of V_6. The solution of the transcendental equation for r_6 is not easy, but we can again approximate the error function by a hyperbolic tangent function. This approximation gives an analytic solution that shows the inter-relation between receiver bandwidth B_6 and pulse width τ. The solution is

$$r_6 = (1/aB_6\tau)(\tfrac{1}{2}c\tau)\,\cosh^{-1}[2 + \cosh(aB_6\tau)]. \qquad (4.30)$$

The range width r_6 is plotted versus $B_6\tau$ in Fig. 4.9. Also shown is a numerical solution of the exact transcendental equation. We see that as the bandwidth–pulse-width product gets large, the 6-dB range width approaches $c\tau/2$. If $f^4(\theta, \phi)$ and $|W^2(r)|$ have Gaussian shapes the resolution volume is a spheroid with maximum angular width $r\theta_1$ and a maximum range extent r_6. Outside this resolution volume, reflectively is weighted by factors $<1/4$ due to either antenna pattern, receiver response, or both. The size of the resolution volume is $0.1\ \mathrm{km}^3$ for $\theta_1 = 1°$, $r = 50\ \mathrm{km}$, $B_6\tau = 1$, and $\tau = 1\ \mu\mathrm{s}$. We must be cautious when ignoring contributions to echo power from regions outside V_6 because hydrometeors have cross sections that vary by many orders of magni-tude. Thus a collection of weak scatterers in V_6 may contribute less power than strong scatterers outside V_6.

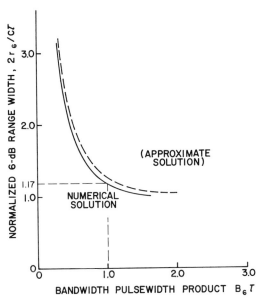

Fig. 4.9 Normalized 6-dB range width, $2r_6/c\tau$, of the resolution volume versus receiver band-width–pulse-width product. The numerical solution to the transcendental equation is the solid line, and the dashed line shows an approximate analytic solution, given by Eq. (4.30).

4.4.5   Reflectivity Factors

Radar meteorologists need to relate the *reflectivity* η, which is general radar terminology for the backscattering cross section per unit volume, to factors that have meteorological significance. If water drops are spherical and have diameters that are small compared to the wavelength (i.e., in the Rayleigh approximation), we can substitute Eq. (3.6) into Eq. (4.10) to obtain

$$\eta = \frac{\pi^5}{\lambda^4}|K_w|^2 Z, \tag{4.31}$$

where

$$Z \equiv \frac{1}{\Delta V}\sum_i D_i^6 = \int_0^\infty N(D, \mathbf{r})D^6\, dD \tag{4.32}$$

is the *reflectivity factor*. Whenever the Rayleigh approximation does not apply, and this is usual for short-wavelength ($\lambda < 10$ cm) radars measuring thunderstorm precipitation, it is accepted practice to write

$$\eta = (\pi^5/\lambda^4)|K_w|^2 Z_e, \tag{4.33}$$

where Z_e is the *equivalent reflectivity factor*. Although the choice of η or Z_e to be used in the radar equation should be equal, radar meteorologists prefer Z_e. Because values of Z commonly encountered in weather observations span many orders of magnitude, radar meteorologists use a logarithmic scale $10 \log_{10} Z$ (where now Z is in units of mm^6/m^3). Precipitation produces Z ranging from near 0 dBZ in cumulus congestus clouds to values somewhat larger than 60 dBZ in regions of heavy rainfall and hail (Chapters 8 and 9). The use of reflectivity factors alone does not really relate radar echo power to meteorologically significant factors such as rainfall rate or liquid water content because, in addition to the phase (i.e., liquid, solid, or a mixture such as melting ice), the particle size distribution needs to be known (Chapter 8).

Incorporating Eq. (4.33) into Eq. (4.16), we have the weather radar equation that gives the mean power, of the weather signal samples, in terms of Z_e (in cubic meters):

$$P(\mathbf{r}_0) = \pi^3 P_t g^2 g_s \theta_1^2 c\tau |K_w|^2 Z_e / 2^{10}(\ln 2)\lambda^2 r_0^2 l^2 l_r, \tag{4.34}$$

where all units are in the MKS system; θ_1 is in radians, and g and g_s are dimensionless. However, it is common to express Z_e in units of (millimeters)^6 per cubic meter, θ_1 in degrees, $\mathbf{r}_0$ in kilometers, λ in centimeters, τ in microseconds, P_t in watts, and $\bar{P}(\tau_s)$ in milliwatts. Thus in units conventional to the radar meteorologist, the mean power is

$$\bar{P}(\text{mW}) = \frac{\pi^5 10^{-17} P_t(\text{W}) g^2 g_s \tau(\mu s)\theta_1^2(\text{deg})|K_w|^2 Z_e(\text{mm}^6/\text{m}^3)}{6.75 \times 2^{14}(\ln 2)r_0^2(\text{km})\lambda^2(\text{cm})l^2 l_r}. \tag{4.35}$$

4.5 Signal-to-Noise Ratio for Distributed Scatterers

The SNR for distributed scatterers is just Eq. (4.16) divided by $kT_{sy}B_n$. Again consistent with the approximation (4.28), we find that the SNR depends on $B_6\tau$ as

$$\text{SNR} = \frac{C_0\eta}{r^2l^2}\left[\coth(aB_6\tau) - \frac{1}{aB_6\tau}\right]\frac{\tau^2}{B_6\tau}, \tag{4.36}$$

where C_0 contains those constants pertaining to the radar. We immediately note that if $B_6\tau$ is a constant, then SNR is proportional to the square of the transmitted pulse width. It is also readily seen that the maximum SNR is obtained as $B_6 \rightarrow 0$. Therefore, unlike for point scatterer measurements, we do not obtain a maximum for the weather SNR at $B_6\tau \approx 1$ (Section 3.5.4).

Even though the SNR increases monotonically as B_6 decreases, the resolution r_6 worsens. If one constrains the resolution to be a constant, then the SNR is an optimum when the filter response is matched to the transmitted pulse. By using a slightly different definition of resolution, one can obtain the optimum signal shape (Zrnic and Doviak, 1978). That is, the resolution can be defined as the square root of the second moment of $|W(r)|^2$. Then, for a given resolution, SNR is a maximum if the transmitted pulse has Gaussian shape matched to the filter's Gaussian impulse response.

As an example, consider a desired resolution of $r_6 = 500$ m, and assume that the filter's impulse response and transmitted voltage pulse are Gaussian with second central moments σ_τ^2. Because the convolution of two Gaussian functions of equal variance is also Gaussian but with double the value of the second moment, it follows that $|W(r)|$ has a second moment σ_w^2 proportional to $2\sigma_\tau^2$ and the range-weighting function $|W(r)|^2$ has a moment proportional to σ_τ^2. Now we can find directly that

$$r_6^2 = 2c^2\sigma_\tau^2 \ln 4. \tag{4.37}$$

Solving for σ_τ (with $r_6 = 500$ m), we find $\sigma_\tau = 1.0$ μs, and the 6-dB pulse width is 2.35 μs. The frequency response of a receiver having a Gaussian impulse response is also Gaussian, and it can be shown by taking the Fourier transform of $h(t)$ that

$$\sigma_f = 1/2\pi\sigma_\tau, \tag{4.38}$$

where σ_f is the second moment of the amplitude spectrum, from which we find the 6-dB width B_6 to be $\sqrt{2\ln 2}/\pi\sigma_\tau = 375$ kHz. If the same calculation is done for a practical radar with rectangular pulse and Gaussian filter, one finds from Fig. 4.9 that $\tau = 2r_6/1.17c = 2.85$ μs and $B_6 = 350$ kHz.

4.6 Correlation of Samples along Range Time

Sample spacing along the range time is usually chosen so that there are independent estimates of reflectivity and velocity along the beam. Both pulse width τ and receiver-filter bandwidth B_6 determine the correlation of these estimates, and sometimes B_6 is deliberately chosen to be small (i.e., not matched to τ) to observe meteorological events in a larger range interval with fewer samples along the range time axis. This approach becomes more advantageous when real-time data processing equipment limits simultaneous observations to few range time samples (as is sometimes the case for real-time Doppler spectral processors) and pulse width cannot be increased. If B_6 is matched to τ or, as in many meteorological radars, is large compared to τ^{-1}, then dimensionally small meteorological events such as tornadic vortices can be missed by samples spaced farther than the range extent of the resolution volume.

In this section we examine the correlation between samples spaced in range time, and determine how receiver bandwidth and transmitter pulse width affect this correlation. In Appendix C we show that the signal at the receiver's input has a correlation R_{xx} given by

$$R_{xx}(\delta\tau_s) = \begin{cases} \bar{P}(\mathbf{r}_0)[1 - |\delta\tau_s|/\tau], & |\delta\tau_s| \le \tau, \\ 0 & \text{otherwise.} \end{cases} \quad (4.39)$$

The signal $V(\tau_s)$, after passing through the filter, has a correlation $R_{vv}(\delta\tau_s)$ given by (Papoulis, 1965, p. 346)

$$R_{vv}(\delta\tau_s) = R_{xx}(\delta\tau_s) \star h^*(-\delta\tau_s) \star h(\delta\tau_s), \quad (4.40)$$

where $h(\delta\tau_s)$ is the unit impulse response of the filter and $\star$ designates the convolution operation. For a Gaussian filter it can be shown that

$$h(t) = 0.5B_6(\pi/\ln 2)^{1/2} \exp[-(\pi B_6 t)^2/4 \ln 2], \quad (4.41)$$

and therefore

$$h^*(-\delta\tau_s) \star h(\delta\tau_s) = \int_{-\infty}^{+\infty} h^*(-\varepsilon)h(\delta\tau_s - \varepsilon)\, d\varepsilon$$

$$= (0.5)^2 B_6 (2\pi/\ln 2)^{1/2}\exp[-(\pi B_6 \delta\tau_s)^2/8 \ln 2]. \quad (4.42)$$

We then find that the samples have a correlation R_{vv} given by the convolution of the input signal correlation R_{xx} and Eq. (4.42):

$$R_{vv}(\delta\tau_s) = \frac{aB_6\tau\bar{P}(\mathbf{r}_0)}{\sqrt{2\pi}} \int_{-1}^{+1} \{(1 - |x|)\exp[-(aB_6\tau)^2(x - \delta\tau_s/\tau)^2/2]\}\, dx, \quad (4.43)$$

where $a = \pi/2\sqrt{\ln 2}$

Equation (4.43) has been evaluated numerically, and the results are plotted in Fig. 4.10. The solution shows that when B_6 is more than twice τ^{-1}, sample

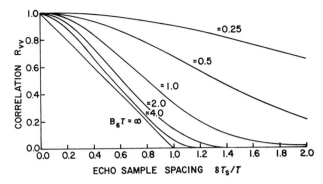

Fig. 4.10 Normalized correlation of weather signal samples spaced by $\delta\tau_s$.

correlation is principally controlled by pulse width, whereas when $B_6 < 0.5\tau^{-1}$ it is controlled by the receiver-filter 6-dB bandwidth. A useful analytic formula for correlation when $B_6 \ll \tau^{-1}$ is

$$R_{vv}(\delta\tau_s) = R_{vv}(0)e^{-(aB_6\,\delta\tau_s)^2/2}, \tag{4.44}$$

where $R_{vv}(0) = aB_6\tau\bar{P}(\mathbf{r}_0)/\sqrt{2\pi}$ and $\delta\tau_s$ is the range time sample spacing.

Problems

4.1 Show that

$$\int_0^\infty P\cdot\mathrm{Prob}(P)\,dP = 2\sigma^2 = \bar{P},$$

where $\bar{P}$ is the mean power of the weather signal samples.

4.2 If a desired range resolution r_6 is 300 meters, determine the width of a Gaussian-shaped transmitted pulse and the 6-dB width of the matched filter receiver that will achieve this resolution. Repeat the problem but with a rectangular pulse and a Gaussian "match" filter.

4.3 Derive a formula for r_6 if the transmit pulse is rectangular of width τ and assume the normalized amplitude transfer function $G(f)$ of the filter is Gaussian (i.e., $\exp(-f^2/2\sigma_f^2)$). First find the filter's impulse response,

$$h(t) = \int_{-\infty}^{+\infty} G(f)\exp(j2\pi ft)\,df,$$

and then determine the range-weighting function.

4.4 Assume hydrometeors have a reflectivity factor of 50 dBZ. Find the reflectivity η for the two wavelengths of 5 and 10 cm. If λ is 10 cm, P_t is 100 kW, total losses are 3 dB, pulse duration is 2 μs, antenna gain is 40 dB, and beamwidth is 1°, compute received power of echoes from hydrometeors at a range of 60 km.

4.5 A 10-cm radar radiates a peak power pulse of $P_t = 10^6$ W and has an antenna gain of 50 dB. A spherical rain drop of diameter 6 mm is located at $r = 20$ km. (a) Compute the level of power density at the drop. (b) Compute the level of power density returned to the antenna. (c) Compute echo power delivered to the receiver. (d) The reflectivity factor (dBZ) of randomly distributed drops, all of 6 mm diameter, is 40 dBZ. Assume that the radar's resolution volume is $0.1 \times 0.5 \times 0.5$ km^3. How many drops are contained in this volume? (e) Assume that all the raindrops, agitated by turbulence, are confined to the resolution volume. Sketch a likely time-dependent trace of echo power. Make sure you indicate on your sketch the important time scales of τ and T_s. (f) Define a sample-time average of echo power and compute it for the conditions stipulated in (d). (g) How else could you define an average of echo power?

4.6   A transmitted pulse leaves an antenna at $t = 0$, and a gating circuit samples, at range-time τ_s, echoes (Fig. P.4.1) from hydrometeors of uniform reflectivity. The receiver response to an echo from a point scatterer that arrives at the antenna at time t, for example $t = 0$, is also sketched in Fig. P.4.1. Determine the range to hydrometeors that contribute most to the echoes sampled at $\tau_s = 100$ μs. If sampling gates are spaced 2 μs apart, what is the spacing of resolution volumes? Describe, in your own words, the physical interpretation of the resolution volume V_6. Which radar parameters determine the angular and range widths of V_6?

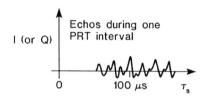 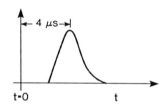

Fig. P.4.1

4.7   A pulsed-Doppler radar has wavelength of 10 cm and is transmitting 1 μs duration pulses at a rate of 1000 Hz. The radar resolution volume at a range of 60 km has a size of 0.1 km^3. (a) What is the size of the resolution volume at 40 km? (b) If the resolution volume is filled with hailstones that are 1 cm in diameter, and their density is 1 per 100 m^3, determine the effective reflectivity factor (in dBZ units). (c) Assuming the one-way attenuation 0.01 dB km^{-1}, find the ratio of the echo powers from hail at ranges of 40 and 60 km. (d) If the hydrometeors are moving at 30 m s^{-1} toward the radar, what Doppler shift (Hz) would they produce, and what velocity would the radar measure?

4.8   There are three raindrops in the radar resolution cell, all located along the beam axis. Raindrop number 1 causes an echo amplitude A. Raindrop number 2 causes an amplitude of $2A$, and raindrop number 3 causes an amplitude of $3A$. The raindrops are in radial motion with random radial velocities. What is the (a) maximum received power, (b) minimum received power, (c) average received power?

4.9   (a) Sketch at least three time frames (i.e., T_s periods) of the I or Q signals for weather echoes as a function of range time τ_s. Assume all weather echoes are received before the next transmitted pulse and that in the period T_s all drops move a distance relative to one another that is much smaller than a wavelength. On your graph indicate the time scale of echo fluctuations. (b) Sketch the time dependence of the sequence of samples obtained at one range time delay. Indicate on the sketch of the I or Q signals where the sample is being taken. Assume whatever Doppler shift you like.

Doppler Spectra of Weather Signals

To measure the power-weighted distribution of velocities, a frequency analysis of $V(\tau_s, mT_s)$ (Fig. 4.1) is needed. This can be accomplished by estimating its power spectrum. We first review the theory of discrete Fourier transforms, then discuss convolution and the correlation of random signals, and show the relation between the correlation function and power spectrum. Throughout this section we omit τ_s from the argument, and the results derived apply to any sequence of samples at fixed τ_s. The very important relationship between the reflectivity and velocity fields and the power spectrum is thoroughly discussed; furthermore the contributions of various physical mechanisms to spectrum broadening are presented.

5.1 Spectral Analysis of Weather Signals

5.1.1 Discrete Fourier Transform

Fourier methods of time series analysis are now commonly used in many branches of science and engineering. Digital computers with associated software or special-purpose machines rapidly compute the Fourier coefficients of lengthy time sequences. The discrete Fourier transform (DFT) of a signal sampled M times at a uniform spacing T_s (Fig. 5.1) is defined as

$$Z(kf_0) = \sum_{m=0}^{M-1} V(mT_s)e^{-j2\pi f_0 T_s mk}, \tag{5.1}$$

where $V(mT_s)$ is the complex voltage of the mth sample and $Z(kf_0)$ is the complex amplitude of the kth spectral coefficient at the frequency $f = kf_0$. The inverse transform (IDFT) is

$$V(mT_s) = \frac{1}{M} \sum_{k=0}^{M-1} Z(kf_0)e^{j2\pi f_0 T_s mk}. \tag{5.2}$$

Note that Eq. (5.1) is an expansion in terms of multiples of the fundamental (lowest) frequency

$$f_0 = 1/MT_s, \tag{5.3}$$

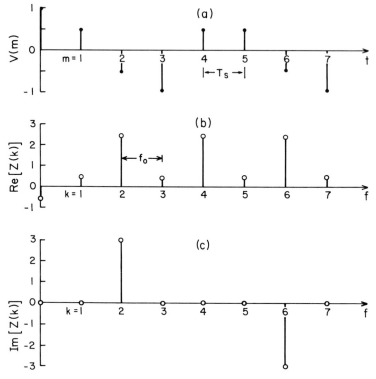

Fig. 5.1 (a) Sequence $V(mT_s)$ of real numbers. (b, c) The complex spectral coefficients $Z(kf_0)$. The number of points $M = 8$. The magnitude $|Z(kf_0)|$ is symmetric with respect to 0 and $4f_0$ (the Nyquist frequency) because the sequence $V(mT_s)$ is real. It can be seen that the sequence $V(mT_s)$ has two dominant sinusoidal components with periods $4T_s$ and $4T_s/3$.

and as such gives M complex Fourier coefficents (amplitudes and phases). We stress that k can also be a continuous variable. Nevertheless, if k is an integer it is easy to prove by substitution that Eq. (5.2) satisfies Eq. (5.1).

$$Z(kf_0) = \frac{1}{M} \sum_{m=0}^{M-1} \sum_{n=0}^{M-1} Z(nf_0)e^{-j2\pi f_0 T_s m(k-n)}. \tag{5.4a}$$

Interchanging the order of summation and summing the geometric progression over m, we get

$$\sum_{m=0}^{M-1} e^{-j2\pi f_0 T_s m(k-n)} = (1 - q^M)/(1 - q), \tag{5.4b}$$

where

$$q = \exp[-j2\pi f_0 T_s(k - n)]. \tag{5.4c}$$

Because Eq. (5.4b) equals M if $k = n$ and 0 otherwise, Eq. (5.4a) is indeed an identity.

We note that $Z(kf_0)$ is periodic in k.

$$Z(kf_0) = Z(lf_0) \qquad \text{if}^1 \quad l = \text{mod}_M(k). \tag{5.5a}$$

Furthermore, if the sequence $V(mT_s)$ is real,

$$Z(-kf_0) = Z^*(kf_0). \tag{5.5b}$$

Thus, for the example in Fig. 5.1 (where $M = 8$), $Z(0) = Z(8)$, $Z(1) = Z(9)$, etc.

Likewise, the sequence $V(mT_s)$ calculated from Eq. (5.2) is periodic in m: $V(mT_s) = V[(m + M)T_s]$. We should not conclude from this property that our original sequence of signal samples is periodic. We have taken a segment, M samples long, of a much longer sequence that, for weather signals, is definitely nonperiodic. It is the property of the transform, however, to repeat periodically the sample values outside the segment MT_s.

The Nyquist frequency $(2T_s)^{-1}$ is the highest frequency that can be unambiguously measured in the sampled sequence because samples of a signal with a frequency $f > (2T_s)^{-1}$ will have the same time dependence as a signal with $f < (2T_s)^{-1}$ (Section 3.6). Because the exponential in Eq. (5.1) is cyclic, $Z(-kf_0) = Z[(M - k)f_0]$. The spectral coefficients are arranged often for display purposes so that there is an equal number $(M/2)$ on each side of $k = 0$. This is done in order to have positive Doppler frequencies to the right and negative ones to the left of the zero ($k = 0$) line. In such a case the power $|Z(Mf_0/2|^2$ is split, so there are actually $M + 1$ lines.

Substituting Eq. (5.3) into Eq. (5.1) and (5.2), we can eliminate f_0 and T_s from the calculation. For simplicity we also drop them from the arguments in Eqs. (5.1) and (5.2) to obtain a compact form of the DFT,

$$Z(k) = \sum_{m=0}^{M-1} V(m)e^{-j(2\pi/M)mk}, \tag{5.6a}$$

and its inverse

$$V(m) = \frac{1}{M} \sum_{k=0}^{M-1} Z(k)e^{j(2\pi/M)mk}. \tag{5.6b}$$

To understand the properties of the DFT, we illustrate how it acts on a sampled complex sinusoid $A \exp[j(2\pi\alpha f_0 mT_s + \psi)]$ of frequency αf_0 and constant phase ψ, where α is a real number.

From Eq. (5.6a) we find the Fourier coefficients

$$Z(k) = A \sum_{m=0}^{M-1} \exp\left[j\left(\frac{2\pi}{M}\alpha m + \psi\right)\right]\exp - j\frac{2\pi}{M}mk. \tag{5.7}$$

---

1. In this notation l is a remainder after k is divided by M.

This finite sum represents a geometric progression, as in Eq. (5.4b), and can be reduced to

$$Z(k) = A \, \frac{\sin[\pi(\alpha - k)]}{\sin[(\pi/M)(\alpha - k)]} \, \frac{\exp[j\pi(\alpha - k)]}{\exp[j(\pi/M)(\alpha - k)]} \, e^{j\psi}.$$

Its magnitude is

$$|Z(k)| = A \left| \frac{\sin[\pi(\alpha - k)]}{\sin[(\pi/M)(\alpha - k)]} \right|. \tag{5.8}$$

If α is an integer, then

$$|Z(k)| = \begin{cases} 0 & \text{for} \quad k \neq \alpha, \\ AM & \text{for} \quad k = \alpha. \end{cases} \tag{5.9}$$

Thus, if the input sinusoid has a period that is exactly contained in the interval length MT_s, its discrete Fourier transform consists of a single coefficient. In other words, the DFT has singled out (filtered) the sinusoid. It is to be noted that although each sample has amplitude A, the amplitude of the spectral coefficient $k = \alpha$ is AM. Samples are therefore said to sum coherently to give a spectral-coefficient amplitude M times the signal amplitude.

Figure 5.2a illustrates a complex sinusoid with period $4T_s$, a sample sequence $8T_s$ in duration ($M = 8$), and $\alpha = 2$. The Fourier coefficients $|Z(k)|$ are zero for $k \neq 2$, and $|Z(2)| = 8A$ (Fig. 5.2b). If k is considered to be a continuous variable the region near the peak of Eq. (5.8) defines the main lobe of the DFT filter, and its width is a measure of the filter's frequency selectivity. The other minor lobes are the filter's sidelobes, which cause the power to be spilled into other frequency bins (k values) if α is not an integer.

If the sinusoid's frequency is not an exact multiple of the fundamental (i.e., α is a noninteger), all of the Fourier coefficients are nonzero. They are strictly a function of $\text{mod}_M (\alpha - k)$ and are largest for coefficients that are closest to $\text{mod}_M \alpha$. Again the DFT filters the sinusoid; but, the magnitude of the spectral coefficient closest to α is less than MA because power has spilled into all harmonics (frequency bins) of f_0 (Fig. 5.3).

The DFT is used to locate and isolate the many sinusoids that compose a signal. Because these sinusoids do not necessarily have periods that are an integer portion of the fundamental MT_s, the power of a large-amplitude sinusoid can spill over into many frequency bins, and this may mask identification of weaker sinusoids. This occurs because the transform acts on the sequence $V(m)$ only over a finite-time or data window; the spillage of power is called the window effect. The illustration in Fig. 5.3 shows an eight-point DFT of a pure sinusoid with frequency $2.5f_0$. Because the sinusoid signal reconstructed from M samples must be periodic outside the interval MT_s, there is a discontinuity in slope at $m = 0$ and $m = 8$. Because of this discontinuity, a single sinusoid

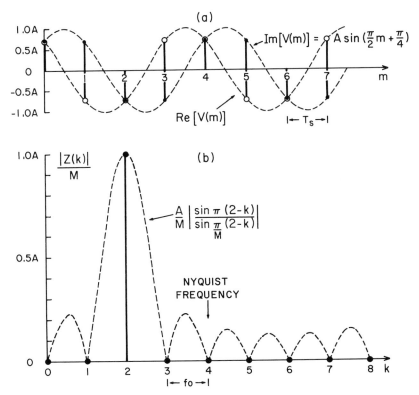

Fig. 5.2 (a) Complex sinusoid of period $4T_s$ (dashed lines). The eight samples, spaced T_s apart, embrace exactly two periods of the sinusoid. (b) The normalized magnitudes of the Fourier coefficients for the sinusoid.

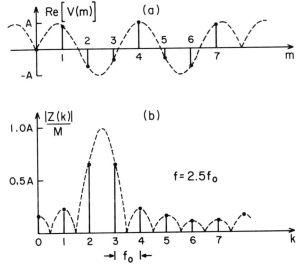

Fig. 5.3 (a) The sinusoid has a period of $3.2T_s$ and hence is not exactly contained in the interval $8T_s$. Only the real component of the complex sinusoid is sketched. (b) The magnitude of the Fourier coefficients.

cannot fit smoothly through the transition, so there must exist power in other frequency bins. Can one reduce this effect? The answer is yes.

Examining Fig. 5.3 and Eq. (5.8), we notice that a spectral coefficient reaches a maximum if one allows k to be a noninteger equal to α. Thus, by continuously changing k in Eq. (5.8), we would have a coherent sum of samples wherever k reached the value equal to the unknown frequency of the sinusoid. However, this does not prevent other sinusoids from possibly spilling their power into the selected frequency bin. Therefore, rather than searching for individual peaks, a more common approach is to weight the time series samples and thus reduce the abrupt transition at the beginning and end of the time sequence. This procedure always broadens the mainlobe of the DFT filter, but the sidelobe amplitudes are decreased so that power spilled into other bins is smaller. We shall examine the window question more thoroughly in Section 5.1.4.

Although efficient methods for calculating the DFT coefficients were discovered at the beginning of this century (Cooley $et\ al.$, 1967), they became widely known only after the 1965 publication by Cooley and Tukey. All such algorithms are appropriately called fast Fourier transforms (FFT). They use advantageously the periodicity of trigonometric functions so that the total number of complex operations for calculating all M complex coefficients is proportional to $M \log_2 M$ if M is a power of 2. The number of complex multiplications is one-half that of the additions (Tretter, 1976). In contrast, the straightforward method, Eq. (5.6a), requires M^2 operations of both kinds. The FFT produces all M coefficients, however. If for some reason only a few coefficients are needed, the regular DFT may be more efficient.

5.1.2 Convolution and Correlation

The Fourier transform has been found to be of great utility in evaluating the convolution and correlation of sample sequences. The output response of any linear system to any input signal can be obtained from the system's impulse response through the application of convolution (Gabel and Roberts, 1973). Let the input sequence be $V(m)$ and the filter impulse response sequence (or weighting function) be $h(m)$. Then the output sequence $Y(l)$ is obtained from the convolution (Papoulis, 1977):

$$Y(l) = V \star h = \sum_{m=-\infty}^{\infty} V(m)h(l-m) = \sum_{m=-\infty}^{+\infty} V(l-m)h(m). \qquad (5.10)$$

The sequence $h(l-m)$ is a mirror image of $h(m)$ about $m = 0$ and shifted to the right by l increments. It is well known in the theory of continuous systems that a convolution has a Fourier transform that equals the product of the transforms of each signal. A similar but not identical relationship holds for discrete periodic signals:

$$Y(l) = \frac{1}{M} \sum_{k=0}^{M-1} H(k)Z(k)e^{j(2\pi/M)lk} = \sum_{m=0}^{M-1} h(m)V(l-m). \qquad (5.11)$$

To prove Eq. (5.11), one should express $h(m)$ and $V(l-m)$ as inverse transforms of $H(k)$ and $Z(n)$. Then, similar to Eq. (5.4a), the sum over m will differ from zero only if $k = n$, resulting in the identity. The difference arises because the product of $H(k)Z(k)$ is a DFT of a circular convolution (Gold and Rader, 1969). Such a convolution (5.11) operates on a periodic extension of either of the sequences $h(m)$ and $V(m)$ (Fig. 5.4).

Two infinite sequences $U(m)$, $V(m)$ can be tested for similarity with the use of a cross correlation function:

$$R_{vu}(l) = E[V(l+m)U^*(m)] \tag{5.12}$$

where $E[\ ]$ denotes the expected value of the ensemble (Papoulis, 1965). The ensemble is a collection or set of random sequences in which $V(m)$ varies from member to member of the ensemble even though m is fixed.

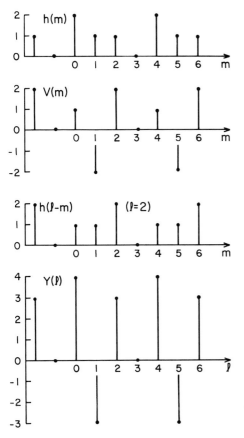

Fig. 5.4   Example of circular convolution. The number of points $M = 4$. The filter weighting function $h(m)$ and the time sequence $V(m)$ have three nonzero terms. Circular convolution involves flipping one of the sequences about $m = 0$, shifting by l increments, cross multiplying the two sequences $V(m)$, $h(l)$, and summing over the M products.

Whereas the convolution operation (5.10) involves shifting the *mirror image* of a sequence and summing products, the correlation calls for shifts and sums only. Therefore, the convolution will equal the correlation if the sequence is symmetric.

A sequence with statistical properties independent of time is called stationary. Strictly speaking, no such sequence can represent a physical process, because all known processes continously change. However, when the time of observation is small compared with the time it takes for significant change to occur in the statistical properties (e.g., mean value, variance, and correlation) of the process, the sequence representing the process can be considered stationary. Weather radar signals are a perfect example of such a stationary process.

In dealing with random signals, the most one can hope to accomplish is to estimate some average statistical parameters. Two useful ones are the mean and autocorrelation; if only they are time invariant, the process is wide-sense stationary.

The autocorrelation of a complex wide-sense-stationary signal $V(m)$ is defined as

$$R(l) = E[V^*(m)V(m+l)] = E[V^*(m-l)V(m)]. \tag{5.13}$$

$R(l)$ is a measure of similarity between the sequence and its shifted conjugate. Noiselike signals have an autocorrelation with a sharp peak at the origin; in fact, the less correlated the signal samples, the sharper is the peak.

The expected signal power is

$$E[|V(m)|^2] = R(0). \tag{5.14}$$

If $V(m)$ has a periodic component, the autocorrelation has it too. If $V(m)$ has a nonzero mean value, its autocorrelation for large lag l equals the square of the mean value.

$$\lim_{l\to\infty} R(l) = E^2[V(m)]. \tag{5.15}$$

This is so because, as l increases, $V^*(m)$ and $V(m+l)$ become less correlated until, in the limit, the correlation of the fluctuating components vanishes and only the mean value contributes to $R(l)$. Although the autocorrelation of uncorrelated samples need not vanish, the autocovariance does. The autocovariance is just the autocorrelation after the mean value $E[V(m)]$ is removed from the signal. For most weather signals, $E[V(m)]$ is zero (Section 4.2), but it is not for echoes from fixed scatterers. Our discussion is illustrated by Fig. 5.5, which shows a possible autocorrelation function of an assumed stationary sequence having nonzero mean. Because the process is stationary, we can clearly see from Eq. (5.13) that $R(-l) = R^*(l)$.

At first sight it appears that the autocorrelation function is simply evaluated. Obviously, acquiring all the realizations of the process is not practical; fortunately, signals we deal with are often stationary, and we may use time averages to deduce the statistical parameters from one realization of a process. For

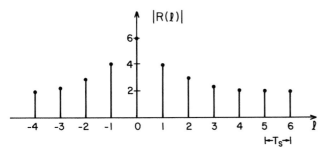

Fig. 5.5 Magnitude of an autocorrelation function. The power is 6, and after about $3T_s$ the time samples are not correlated. The magnitude of the mean $E[V(m)]$ is $\sqrt{2}$.

example, statistical properties of scatterers (e.g., reflectivity) are stationary during the period of observation (typically <1 s), and scatterers produce one member of the ensemble of all possible sample sequences that could occur under similar conditions. It is more practical to examine one member and deduce its statistical properties. If the statistical properties of the ensemble can be deduced from sample time averages, the ensemble members $V(m)$ are said to be ergodic. Only then can the properties of an ensemble be inferred from time averages of one member. Sample sequences produced by scatterers with statistically homogeneous properties are members of an ergodic ensemble (if one ignores beam spreading and attenuation). We shall assume all our signals to be ergodic. Thus, the ensemble average (5.13) is equivalent to

$$R(l) = \lim_{M \to \infty} \frac{1}{M} \sum_{m=0}^{M-|l|-1} V^*(m)V(m+l). \qquad (5.16)$$

Because there is usually a finite number of samples we can only form estimates $\hat{R}(l)$ (the caret is used to denote estimates), such as

$$\hat{R}(l) = \begin{cases} \dfrac{1}{M} \displaystyle\sum_{m=0}^{M-|l|-1} V^*(m)V(m+l) & \text{for} \quad |l| \le M-1, \\ 0 & \text{otherwise.} \end{cases} \qquad (5.17)$$

If $R(l)$ vanishes at small l and remains small, $V(m)$ changes significantly from sample to sample.

5.1.3 Power Spectrum of Random Sequences

Besides being useful in its own right for determining the rapidity of signal change, the mean power, periodicity, noisiness, etc., the autocorrelation enables us to find, through its Fourier transform, the frequency distribution of the random signal power. The power spectrum $S(f)$ is defined as the DFT of the autocorrelation function.

$$S(f) \equiv \lim_{M \to \infty} T_s \sum_{l=-(M-1)}^{M-1} R(l)e^{-j2\pi f T_s l}. \qquad (5.18)$$

We can apply the theory of Fourier series to obtain the inverse relation

$$R(l) = \int_{-1/2T_s}^{1/2T_s} S(f)e^{j2\pi f T_s l}\, df. \qquad (5.19)$$

The power spectrum has a repetitive cycle equal to T_s^{-1}, the unambiguous interval. This can be seen on substituting $f = f + T_s^{-1}$ in Eq. (5.18). Equation (5.18) and its inverse (5.19) form the Fourier transform pair and uniquely relate $R(l)$ and $S(f)$. They are completely analogous to a regular Fourier series representation of a periodic function; $S(f)$ is a *continuous* function with period T_s^{-1}, and the $R(l)$ are its complex Fourier series coefficients spaced T_s apart.

One may question how well the transform pair (5.18), (5.19) represents the properties of the actual weather signal since $V(t)$ is sampled at increments spaced T_s apart (Section 4.1). To answer this question, we resort to the sampling theorem: *If a function $V(t)$ contains no frequencies higher than $Mf_0/2$ cycles per second, it is completely determined by its ordinates at a series of points spaced $(Mf_0)^{-1} = T_s$ apart, the series extending throughout the time domain* (Gabel and Roberts, 1973). Thus the sequence $R(l)$ completely determines the signal power spectrum so long as the Nyquist frequency $(2T_s)^{-1}$ is larger than the highest frequency contained in $V(t)$. The relations (5.18) and (5.19) are useful even if $V(t)$ has frequencies beyond the unambiguous interval. Then $S(f)$ is an aliased spectrum that may be considerably different from the true one. Nevertheless, useful information can often be obtained from aliased power spectra if aliasing is recognized.

$S(f)$ represents the power per hertz, from which the name power spectral density derives. From Eq. (5.13), $R(-l) = R^*(l)$; then, because the summation in Eq. (5.18) is symmetric about $l = 0$, it can be shown that $S(f)$ *is real and positive*. The power spectrum is only one of the statistical parameters of a signal, but it is a very useful one. Nevertheless, it is not a complete measure or description of the random process.

A process that generates a particularly simple power spectrum is white noise, for which

$$S(f) = NT_s$$

and

$$R(l) = \begin{cases} N & \text{for} \quad l = 0, \\ 0 & \text{otherwise.} \end{cases} \qquad (5.20)$$

N is the total noise power in the frequency interval between $-1/2T_s$ and $1/2T_s$. In the absence of echoes, the output of a radar receiver can be described as white noise. Sources of white noise are thermally induced random motions of electrons in the rf amplifier and the synchronous detectors of the receiver, and cosmic and atmospheric radiation. Quantization from analog-to-digital conversion (Section 7.7.1) is white only if the quantization error (i.e., the difference between the quantized and actual values) is uncorrelated from sample to sample. In addition to being white, the in-phase and quadrature components of thermal noise and radiation have Gaussian amplitude distributions (very much like weather signals). That is, the I and Q values of receiver noise are normally distributed about zero, and the variance of the distribution is equal to the noise power N (Section 4.3; $P = N$ for thermal noise). This is not the case for quantization noise, which may have values uniformly distributed over the quantization interval. Thus white noise, which relates to the uniformity of the power density as a function of f, and the Gaussian amplitude distribution of Is and Qs are independent characteristics of the signal.

We shall examine two power spectrum estimates and prove their equivalence. First, one can compute the spectrum estimate using the definition (5.18) and the estimate $\hat{R}(l)$.

$$\hat{S}_1(f) = T_s \sum_{l=-(M-1)}^{M-1} \hat{R}(l)e^{-j2\pi fT_s l}. \tag{5.21}$$

Second, it is often computationally more efficient to use the FFT to compute $\hat{S}_2(f)$ directly from the data:

$$\hat{S}_2(f) = |Z(f)|^2 T_s/M, \tag{5.22}$$

or, using Eq. (5.1)

$$
\begin{aligned}
\hat{S}_2(f) &= \frac{T_s}{M} \left[\sum_{m=0}^{M-1} V^*(m)e^{j2\pi fT_s m} \sum_{n=0}^{M-1} V(n)e^{-j2\pi fT_s n} \right] \\
&= \frac{T_s}{M} \left\{ \sum_{m,n}^{M-1} V^*(m)V(n)e^{-j2\pi fT_s l} \right\},
\end{aligned} \tag{5.23}
$$

where $l = n - m$ and the double sum has been abbreviated. The double sum is really an addition of the product terms in an $M \times M$ matrix (Fig. 5.6). The diagonal is a sum of M samples of $\hat{R}(0)$, and the terms along the lines on either side of the diagonal are the $M - |l|$ correlation estimates at different lags l [multiplied by $\exp(-j2\pi fT_s l)$], as shown in Fig. 5.6. Thus, we can evaluate Eq. (5.23) by summing the various estimates $\hat{R}(l)$ along l. This is equivalent to

$$\sum_{m,n}^{M-1} V^*(m)V(n)e^{-j2\pi fT_s l} = \sum_{l=-(M-1)}^{M-1} e^{-j2\pi fT_s l} \sum_{m=0}^{M-|l|-1} V^*(m)V(m+l). \tag{5.24a}$$

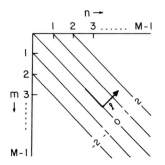

Fig. 5.6 Matrix showing how a double sum can be reduced to a single sum of the correlation estimates $\hat{R}(l)$.

The inner sum of the right side of Eq. (5.24a) is a sum along lines parallel to the diagonal and is, from Eq. (5.17), equal to $M\hat{R}(l)$. Therefore,

$$\frac{T_s}{M} \sum_{m,n}^{M-1} V^*(m)V(n)e^{-j2\pi f T_s l} = T_s \sum_{l=-(M-1)}^{M-1} \hat{R}(l)e^{-j2\pi f T_s l}, \qquad (5.24b)$$

and hence

$$\hat{S}_2(f) = \hat{S}_1(f), \qquad (5.25)$$

which completes the proof. The estimate $\hat{S}_2(f)$ has the descriptive name *periodogram* because it was used by geophysicists to find periodicities in data.

5.1.4 Bias, Variance, and the Window Effect

Estimation theory has established several attributes of estimators. Two important ones are the bias and the variance. An estimate is said to be unbiased if its mean value equals the true value of the parameter that is estimated. To test if $\hat{S}(f)$ is biased, we examine the ensemble average of Eq. (5.23). Before doing this it is convenient to introduce an artifice: We represent $V_M(m)$, the finite segment of our complex sequence, by the product

$$V_M(m) = V(m)d(m), \qquad (5.26)$$

where

$$d(m) = \begin{cases} 1 & \text{for} \quad 0 \leq m \leq M-1, \\ 0 & \text{otherwise}. \end{cases} \qquad (5.27)$$

The weighting sequence $d(m)$ in Eq. (5.26) is referred to as the data window because it reveals only a finite portion (of length M) of the otherwise infinite sequence $V(m)$. In other words, an observer sees through the window $d(m)$ the truncated series $V_M(m)$.

With $V_M(m)$ expressed by Eq. (5.26), the expectation of Eq. (5.23) is

$$E[\hat{S}(f)] = \frac{T_s}{M} \sum_{l=-(M-1)}^{M-1} \sum_{m=0}^{M-1-|l|} d^*(m)d(m+l)E[V^*(m)V(m+l)]e^{-j2\pi f T_s l}$$

$$= T_s \sum_{l=-(M-1)}^{M-1} R(l)e^{-j2\pi f T_s l} \sum_{m=0}^{M-1-|l|} \frac{d^*(m)d(m+l)}{M}$$

$$= T_s \sum_{l=-(M-1)}^{M-1} w(l)R(l)e^{-j2\pi f T_s l}. \tag{5.28}$$

To arrive at Eq. (5.28) we have assumed $V(m)$ to be stationary, which means that $E[V(m)V^*(m+l)]$ is independent of m. Therefore only the data window product is summed over m. The sum of data window products is a correlation of the data window samples and is called the lag window $w(l)$. From Eq. (5.28) the lag window $w(l)$ is

$$w(l) = \frac{1}{M} \sum_{m=0}^{M-1-|l|} d^*(m)d(m+l). \tag{5.29}$$

In arriving at Eqs. (5.28) and (5.29) we did not specify $d(m)$. Therefore Eq. (5.29) is quite general and valid for any window type. A rectangular data window produces a triangular (i.e., Bartlett) lag window, given by

$$w(l) = \begin{cases} 1 - \dfrac{|l|}{M}, & -M \le l \le M, \\ 0 & \text{otherwise,} \end{cases} \tag{5.30}$$

which, when introduced into Eq. (5.28), results in the following mean value of the spectrum estimate:

$$E[\hat{S}(f)] = T_s \sum_{-(M-1)}^{M-1} \left(1 - \frac{|l|}{M}\right)R(l)e^{-j2\pi f T_s l}. \tag{5.31}$$

Note that Eq. (5.31) equals the true spectrum (5.18) only if $M \to \infty$. Therefore *the periodogram is a biased estimate of the true power spectrum*.

To summarize, we have shown that a rectangular data window is equivalent to the triangular lag window. If a periodogram is obtained from a uniformly weighted time series, the result, in the mean, is the same as if the autocorrelation is weighted with a triangular window as in the transform (5.31). The effect of the lag window on the correlation is illustrated in Fig. 5.7, where one can see that the largest deviation from the true autocorrelation occurs at larger values of lag l.

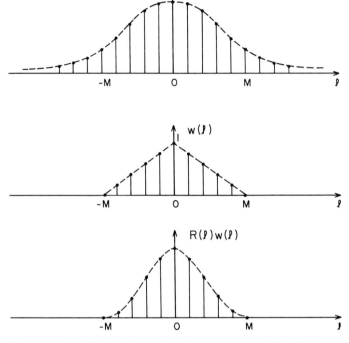

Fig. 5.7 True $R(l)$ and the windowed autocorrelation $R(l)w(l)$ functions for a rectangular data window.

5.1.5 Expressing Spectral Estimates in Terms of the True Spectrum

An alternative and more illustrative method to demonstrate that the periodogram is a biased estimate of the true spectrum involves an expression of $E[\hat{S}(f)]$ in terms of the true $S(f)$. We shall prove that

$$E[\hat{S}(f)] = T_s \sum_{l=-\infty}^{\infty} w(l)R(l)e^{-j2\pi fT_s l} = \int_{-(2T_s)^{-1}}^{(2T_s)^{-1}} S(f')W(f-f')\,df', \quad (5.32)$$

where the spectral window $W(f)$ and the lag window $w(l)$ are Fourier transform pairs like Eqs. (5.18) and (5.19).

$$W(f) = T_s \sum_{l=-(M-1)}^{M-1} w(l)e^{-j2\pi fT_s l}, \quad (5.33)$$

$$w(l) = \int_{-(2T_s)^{-1}}^{(2T_s)^{-1}} W(f)e^{j2\pi fT_s l}\,df. \quad (5.34)$$

We now apply the inverse Fourier transform (5.19) to both sides of (5.32). Because the left side is a Fourier transform, its inverse simply retrieves the product

$$w(l)R(l) = \int_{-(2T_s)^{-1}}^{(2T_s)^{-1}} \int_{-(2T_s)^{-1}}^{(2T_s)^{-1}} S(f')W(f-f')e^{j2\pi fT_s l}\,df'\,df. \qquad (5.35)$$

After the change of variable $f - f' = \varepsilon$, and because $W(f)$ has a period T_s^{-1}, we obtain

$$w(l)R(l) = \int_{-(2T_s)^{-1}}^{(2T_s)^{-1}} W(\varepsilon)e^{j2\pi\varepsilon T_s l}\,d\varepsilon \int_{-(2T_s)^{-1}}^{(2T_s)^{-1}} S(f')e^{j2\pi f' T_s l}\,df', \qquad (5.36)$$

which is an identity. Therefore, Eq. (5.32) reveals that the expected value of a periodogram is a circular convolution of a true spectrum $S(f')$ with the spectral window $W(f')$. Because the convolution is an integration, it broadens the true spectrum and produces spillage of the sharp spectral peak into the sidelobes (Fig. 5.8). The lower curve in Fig. 5.8 is the expected value of the estimated power spectrum; a value of $E[\hat{S}(f_k)]$ is obtained by multiplying the upper two curves point by point, integrating the result, and repeating these steps for various values of f_k. The convolution is circular; as the window peak moves to the right, the sidelobes from its periodic extension appear in the interval $\pm 1/2T_s$. It is best to visualize the points $-1/2T_s$ and $1/2T_s$ as being tied together so that all frequencies lie on a circle; thus the convolution operation is the same as moving average on a circle. Because the lag window is a correlation of data window samples [Eq. (5.29)], it follows that the amplitude spectral window $D(f)$ can be computed from the Fourier transform of the data window.

The various relations and transforms between windows and their spectra are summarized in Table 5.1 and schematically depicted in Fig. 5.9. As an example, consider a rectangular window (i.e., a sequence of uniform unit weights). The magnitude of its transform $|D(k)|$ is equal to Eq. (5.8) with $\alpha = 0$.

$$|D(f)| = \left| \frac{\sin(\pi T_s M f)}{\sin(\pi T_s f)} \right|, \qquad (5.37)$$

where $f = kf_0$ was substituted into Eq. (5.8).

Figure 5.10 illustrates spectra of a weather signal weighted with a rectangular window, and one with a von Hann window (Table 5.2); considerable difference is in the spectral domain especially in spectral skirts. In the periodogram obtained with uniform weight applied to data samples (Fig. 5.10 RECT) we notice the slower decay [proportional to the square of Eq. (5.37)] of spectral coefficients in the skirts. Since the von Hann window has a gradual transition between zero and unit weight, its spectral window has a less concentrated main lobe and significantly lower sidelobes. The resulting spectrum retains these properties and enables one to observe the spectra of signals as weak as 40 dB below the main peak. This is very significant when one is trying to

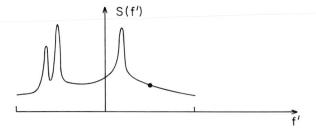

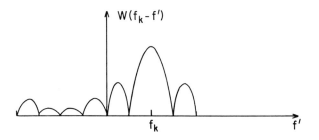

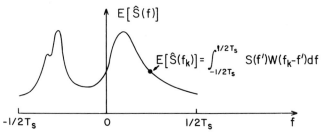

Fig. 5.8 Spectral window effect. The upper plot is a true spectrum $S(f')$. The middle is a spectral window centered at f_k, whereas the lower graph is the smeared spectrum estimate.

Table 5.1
Relations between Windows and Their Spectra[a]

Amplitude spectral window:	$D(f) = \sum\limits_{m=0}^{M-1} d(m)e^{-j2\pi f T_s m}$
Data window:	$d(m) = \dfrac{1}{M} \sum\limits_{m=0}^{M-1} D(f)e^{j2\pi f T_s m}$
Lag window:	$w(l) = \dfrac{1}{M} \sum\limits_{-(M-1)}^{M-1} d^*(m)d(m+l) = \displaystyle\int_{-(2T_s)^{-1}}^{(2T_s)^{-1}} W(f)e^{j2\pi f T_s l}\,df$
Power spectral window:	$W(f) = |D(f)|^2 T_s/M = T_s \sum\limits_{l=-(M-1)}^{M-1} w(l)e^{-j2\pi f T_s l}$

[a] The data window is a sequence of weights $d(mT_s)$ spaced T_s apart.

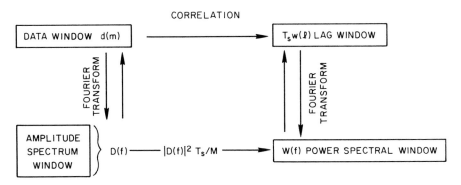

Fig. 5.9 Various relations between windows.

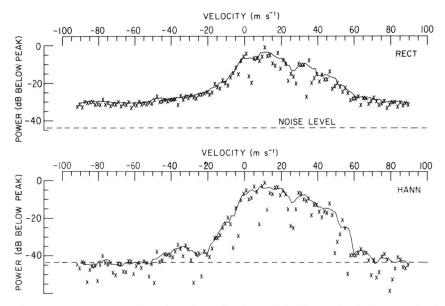

Fig. 5.10 Power spectra of weather echoes, showing statistical fluctuations in the spectral estimates (denoted by x). RECT signifies the spectra of uniformly weighted weather signal samples, whereas HANN signifies samples weighted by a von Hann window. Total number of points is 128. Solid curves are five-point running averages of the spectral power. This spectrum is from a small tornado that touched down on 20 May 1977 at 18:53:50 in Del City, Oklahoma at about 35 km from the radar. Azimuth: 6.1°; elevation: 3.1°; altitude: 1.952 km; SNR = 31 dB.

Table 5.2
Spectra of Several Data Windows

Data window type	Weighting function[a]	Amplitude spectrum window
Rectangular	$d(m) = 1,\ 0 \le m \le M-1$	$\|D(f)\| = \left\|\dfrac{\sin(\pi T_s M f)}{\sin(\pi T_s f)}\right\|$
Triangular	odd M $\ \ d(m) = \begin{cases} m+1, & 0 \le m \le (M-1)/2 \\ M-m, & (M-1)/2 \le m \le M-1 \end{cases}$	$\|D(f)\| = \dfrac{\sin^2[\pi T_s(M+1)f/2]}{\sin^2(\pi T_s f)}$
	even M $\ \ d(m) = \begin{cases} m+1, & 0 \le m \le M/2 - 1 \\ M-1, & M/2 \le m \le M-1 \end{cases}$	$\|D(f)\| = \dfrac{\sin^2(\pi T_s M f/2)}{\sin^2(\pi T_s f)}$
von Hann[b]	$d(m) = a + b\cos\left[\left(m - \dfrac{M-1}{2}\right)\dfrac{2\pi}{M}\right]$	$\|D(f)\| = \left\|\dfrac{\sin(\pi T_s M f)}{\sin(\pi T_s f)}\right\|$ $\times \left[a - b\,\dfrac{\sin^2(\pi T_s f)\cos(\pi/M)}{\sin^2(\pi T_s f) - \sin^2(\pi/M)}\right]$
Hamming[b]		

[a] Sometimes the listed weighting functions are used as lag window coefficients $w(l)$.
[b] For the von Hann window $a = b = 0.5$ and for the Hamming window $a = 0.54$, $b = 0.46$.

estimate the peak winds of tornados or other severe weather within the resolution volume; the power from scatterers moving at the highest velocities is rather weak (because there are few of them) and would be masked by strong spectral peaks (associated with the majority of scatterers) seen through the window sidelobes unless a suitable window were applied. The apparent lack of randomness in the coefficients in the spectral skirts for the rectangularly weighted data is due to the larger correlation between coefficients. This correlation is attributed to the strong spectral powers seen through the nearly constant level-window sidelobes [Zrnic, 1980].

The rectangular window power spectrum $|D(f)|^2 T_s/M$ [Eq. (5.37)] is readily apparent at negative velocities (Fig. 5.10). The dynamic range [i.e., the decibel difference between the mainlobe and sidelobe value of $|D(f)|^2$] for the most distant sidelobe is about 30 dB. This is in contrast to the 45-dB dynamic range obtained with the von Hann window, which better defines the true spectrum and the maximum velocity (60 m s^{-1}). The receiver noise level is indicated by the dashed lines in Fig. 5.10. It is apparent that this level is masked by the powers seen through the sidelobes of the rectangular window.

Much effort was spent by investigators in devising window functions. Two that are simple to use are the von Hann data window and the Hamming window. Table 5.2 lists four window types and their transforms. Some general rules governing window use are

1. Windows with a smooth transition between zero and unit weight have spectra $W(f)$ with lower sidelobes.
2. The window effect is negligible when the window length is large compared to the lag that is required to decorrelate the data.
3. It is not possible to reduce spectral window sidelobes without increasing the width of the mainlobe.
4. Data windows usually have even symmetry; that is,

$$d(M - 1 - m) = d(m) \qquad \text{for} \quad m = 0, 1, \ldots, M - 1$$

5.1.6 Variance of the Periodogram

The probability distribution of the in-phase and quadrature components and the window type determine the variance of the periodogram. We consider in-phase and quadrature components that are zero-mean Gaussian distributed (weather signals) and for which the fourth moment of the complex signal is (Reed, 1962)

$$E[V^*(m)V(n)V^*(k)V(l)] = E[V^*(m)V(n)]E[V^*(k)V(l)]$$
$$+ E[V^*(m)V(l)]E[V(n)V^*(k)]. \qquad (5.38)$$

For this condition it can be shown (Zrnic, 1980) that the variance of a periodogram is exactly given[2] by

$$\text{var}[\hat{S}(f)] = E^2[\hat{S}(f)]. \tag{5.39}$$

Moreover, power of spectrum coefficients at any frequency f is exponentially distributed regardless of the spectral shape, window type, or number of points in the transform. The last conclusion is bothersome because it means that the estimate does not improve with longer periodograms; only the resolution increases. It is natural to expect that the estimate could be improved by increasing the number of data points. Two roughly equivalent methods that reduce the variance are (1) averaging periodograms from short sequences of time samples and (2) making weighted running averages on a periodogram derived from a long sequence.

Although no comprehensive treatment of the digital processing of complex signals is available, several books deal extensively with the processing of real signals (e.g., Tretter, 1976). Many of the results valid for real signal carry over to complex signals; however, there are differences, some of which we have highlighted in the first part of this chapter.

We have emphasized the periodogram approach to power spectral estimation, which is especially suitable for signals that have an unknown spectral shape. But the reader is cautioned that optimum spectral estimation is a complex problem and depends heavily on the type of signals and the desired results. For example, if spectra have a shape (e.g., Gaussian; Section 5.2) that can be described with few parameters, it is advantageous to fit such shapes either to the periodogram (Waldteufel, 1976) or to the autocorrelation function (Sato and Woodman, 1982). Spectra that are well modeled as outputs of linear filters driven by white noise can be estimated from the filter coefficients (Cadzow, 1982). Although models based on parametric fitting allow extensions of the estimated autocorrelation function beyond the limits imposed by the finite-time window, the amount of extractable information is not increased.

5.2 Weather Signal Spectrum and Its Relation to Reflectivity and Radial Velocity Fields

The power spectrum of weather signal, often referred to as the Doppler spectrum, is a power-weighted distribution of the radial velocities of the scatterers. The power weight depends not only on the reflectivity of the scatterers, but also on the weights given to them by the antenna pattern, the transmitted pulse shape, and the receiver's response to it. Consider scatterers that produce a

---

2. Analogous results for a real Gaussian signal hold only in the limit of a large number of samples.

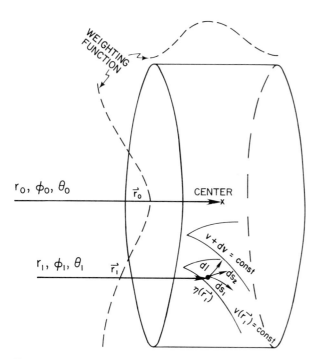

Fig. 5.11 Parameters and geometry that contribute to the weather signal power spectrum; $\eta(\mathbf{r}_1)$ is the reflectivity and $v(\mathbf{r}_1)$ the radial velocity field; $\mathbf{r}_0$ is the center of resolution volume. The weighting functions in angle and range are indicated by dashed lines.

reflectivity field $\eta(\mathbf{r}_1)$ and follow a radial velocity field $v(\mathbf{r}_1)$.[3] Let the center of the resolution volume V_6 be at a location $\mathbf{r}_0$ (Fig. 5.11) so the scatterers' cross sections are weighted by

$$I(\mathbf{r}_0,\mathbf{r}_1) = \frac{Cf^4(\theta - \theta_0, \phi - \phi_0)|W(\mathbf{r}_0,\mathbf{r}_1)|^2}{l^2(\mathbf{r}_1)r_1^4}, \qquad (5.40)$$

where

$$C = \frac{P_t g^2 \lambda^2}{(4\pi)^3}$$

and $\mathbf{r}_1$ is the radius vector to the elemental backscattering volume. $I(\mathbf{r}_0,\mathbf{r}_1)$ is a composite weighting function that henceforth will be designated simply as the *weighting function*.

3. Here and in subsequent chapters we shall omit wherever possible the subscript r and the adjective *radial*, letting v always designate the radial velocity except in those sections where it is needed to denote a Cartesian component of the wind. (The subscript r will be used for the radial Doppler velocity when necessary to avoid confusion.)

Assume a stationary velocity field $v(\mathbf{r}_1)$ and locate a surface of constant velocity [i.e., $v(\mathbf{r}_1) = $ constant]. We seek to determine the expected weather signal power from scatterers that pass through or move within the space between the two surfaces of constant velocity v and $v + dv$. This power will obviously be a sum of expected powers from elemental volumes between the two surfaces. Thus it is convenient to choose elemental volumes to be the product $ds_1\, ds_2\, dl$, where ds_1 and ds_2 are two orthogonal differential lengths, at the point $\mathbf{r}_1$ and tangent to the surface $v(\mathbf{r}_1) = $ constant (Fig. 5.11). The third element of length dl is perpendicular to the surface of constant v.

$$dl = |\mathbf{grad}\ v(\mathbf{r}_1)|^{-1}\, dv. \tag{5.41}$$

The elemental volume contributes an increment of expected power $d\bar{P}$ in the velocity interval v, $v + dv$ equal to

$$d\bar{P}(v) = \eta(\mathbf{r}_1)I(\mathbf{r}_0,\mathbf{r}_1)|\mathbf{grad}\ v(\mathbf{r}_1)|^{-1}\, ds_1\, ds_2\, dv. \tag{5.42}$$

Finally, the integral over the surface A of constant v gives the expected power $\bar{P}(\mathbf{r}_0, v)$ which, by definition is the product of the power spectrum and dv.

$$\bar{P}(\mathbf{r}_0, v) = S(\mathbf{r}_0, v)\, dv = \left[\iint\limits_A \eta(\mathbf{r}_1)I(\mathbf{r}_0,\mathbf{r}_1)|\mathbf{grad}\ v(\mathbf{r}_1)|^{-1}\, ds_1\, ds_2 \right] dv. \tag{5.43}$$

Equation (5.43) is fundamental and worthy of more discussion. The area A consists of all *isodop* surfaces (surfaces of constant Doppler velocity); thus, it is a union of such surfaces. At each point $\mathbf{r}$ on the surface, the reflectivity is multiplied with a corresponding value of the weighting function. The gradient term adjusts the isodops' contribution according to their density; the closer the isodop surfaces, the smaller the number of scatterers in the velocity interval between two isodops. We emphasize that although velocity and reflectivity fields uniquely specify $S(\mathbf{r}_0, v)$, the converse is not true; a variety of reflectivity–velocity combinations may yield identical power spectra.

Because the hydrometeors between the isodop surfaces v, $v + dv$ are continually being replaced by new ones, wobble, and move randomly within the surfaces, *the estimates $\hat{S}(\mathbf{r}_0, v)$ of Eq. (5.43) will change randomly, even though the velocity field might not be changing.* If, within these isodop surfaces, there are many hydrometeors of similar cross sections (i.e., no single scatterer dominates), $\hat{S}(\mathbf{r}_0, v)$ has an exponential distribution because then the central-limit theorem applies to the sum of the voltages (i.e., the voltages I and Q have Gaussian distributions, Section 4.3). Thus, the periodograms have spectral coefficients (e.g., denoted by x in Fig. 5.10) which randomly fluctuate (i.e., from periodogram to periodogram) and are exponentially distributed. By averaging a large number of these randomly varying periodograms we obtain, in the limit, a spectrum equal to $S(\mathbf{r}_0, v)$ convolved with the window spectrum [e.g., Eq. (5.32)]. The five-point running average (solid line on Fig. 5.10) gives a better estimate of $S(\mathbf{r}_0, v)$ at the expense of velocity resolution.

If the velocity field changes with time, so does $S(\mathbf{r}_0, v)$. When turbulence is superimposed on an otherwise deterministic velocity field, $\hat{S}(\mathbf{r}_0, v)$ fluctuates both because hydrometeors within the volume between v and $v + dv$ are constantly being replenished, and because the velocity field is altered by turbulence. In this case $S(\mathbf{r}_0, v)$ given by Eq. (5.43) can be considered to be the expected value *for one configuration* of the deterministic velocity field distorted by turbulence.

To calculate the mean velocity and the spectrum width, the normalized $S_n(\mathbf{r}_0, v)$ version of Eq. (5.43) is used:

$$S_n(\mathbf{r}_0, v) = S(\mathbf{r}_0, v) \bigg/ \int_{-\infty}^{\infty} S(\mathbf{r}_0, v)\, dv. \qquad (5.44)$$

Note that the integral in the denominator is the total power and can be obtained from the volume integral of Eq. (5.42):

$$\bar{P}(\mathbf{r}_0) = \iiint \eta(\mathbf{r}_1) I(\mathbf{r}_0, \mathbf{r}_1)\, dV_1, \qquad (5.45)$$

where $dV_1 = ds_1\, ds_2\, dl$.

The volume-averaged reflectivity $\bar{\eta}(\mathbf{r}_0)$ is obtained by dividing Eq. (5.45) by the volume integral of the weighting function $I(\mathbf{r}_0, \mathbf{r}_1)$[4]:

$$\bar{\eta}(\mathbf{r}_0) = \iiint \eta(\mathbf{r}_1) I_n(\mathbf{r}_0, \mathbf{r}_1)\, dV_1, \qquad (5.46a)$$

where the normalized weighting function is defined as

$$I_n(\mathbf{r}_0, \mathbf{r}_1) = I(\mathbf{r}_0, \mathbf{r}_1) \bigg/ \iiint I(\mathbf{r}_0, \mathbf{r}_1)\, dV_1. \qquad (5.46b)$$

The differential volume dV_1 no longer needs to be tied to the coordinates s_1, s_2 and l. Hence, dV_1 can now be $r_1^2\, dr_1 \sin\theta\, d\theta\, d\phi$, for example. The integral value in the denominator of Eq. (5.46) can be found from Eqs. (4.13) and (4.15) for weather radar parameters usually met in practice.

We now turn to the mean Doppler velocity, defined as

$$\bar{v}(\mathbf{r}_0) = \int_{-\infty}^{\infty} v S_n(\mathbf{r}_0, v)\, dv, \qquad (5.47)$$

which depends on reflectivity and $I(\mathbf{r}_0, \mathbf{r}_1)$. The relationship between the point velocities $v(\mathbf{r}_1)$ and the power weighted moment $\bar{v}(\mathbf{r}_0)$ is obtained by substituting

4. Overbar denotes also the spatial average of expected quantities.

Eqs. (5.41), (5.43), and (5.44) into Eq. (5.47) to obtain

$$\bar{v}(\mathbf{r}_0) = \iiint v(\mathbf{r}_1)\eta(\mathbf{r}_1)I(\mathbf{r}_0,\mathbf{r}_1)\,dV_1 \Big/ \iiint \eta(\mathbf{r}_1)I(\mathbf{r}_0,\mathbf{r}_1)\,dV_1. \qquad (5.48)$$

Unlike the average reflectivity, Eq. (5.48) is the average of point velocities weighted by both reflectivity and the illumination function. In the special case of uniform reflectivity, the $I(\mathbf{r}_0,\mathbf{r}_1)$-weighted mean velocity $\bar{v}_I(\mathbf{r}_0)$

$$\bar{v}_I(\mathbf{r}_0) = \iiint v(\mathbf{r}_1)I_n(\mathbf{r}_0,\mathbf{r}_1)\,dV_1. \qquad (5.49)$$

equals the first moment of $S_n(v)$. Like the reflectivity this is a spatial average of point velocities weighted by the normalized weighting function. Thus $\bar{v}(\mathbf{r}_0)$ cannot in general be equated to a spatial mean velocity. But if reflectivity and illumination are symmetrical about the resolution volume center, and if radial wind changes linearly across V_6, then $\bar{v}(\mathbf{r}_0)$ is the true spatial average of the radial wind component (Section 5.2.1).

The velocity spectrum width $\sigma_v(\mathbf{r}_0)$ is defined by

$$\sigma_v^2(\mathbf{r}_0) = \int_{-\infty}^{\infty} [v - \bar{v}(\mathbf{r}_0)]^2 S_n(\mathbf{r}_0,v)\,dv. \qquad (5.50)$$

Similarly, the spectrum width can be related to the point velocities to obtain

$$\sigma_v^2(\mathbf{r}_0) = \iiint v^2(\mathbf{r}_1)\eta(\mathbf{r}_1)I(\mathbf{r}_0,\mathbf{r}_1)\,dV_1 \Big/ \iiint \eta(\mathbf{r}_1)I(\mathbf{r}_0,\mathbf{r}_1)\,dV_1 - \bar{v}^2(\mathbf{r}_0) \quad (5.51)$$

and is a weighted deviation of velocities from the weighted mean velocity.

5.2.1 Power Spectrum for Uniform Shear and Reflectivity

A simplification of Eq. (5.43) occurs if the radial velocity field and the reflectivity are height invariant. Then the power spectrum reduces to

$$S(v,\mathbf{r}_0) = \iint \left[\int \eta(x_1,y_1)I(\mathbf{r}_0,x_1,y_1,z_1)|\mathbf{grad}\,v(x_1,y_1)|^{-1}\,ds \right] dz_1. \qquad (5.52)$$

Assuming that I is product separable in the variables x_1, y_1, z_1, the inner integral is two-dimensional and sums the contributions along the line $s = s(x_1,y_1)$ on which $v(x_1,y_1)$ is constant. Now, $ds = (dx_1^2 + dy_1^2)^{1/2}$ so Eq. (5.52) is a surface integral with area element $ds\,dz_1$. Both $\eta(x_1,y_1)$ and $\mathbf{grad}\,v(x_1,y_1)$ are independent of z, but the weighting function may not be. The formulation in Eq. (5.52) is useful in simulating the spectra of tornado vortices, because in this case both velocity and reflectivity can be nearly independent of height, at least over the resolution volume. At each point x_1, y_1 along an infinitesimal strip of constant velocity at z_1, the reflectivity is multiplied by the corresponding weighting function. To account for the contributions of other infinitesimal strips within the resolution volume, integration is performed along the third (z-axis) dimension. Equation (5.52) was used to compute the spectra of model tornados and mesocyclones. These compare well with actual measurements (Chapter 9).

It can be shown using Eq. (5.52) that if wind shear and η are constant across the resolution volume, the power spectrum follows the weighting function shape. A simple example illustrates these statements.

Let us obtain the power spectrum of weather signals from scatterers with uniform reflectivity, carried by a radial wind having uniform shear k_ϕ in a direction perpendicular to the beam axis (Fig. 5.12). For simplicity we consider

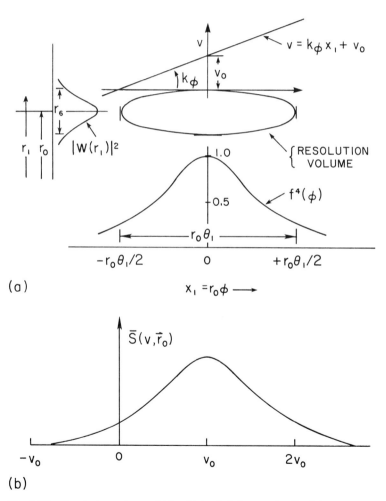

(a)

(b)

Fig. 5.12 Resolution volume within which wind shear and reflectivity are assumed to be uniform. The beamwidth (3 dB) at range r_0 is $r_0\theta_1$, and $f^4(\phi)$ is the two-way antenna pattern function. (b) The power spectrum $S(v, r_0)$ in this example follows the shape of the weighting function.

only the inner integral along iosdops (a two-dimensional problem). Furthermore, we assume the beamwidth is sufficiently narrow that Cartesian distances approximate well the azimuthal arcs over the resolution volume V_6 (i.e., $x_1 = r_0\phi$). Therefore, the pattern weight $f^4(\phi)$ in terms of x_1, Eq. (C.18)], is

$$f^4(x_1) = \exp\{-[4 \ln 4/(r_0\theta_1)^2]x_1^2\}, \qquad (5.53)$$

where θ_1 is the one-way 3-dB beamwidth. The ellipse (Fig. 5.12) represents the cross section in space where range and antenna pattern weights are significant; in this example both weighting functions have a Gaussian shape, hence the resolution volume is a spheroid with the symmetry axis along the beam. Clearly here $|\mathbf{grad}\ v| = |k_\phi|$. Applying Eq. (5.52), we have

$$S(v, \mathbf{r}_0) = (\eta/k_\phi)\ dz_1 \int I(\mathbf{r}_0, x_1, r_1)\ ds. \qquad (5.54)$$

The integration along a contour of constant v constrains x_1. In general, both x_1 and r_1 are linked by $v = $ const. Thus, in terms of $v = $ const, $ds = dr_1$ and

$$x_1 = (v - v_0)/k_\phi. \qquad (5.55)$$

Therefore, from Eq. (5.40), the weighting function is

$$I(\mathbf{r}_0, \mathbf{r}_1, v) = Cf^4\left(\frac{v - v_0}{k_\phi}\right)|W(r_0, r_1)|^2/r_0^4. \qquad (5.56)$$

Substituting this in Eq. (5.54), we obtain

$$S(v, \mathbf{r}_0) = \frac{\eta\ dz_1 C}{r_0^4 k_\phi} \exp\left\{-\left[\frac{4 \ln 4}{(r_0\theta_1)^2}\left(\frac{v - v_0}{k_\phi}\right)^2\right]\right\} \int |W(r_0, r_1)|^2\ dr_1. \qquad (5.57)$$

Thus, the power spectrum (Fig. 5.12b) follows the antenna pattern shape in this not very rare situation of uniform shear.

5.2.2 Contributions of Independent Meteorological Mechanisms to the Power Spectrum

The previous section showed that if shear and reflectivity fields are uniform, the power spectrum shape follows the shape of the weighting function [Eq. (5.40)]. Because the principal part of the weighting function for weather radars can, in most cases, be approximated by a Gaussian function, the power spectrum will have, to the extent of this approximation, a Gaussian shape. It has been found that over 75% of observed power spectra are Gaussian shaped (Jansen and Van der Speak, 1985) and we might infer that the reflectivity and radial velocity shear are then uniform within the resolution volume V_6. But power spectra are also well approximated by a Gaussian shape if several broadening mechanisms (e.g., shear, turbulence, etc.) act independently. Because of independence the resultant spectral shape can be deduced by considering each mechanism separately, as we now demonstrate.

Scatterer velocity is a linear sum of velocities associated with various kinds of motion (e.g., uniform wind, shear, turbulence, etc.) and each contributes to the spread of the power spectrum (the uniform wind contributes to the spread because radial velocities vary across V_6; constant wind also brings new configurations of scatterers into the resolution volume, thus changing echo amplitude and phase even though reflectivity might be uniform). Shear produces a spread (proportional to shear magnitude) of Doppler velocities about the mean radial velocity. Likewise, turbulence provides a Doppler spectrum of echo power with a spread proportional to the intensity of turbulence. We now show that the observed power spectrum is obtained from the convolution of the individual spectra, irrespective of whether they are associated with real or apparent velocity distributions.

The in-phase and quadrature phase signal components at sample time 0 are

$$V(\tau_s, 0) = I_0 + jQ_0 = \sum_i A_i(0)F_i(0)e^{-j\phi_i} \tag{5.58a}$$

where $\phi_i \approx 4\pi r_i / \lambda$ and $|F_i(0)|^2$ is proportional to the weight [Eq. (5.40)] impressed on the ith scatterer.[5] These components for subsequent signal samples at the same range time τ_s, but mT_s later, are

$$V(\tau_s, mT_s) = I_m + jQ_m = \sum_i A_i(mT_s)F_i(mT_s)e^{-j\zeta_i} \tag{5.58b}$$

where $\zeta_i \equiv \phi_i + 4\pi v_i mT_s / \lambda$. It is important to keep in mind that the phases ϕ_i are uniformly distributed over the interval 2π (Section 4.3), whereas the velocities v_i are usually concentrated around a mean velocity. Strictly, v_i is an average radial velocity that takes the ith scatterer to its new position in the time mT_s. Furthermore, we assume that the A_is are independent of the phases ϕ_i and velocities v_i (i.e., the scatterer's cross section and phase shift ψ_{si} due to scattering should not depend on its position or velocity), and the value of the weighting function applied to any scatterer does not change appreciably [i.e., $F_i(0) \approx F_i(mT_s)$] while the scatterer moves during the time of spectral estimation.

A_i varies randomly in time because the scatterer (e.g., a drop) may oscillate or change its canting angle (i.e., the angle between the vertical and the scatterer's axis of symmetry). Thus, the autocorrelation of signal samples is

$$
\begin{aligned}
R(mT_s) &= E[V^*(\tau_s, 0)V(\tau_s, mT_s)] \\
&= \sum_i \sum_k F_i^* F_k E[A_i^*(0)A_k(mT_s)] \\
&\quad \times E[\exp\{j(\phi_i - \phi_k - 4\pi v_k mT_s / \lambda)\}] \\
&= \sum_k R_k(mT_s)|F_k|^2 E[\exp\{-j4\pi v_k mT_s / \lambda\}]
\end{aligned} \tag{5.59a}
$$

5. The form (5.58a) is the same as Eq. (4.1) except both the antenna pattern and range weights are contained in F_i.

where

$$R_k(mT_s) \equiv E[A_k^*(0)A_k(mT_s)].$$ (5.59b)

The double summation reduces to a single one because the expectations with respect to the exponential argument are zero, except at $i = k$. As we can see, the autocorrelation is independent of the initial phases ϕ_i. Because $R(0)$ is proportional to the mean power $\bar{P}$, and because

$$\bar{P} = \sum_k \sigma_{bk} I(\mathbf{r}_0, \mathbf{r}_k)$$ (5.59c)

[i.e., from Eq. (4.11)], where σ_{bk} is the backscattering cross section of the kth hydrometeor, it follows that $R_k(0)$ is proportional to σ_{bk} and thus $R_k(mT_s)$ gives the loss in signal correlation due to temporal variations in backscattering cross sections.

We can show, for Rayleigh scatter, horizontal propagation, and vertically polarized waves, that the average backscattering cross section of a vibrating drop is larger than that for a nonvibrating drop of equal volume (Zrnić and Doviak, 1989). On the other hand, if waves are horizontally polarized, average backscattering cross section decreases when drops vibrate.

We can now invoke Eq. (5.18) to obtain the power spectrum $S(f)$. The spectral broadening mechanisms in Eq. (5.59a) act through product terms. For example, the kth scatterer's velocity v_k is the sum of the velocities due to mean wind, shear, and turbulence, which move the scatterer from one range position to the next in the interval mT_s, Because v_k appears in the exponent of Eq. (5.59a), the velocities associated with mean wind, shear, and turbulence can each be placed into separate exponential functions that multiply one another, and then the transform of Eq. (5.59a) can be expressed as convolution of the spectrum associated with each function. It is shown that, under certain general conditions, the convolution of a number n of spectra approaches a Gaussian function as n increases, independent of the shape of each spectrum (Papoulis, 1965, p. 267). Because several mechanisms contribute to the Doppler spectrum, we see why the observed spectra often appear to be Gaussian in shape.

It should be noted that, because we have assumed $F_i(0) \approx F_i(mT_s)$, spectral broadening due to replenishment of scatterers in V_6 (i.e., either due to beam scanning or to antenna motion if, for example, the radar is airborne) has been ignored. For most storm observations from ground-based radars this contribution is small. For example, a 10-cm radar with a one-degree beamwidth scanning an array of fixed scatterers at a rate of 5 RPM would generate a Doppler spectral width of about 0.5 m s^{-1} [Eq. (C.23)], significantly smaller than the average (4 m s^{-1}) observed for severe storms (e.g., Fig. 10.11). Replenishment of scatterers in V_6 could be a dominant contributor to spectral width if the spectrum width due to turbulence and shear is negligible.

Examination of Eq. (5.59a) reveals that there is also a spectral broadening mechanism due to the change in hydrometeor shape or orientation with respect to the incoming electric field. This change is described by the autocorrelation $R_k(mT_s)$, which produces a narrower $R(mT_s)$, and consequently a broader Doppler spectrum than without change in hydrometeor shape or orientation.

5.2.3   Probability Distribution of Turbulent Velocities Related to the Power Spectrum

We now demonstrate that if hydrometeors are perfect tracers of turbulent motion (i.e., they do not vibrate, wobble, or move relative to the air), the normalized power spectrum

$$S_n(r_0, v) = \frac{S(r_0, v)}{\displaystyle\int_{-\infty}^{\infty} S(r_0, v)\, dv} \tag{5.60}$$

is equal to the probability density function of radial velocities v of the turbulent flow.

If hydrometeors did not vibrate or wobble, then the autocorrelation would satisfy

$$R_k(mT_s) = R_k(0). \tag{5.61}$$

If a scatterer, selected randomly, has the same probability distribution of turbulent velocities $p(v)$ as any other scatterer (i.e., the turbulent flow is statistically homogeneous), we can express the expectation, of the exponential in Eq. (5.59a) as

$$E[\exp(-j4\pi v_k mT_s/\lambda)] = \int_{-\infty}^{\infty} p(v)\, \exp(-j4\pi v m T_s/\lambda)\, dv. \tag{5.62}$$

Substituting Eqs. (5.61) and (5.62) into Eq. (5.59a), we obtain the normalized correlation function (i.e., correlation coefficient)

$$\rho(mT_s) \equiv \frac{R(mT_s)}{R(0)} = \int_{-\infty}^{\infty} p(v)\, \exp(-j4\pi v m T_s/\lambda)\, dv, \tag{5.63}$$

where

$$R(0) \equiv \sum_k R_k(0)|F_k|^2.$$

Because the normalized correlation function is related to the normalized power spectrum through Eq. (5.19),

$$\rho(mT_s) = \int_{-1/2T_s}^{1/2T_s} S_n(f) e^{j2\pi f m T_s}\, df, \tag{5.64}$$

we can equate Eq. (5.63) to Eq. (5.64) and use

$$S(-v) = \frac{1}{\lambda} S(f) \qquad (5.65)$$

and the Doppler shift relation, $f = -2v/\lambda$, to obtain

$$p(v) = S_n(-v). \qquad (5.66)$$

Thus, for homogeneous turbulence, the normalized power spectrum is equal to the velocity probability distribution; moreover it is independent of the angular and range weighting function. Although this derivation did not require that reflectivity is uniform, our assumption that the hydrometeor's size does not depend on its position or velocity implicity assumes reflectivity to be uniform.

5.3 Velocity Spectrum Width

The velocity spectrum width (i.e., the square root of the second spectral moment about the mean velocity) is a function both of radar system parameters such as beamwidth, bandwidth, and pulse width and the meteorological parameters that describe the distribution of hydrometeor density and velocity within the resolution volume. Relative radial motion of hydrometeors broadens the spectrum. For example, turbulence produces random relative radial motion of scatterers. Wind shear can cause relative radial motion, as can differences in the speed of fall of various size hydrometeors. There is also a contribution to spectrum width caused by the V_6 sweeping through space (i.e., contributions from scatterers to successive signal samples are not identically weighted). This change in the location of the resolution volume V_6 from PRT to PRT results in a decorrelation of signal samples and a consequent increase in the spectrum width σ_v (Appendix C). The signal samples decorrelate quicker (independent of particle motion inside V_6) if V_6 moves faster. Thus spectrum width increases in proportion to the antenna motion. Because the cited spectral broadening mechanisms are independent of one another, the square of the velocity spectrum width σ_v^2 can be considered as a sum of contributions by each. That is,

$$\sigma_v^2 = \sigma_s^2 + \sigma_\alpha^2 + \sigma_d^2 + \sigma_o^2 + \sigma_t^2, \qquad (5.67)$$

where σ_s^2 is due to shear, σ_α^2 to antenna motion, σ_d^2 to different speeds of fall for different sized hydrometeors, σ_o^2 to change in orientation or vibration of hydrometeors, and σ_t^2 to turbulence. Because broadening caused by the mean relative motion between the antenna and scatterers is small except in the case of an airborne radar, σ_α principally depends on antenna rotation.

We ignore the spectrum broadening caused by the window effect because it is related to the method of spectral moment analysis and furthermore does not contribute to signal sample decorrelation, as do the mechanisms considered in

Eq. (5.67). These mechanisms are related to the changes in scatterers' velocities, shapes, or the weighting applied to them. Methods other than DFT spectral width estimation—for example, the covariance method (Chapter 6)—would produce different width estimates (i.e., biases). The significance of the width σ_v for weather radar design is discussed in Chapter 7.

The contribution σ_d^2 due to the different radial components of fall speed of the assorted-size drops, is related to the radar and meteorological parameters:

$$\sigma_d^2 = (\sigma_{d0} \sin \theta_e)^2. \tag{5.68}$$

The width σ_{d0} is caused by the spread in terminal velocity of various size hydrometeors falling relative to the air. Lhermitte (1963) has shown that for rain $\sigma_{d0} \simeq 1.0$ m s^{-1} and is nearly independent of the drop size distribution. The elevation angle $\theta_e = (\pi/2) - \theta_0$ is measured with respect to the beam center. If the antenna pattern is Gaussian with a one-way half-power width θ_1 and rotates at an angular velocity α, the spectrum width due solely to resolution volume displacement is (Appendix C)

$$\sigma_\alpha^2 = (\alpha\lambda \cos \theta_e/2\pi\theta_1)^2 \ln 2. \tag{5.69}$$

We shall prove that the wind shear width term σ_s is composed of three contributions:

$$\sigma_s^2 = \sigma_{s\theta}^2 + \sigma_{s\phi}^2 + \sigma_{sr}^2, \tag{5.70}$$

where the terms are due to *radial velocity* shear along three orthogonal directions through $\mathbf{r}_0$. The assumptions behind (5.70) are that shear is uniform and the weighting function and reflectivity are product separable along the orthogonal directions (e.g. x, y, z). Because we use a spherical coordinate system (5.70) automatically includes spectral broadening caused by uniform wind across a finite width beam (i.e., the so-called beam broadening term; Nathanson, 1969).

If the wind is linear about v_0, the speed v can be expressed as

$$v - v_0 = k_x x + k_y y + k_z z, \tag{5.71}$$

where the ks are the components of shear along the various axes. Let us orient the coordinate system so that y is in the elevation direction, x transverse and z parallel to the beam axis; these coordinates are orthogonal components of $\mathbf{r}_1 - \mathbf{r}_0$ with the origin at the tip of $\mathbf{r}_0$ and $|\mathbf{r}_1 - \mathbf{r}_0| \ll r_0$. From (5.71) it follows that the mean velocity $\bar{v} = v_0$ at $\mathbf{r}_1 = \mathbf{r}_0$. If the weighting function is product separable,

$$I(x, y, z) = C|W(z)|^2 f_\phi^4(x) f_\theta^4(y)/r_0^2 \tag{5.72}$$

(and also the reflectivity), substitution of (5.71) and (5.72) into (5.51) produces

$$\sigma_s^2 = \sigma_x^2 k_x^2 + \sigma_y^2 k_y^2 + \sigma_z^2 k_z^2, \tag{5.73}$$

where σ_x^2, σ_y^2, and σ_z^2 are second moments of $f_\phi^4(x)$, $f_\theta^4(y)$, and $|W(z)|^2$, respectively at range $\mathbf{r}_0$. Because resolution volume dimensions are small compared to their range, distances transversed to the beam axis can be approximated by arclengths. If k_θ, k_ϕ, k_r are the shears in the directions θ, ϕ, r, then

$$\sigma_s^2 = \sigma_{s\theta}^2 + \sigma_{s\phi}^2 + \sigma_{sr}^2 = (r_0\sigma_\theta k_\theta)^2 + (r_0\sigma_\phi k_\phi)^2 + (\sigma_r k_r)^2, \qquad (5.74)$$

where σ_θ^2 and σ_ϕ^2 are defined as the second central moments of the two-way antenna power pattern in the indicated directions and $\sigma_z^2 = \sigma_r^2$ is the second central moment of the weighting function $|W(r)|^2$. A circularly symmetric Gaussian pattern has

$$\sigma_\theta^2 = \sigma_\phi^2 = \theta_1^2/(16 \ln 2). \qquad (5.75)$$

For a rectangular transmitted pulse and a Gaussian receiver frequency response under matched conditions (i.e., $B_6\tau = 1$).

$$\sigma_r^2 = (0.35c\tau/2)^2 = (0.30r_6)^2. \qquad (5.76)$$

To estimate the shears needed in Eq. (5.74), one may use differences between radial velocities at adjacent elevations and azimuths in the spherical system with the radar at its center. The weather radar usually scans conically, with the vertical axis along the cone's axis. Thus it becomes difficult, because of the finite beamwidth and errors in velocity estimates, to measure accurately the shears k_ϕ, k_θ at the highest elevation angles. For example, as $\theta_e \rightarrow \pi/2$ one should use four measurements $90°$ apart in azimuth and slightly displaced from the zenith to deduce the two orthogonal shear components.

Spectral broadening σ_0 due to changes in orientation or vibrations of hydrometeors is small and can be obtained from Eq. (5.59b). For example if drop oscillations produce an rms change in axis ratio of 10% (see Chapter 8 for drop shape) there would be a 10% increase in spectral width (Zrnic and Doviak 1989).

The width σ_t due to turbulence is somewhat more difficult to model. For turbulence that is homogeneous and isotropic within V_6, Frisch and Clifford (1974) have shown that σ_t^2 is related to the eddy dissipation rate ε (Chapter 10). A detailed discussion of the contribution of turbulence and shear to the spectrum width is presented in Chapter 10, where examples of data fields are also given.

Problems

5.1   Show that if the reflectivity field is uniform, the reflectivity weighted mean Doppler velocity $\bar{v}(\mathbf{r})$,

$$\bar{v}(\mathbf{r}) = \int v \cdot S_n(\mathbf{r}, v)\, dv$$

(where $S_n(\mathbf{r}, v)$ is the normalized power spectrum) is equal to the illumination weighted mean velocity $\bar{v}_I(\mathbf{r})$ where

$$\bar{v}_I(\mathbf{r}) \equiv \frac{\iiint v(\mathbf{r}) \cdot I(\mathbf{r}, \mathbf{r}_1)\, dV}{\iiint I(\mathbf{r}, \mathbf{r}_1)\, dV},$$

dV is an elemental volume, $\mathbf{r}$ is the location of the resolution volume, and $\mathbf{r}_1$ is a variable of integration.

5.2   A discrete Fourier transform (DFT) is performed on white noise. Assume the noise power N is $2\sigma^2$ and the number of samples is M. (a) Determine the expected noise power for any spectral coefficient. (b) Compare the noise level in the spectral domain with the power level of the spectral coefficient for a time-domain complex sinusoid having a signal power equal to noise power and a frequency equal to a multiple of the fundamental $(MT_s)^{-1}$. (c) Calculate the signal-to-noise level in the spectral domain.

5.3   The Doppler spectrum of signal plus noise calculated from an 8-sample DFT is plotted in Fig. P.5.1. (a) Discuss why nine lines are shown in Fig. P.5.1. (b) A 10-cm radar operating at a PRF of 1000 Hz measures a spectral integrated noise power of 12 units (i.e., without echoes). For the spectrum plotted in Fig. P.5.1 find the mean velocity and spectral width. The mean velocity and spectral width estimates could be biased by the noise power. Unbias the estimates and recalculate the mean velocity and spectral width. (c) If the data window is rectangular, how much is the additional bias that one should remove from the width estimate?

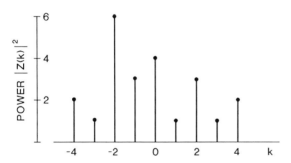

Fig. P.5.1

5.4   Calculate the Doppler power spectrum (W s per unit height interval) of a simply rotating (i.e., no divergence) mesocyclone that is centered at an azimuth $\phi = 0$ and range r_0 when a sampling gate is placed such that its corresponding resolution volume is also centered at r_0. Treat a simplified one dimensional problem by making the following assumptions: (i) Reflectivity is uniform. (ii) The antenna gain pattern is Gaussian in azimuth and uniform in elevation, so its two-way normalized pattern is

$$f^4(\phi - \phi_0) = \exp\{-4 \cdot \ln 4 \cdot (\phi - \phi_0)^2/(\theta_1)^2\},$$

where ϕ_0 is the azimuth angle of the beam axis and θ_1 is the one-way 3-dB width of the pattern. (iii) The beam elevation angle is zero and $\theta_1 \ll \pi$. (iv) The range-weighting function $|W(r)|^2$ is much narrower than the diameter of maximum velocities of the mesocyclone. (v) The tangential velocities v_t are given by

$$v_t = \frac{v_m x}{x_m}, \qquad x \approx r\phi \leq x_m,$$

and

$$v_t = \frac{v_m x_m}{x}, \qquad x \geq x_m,$$

where x_m is the radius of maximum tangential wind. (a) Plot the Doppler velocity isotachs (i.e., isodops) versus azimuth angle and range assuming that the circle of peak tangential wind subtends a maximum azimuthal angle $\phi_m \ll \pi$. (b) Plot the power spectrum for the two cases:

$$(1) \quad \phi_0 = \phi_m = \theta_1 \qquad (2) \quad \phi_0 = 0; \quad \phi_m = \theta_1$$

5.5   Because signal power from scatterers having different velocities may vary considerably, spectra are often plotted on a logarithmic (dB) scale. Assume that the equation for a Gaussian-shaped power spectrum is

$$10 \log[S(v)/S_0] = -(v - 10 \text{ m s}^{-1})^2/4 \text{ m}^2 \text{ s}^{-2}$$

(a) Sketch this spectrum on a dB scale. (b) What is the mean Doppler velocity and what is the Doppler frequency (Hz) if the radar wavelength is 10 cm? (c) Find the Doppler spectral width. (d) Assume that the radar equation is

$$P_r = F \cdot Z \text{ (mm}^6 \text{ m}^{-3}) \qquad \text{where} \quad F = 10^{-16} \text{ mm}^{-6} \text{ m}^3 \text{ mW}.$$

Also, assume that $S_0 = 10^{-10}$ mW m^{-1} s. List at least three (3) factors of which F is a function. Find Z in units of dBZ.

5.6   A sequence of complex voltages $V(m)$ represents the sampled radar receiver *noise* output. It consists of independent zero mean Gaussian random variables $I(m)$, $Q(m)$ for which

$$V(m) = I(m) + jQ(m); \quad E[V^*(m)V(n)] = 0, \qquad m \neq n$$
$$E[I^2(m)] = E[Q^2(m)] = \sigma^2; \qquad\qquad\qquad m = n$$

(a) Use Eq. (4.5) to prove that

$$E[I(m)Q(m)] = 0.$$

Note that the expectations of a product of two independent random variables is equal to the product of the expectations of each. (b) A power estimate $\hat{P}$ is formed from M samples

$$\hat{P} = \frac{1}{M} \sum_{m=0}^{M-1} V(m)V^*(m).$$

Find the expected value $E[\hat{P}]$ and the variance $\text{VAR}[\hat{P}]$ of the estimate $[\hat{P}]$ in terms of σ^2 and M. (c) Repeat the problem, but consider correlated signal samples, that is

$$E[V^*(m)V(n + l)] = 2\sigma^2 \rho(l),$$

where $\rho(l)$ is the correlation coefficient:

$$\rho(\ell) = \frac{R(\ell)}{2\sigma^2}.$$

Compare VAR $[\hat{P}]$ with that derived in (b).

5.7   Very often the Doppler spectrum of a weather signal has Gaussian shape. Receiver and quantization noises are white. A DFT of a *long time* sequence is taken and a composite (signal and noise) spectrum is computed. Then several spectra are averaged so that the estimate is smooth in v with closely spaced lines (Fig. P.5.2 gives the expected values). Assume that the total power P of the composite spectrum is known, the total noise power N is known, the unambiguous velocity interval $2v_{max}$ is given, and the width of the weather signal σ_v is known. Calculate (a) the level of noise spectral density $N(v)$, (b) the peak spectral-to-noise density ratio, and (c) the width of the composite spectrum. In all calculations use the fact that $\sigma_v \ll 2v_{max}$.

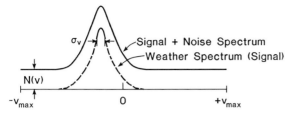

Fig. P.5.2

5.8   A Doppler radar is observing a front that has uniform reflectivity and a velocity component in the radar direction, as shown in Fig. P.5.3. The beam is centered on the shear discontinuity and the resolution volume is 60 km from the radar. (a) Sketch the Doppler spectrum and indicate its important parameters. (b) Find the mean Doppler velocity.

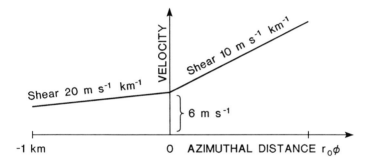

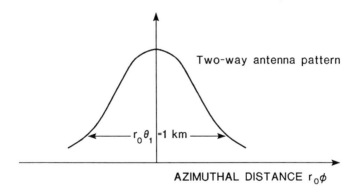

RELATIVE POSITION OF ANTENNA
PATTERN WEIGHTING FUNCTION

Fig. P.5.3

6

Weather Signal Processing

In this chapter proven methods of weather signal processing are presented, with emphasis on obtaining the first three spectral moments. A section on methods to obtain simultaneously spectral moments and polarimetric measurements is also included. Weather signals are sampled at a large number of range time locations (e.g., approximately 1000), and hence the optimum estimation algorithms, which require much processing, may be impractical. Simpler methods, with fewer calculations and smaller memory storage, are used.

6.1 Spectral Moments

Signal processors supply the three most important spectral moment estimates. These are

1. Weather signal power or the zeroth moment of the Doppler spectrum [Eq. (5.45)]. This can be related to liquid water content or precipitation rate in the resolution volume.
2. Mean Doppler velocity or the first moment of the power-normalized spectra [Eq. (5.47)]. This is for near-horizontal antenna orientations, essentially the air motion toward or away from the radar.
3. Spectrum width σ_v, the square root of the second moment about the first of the normalized spectrum [Eq. (5.50)]. This is a measure of the velocity dispersion, that is, shear or turbulence within the resolution volume.

Moments are estimated from samples of a randomly varying signal. In the case of weather signals, few sample estimates have too large a statistical uncertainty to yield meaningful data interpretation. Thus, a large number of signal samples (acquired during a few tens of milliseconds) must be processed to provide the required accuracy. The actual number required depends on both radar system characteristics and meteorological conditions. These include the signal-to-noise ratio, the distribution of velocities within the resolution volume, and the receiver transfer function (i.e., linear, logarithmic, etc.).

Whereas reflectivity estimation requires only power sample averaging to reduce statistical fluctuations, mean Doppler velocity estimation involves the

Fourier transform or complex covariance calculation, both of which are more complicated. The need to obtain the principal Doppler moments economically, with minimum uncertainty, and in digital format (to facilitate further processing and analysis with electronic computers) has prompted the use of the covariance calculations. The advantages of a covariance processing coupled with the new technology (e.g., medium-scale integrated circuits) have made possible simultaneous estimates of the spectral moments at hundreds of resolution volumes along the radar beam.

It is the very nature of weather signals that imposes limitations and tradeoffs. Weather signals easily span 80 dB of dynamic range and are random. Furthermore, because weather is distributed quasicontinuously over large spatial regions (from tens to hundreds of kilometers), measurements need to be made at millions of locations in a few minutes.

6.2 Weather Signals in a Receiver

The block diagram in Fig. 6.1 represents a typical meteorological Doppler radar receiver. The mixer has exactly the same function as the synchronous detector in

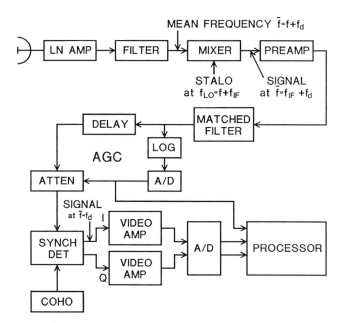

Fig. 6.1 Block diagram of the Doppler weather radar receiver. LN AMP is a low-noise amplifier, COHO is the coherent oscillator that generates the continuous IF, and AGC is the automatic gain control circuit.

Fig. 3.1 except that here the output has a carrier frequency at the intermediate frequency (IF) whereas in Fig. 3.1 the output has no carrier (it is a video signal).

The automatic gain control (AGC) is a feed-forward instantaneous type. This gain control circuit matches the signal's rms value to the range of the analog-to-digital (A/D) converter (at the output of the video amplifiers); otherwise many more bits would be required in the A/D converter, with the severe consequence of limiting the time of conversion from the analog level to the digital word. Matching is accomplished by splitting the signal and delaying a part of it (Fig. 6.1). The part not delayed is passed through a logarithmic amplifier–detector (LOG in Fig. 6.1) which has a large dynamic range. The delay is needed to allow enough time to switch attenuator settings. The attenuator setting is switched at a megahertz rate to match the typical 1-μs spacing of range gates. After detection, it is converted to digital form; thus the digital value of signal strength controls the attenuator setting. The attenuation setting is also supplied to the processor so that echo power can be computed from the attenuated I and Q samples.

The radio frequency (rf) filter (after the LN AMP) serves the purpose of rejecting interference from other transmitters, as well as rejecting image noise, so that radar sensitivity is improved. The matched filter is either in the video (as in Fig. 3.1) or in the IF circuits (as in Fig. 6.1) of the receiver; direct current (dc) offset control and gain balance circuits must be available in the two video amplifiers. The synchronous detector operates in exactly the same way as the one in Fig. 3.1 except that it mixes the IF output of the COHO with the IF carrier that has an amplitude and phase modulation identical to the microwave carrier. Ground clutter suppression and the calculation of spectral moments are done in the processor, which contains a special-purpose computer. The echo power estimate is sometimes obtained from the output of the logarithmic amplifier.

Each and every range gate produces a sequence of complex video samples $V(kT_s)$ consisting of a signal part $s_k \exp(j\omega_d kT_s)$ and a white noise part n_k; that is,

$$V(kT_s) = s_k e^{j\omega_d kT_s} + n_k, \qquad k = 0, 1, \ldots, M-1. \qquad (6.1)$$

The signal sample is written as a product of two terms for later convenience. The multiplication of a time signal by $\exp(j\omega_d t)$ shifts the spectrum by f_d. Thus the spectrum of s_k is assumed to be centered about zero frequency (i.e., its power-weighted mean frequency is zero), whereas the spectrum of $V(kT_s)$ is centered on the Doppler frequency f_d. Both n_k and s_k are zero-mean Gaussian processes (i.e., their average value is zero), but s_k is narrowband compared to the receiver bandwidth. It is desirable to have the unambiguous interval squared, T_s^{-2}, large compared to the second spectral moment of s_k to improve the accuracy of the various estimators. This is because, for a fixed dwell time MT_s (collection time of M samples for one estimate), the unambiguous interval expands as the number of samples M increases and this reduces the noise contamination of the signal spectrum. But other considerations such as multiple trip signals require sample separation (T_s) to be large, so the second moments cannot be made arbitrarily small compared to T_s^{-2}.

Let the average signal power be S. The magnitude of the normalized correlation function ρ at lag mT_s depends on the sample spacing and σ_v. Thus the autocorrelation function of the signal samples $V(kT_s)$ is

$$R(mT_s) = R_m = E\{V^*(kT_s)V[(k + m)T_s]\} = S\rho(mT_s)e^{j\omega_d mT_s} + N\delta_m, \quad (6.2)$$

where N is defined as the mean white noise power and δ_m is 1 for $m = 0$ and zero otherwise. We can deduce from Eq. (5.18) that if the correlation coefficient $\rho(mT_s)$ is real the power spectrum is symmetric about $f = 0$. Conversely, $\rho(mT_s)$ of the s_k is real if its power spectrum is symmetric about $f = 0$. Nevertheless, the correlation of the $V(kT_s)$ is complex, and its power spectrum $S(f)$ is centered about f_d.

It was explained in Chapter 5 that weather signals from regions of uniform reflectivity should have spectra closely resembling a Gaussian function because uniform shear and many broadening mechanisms independently contribute to the spectrum shape. Thus in analyzing the statistical properties of various estimators, it is convenient to assume a Gaussian power spectrum,

$$S(v) = \frac{S}{(2\pi)^{1/2}\sigma_v}\exp[-(v - \bar{v})^2/2\sigma_v^2] + \frac{2NT_s}{\lambda}, \quad (6.3)$$

with a corresponding autocorrelation at lag mT_s

$$R(mT_s) = S\exp[-8(\pi\sigma_v mT_s/\lambda)^2]e^{-j4\pi\bar{v}mT_s/\lambda} + N\delta_m. \quad (6.4)$$

The magnitude of the normalized signal correlation (i.e., the correlation coefficient) is then

$$\rho(mT_s) = \exp[-8(\pi\sigma_v mT_s/\lambda)^2]. \quad (6.5)$$

Each of the three parameters, power S, mean velocity $\bar{v} = -\lambda f_d/2$, and spectrum width σ_v, are related to specific meteorological fields, hence accurate estimates are required for meaningful interpretation.

6.3 Signal Power Estimation

6.3.1 Sample Time Averaging

The sampled signal is a weighted sum of echoes returned from individual scatterers, which are moving relative to each other. This relative movement produces fluctuation in the samples S_k of signal power. Added to this is the sampled receiver noise power N_k, which also fluctuates like the weather signal. To obtain a quantitative estimate of the total mean power $\bar{P} = \bar{S} + \bar{N}$, samples P_k must be averaged over a period long enough that the uncertainty in the mean power estimate is reduced to a tolerable level. The mean and variance of the P population are completely described by its probability density (Section 4.3).

6.3.1.1 Statistical Properties of Signal Samples from Linear and Nonlinear Receiver–Detectors

In the sequel we consider common receiver–detector combinations that act on the complex signal $V(k)$. Detectors in an incoherent receiver (Section 3.4.2) operate on the IF (or rf) signals and generate at their output, after filtering, low-frequency envelopes of $V(k)$. Those envelopes are nonlinearly related, in general, to the magnitude $|V(k)|$; i.e., the output signal Q of the incoherent detector can have one of many dependences on the input signal power. Three common transfer functions are

1. square law, in which case the output sample Q is proportional to the input power P_k [i.e, $Q_k \propto P_k \propto |V(k)|^2$]
2. linear amplitude, so that $Q_k \propto P_k^{1/2} \propto |V(k)|$
3. logarithmic, or $Q_k \propto \log P_k \propto \log |V(k)|$

It should be evident that any power transfer function can be obtained by operating on the magnitudes $|V(k)|$ at the output of the coherent (synchronous) detector. In modern radar systems (e.g., WSR-88D) the square law characteristic is obtained from the sum of the squares of the in-phase and quadrature components.

To improve the accuracy of power estimates, Q_k samples are averaged and the sample mean probability density is derived from the probability density of P_k and the receiver–detector transfer function (Marshall and Hitschfeld, 1953). A uniform average,

$$\hat{P} = \frac{1}{M} \sum_{k=0}^{M-1} P_k, \tag{6.6}$$

usually gives very good estimates of $S + N$. The M samples in Eq. (6.6) are spaced by T_s.

Reflectivity, from which liquid water content and the rainfall rate can be estimated, is proportional to $\bar{P}$. The $\bar{P}$ values of meteorological interest can easily span a range of 10^8, and often the choice of receiver type hinges on the cost of meeting this large-dynamic-range requirement. The problem is to estimate $\bar{P}$ from sample averages of Q. The estimation is complicated if Q is not linearly related to P. That is, when one operates on the mean output estimates $\hat{Q}$,

$$\hat{Q} = \frac{1}{M} \sum_{k=0}^{M-1} Q_k, \tag{6.7}$$

with the inverse of the transfer function (i.e., Q versus P) to obtain estimates $\hat{P}$, biases are generated and the uncertainty is larger than if one averages P_k directly (Sirmans and Doviak, 1973). [An estimate is biased if its expected value $E(P)$

differs from the mean of the variable being estimated (i.e., $\bar{P}$).]. The output mean for linear and logarithmic receiver–detectors is related to $\bar{P}$. $\bar{P}$ estimate bias and standard deviation for the three transfer functions are given in (Zrnic, 1975a and Doviak and Zrnic, 1984a). Only the square law transfer function generates unbiased estimates of power.

Although the uniform block average [Eq. (6.6)] is conceptually simple, it does not provide a continuous update of the output (i.e., an update with each new input sample). One may choose a uniformly weighted running average, which is more complex to implement and requires considerable memory storage, or use a first-order recursive filter (an exponentially weighted running average), for which the estimate is

$$\hat{Q}_k = (1 - b)\hat{Q}_{k-1} + b Q_k, \tag{6.8}$$

where b is a number between zero and 1. It has been shown that the means and variances of the input power estimates obtained from Eq. (6.8) for various transfer functions (e.g., log or square law) are almost identical to those obtained from a uniform average of M samples if $M = (2 - b)/b$ (Zrnic, 1977a).

6.3.1.2 Number of Equivalent Independent Samples

The total number M of samples (for a resolution volume) is determined by the PRT and dwell time. Because considerable correlation may exist from sample to sample, however, we determine the equivalent number M_I of independent samples in order to express simply the reduction in estimate variance achieved by averaging. The degree of correlation between samples is a function of the radar parameters (e.g., wavelength, PRT, beamwidth, and pulse width) and the spread of velocities in the resolution volume.

If the estimate $\hat{Q}$ of the true output mean $\bar{Q}$ is derived from an average of M_I uniformly weighted independent samples (i.e., $M = M_I$), the output single sample estimate variance σ_Q^2 is reduced by a factor of $1/M_I$. That is,

$$\sigma_{\bar{Q}}^2 = \sigma_Q^2/M_I. \tag{6.9}$$

On the other hand if M samples are correlated, the estimate variance $\sigma_{\bar{Q}}^2$ does not vary as $1/M$. Instead, for a stationary process and equally spaced samples, the estimate-variance reduction factor for the M-sample average is given by (Papoulis, 1965)

$$\frac{\sigma_{\bar{Q}}^2}{\sigma_Q^2} = M_I^{-1} = \sum_{m=-(M-1)}^{M-1} \frac{M - |m|}{M^2} \rho_Q(mT_s). \tag{6.10}$$

The normalized correlation ρ_Q of the output samples can be expressed in terms of the Doppler spectrum; the parameters of this spectrum can be related to atmospheric and radar system parameters. To determine rigorously the autocorrelation of the output Q, we must transform this spectrum by the transfer function (which is nonlinear for the three detectors). For example, for a square law receiver–detector the output correlation ρ_s equals the square of the input correlation ρ (Papoulis, 1965). Correlation functions for the output of the other receiver–detectors are more complicated, and the reader is referred to other works for details (Kerr, 1951; Davenport and Root, 1958).

Assuming the input power spectrum to be Gaussian [Eq. (6.3)] with a corresponding correlation coefficient [Eq. (6.5)] and zero noise, the correlation of the output of a square law receiver–detector is

$$\rho_s(mT_s) = \exp[-(2\pi m\sigma_{vn})^2], \tag{6.11a}$$

where for convenience

$$\sigma_{vn} \equiv \sigma_v/2v_a = 2\sigma_v T_s/\lambda \tag{6.11b}$$

is the normalized (to the unambiguous interval) spectrum width, and $v_a = \lambda/4T_s$ is the unambiguous velocity (Section 3.6).

Combining Eqs. (6.10) and (6.11a) gives the variance reduction factor for the square law receiver–detector:

$$M_1^{-1} = \sum_{m=-(M-1)}^{M-1} \frac{M - |m|}{M^2} \rho_s(mT_s). \tag{6.12}$$

For large M and $\sigma_{vn} \ll 1$, $M_I = 2M\sigma_{vn}\pi^{1/2}$. The variance of power estimated from M samples is calculated using the distribution of P [Eq. (4.7)] which gives σ_Q^2, and Eq. (6.12). Thus, the standard deviation (SD) σ_Q of single-sample power estimate equals $\bar{P}$.

To estimate S we need to subtract an estimate $\hat{N}$ (which may depend on beam position; Section 3.5.1) of receiver noise power from $\hat{P}$. Note that $\hat{S}$ is derived from the difference of two independent random variables with similar probability distributions; therefore the SD of $\hat{S}$ is proportional to $\sqrt{\bar{P}^2 + \bar{N}^2}$. Because power and, proportionately, reflectivity factor Z are usually expressed in dB units, it is desirable to express SD also in dB units. For large SNR the SD of reflectivity factor σ_z is approximately given by

$$\sigma_z \approx 10 \log[1 + \text{SD}(\hat{S})/\bar{S}], \tag{6.13}$$

which is valid for $\text{SD}(\hat{S})/\bar{S} \ll 1$. The values of σ_z versus M with σ_{vn} as a parameter are plotted in Fig. 6.2. For small M the approximation is no longer valid, and for $M \to 1$, $\sigma_z \to 5.6$ dB independent of σ_{vn}.

6.3.1.3 Independent Samples Due to Shifts in Resolution Volume Location

Acquisition of an ensemble sample series while the antenna is moving or rotating (at a rate α) results in a continuous alteration in resolution volume location. The signal samples for resolution volumes corresponding to two beam positions are correlated if the two resolution volumes overlap and if there is no

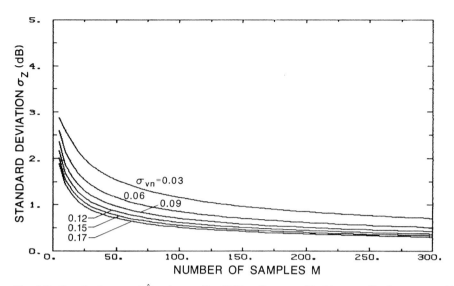

Fig. 6.2 Standard error of $\hat{Z}$ estimates (for SNR $\gg 1$) versus M with normalized spectrum width σ_{vn} as a parameter.

significant decorrelation from reshuffling of the scatterers' positions. In Appendix C we show that the correlation of signal samples taken at different beam positions equals the correlation of the one-way radiation pattern lagged by the beam spacing. For many types of meteorological measurements, the antenna mainlobe can be approximated by a Gaussian function; then the number of analysis similar to that for sample time averaging (Doviak and Zrnić, 1984, Section 6.3.1.3).

6.3.2 Range Time Averaging

Estimate variance can be reduced by averaging along the range time τ_s as well as along the sample time. Because the range extent of the resolution volume is usually small compared to its size perpendicular to the beam axis, averaging over a range interval Δr results in a more symmetric spatial resolution volume with no significant degradation of the spatial resolution; for the WSR-88D Δr is 1 or 2 km (Table 6.1). The incremental spacing between samples multiplied by the number of range samples being averaged gives Δr. Range time averaging the output of a linear or logarithmic receiver–detector introduces a systematic bias of the estimate caused by reflectivity gradients, and this bias limits the maximum Δr useful for averaging (Rogers, 1971). There are no such limitations with the square law receiver–detector. The sampling increment is chosen by considering the autocorrelation of the consecutive range samples of the return signal plus receiver noise. Averages containing range-correlated samples have

Table 6.1
WSR-88D Signal Processor

Reflectivity factor calculator	
Algorithm	Power average
Estimate accuracy	1 dB
Number of samples averaged	in range, 4 or 8;
	in sample time, 6 to 64
Range increment	1 or 2 km
Doppler velocity calculator	
Algorithm	Autocovariance processing
Estimate accuracy	1 m s^{-1} at SNR>8 dB, $\sigma_v = 4$ m s^{-1}
Number of samples processed	40 to 200
Range increment	0.25 km
Spectrum width calculator	
Algorithm	Single lag autocovariance
Estimate accuracy	1 m s^{-1} at SNR>10 dB, $\sigma_v = 4$ m s^{-1}
Number of samples processed	40 to 200
Range increment	0.25 km

an estimate variance that depends not only on the number of samples, but also on the sample correlation.

Quite often the bandwidth of the receiver may be about two to three times the reciprocal of the transmitter pulse width τ; there is then significantly less correlation of signal samples spaced a pulse width τ apart along τ_s (Fig. 4.10). For matched filter receivers, samples become significantly correlated only if the product $\delta\tau_s B_6$ of range sampling increment and receiver bandwidth is less than about 0.5 (Section 4.6). This suggests use of a range sample increment smaller than required for complete independence (i.e., $\delta\tau_s \geq \tau$ and $B_6 \rightarrow \infty$). The reduction in estimate variance in the case of correlated samples is given by Eq. (6.10), where T_s is replaced by the range time increment (Appendix C) and ρ_Q is now the correlation along the range time axis τ_s (Section 4.6).

6.4 Mean Frequency Estimators

As for reflectivity estimators, we shall consider only those methods of mean frequency extraction that have been implemented on meteorological Doppler radars; they are based either on the autocovariance or on the DFT of I and Q samples. In applications where signal samples are tightly correlated, savings in computations can be achieved by summing several consecutive $V(kT_s)$s prior to estimation of Doppler spectral parameters. Enhancement of signal-to-white-noise ratio is proportional to the number of these "coherently" summed signals. Such a technique is used in wind profiling radars (Section 11.8.1) that have long wavelengths and therefore signals with narrow spectral widths relative to the unambiguous interval (Strauch *et al.*, 1984). Doppler processing in polarimetric radars requires special considerations, which are discussed in Section 6.8.

6.4.1 Autocovariance Processing: The Pulse Pair Processor

The pulse pair estimator calculates the first two moments of the Doppler spectrum from estimates of the autocovariance function at lag T_s. Using Eq. (5.19), we can express $R(T_s)$ in terms of the power-weighted mean Doppler frequency f_d.

$$R(T_s) = e^{j2\pi f_d T_s} \int_{-1/2T_s}^{1/2T_s} S(f) e^{j2\pi T_s(f-f_d)} \, df. \tag{6.14}$$

Note that f_d can be calculated from the argument, $2\pi f_d T_s$, of $R(T_s)$ if the integral in Eq. (6.14) is real, which implies that the imaginary part of the integral is zero; that is,

$$\varepsilon_s = \int_{-1/2T_s}^{1/2T_s} S(f) \sin[2\pi T_s(f - f_d)] \, df = 0. \tag{6.15}$$

Thus if the spectrum $S(f)$ is symmetric with respect to f_d, the argument of $R(T_s)$ yields the mean frequency; otherwise there is a bias error related to ε_s.

Consider the following commonly used definition of the estimate for the power-weighted mean Doppler frequency $\hat{f}_d$:

$$\hat{f}_d = \int_{-1/2T_s}^{1/2T_s} f \hat{S}_n(f) \, df, \tag{6.16}$$

where $\hat{S}_n(f)$ is the normalized estimate of the power spectrum (Eq. 5.44). If the spectrum $S(f)$ has spectral components that exceed the Nyquist limits $\pm(2T_s)^{-1}$, these components are undersampled and thus aliased with a net result that $\hat{f}_d$ is a biased estimate of the mean Doppler frequency f_d (Sirmans and Bumgarner, 1975). This bias can be eliminated, however, by additional processing (Section 6.4.2). In contrast to this, some thought reveals that if $S(f)$ is symmetric about f_d but has some frequency components that exceed the Nyquist limit, the argument of $R(T_s)$ still yields f_d without bias. Another way of seeing this is to change variable in Eq. (6.15) to get the following instructive formula:

$$\varepsilon_s = \int_{-1/2T_s-f_d}^{1/2T_s-f_d} S(f' + f_d) \sin(2\pi f' T_s) \, df', \tag{6.17}$$

where $S(f' + f_d)$ is centered at the origin $f' = f - f_d = 0$. Because $S(f)$ is periodic with period T_s^{-1}, we can add f_d to the limits of the integral (6.17), making them symmetric about $f' = 0$. Note that $\sin 2\pi(f - f_d)T_s$ has period T_s^{-1} and is odd with respect to f_d. Therefore ε_s is zero whenever $S(f)$ is symmetric about f_d (Fig. 6.3).

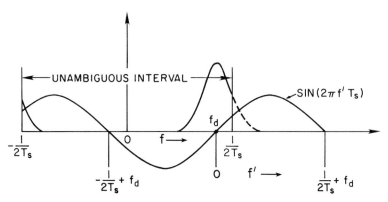

Fig. 6.3 The autocovariance estimate of the mean Doppler frequency is not biased by the frequency components that are undersampled if $S(f)$ is symmetric. —— sampled power spectrum; ---, undersampled components of spectrum $S(f)$.

    With reference to Fig. 6.3, we deduce that, even if the spectra are not symmetric, the argument of $R(T_s)$ yields an almost unbiased estimate provided that $S_n(f)$ has its frequencies contained in a band small compared to $(T_s)^{-1}$. In view of the foregoing, we also conclude that white noise (because it has a symmetric spectrum) will not bias the autocovariance estimator.

    If a sequence of $M + 1$ uniformly spaced pulses is transmitted, the autocorrelation at a sample time lag T_s (for weather signals sampled at range delay τ_s) can be estimated from the sum

$$\hat{R}(T_s) = \frac{1}{M} \sum_{m=0}^{M-1} V^*(m)V(m + 1), \tag{6.18}$$

so that the mean velocity estimate becomes

$$\hat{v} = -(\lambda/4\pi T_s)\mathrm{arg}\,\hat{R}(T_s). \tag{6.19}$$

The argument of $\hat{R}(T_s)$ is in radians, and the negative sign signifies that positive Doppler shifts create negative velocities, in accordance with Eq. (3.30).

    It should be clear that pulses need not form a uniform sequence for $R(T_s)$ to be estimated. In fact, pulses can be transmitted in pairs such that the spacing between the pairs is much larger than the intrapair period (Fig. 6.4). In this case

Fig. 6.4 Spaced sample pairs for calculating the mean velocity or spectrum width. The spacing between the pulses of a pair is T_s and the pair separation is T.

$R(T_s)$ is the covariance for the time lag T_s, which is equal to the intrapair period, and the sample estimates $V^*(m)V(m+1)$ have themselves a correlation that depends on the interpair spacing T. Spaced pairs are advantageous in alleviating the contamination of velocity estimates by range-overlaid echoes (Section 7.4.2).

Perturbation analysis has been successfully used to derive the variance of the mean velocity estimated from pulse pairs (Zrnić, 1977b). Nevertheless, such an analysis is not without flaws: For very narrow spectrum widths or low signal-to-noise ratios, it matches experimental results only if a very large number of samples is used. Briefly, the following two conditions are necessary for the perturbation analysis to be valid:

$$2\pi M \sigma_{vn} \gg 1, \tag{6.20a}$$

$$\rho^2(T_s)M \gg (N/S+1)^2. \tag{6.20b}$$

Condition (6.20a) expresses the requirement for a large number of independent samples, and (6.20b) ensures that the argument of $\hat{R}(T_s)$ has a distribution width small compared to 2π. With these assumptions, an expression for the mean frequency variance of correlated but spaced pairs has been derived (Zrnić, 1977b):

$$\mathrm{var}(\hat{v}) = \lambda^2 [32\pi^2 T_s^2 \rho^2(T_s)]^{-1} \{ M^{-2}[1-\rho^2(T_s)] \sum_{m=-(M-1)}^{M-1} \rho^2(mT)(M-|m|)$$

$$+ N^2/MS^2 + (2N/MS)[1+\rho(2T_s)(1/M-1)\delta_{T-T_s,0}]\}, \tag{6.21}$$

where now M is the number of sample pairs, the spacing between pairs is T, and $\delta_{T-T_s,0}=1$ for $T=T_s$ and zero otherwise. This expression for the variance is valid when sample pairs are contiguous (i.e., share a common sample when transmitted pulses are uniformly spaced) or when they are independent (which can be achieved by increasing the separation T or by changing the transmitted frequency between pairs). The plot of Eq. (6.21) in Fig. 6.5 reveals that the variances are not greatly different for independent and contiguous pairs. At narrow widths and low SNR, contiguous pairs have a lower standard deviation of the estimates; otherwise the opposite is true.

Superimposed in Fig. 6.5 are results from simulations for contiguous pairs. Several 8192-point periodograms were generated with Gaussian-shaped spectra (Zrnić,1975b). The inverse of such periodograms produced long-time records from which chunks of 64 samples were processed. Altogether 1024 chunks were processed to estimate the mean and variance. Because the number of processed samples is much less than the total number in the sequence, the method effectively introduces a rectangular window. The theoretical predictions at narrow widths and low SNR deviate from the simulation results because only 63 samples were processed for each estimate, and thus Eq. (6.20) was not satisfied. With more samples this deviation decreases. Torp (1992) has obtained a probability

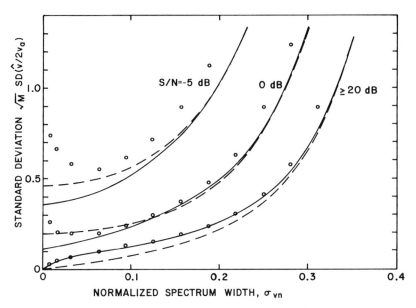

Fig. 6.5 Standard deviations of the mean frequency estimate (autocovariance processing). Spectrum width is normalized to the unambiguous interval $2v_a$. Note a gradual increase between 0 and 0.2 and an exponential rise thereafter. ———, contiguous pairs; ———, independent pairs; $\bigcirc$, simulation.

density function that produces a standard deviation consistent with simulation results at small spectrum widths and low SNR.

Because of condition (6.20a) the sum in Eq. (6.21) can be replaced with an integral to yield

$$\text{var}(\hat{v}) \approx \lambda^2 [32\pi^2 M\rho^2(T_s)T_s^2]^{-1}\{[1 - \rho^2(T_s)]T_s/2\sigma_{vn}T\sqrt{\pi} \\ + N^2/S^2 + 2(N/S)[1 - \rho(2T_s)\delta_{T-T_s,0}]\}. \tag{6.22a}$$

If sample pairs are independent, relation (6.22a) reduces to

$$\text{var}(\hat{v}) \approx \lambda^2 [32\pi^2 M\rho^2(T_s)T_s^2]^{-1}[(1 + N/S)^2 - \rho^2(T_s)]. \tag{6.22b}$$

The direct proportionality to ρ^{-2} signifies that the variance is heavily dependent on σ_{vn}^2 [and hence on the exponent in Eq. (6.5)]. Thus the effect of increasing T_s or σ_v is an exponential growth in the variance (Fig. 6.5).

For large signal-to-noise ratios, narrow spectrum widths, and contiguous pairs, relation (6.22a) becomes

$$\text{var}(\hat{v}) \approx \sigma_v \lambda/(8MT_s\sqrt{\pi}). \tag{6.23}$$

6.4.2   Spectral Processing

The advent of the FFT and the subsequent decrease in the cost of the circuits for the required computation make direct spectral methods attractive for mean frequency calculations. Still, the complexity and cost of FFT processors are an order of magnitude larger than those of their autocovariance counterparts. To calculate the mean frequency, one first forms the periodogram (5.22), which is an estimate of the power spectrum (it may contain aliases if there are frequency components outside the unambiguous interval). The next step is to obtain a rough mean frequency estimate k_m/MT_s, where k_m $(-M/2 \leq k_m \leq M/2)$ could be the index of the strongest Fourier coefficient. Then the mean velocity estimate becomes

$$\hat{v} = -\frac{\lambda}{2M}\left\{\frac{k_m}{T_s} + \frac{1}{\hat{P}T_s}\sum_{k_m-M/2}^{k_m+M/2}(k-k_m)\hat{S}[\mathrm{mod}_M(k)]\right\}, \qquad (6.24)$$

where $\hat{P}$ is the total power in the periodogram and $\mathrm{mod}_M(k)$ is the remainder on dividing k by M. This form, in contrast to that given in Eq. (6.16), eliminates biases due to aliasing if spectra are symmetric. Modifications of Eq. (6.24) are used to improve accuracy; among the simplest is the elimination of noise and weak signal spectral components by choosing a spectral density threshold above which spectral powers are included in the estimator. The standard deviations of the estimates if thresholds are used are listed by Sirmans and Bumgarner (1975). Berger and Groginsky (1973) have derived the variance of a power-weighted mean velocity estimate in which the noise spectral density is subtracted from the periodogram and the summation is replaced with the integral over the unambiguous interval. The variance reads

$$\mathrm{var}(\hat{v}) = \frac{\lambda^2}{4MT_sS^2}\int_{-1/2T_s}^{1/2T_s} f^2S^2(f+f_d)\,df. \qquad (6.25)$$

If Gaussian signal spectra [Eq. (6.3)] of narrow width [i.e., $\rho^2(T_s) \approx 1$] are substituted into Eq. (6.25), the variance simplifies to

$$\mathrm{var}(\hat{v}) = \frac{\lambda^2}{4MT_s^2}\left[\frac{\sigma_{vn}}{4\sqrt{\pi}} + 2(\sigma_{vn})^2\frac{N}{S} + \frac{1}{12}\left(\frac{N}{S}\right)^2\right]. \qquad (6.26)$$

It can be shown that the first term of Eq. (6.26) is equal to the corresponding term of the first-order expansion in the pulse pair variance (6.22a); the other terms in Eq. (6.26) are larger than their counterparts in (6.22a). Consequently, at small SNR and narrow widths $(\sigma_{vn} < 1/2\pi)$, the autocovariance estimator performs better. This is not the case at larger widths. Simulations show that the Fourier-derived mean has a variance that increases much more slowly than the exponential increase associated with the autocovariance method. Another advantage offered by spectral processing is the ease of editing data so that anomalies and system malfunctions can be identified and often eliminated.

6.5 Estimators of the Spectrum Width

Whereas only coherent radars can provide mean frequency information, the Doppler spectrum width can also be obtained from incoherent radars (Bello, 1965). One of the early devices, the R meter, estimates the width from the rate at which the detected weather signal envelope [i.e., some function of $|V(k)|$] crosses a threshold (Rutkowski and Fleisher, 1955). Neither the R meter nor any other method based on measurement of the envelope is widely used in radar meteorology. Although the width of the detected envelope spectrum is uniquely related to the Doppler spectrum width, the envelope spectrum is broader, hence the estimate accuracy would be degraded. Nevertheless, methods used to extract the width in coherent radars can also be employed on the signal envelopes from incoherent radars. Computation of spectral width in polarimetric radars requires special considerations (Section 6.8).

6.5.1 Autocovariance Processing

If weather signal spectra closely follow a Gaussian shape, it is natural to estimate the spectrum width from estimates of the autocorrelation coefficient [Eq. (6.5)]. One first needs to estimate $|R_1| = |R(T_s)|$ and $\hat{S}$, after which the logarithm of their ratio, from Eq. (6.4), will retrieve $\hat{\sigma}_v$.

$$\hat{\sigma}_v = \frac{\lambda}{2\pi T_s \sqrt{2}} \left| \ln\left(\frac{\hat{S}}{|\hat{R}_1|}\right) \right|^{1/2} \mathrm{sgn}\left[\ln\left(\frac{\hat{S}}{|\hat{R}_1|}\right) \right]. \tag{6.27}$$

The signal power estimate is obtained by subtracting the known noise power from the average of the squares of the magnitudes.[1]

$$\hat{S} = \frac{1}{M} \sum_{k=0}^{M-1} |V(k)|^2 - N. \tag{6.28}$$

The autocorrelation estimate $\hat{R}_1$ at lag T_s is given by Eq. (6.18).

The sgn term in Eq. (6.27) warrants some discussion. Because $\hat{S}$ and $|\hat{R}_1|$ are estimates, it is possible at narrow spectrum widths and low SNR for the logarithm in Eq. (6.27) to become negative. The sgn term tags these negative cases so that they can be eliminated from data analysis (or assigned to zero).

A related estimator is obtained by expanding the logarithm in Eq. (6.27) for values $|\hat{R}_1|$ close to $\hat{S}$, values that occur when $\sigma_{vn} \ll 1$.

$$\hat{\sigma}_v = \frac{\lambda}{2\sqrt{2}\pi T_s} \left| 1 - \frac{|\hat{R}_1|}{\hat{S}} \right|^{1/2} \mathrm{sgn}\left(1 - \frac{|\hat{R}_1|}{\hat{S}} \right). \tag{6.29}$$

At large widths Eq. (6.29) has an asymptotic ($M \to \infty$) bias, (Zrnić, 1977b) whereas Eq. (6.27) does not (provided the spectrum is Gaussian).

It can be shown that the variances of the spectrum width estimates Eqs. (6.27) and (6.29) are equal when the spectra are Gaussian and the asymp-

---

1. To avoid occurrence of negative estimates a multiplicative correction $\mathrm{S\hat{N}R}/(\mathrm{S\hat{N}R} + 1)$ to the sum is sometimes used.

totic bias is removed from Eq. (6.29). Under those conditions, perturbation expansion yields the following expression for the variance (Zrnić, 1977b, 1979a):

$$
\begin{aligned}
\mathrm{var}(\hat{\sigma}_v) = \frac{\lambda^2}{128 M \pi^4 \sigma_{vn}^2 \rho^2(T_s) T_s^2} &\Big\{ 2[1 - (1 + \delta_{T-T_s,0})\rho^2(T_s) \\
&+ \delta_{T-T_s,0}\rho^4(T_s)]N/S + [1 + (1 + \delta_{T-T_s,0})\rho^2(T_s)]N^2/S^2 \\
&+ \rho^2(T_s) \sum_{m=-(M-1)}^{M-1} \{2\rho^2(mT) + \rho^2(mT)\rho^{-2}(T_s) \\
&+ \rho^2[mT + T_s(1 - \delta_{T-T_s,0})] \\
&- 4\rho(mT + T_s)\rho(mT)\rho^{-1}(T_s)\}(1 - |m|/M) \Big\}.
\end{aligned}
\tag{6.30a}
$$

With independent pairs Eq. (6.30a) simplifies to

$$
\begin{aligned}
\mathrm{var}(\hat{\sigma}_v) = \frac{\lambda^2}{128 M \pi^4 \sigma_{vn}^2 \rho^2(T_s) T_s^2} &\{[1 - \rho^2(T_s)]^2 \\
&+ 2[1 - \rho^2(T_s)]N/S + [1 + \rho^2(T_s)]N^2/S^2\}.
\end{aligned}
\tag{6.30b}
$$

Because Eqs. (6.30) are obtained from a perturbation analysis of the expansions around the mean autocovariance $\bar{R}(T_s)$ and the $\bar{\sigma}_{vn}$, it is well to emphasize that at very narrow widths ($\sigma_{vn} < 0.01$) or if the signal-to-noise ratio and the number of pairs are small [e.g., condition (6.20) is violated], the results become erroneous. Nevertheless, the results are valid for common conditions.

Equation (6.30a) for contiguous pairs and Eq. (6.30b) for independent pairs are plotted in Fig. 6.6. There is little difference between the two, and it can be seen that the estimators are good (the standard deviation is small) as long as $0.02 < \sigma_{vn} < 0.2$ and SNR > 5 dB. At larger widths the estimate, like the mean frequency estimate, degrades exponentially. Simulated results (also included in Fig. 6.6) agree well with the theory for a wide range of widths and SNRs, but in regions where the conditions (6.20) are violated, deviations from the plotted curves occur.

Aliasing from undersampling ($\sigma_{vn} > 1/2\pi$) does not bias the estimator. Sample-dependent bias is inherently present, but because it is proportional to M^{-1} it can be neglected compared to the standard deviation (Zrnić, 1977b). Probably the most serious defect of the estimator is that it is not adaptable to editing, that is, to identifying errors arising from spurious spectral peaks. If $\sigma_{vn} < (2\pi)^{-1}$, the approximate variances for contiguous pairs are

$$
\mathrm{var}\,\hat{\sigma}_v \approx (3\lambda^2/128\sqrt{\pi}MT_s^2)\sigma_{vn}
\tag{6.31a}
$$

for high SNR and

$$
\mathrm{var}\,\hat{\sigma}_v \approx 3\lambda^2 N^2/128\pi^4\sigma_{vn}^2 T_s^2 M S^2
\tag{6.31b}
$$

for low SNR.

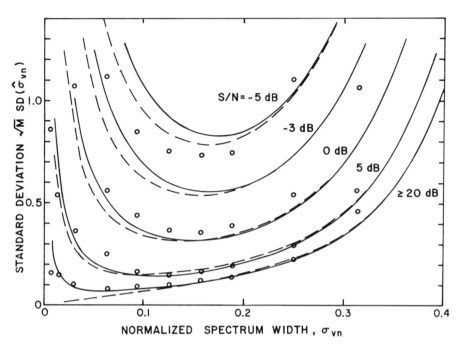

Fig. 6.6 Standard deviation of the width estimate (autocovariance processing). The left portion of the curve cannot be extended toward the origin because the standard deviation there follows an $M^{-1/4}$ law. ——, contiguous pairs; ---, independent pairs; ○, simulation.

There is one more estimator of spectral width that has found applications in weather radars. It is suitable for contiguous samples and is obtained from the ratio of autocovariances at lag 1 and 2. As in Eq. (6.27), a Gaussian shape of the spectrum permits the σ_v estimate to be written as (Srivastava *et al.*, 1979)

$$\hat{\sigma}_v = (\lambda/2\pi T_s 6^{1/2})| \ln |\hat{R}_1/\hat{R}_2||^{1/2} \, \text{sgn}[\ln |\hat{R}_1/\hat{R}_2|]. \tag{6.32}$$

Again the perturbation expansion (Zrnić, 1979b) is needed to obtain the variance of Eq. (6.32)

$$\text{var}(\sigma_v) = [\lambda^2/576\pi^4 T_s^2 \sigma_{vn}^2](A + B + C), \tag{6.33}$$

where A, B, and C are as follows:

$$A = (1/2M^2) \sum_{-(M-1)}^{M-1} [\rho^2(m)/\rho^2(1) + \rho(m+1)\rho(m-1/\rho^2(1)](M - |m|)$$
$$+ [1/2M\rho^2(1)](N/S)^2 + [1 + M\rho^2(1)]N/S$$
$$+ [(M-1)\rho(2)/M^2\rho^2(1)]N/S$$

$$B = -(1/M^2) \sum_{-(M-1)}^{M-1} \{[\rho(m+2)\rho(m-1) + \rho(m)\rho(m+1)]/\rho(1)\rho(2)\}(M-|m|)$$

$$-\{[(2M-3)\rho(3) + (2M-1)\rho(1)]/M^2\rho(1)\rho(2)\}N/S$$

$$C = (1/2M^2) \sum_{-(M-1)}^{M-1} [\rho^2(m)/\rho^2(2) + \rho(m+2)\rho(2-m)/\rho^2(2)](M-|m|)$$

$$+ [1/2M\rho^2(2)]N^2/S^2 + [1/M\rho^2(2)]N/S$$

$$+ [(M-2)\rho(2)/M^2]N/S$$

Note that T_s has been omitted from the arguments of ρ.

The useful range of normalized spectral widths for this estimator (Fig. 6.7) is decreased by a factor of two compared to the estimator (6.27) shown in Fig. 6.6. But Eq. (6.32) does not require separate estimation of white noise power; this is an advantage in systems where incoherent echoes (second or higher trip) are overlaid on coherent echoes (such as the case with magnetron transmitters). Also at SNR less than 0 dB there is a region where the estimate variance (6.33, Fig. 6.7) is smaller than the one in Fig. 6.6. An M-dependent bias is present with this estimator, but it is not significant because it is small compared to the standard deviation (Zrnić, 1979b).

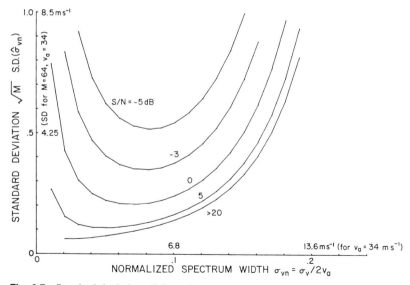

Fig. 6.7 Standard deviation of the estimator (6.32) that uses autocovariances at lag 2 and lag 1.

6.5.2 Spectral Processing

Width estimated using the algorithm

$$\hat{\sigma}_v^2 = \frac{\lambda^2}{4\hat{P}T_s^2} \sum_{k_m-M/2}^{k_m+M/2} \left(\frac{k}{M} + \frac{2\hat{v}T_s}{\lambda}\right)^2 \hat{S}[\mathrm{mod}_M(k)]. \qquad (6.34)$$

avoids bias due to aliasing provided the width is small compared to the unambiguous interval. The computed width contains a bias σ_b^2 due to the window effect. In general, this bias is difficult to compute exactly. One may argue that both the mean frequency [Eq. (6.24)] and the width [Eq. (6.34)] should be obtained from integrals of the power spectrum (using a continuum of k from $-M/2$ to $M/2$), rather than summations, to account exactly for the bias introduced by the window. In practice this would require an enormous amount of computations; furthermore, the sum in Eq. (6.34) is close to the integral when spectra are broad compared to the Doppler frequency resolution, $1/MT_s$. For such cases the bias σ_b^2 can be computed exactly because the measured spectrum is equal to a convolution [Eq. (5.32)] of the true spectrum with the lag window transform, thus

$$\sigma_b^2 = \int_{-v_a}^{v_a} v^2 D^2(v)\, dv \Big/ \int_{-v_a}^{v_a} D^2(v)\, dv, \qquad (6.35)$$

where $D(v)$ represents the Fourier transform of the data window (Table 5.2).

It should be emphasized that the estimated width [Eq. (6.34)] is biased even at large SNR if spectra have width comparable to the unambiguous interval; in this respect the autocovariance method is superior to Fourier processing. To maintain accuracy, however, the radar systems must be designed to have maximum expected spectrum widths reasonably smaller than T_s^{-1}. For narrow spectra both autocovariance and spectral estimators generate biases that can be an appreciable fraction of the estimate (Zrnić, 1979a and b).

Next we present a formula for the variance of the width assuming that the integral form of Eq. (6.34) is used and the noise spectral density is subtracted (Berger and Groginsky, 1973):

$$\mathrm{var}(\hat{\sigma}_v) = \frac{T_s}{2\lambda M \sigma_{vn}^2 S^2} \int_{-v_a}^{v_a} (v^2 - \sigma_v^2)^2 S^2(v + v_d)\, dv. \qquad (6.36a)$$

For narrow Gaussian spectra in white noise this formula becomes

$$\mathrm{var}(\hat{\sigma}_v) = \frac{\lambda^2}{4MT_s^2}\left[\frac{3\sigma_{vn}}{32\sqrt{\pi}} + \sigma_{vn}^2 \frac{N}{S} + \left(\frac{1}{320\sigma_{vn}^2} - \frac{1}{24} + \frac{\sigma_{vn}^2}{4}\right)\frac{N^2}{S^2}\right]. \qquad (6.36b)$$

Unlike the autocovariance estimator of the spectrum width, the variance here does not increase exponentially with width. Note that the first term of Eq. (6.36b) is the right side of Eq. (6.31a); i.e, the Fourier method is equivalent to the pulse pair method at large SNR. In contrast, at low SNR the pulse pair variance, Eq. (6.31b), is about three times smaller than the corresponding term of Eq. (6.36b).

It is significant that mean frequency and width estimators have variances comprising a signal contribution, a noise part, and a cross product of signal and noise. At large SNR (20 dB or so) the signal contribution dominates and the accuracy is solely a function of spectrum width. In such a case, averaging estimates over range and azimuth decreases the error at the expense of resolution. At low SNR the variance is inversely proportional to SNR and a much more efficient method of reducing it is to increase the SNR rather than to average the estimates along range. This can be accomplished by increasing the pulse length (Zrnić and Doviak, 1978; Section 4.5).

6.6 Minimum Variance Bounds

It is characteristic of maximum-likelihood estimators that they provide estimates with minimum variance if the number of samples M is large. Although these estimators may be quite complex and their algorithms are often unknown, we can nevertheless compute their variance, which establishes a lower bound to be compared with the variance achieved using simpler algorithms. A comprehensive treatment of the Cramer–Rao bounds is that of Zrnić (1979a), from which we have extracted the principal results. The minimum variance for the estimates of mean velocity, assuming a pure sinusoid in white Gaussian noise, is

$$\min \operatorname{var}(\hat{v}) = 3\lambda^2 N / 8M^3 \pi^2 T_s^2 S. \tag{6.37}$$

For weather signals with Gaussian-shaped spectra, the following two approximate expressions describe the minimum variance well:

$$\min \operatorname{var}(\hat{v}) = 3\lambda^2 \sigma_{vn}^4 / MT_s^2 (1 - 12\sigma_{vn}^2) \tag{6.38}$$

in the case of large SNR, and

$$\min \operatorname{var}(\hat{v}) = (\sqrt{\pi}\lambda^2 \sigma_{vn}^3 / MT_s^2)(N^2/S^2) \tag{6.39}$$

for small SNR.

Discussion of these bounds and comparison with the pulse pair algorithm follows. We shall constrain the analysis to a constant dwell time $T_d = MT_s$ because (1) this is a parameter that, together with the antenna rotation rate, determines the effective antenna beamwidth (Section 7.8); and (2) a comparison of estimators is meaningful if they operate on a time series of constant length. For weak signal power (SNR $\ll$ 1) and contiguous samples, we find from Eq. (6.22a) the value of T_s that minimizes the variance of the pulse pair velocity estimates.

$$T_s(\text{optimal})_{pp} = \lambda / 4\pi\sigma_v \sqrt{2}, \qquad N \gg S. \tag{6.40}$$

The reason for a minimum lies in two opposing effects. First, because the pulse pair processor measures the phase from which the frequency (velocity) is computed, it follows that, for any given error in phase, the error in the velocity decreases with an increase in T_s. But, as T_s increases, the correlation between phase samples decreases, which in turn increases the phase errors. The corresponding minimum variance of the pulse pair velocity estimates is

$$\text{var}(\hat{v}) = (\lambda \sigma_v \sqrt{2e}/8T_d \pi)(N^2/S^2), \tag{6.41}$$

and the Cramer–Rao bound (6.39) for T_s given by Eq. (6.40) becomes

$$\min \text{var}(v) = (\lambda \sigma_v \pi^{-3/2}/4T_d)(N^2/S^2). \tag{6.42}$$

Thus, if signal-to-noise ratio is low, the pulse pair algorithm produces mean velocity estimates with a variance that is only $\sqrt{e\pi/2} \approx 2$ times larger than the best possible.

At a 10-cm wavelength and $\sigma_v = 4 \text{ m s}^{-1}$ we find $T_s(\text{optimal}) \approx 1.4 \text{ ms}$, which is about 40% larger than the values typically used. Other factors (Chapter 7) have motivated the use of the shorter pulse repetition times.

Expressions similar to Eqs. (6.38) and (6.39) are available for the minimum variance of the Doppler spectrum width estimate. These are

$$\min \text{var}(\hat{\sigma}_v) = 45\lambda^2 \sigma_{vn}^6/MT_s^2 \tag{6.43}$$

in the case of large SNR, and

$$\min \text{var}(\hat{\sigma}_v) = \sqrt{\pi}\lambda^2 \sigma_{vn}^3 N^2/MT_s^2 S^2 \tag{6.44}$$

in the case of small SNR.

In summary, the Fourier method and autocovariance processing give comparable accuracy in spectral moments. Because the Fourier transform makes use of more information, one would expect it to be always better, but this is the case only in some situations, such as a pure sinusoid in white noise. The optimum processor, then, is a parallel bank of narrowband filters (these can be FFT coefficients), and the mean velocity is associated with the filter having the largest output (Whalen, 1971). Surprisingly, autocovariance (pulse pair) processing of a correlated sequence produces a maximum-likelihood estimate of mean velocity if the signal has an exponential correlation function (Zrnić, 1979a). Thus in general the underlying signal statistics dictate the optimum processing, and for a range of signal-to-noise ratios and spectrum widths there is no unique solution. Nevertheless, it is gratifying that relatively simple algorithms perform very well; they approach and even achieve the minimum variance bound for some class of signals. Therefore most weather radars, including the WSR-88D (Table 6.1), use autocovariance processing.

6.7 Performance on Data

A typical Doppler spectrum for a resolution volume within a thunderstorm is plotted in Fig. 6.8. This spectrum is obtained from a Fourier transform of 64 samples weighted with a von Hann window. The unambiguous velocity in this example is 28.5 m s^{-1}, therefore the spacing of the spectral coefficients is 0.9 m s^{-1}. Both autocovariance and spectral analysis (with a threshold 15-dB below the spectral peak) were used to estimate the mean velocity and spectrum width. The two moments computed from the autocovariance are $\hat{v} = 14.9 \text{ m s}^{-1}$ and $\hat{\sigma}_v = 2.5 \text{ m s}^{-1}$. The Fourier method yields $\hat{v} = 15.4 \text{ m s}^{-1}$ and $\hat{\sigma}_v = 2 \text{ m s}^{-1}$. To find the uncertainty in the estimates derived from the autocovariance, one can use Figs. 6.5 and 6.6 or Eqs. (6.23) and (6.31a). Because we do not know the true spectrum width, let us take the average, 2.25 m s^{-1}, of the two measurements. $\text{SNR} = 17 \text{ dB}$, and one can safely use Eq. (6.23) (with $\lambda/4T_s = 28.5 \text{ m s}^{-1}$ and $M = 64$) to find for $\hat{v}$ a standard deviation of 0.53 m s^{-1}. The same result is obtained for the Fourier-derived mean velocity because at large SNR Eq. (6.23) is equivalent to Eq. (6.25). The standard deviation of the spectrum width estimate $\hat{\sigma}_v$ is found from Fig. 6.6, and it is 0.57 m s^{-1}. For

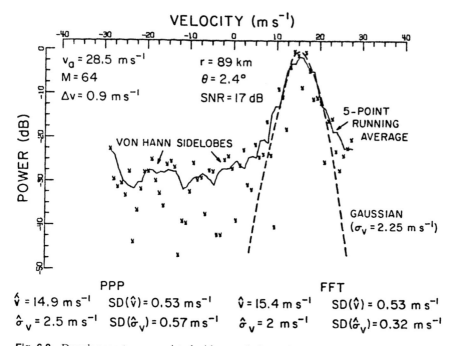

PPP		FFT	
$\hat{v} = 14.9 \text{ m s}^{-1}$	$\text{SD}(\hat{v}) = 0.53 \text{ m s}^{-1}$	$\hat{v} = 15.4 \text{ m s}^{-1}$	$\text{SD}(\hat{v}) = 0.53 \text{ m s}^{-1}$
$\hat{\sigma}_v = 2.5 \text{ m s}^{-1}$	$\text{SD}(\hat{\sigma}_v) = 0.57 \text{ m s}^{-1}$	$\hat{\sigma}_v = 2 \text{ m s}^{-1}$	$\text{SD}(\hat{\sigma}_v) = 0.32 \text{ m s}^{-1}$

Fig. 6.8 Doppler spectrum associated with a resolution volume in a storm.

Fourier-derived widths we use the first term of Eq. (6.36b) and find 0.32 m s^{-1}.
Both of these errors are typical, and the accuracy is adequate for most measurements in storms.

Finally, a real-time display on an oscilloscope of the three moment estimates versus range demonstrates what is to be expected from uniform precipitation (Fig. 6.9). Data were collected while the antenna was stationary but elevated to 4°. Fairly uniform rain caused echoes that extend in range until the beam exceeds the precipitation top (4 km at a range of about 60 km). Two traces (d) and (c), respectively, are nonintegrated and integrated logarithmic powers. There are 762 range gates spaced 1 μs, and the recursive first-order filter (digital integrator) operates on logarithmic samples of power with a time constant of 24 ms for this example, which is equivalent, in terms of variance reduction, to $M = 49$ uniformly weighted samples. With spectrum widths of 3 m s^{-1} there are about 10 independent samples [Eq. (6.12)].

The mean velocities (b) and spectrum widths (a) are analog representations of values digitally calculated using the autocovariance algorithms, Eqs. (6.19) and (6.29). The number of samples (from each resolution volume) for spectrum width and velocity calculations is 64. The continuity and low spread in the width and velocity estimates up to 60 km are indicative of a large SNR. Note the

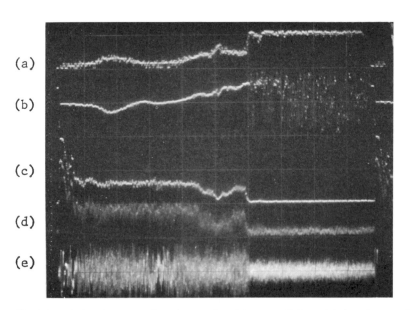

Fig. 6.9 Displayed as a function of range time (along a radial) are (a) $\hat{\sigma}_v$ estimates with a vertical scale of 15 m s^{-1}/div; (b) $\hat{v}$ estimates with a scale of 34 m s^{-1}/div; (c) $\hat{P}$ estimates in log units (40 dB/div); (d) nonintegrated power samples; and (e) one component of the Doppler channel (either I or Q). The horizontal scale is 11.5 km/div.

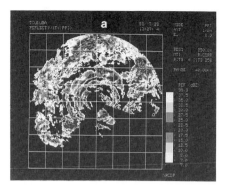
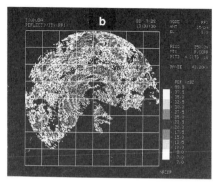

Fig. 6.10 Reflectivity from a storm complex. (a) The number of samples per estimate is 256. (b) The number of samples per estimate is 16. Figure provided by Masayuki Maki, National Center for Disaster Prevention, Japan.

uniform spread of velocities due to noise in the region of no echoes. Because the width is computed from Eq. (6.29) without noise removal (i.e., N is not subtracted from the power estimate $\hat{P}$), it is biased and therefore shows a sharp discontinuity at 60 km and large values at places where there is no weather signal.

To illustrate the effects of the number of integrated samples on the power estimates, two fields of the reflectivity factor from the same storm complex are presented in Fig. 6.10a and b. The smoother field (Fig. 6.10a) was obtained from 256 samples and antenna rotation rate of 1 rpm; the noisier field (Fig. 6.10b) was obtained with 16 samples at 15 rpm. Thus the azimuthal resolution for both examples is nearly identical and the difference in appearance is solely due to the number of integrated samples.

6.8 Signal Processing for Coherent Polarimetric Radar

A radar capable of transmitting two linear polarizations provides several additional parameters of interest to the meteorologist (Chapter 8). We are concerned here with only three of these, namely the differential reflectivity (Z_{DR}), the specific differential phase (K_{DP}), and the correlation coefficient $\rho_{hv}(0)$. The alternate polarization switching that enables estimation of these three additional polarization dependent parameters affects the estimates of the first three spectral moments if the algorithms previously presented are not modified. A fourth parameter, the differential phase shift ϕ_{DP}, is linearly related to K_{DP} and is needed for velocity estimation. In the following, the estimation of each of these radar observables is discussed.

6.8.1 Reflectivity and Differential Reflectivity

The primary intended use of a polarimetric radar is for identification of hail and accurate rainfall estimation. To estimate rainfall rate more accurately than is possible by an R, Z relation (Section 8.4.1), errors in differential reflectivity Z_{DR} must be less than 0.1 dB, and errors in reflectivity factor must be less than 1 dBZ (Sachidananda and Zrnić, 1985). Therefore, the number of samples needed for averaging is dictated by these accuracy requirements.

The mean sample powers $\hat{S}_h$ and $\hat{S}_v$ for the two polarizations must be calculated separately to estimate Z_{DR}. If samples from the linear receiver are used, Z_h (the reflectivity factor for horizontal polarization) and Z_{DR} are estimated from

$$\hat{S}_h = \frac{1}{M} \sum_{i=1}^{M} |H_{2i}|^2 - \hat{N}_h, \tag{6.45a}$$

$$\hat{S}_v = \frac{1}{M} \sum_{i=1}^{M} |V_{2i+1}|^2 - \hat{N}_v, \tag{6.45b}$$

and

$$\hat{Z}_{DR} \equiv 10 \log\left(\frac{\hat{S}_h}{\hat{S}_v}\right) \quad \text{(dB)}. \tag{6.46}$$

Here, M is the number of H or V samples; these are the weather signals received when horizontally or vertically polarized waves are transmitted. Subscripts indicate the sample's position in time ($2iT_s$ or $2iT_s + T_s$). $\hat{N}_h$ and $\hat{N}_v$ are white noise powers when the receiver is tuned to either horizontally or vertically polarized signals. It may be noted that the number of samples available for $\hat{S}_h$ or $\hat{S}_v$ estimation is half of the total available in a monopolarized radar, over the same time interval. But, the increase in the standard deviation due to this reduced number is usually insignificant because the standard deviation is principally a function of the dwell time $2MT_s$ and the spectrum width. The SD of S_h or S_v can be determined from Fig. 6.2 by realizing that the normalization of spectrum width is for sample spacing of $2T_s$.

Figure 6.11 shows the standard deviation σ_{DR} of $\hat{Z}_{DR}$ as a function of the number of samples averaged. This error is computed from the following approximate equation valid for $\text{SD}(\hat{S}_h/\hat{S}_v)/(\overline{S}_h/\overline{S}_v) \ll 1$ (Sachidananda and Zrnić, 1985):

$$\sigma_{DR} = 10 \log[1 + \text{SD}(\hat{S}_h/\hat{S}_v)/(S_h/S_v)] \quad \text{(dB)}. \tag{6.47}$$

The variance of $\hat{S}_h/\hat{S}_v$ needed in Eq. (6.47) is given by Sachidananda and Zrnić (1985) who show it to be

$$\text{var}(\hat{S}_h/\hat{S}_v) = 2\left(\frac{S_h}{S_v}\right)^2 \sum_{m=-(M-1)}^{M-1} \frac{(M-|m|)}{M^2} \{|\rho(mT_s)|^2 - |\rho[(2m+1)T_s]|^2|\rho_{hv}(0)|^2\}$$

$$+ \frac{1}{M}\left(\frac{S_h}{S_v}\right)^2 \left(\frac{2N_h}{S_h} + \frac{N_h^2}{S_h^2} + \frac{2N_v}{S_v} + \frac{N_v^2}{S_v^2}\right) \tag{6.48}$$

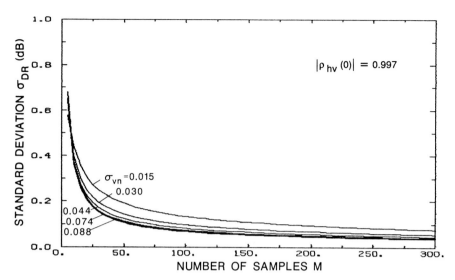

Fig. 6.11   Standard error of Z_{DR} estimates. $\rho_{hv}(0)$ is the correlation coefficient between horizontally and vertically polarized weather signals at lag 0. (Sachidananda and Zrnić, 1985.)

where $\rho_{hv}(0)$ is the correlation coefficient of the horizontally and vertically polarized samples that would be obtained if they were acquired simultaneously. High values of $|\rho_{hv}(0)|$ (Section 6.8.6) are expected (Sachidananda and Zrnić, 1986) for rain; measurements give a mean of about 0.98 (Fig. 8.26a). It is obvious from what was said at the beginning of the section and a comparison of Fig. 6.2 and 6.11 that the standard error in Z_{DR} dictates the number of samples required for improved rain-rate estimation or discrimination of hail.

6.8.2   Mean Radial Velocity and Differential Propagation Phase

The mean radial velocity and the differential propagation phase are estimated from polarization-dependent autocovariances. This is in contrast to estimator (6.18), which uses signals from monopolarized radars. Estimator (6.18) does not perform satisfactorily when applied to alternately polarized radar signals. The reason for this becomes clear from an examination of spectra and autocovariances of sequences of alternately polarized signals.

6.8.2.1   Spectra of Signals with Alternate Polarizations

The effects of alternating polarizations (Fig. 6.12) on power spectra has been analyzed by Sachidananda and Zrnić (1988) and we follow their explanation. Figure (6.13a) contains a typical spectrum obtained from the time series record of the NSSL radar with alternating polarization. The weather signal is from a region of high reflectivity (50 dBZ) and large signal-to-noise ratio (SNR > 20 dB) so that the weather spectral powers are well above the nearly

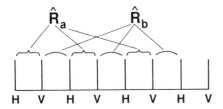

Fig. 6.12 A schematic of a sequence of alternate horizontally (H) and vertically (V) polarized weather signal samples. Two autocovariance estimates that are needed for simultaneous retrieval of Doppler and differential phase are indicated.

zero noise level. The number of samples used in a discrete Fourier transform is 256, and the magnitudes of the spectral coefficients are plotted with a normalized linear vertical scale.

We see that the spectrum has two halves separated by ν_a. The radar cross section of the hydrometeor scatterers is different for the two polarizations; thus the weather signal is amplitude modulated at the sampling frequency T_s^{-1}, which causes the appearance of a secondary spectrum. To recover the mean velocity, we need to know which one is the primary spectrum. Similarly, the total propagation phase shift of the signal from the radar antenna to the resolution volume and back to the radar, as well as through the radar system, is different for the two polarizations, giving rise to a phase modulation.

If we use spectral processing, Eqs. (6.24) and (6.34), to recover the mean radial velocity and the spectral width, we would get either an unbiased mean velocity ν or $\nu + \nu_a$ depending on which half is larger; the width can be recovered if a provision is made to recognize the two distinct halves. In Fig. 6.13b, we show typical spectra of H and V samples taken separately. Note that because the PRT for either sequence of equal polarization is twice that for the combined signal, the unambiguous interval is reduced by a factor of 2, and so is the number of data points. The remarkable similarity of the shape of H and V spectra indicates that the two sequences have a very high correlation. The same shape is nearly reproduced in the combined spectra (Fig. 6.13a). It is possible to estimate the mean velocity and spectrum width using only the H or V data, but the unambiguous interval would be half as wide and that would cause more frequent ambiguity of the estimated mean velocity. Additional information is needed to resolve the ambiguity.

6.8.2.2 Autocovariance Processing

With alternate polarization the autocorrelation estimate using the conventional pulse pair processor can be expressed as

$$\hat{R}(T_s) = \frac{1}{2M} \sum_{i=1}^{M} (H_{2i}^* V_{2i+1} + V_{2i+1}^* H_{2i+2}). \tag{6.49}$$

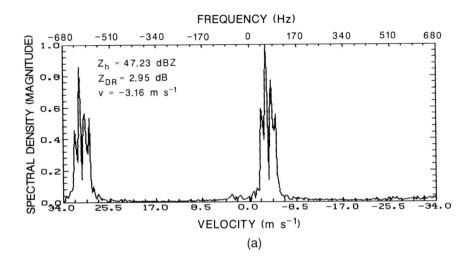

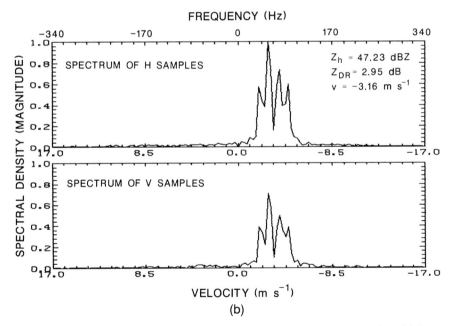

Fig. 6.13   (a) Normalized power spectrum of a signal with alternate polarization. (b) Spectra of separated H and V samples for the same data as in (a).

Summing the two products in Eq. (6.49) separately, and representing them by $\hat{R}_a$ and $\hat{R}_b$, we have

$$\hat{R}_a(T_s) = \frac{1}{M} \sum_{i-1}^{M} H_{2i}^* V_{2i+1}, \tag{6.50}$$

$$\hat{R}_b(T_s) = \frac{1}{M} \sum_{i=1}^{M} V_{2i+1}^* H_{2i+2}, \tag{6.51}$$

and

$$\hat{R}(T_s) = \tfrac{1}{2}[\hat{R}_a + \hat{R}_b]. \tag{6.52}$$

True values of estimates $\hat{R}_a$ and $\hat{R}_b$ are the expected values of products $H_n^* V_{n+1}$ and $V_{n+1}^* H_{n+2}$, respectively ($n = 2i$). It can be shown (Sachidananda and Zrnić, 1988) that the phase of R_a is the sum of phases due to the Doppler shift $\psi_d = -2T_s(k_0 + k_h)v$ and the two-way differential propagation phase shift,

$$\phi_{DP} = 2 \int_0^{r_0} [k_h(r) - k_v(r)]\, dr + \phi_{DS},$$

where ϕ_{DS} is the differential phase upon scattering; k_h and k_v are increments (for horizontal and vertical polarizations) to the free space propagation constant k_0 due to the presence of hydrometeors. The phase of R_b is the Doppler phase ψ_d minus the ϕ_{DP}. Although k_h, k_v are small compared to k_0 (e.g., at a 10-cm wavelength and rain rate of 100 m h^{-1}, $k_h = 24.4$ deg km^{-1}, $k_v = 20.7$ deg km^{-1}, and $k_0 = 3.6 \cdot 10^6$ deg km^{-1}) the cumulative effects (rk_h) are measurable. We can write the expected (or true) $R(T_s)$ as

$$R(T_s) = \tfrac{1}{2}[|R_a|e^{j(\psi_d + \phi_{DP})} + |R_b|e^{j(\psi_d - \phi_{DP})}]. \tag{6.53}$$

The magnitudes of R_a and R_b are equal, therefore we can simplify Eq. (6.53) to

$$R(T_s) = |R_a| \cos(\phi_{DP})e^{j\psi_d}. \tag{6.54}$$

This equation holds approximately for the estimates as well. Thus, we can unambiguously estimate the Doppler phase shift ψ_d using the conventional autocovariance processor only if ϕ_{DP} is less than $90°$.

Sachidananda and Zrnić (1988) have developed a modified estimator to extract the correct velocity along each radial even if $\phi_{DP} \geq 90°$. It not only enables one to resolve the velocity ambiguity caused by ϕ_{DP} but it also removes a noisy band in the velocity estimate if ϕ_{DP} is near $90°$. Before we can apply the modified estimator, we need to estimate ϕ_{DP}. An estimator for ϕ_{DP} suggested by Mueller (1984) is

$$\hat{\phi}_{DP} = \tfrac{1}{2} \arg(\hat{R}_a \hat{R}_b^*). \tag{6.55}$$

The variance of the estimator has been obtained by Sachidananda and Zrnić (1986), but a simpler equation can also be calculated from Zrnić (1977b, Eq. 25). In either case the fluctuations of ϕ_{DP} due to temporal changes along the propagation path are ignored, so that the contributors to errors are associated with scatterers within the resolution volume and with white noise. Following Zrnić (1977) the relatively simple equation for $\text{var}(\hat{\phi}_{\text{DP}})$ is

$$\text{var}(\hat{\phi}_{\text{DP}}) = \tfrac{1}{2}\,\text{var}(\hat{\psi}_{\text{d}})[1 - \text{cor}(\hat{\psi}_{\text{a}}, \hat{\psi}_{\text{b}})], \qquad (6.56)$$

where $\hat{\psi}_{\text{d}} = \hat{\omega}_{\text{d}}T_{\text{s}}$ is the Doppler phase shift estimate; $\hat{\psi}_{\text{a}}$, $\hat{\psi}_{\text{b}}$ are the Doppler phase shift estimates but calculated from $\hat{R}_{\text{a}}$ and $\hat{R}_{\text{b}}$. The var $\hat{\psi}_{\text{d}}$ is obtained by multiplying Eq. (6.21) or (6.22a) with $(4\pi T_{\text{s}}/\lambda)^2$ and setting $T = 2T_{\text{s}}$; also, $\rho^2(T_{\text{s}})$ in Eq. (6.21) or (6.22a) must be multiplied with $|\rho_{\text{hv}}(0)|$; M is the number of HV (or VH) consecutive pairs (i.e., one-half of the total number of samples). The correlation coefficient between $\hat{\psi}_{\text{a}}$ and $\hat{\psi}_{\text{b}}$ estimates can be derived using correlations of Doppler shifts given by Zrnić (1977)

$$
\begin{aligned}
\text{cor}(\psi_{\text{a}}, \psi_{\text{b}}) &= \frac{|\rho_{\text{hv}}(0)|^2 \displaystyle\sum_{m=-(M-1)}^{M-1} \rho^2(2mT_{\text{s}} + T_{\text{s}}) - 2\rho(2T_{\text{s}})N/S}{\displaystyle\sum_{m=-(M-1)}^{M-1} \rho^2(2mT_{\text{s}}) + 2N/S + N^2/S^2} \\[2mm]
&\approx \frac{\pi^{3/2}\sigma_{\text{vn}} - 2\rho(2T_{\text{s}})N/S}{\pi^{3/2}\sigma_{\text{vn}} + N^2/S^2 + 2N/S}.
\end{aligned}
\qquad (6.57)
$$

To arrive at the approximate formula, sums were replaced by integrals and a Gaussian correlation [Eq. (6.5)] has been assumed. Figure 6.14 gives the standard deviation of ϕ_{DP} versus the number of sample pairs averaged (Sachidananda and Zrnić, 1986).

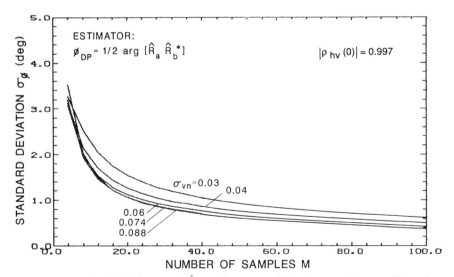

Fig. 6.14   Standard deviation σ_ϕ of $\hat{\phi}_{\text{DP}}$ estimates. $\rho_{\text{hv}}(0)$ is the correlation coefficient between horizontally and vertically polarized signals at lag 0, and M is the number of HV or VH sample pairs.

Differential phase shift computed from Eq. (6.55) is ambiguous if the actual value is outside a 180° interval, but it is easy to resolve this ambiguity knowing that ϕ_{DP} is always positive in the rain medium, and is a monotonically increasing function of range. As a matter of fact, the majority of hydrometeor types would produce either positive or zero differential phase shift. Only vertically oriented particles produce negative ϕ_{DP}. If the differential phase shift in the radar system is adjusted to zero, ϕ_{DP} will be small (i.e., equal to ϕ_{DS}) for a path void of hydrometeors and will increase only if the propagation path encounters precipitation. Thus, it is tempting to use continuity of ϕ_{DP} (from 0° to 180°) in range to correct the ambiguity whenever two consecutive values differ by a large amount (e.g., more than 90°). But at close range ϕ_{DP} is small or zero, and statistical uncertainty may produce negative values that would appear close to 180°. Thus, range-dependent correction procedures should be used on ϕ_{DP} data.

The mean velocity should be computed from

$$\hat{v}_p = -(\lambda/4\pi T_s) \, \arg[\hat{R}_a \exp(-j\hat{\phi}_{DP})], \tag{6.58}$$

where the subscript p is used to indicate that this velocity estimate is from an alternately polarized sequence and $\hat{\phi}_{DP}$ is the corrected (unambiguous) value. Multiplication with the exponent in the argument permits automatic adjustment of the phase of $\hat{R}_a$ so that even for angles larger than 180° a correct (within $\pm v_a$) velocity is obtained.

The variance of Eq. (6.58) can be obtained by following the procedure in Zrnić (1977), and the result is

$$\text{var}(\hat{v}_p) = \frac{\text{var}(\hat{v})}{2}[1 + \text{cor}(\psi_a, \psi_b)], \tag{6.59}$$

where $\text{var}(\hat{v})$ is given by Eq. (6.21) or (6.22a), with $T = 2T_s$ (again $\rho^2(T_s)$ in Eq. (6.21) or (6.22a) must be multiplied with $|\rho_{hv}(0)|^2$); $\text{cor}(\hat{\psi}_a, \hat{\psi}_b)$ is computed according to Eq. (6.57).

In Fig. 6.15 we show plots of consecutive complex product vectors $(H_n^* V_{n+1})$ and $(V_{n+1}^* H_{n+2})$, for one time series record with 256 samples. Each point corresponds to the tip of a product vector. The figure indicates clearly that the product vectors form two distinct groups. In fact, all the vectors in the upper half are $(H*V)$ products and the ones in the lower half are $(V*H)$ products; the means of the component vectors represent $\hat{R}_a$ and $\hat{R}_b$, respectively. The angular spread of the component vectors in each group is a measure of the spectrum width of the signal. The remarkable similarity in the pattern of distribution of the component vectors of $\hat{R}_a$ and $\hat{R}_b$ is due to the high correlation between H and V samples ($|\rho_{hv}(0)| \approx 0.98$). The inference that the mean velocity estimate from Eq. (6.49) would be noisy if ϕ_{DP} is near 90° becomes obvious after examining Fig. 6.15. That is if $\phi_{DP} = 90°$, the two mean vectors $\hat{R}_a$ and $\hat{R}_b$ exactly oppose each other, making the sum nearly zero, and because of noise the argument of $\hat{R}$ would be almost uniformly distributed over 2π.

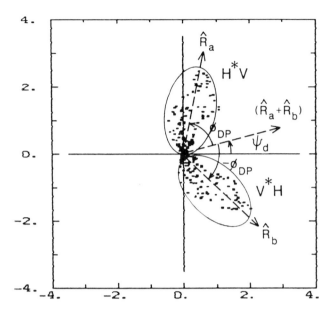

Fig. 6.15 Complex pulse-pair product vectors from a time series record. Each ellipse encircles either the H^*V or V^*H complex product vectors from which R_a or R_b (magnified for clarity) are estimated. x, y axes are in arbitrary units.

From our analysis we can infer that, if the differential propagation phase shift is near 90°, there will be a region of discontinuity in the velocity estimates along a radial. This is clearly shown in Fig. 6.16 in which we have plotted (1) the conventional pulse pair velocity estimate (6.19), (2) the velocity derived from a modified estimator (6.58), and (3) the corresponding differential phase shift $\hat{\phi}_{DP}$, along a radial; all parameters have been calculated from time series records. Note that the conventional pulse pair estimator and the modified estimator give the same velocity value as long as $\hat{\phi}_{DP}$ is less than 90°; if $\hat{\phi}_{DP} \approx 90°$, conventional pulse pair estimates become noisy; and for $\hat{\phi}_{DP} > 90°$, the velocity shifts by v_a, the Nyquist velocity.

6.8.3 Specific Differential Phase

Specific differential phase K_{DP} is defined as a difference between propagation constants for horizontally and vertically polarized electromagnetic waves.[2] In a homogeneous medium, K_{DP} can be directly obtained from differential phase

---

2. The specific differential phase defined here is for one way propagation; note that Sachidananda and Zrnić (1986, 1987) define K_{DP} for two-way propagation.

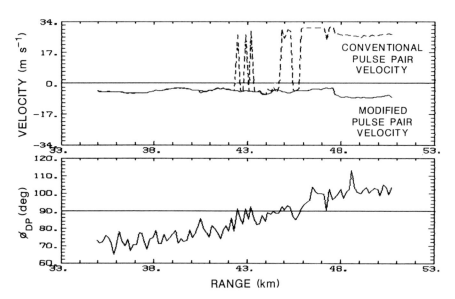

Fig. 6.16 A comparison of velocity estimates from the conventional pulse-pair estimator and the modified estimator. The corresponding ϕ_{DP} estimates are also shown. Discrete data are spaced in range by 150 m, but are connected for visual clarity.

shifts at two range locations (Fig. 6.17),

$$K_{DP} = \frac{\phi_{DP}(r_2) - \phi_{DP}(r_1)}{2(r_2 - r_1)}. \tag{6.60}$$

To reduce errors, K_{DP} should be estimated from measurements spaced over several resolution volumes. Further decrease in error can be obtained by averaging several contiguous estimates. Although this serves to decrease the errors of estimates, a least-squares fit is recommended for estimating K_{DP} because in a homogeneous medium $\phi_{DP}(r_i)$ is linearly related to r_0. Thus, at a center range r_0 given by

$$r_0 = \sum_i r_i / L, \tag{6.61}$$

Fig. 6.17 Differential phases corresponding to resolution volume locations at r_1 and r_2 needed for estimation of specific differential phase over the distance $r_2 - r_1$.

where L is the number or range locations, the least squares solution for $\hat{K}_{DP}(r_0)$ is

$$\hat{K}_{DP}(r_0) = \frac{\sum [\phi_{DP}(r_i) - \bar{\phi}_{DP}(r_i)](r_i - r_0)}{2\sum_i (r_i - r_0)^2}. \tag{6.62}$$

Note that the overbar in Eq. (6.62) denotes the average value over the L range locations. Now the variance of $\hat{K}_{DP}$ is linearly related to the variance σ_ϕ^2 of ϕ_{DP} by

$$\text{var}(\hat{K}_{DP}) = \frac{\sigma_\phi^2}{4\sum_i (r_i - r_0)^2}. \tag{6.63}$$

6.8.4   Spectrum Width

Because, in a radar with alternating polarization, samples of equal polarization are available only at a PRT of $2T_s$, the autocorrelation $R(mT_s)$ for even m is unaffected by switching. Thus, the spectrum width σ_v can be estimated from $\hat{R}(2T_s)$.

We can express the correlation coefficient $\rho(2T_s)$ as

$$\hat{\rho}(2T_s) = \frac{\left| \sum_1^M (H_{2i}^* H_{2i+2} + V_{2i+1}^* V_{2i+3}) \right|}{M(\hat{S}_h + \hat{S}_v)} \tag{6.64}$$

where the mean power estimates $\hat{S}_{h,v}$ are given by Eqs. (6.45a) and (6.45b). The spectrum width can now be estimated as

$$\hat{\sigma}_v = \lambda[-0.5 \ln \rho(2T_s)]^{1/2}/(4\pi T_s). \tag{6.65}$$

If $\hat{R}(2m)$ and $\hat{R}(2m+1)$ are highly correlated the variance of Eq. (6.65) can be obtained from Fig. 6.5, assuming that only $\hat{R}(2m)$ is used for $\hat{\sigma}_v$.

6.8.5   Correlation Coefficient

The correlation coefficient between H and V at zero lag $\rho_{hv}(0)$ depends on the shape, oscillation, wobbling, and canting angle distribution of hydrometeors (Sachidananda and Zrnić, 1985). Because simultaneous H and V samples are not available we use the correlation $\rho_{hv}(T_s)$ which, however, mainly depends on the spread of radial velocities. Two assumptions are needed to estimate $|\rho_{hv}(0)|$. First, some *a priori* model for the power spectrum shape (e.g., Gaussian) is needed. Second, the correlation at a lag $(2m+1)T_s$ is assumed to contain independent contributios from the distribution of radial velocities, and the distributions of canting angles and shapes; therefore the correlation coefficient $\rho_{hv}[(2m+1)T_s]$ can be expressed as a product $\rho(2m+1) \cdot |\rho_{hv}(0)|$.

An estimate of $|\rho_{hv}(T_s)|$

$$|\hat{\rho}_{hv}(T_s)| = \frac{|\hat{R}_a| + |\hat{R}_b|}{2\sqrt{\hat{S}_h \hat{S}_v}} \qquad (6.66)$$

is obtained from Eqs. (6.45a), (6.45b), (6.50), and (6.51). The correlation due to the radial velocity distribution at lag $2T_s$ is given by Eq. (6.64), and the assumption of Gaussian spectral shape allows equating $\hat{\rho}(T_s)$ to $\hat{\rho}(2T_s)^{0.25}$ so that the correlation coefficient magnitude $|\hat{\rho}_{hv}(0)|$ is directly computed.

$$|\hat{\rho}_{hv}(0)| = |\hat{\rho}_{hv}(T_s)| / [\hat{\rho}(2T_s)]^{0.25}. \qquad (6.67)$$

The normalized variance of $|\hat{\rho}_{hv}(0)|$ can be written using a perturbation expansion.

$$\begin{aligned}
\frac{\operatorname{var}(|\hat{\rho}_{hv}(0)|)}{|\rho_{hv}(0)|^2} &= \frac{\operatorname{var}(|\rho_{hv}(T_s)|)}{|\rho_{hv}(T_s)|^2} \\
&+ \frac{\operatorname{var}[\rho^2(2T_s)]}{16\rho^2(2T_s)} - \frac{\operatorname{cov}[|\rho_{hv}(T_s)|, \rho(2T_s)]}{2|\rho_{hv}(T_s)|\rho(2T_s)}
\end{aligned} \qquad (6.68)$$

From (6.68) the normalized standard deviation is plotted in Fig. 6.18. It is valid for a slightly simpler estimator in which the terms for vertical polarization in (6.64) are omitted and the R_b in (6.66) is not used; inclusion of these terms would reduce the variance, and thus we can consider the plots in Fig. 6.18 as an upper limit. It is evident that a system with an unambiguous velocity of 25 m s^{-1} would need less then 50 pairs (dwell time of 100 T_s) to lower the standard

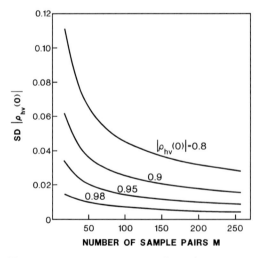

Fig. 6.18 Standard deviations of $|\hat{\rho}_{hv}(0)|$ estimates; M is the number of sample pairs (HV or VH) and $\sigma_{vn} = 0.06$. (Courtesy of Liu Li, Colorado State University.)

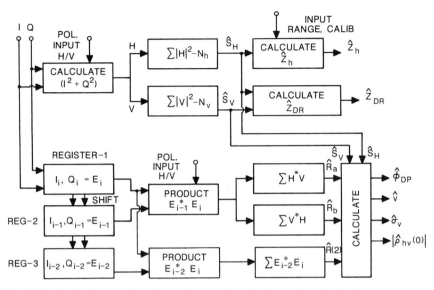

Fig. 6.19   A signal processing scheme for a radar with alternating polarizations.

deviation below 0.1, if the spectrum width is 3 m s^{-1} and the $|\rho_{hv}(0)| > 0.8$. The bias in the estimate is an order of magnitude smaller than the standard deviation (Balakrishnan and Zrnić, 1990b).

6.8.6   A Signal Processing Scheme for Echoes with Alternating Polarization

A signal processing schematic for a radar with alternating polarization is shown in Fig. 6.19. A linear receiver needs adequate dynamic range so that there are no distortions of the I and Q components. The reflectivity and differential reflectivity are derived from I and Q samples. Note the additional accumulator in the reflectivity processor and in the pulse pair processor, to store the values for the two polarizations separately. The mean velocity and the spectrum width estimators are given by Eqs. (6.58) and (6.65).

6.9   Concluding Remarks

In summary, the first three spectral moments and polarimetric variables can be routinely obtained at over 1000 contiguous range locations at speeds commensurate with real-time applications (i.e., antenna rotation rate of a few rpm). The speed and convenience of velocity and width calculations were improved by the advent of the pulse pair processor and associated digital circuitry. As these circuits become even more accessible, Fourier methods of spectral processing may find wider use in weather radars, especially in situations where signals are contaminated by spectral artifacts (sidelobes, power-supply ripple frequencies, etc.). Although spectral processing that uses the classical definition of moments is inferior to the autocovariance method at low SNR and narrow spectrum widths, the estimators based on the autocovariance for broader spectra have an exponential increase in the standard error; this is not the case for the Fourier method, which creates finite errors in both moment estimates and merely develops a bias in the width estimate. More important advantages of the Fourier method are the absence of bias due to nonsymmetric spectra and the feasibility of eliminating anomalous spectral powers.

We emphasize that an estimate of the power, mean velocity, or spectrum width for a single resolution volume might not have meteorological meaning; there remain statistical uncertainty and possible contamination from range overlaid echoes, anomalies, ground clutter, etc. Before accepting spectral moment data, one must check them for meteorological consistency over contiguous range and azimuthal locations.

Problems

6.1   Echo decorrelation time may be defined as the time it takes the argument of the exponential in Eq. (6.5) to reach -1. Find the decorrelation times for a 10-cm and a 5-cm radar wavelength if the spectrum width is 1 m s^{-1}.

6.2   A power transfer function of a receiver is given by $Q = \alpha \log_\beta P$ where Q is the output and P the input and α, β are constants. Assume that P is exponentially distributed with mean $\overline{P}$. Determine the mean and variance of Q. You may find the following two integrals useful:

$$\int_0^\infty e^{-\mu x} \ln x \, dx = -\frac{1}{\mu}(c + \ln \mu)$$

$$\int_0^\infty e^{-\mu x} \ln^2 x \, dx = \frac{1}{\mu}\left[\frac{\pi^2}{6} + (c + \ln \mu)^2\right]$$

where $c = 0.577215$ is Euler's constant.

6.3   The exponentially weighted estimates of power after square-law detection is

$$\hat{Q}_i = (1 - b)\hat{Q}_{i-1} + bQ_i.$$

Assume Q_i's are independent and exponentially distributed with mean P_0 and that the estimator has reached steady state; in other words, the recursion has been going on for a long time so that the moments $E(\hat{Q}^n) = E(\hat{Q}_{i-1}^n)$ for all n. Find the mean and variance of the estimate $\hat{Q}$ and compare with the one obtained by averaging M uniformly weighted samples. Under what conditions would the two be equal?

6.4   (a) Find the dwell time (MT_s) needed to reduce the standard error of velocity estimates to 1 m s^{-1} if the Doppler spectral width is 2 m s^{-1} and radar wavelength is 10 cm. Assume that this spectral width is narrow compared to the unambiguous velocity, that signal-to-noise ratio is large, and that the autocovariance algorithm for contiguous pairs is employed to estimate the mean velocity. (b) Repeat for the spectral width estimate.

6.5   (a) From Eq. (6.22) derive an approximate expression for velocity estimate variance if the signal-to-noise ratio is very low. (b) Let $T_s = 1$ ms and $\sigma_v = 2$ m s^{-1}, $\lambda = 0.1$ m and SNR $= 0.1$; use your expression to find the dwell time for which the standard deviation of the estimate is less than 1 m s^{-1}. (c) Repeat for the spectral width estimate.

6.6   Assume the effective white noise power at the receiver input is 10^{-14} W, and that a signal-to-noise ratio is -6 dB. (a) Plot the magnitudes of the signal plus noise autocorrelation function $R(0)$ and $R(1)$ for two normalized spectral widths: $\sigma_{vn} = 0.1$, and $\sigma_{vn} = 0.9$. (b) S is always larger than $R(1)$; nevertheless, for the respective estimates this may not hold. From your plot deduce which of the two widths will be more likely to make the estimate $\hat{R}(1) > \hat{S}$.

6.7   The complex voltage signal from a moving scatterer is given by

$$V(nT_s) = 5[\cos(2000\pi n T_s) + j\,\sin(2000\pi n T_s)].$$

Assume that the radar wavelength is 10 cm. (a) Find the autocovariance at lag T_s and from it calculate the mean velocity. (b) What is the Doppler spectral width of this signal?

6.8   A 10-cm pulsed-Doppler radar has a pulse repetition interval $T_s = 1$ ms. The signal-to-noise ratio is very large, the total signal power is 1, and there is no velocity aliasing. The auto-covariance has a Gaussian shape and, at lag $T_s = 1$ ms, $R(T_s) = 0.5 + j0.5$. Give an analytic expression for the power spectrum and numerical values for the mean velocity and spectral width.

6.9   A 10-cm weather radar transmits at a rate of 1000 pulses per second. It receives echoes from a storm. An estimate of the autocorrelation magnitude is made; its value at lag 0 is $3V^2$ and at lag 1 it is $2V^2$. The white noise power is $0.5V^2$ and the mean Doppler velocity is 5 m s^{-1}. (a) Write an expression for the autocorrelation. (b) Estimate the Doppler spectral width. (c) Write an expression for the Doppler spectrum.

6.10 A 10-cm radar receives alternately polarized signals along a 100-km path containing uniform rain. The specific differential phase is 1°. The radial velocity, in a resolution volume 100 km away, is 10 m s^{-1} toward the radar and the PRT is 1 ms. Find the differential phase. If the correlation coefficient magnitude, measured by radar, is 0.95 at lag T_s, what is the value of the complex correlation coefficient if autocovariance processing is used. Find also the correlations of $H_n^* V_{n+1}$ and $V_{n+1}^* H_{n+2}$. If the spectrum width is 2 m s^{-1} and the spectrum has Gaussian shape find $|\rho_{hv}(0)|$.

6.11 Two consecutive samples of a complex signal are $V(mT_s) = 0.9 + j0.1$ and $V[(m + 1)T_s] = 0.5 + j0.2$ volts. Find an estimate of power through a 1-Ω resistor. Find the mean velocity if $T_s = 1$ ms and $\lambda = 0.1$ m. Determine the Doppler spectral width.

Considerations in the Observation of Weather

This chapter examines limitations in pulsed-Doppler radar observations of weather caused by range–velocity ambiguities, weather signal decorrelation, ground clutter, and antenna sidelobes and rotation. Furthermore, because of the statistical uncertainties associated with Doppler spectral moment estimation, sometimes an undesirably long dwell time is required for acceptable measurement accuracy. Various techniques to mitigate these restrictions are described. Finally, we briefly discuss how radar hardware affects measurement accuracy.

7.1 Range Ambiguities

To illustrate range ambiguities, Fig. 7.1 shows a conglomerate of thunderstorm cells, to ranges beyond 300 km as seen by a (WSR-57) radar having a long PRT (6 ms, with an unambiguous range of 900 km). One of the storm cells produced a tornado that was tracked with a Doppler radar located near the WSR-57. The Doppler radar's PRT is significantly shorter in order to have a reasonably large (34 m s^{-1}) unambiguous velocity, and consequently its unambiguous range is only 115 km; Fig. 7.2 shows the range-ambiguous storms as observed by the Doppler radar. Not only are the ranges ambiguous, but because there were storms within as well as beyond 115 km, a tornado-producing cell at 150 km (within the Doppler radar's second-trip $cT_s/2$ interval in this particular case) is partially overlaid with (obscured by) echoes from storms in other trips (first and third $cT_s/2$ intervals). This could make accurate spectral moment estimation difficult.

Doppler velocity and spectrum width measurements are possible only within unobscured regions of this second-trip storm when the Doppler radar, as in this case, is fully coherent from pulse to pulse so that phase information is preserved for all echoes. However, if echoes happen to be overlaid, then moment estimation is still possible with a fully coherent radar for the echo that has significantly more power ($\geq$10 dB for velocity and 15 dB for width) than the sum of the other

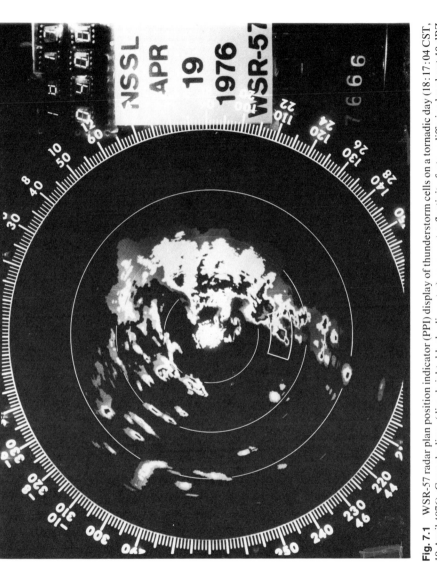

Fig. 7.1 WSR-57 radar plan position indicator (PPI) display of thunderstorm cells on a tornadic day (18:17:04 CST, 19 April 1976). Gray shadings (dim, bright, black, dim, etc.) represent reflectivity factors differing by about 10 dBZ starting at 17 dBZ. Constant-range circles are 100 km apart; elevation angle is 0.0°. The unambiguous range is 900 km. The boxed area outlines a tornadic storm cell; its mesocyclone signature was detected in real time by NSSL's Norman Doppler radar, which was about 100 m from the WSR-57.

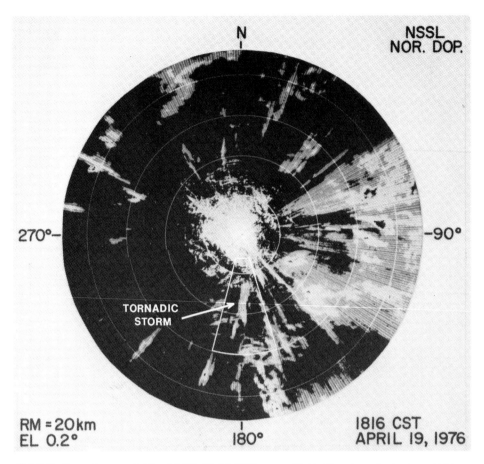

Fig. 7.2 Same storms as in Fig. 7.1 (18:16:35 CST), seen with the Doppler radar having a 115-km unambiguous range (only the first 95 km are shown). 10 log Z brightness categories (dim, bright, etc.) start at 10 dBZ and are incremented in 10-dBZ steps. The 10 log Z scale applies only to first-trip echoes. Some range-overlaid storms can be recognized by their radially elongated shape. The box outlines the same areas as in Fig. 7.1. Constant range circles are 20 km apart. Part of the tornadic storm is obscured by ground clutter and a nearby (30–60 km range) first-trip storm.

trip echoes being overlaid. If a phase diversity radar is used with appropriate filtering (Section 7.4.1), then echo power need not be greater than the sum of other echoes to obtain reasonably accurate velocity measurements.

Range-overlaid storm cells can obscure radar signatures associated with significant meteorological phenomena, such as tornado cyclones and down-drafts, that contain wind shears hazardous to aircraft in their descent into

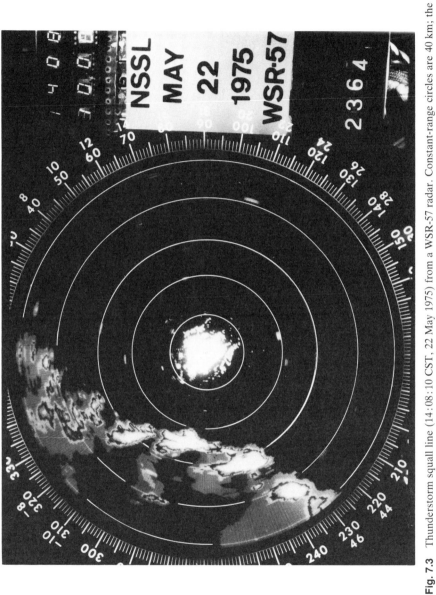

Fig. 7.3 Thunderstorm squall line (14:08:10 CST, 22 May 1975) from a WSR-57 radar. Constant-range circles are 40 km; the unambiguous range is 900 km.

or departure from an airport. The characteristics of signatures are related to meteorological phenomena in Chapter 9. Density currents produced by strong downdrafts can extend far beyond the precipitation areas, where their effective reflectivity factors could be as low as 0 dBZ and sometimes even less (Table 9.2 and Fig. 9.38a); yet they can harbor sudden wind-shift changes at their fronts (i.e., gust fronts) that can be hazardous. Tornado cyclone reflectivities are usually significantly higher (>10 dBZ).

Geometric considerations have lead Doviak *et al.*, (1978) to a conclusion that weather Doppler radars having unambiguous range r_a large compared to storm cell diameters should infrequently experience obscuration of first-trip phenomena by randomly distributed cells. However when cells are organized along a line as is the case for a squall line (Fig. 7.3), the probability of obscuration may increase significantly. It is evident that the increase would occur if the line is passing over the radar site. We can deduce that phenomena located in the second- (or higher-order) trip zones would experience a larger probability of obscuration. This is because storm cells in the first trip occupy a larger fraction of the first-trip area than the storms from the second trip. Furthermore, first-trip echo power is usually larger because it has an r^{-2} advantage.

7.2 Velocity Ambiguities

Hydrometeor velocities become ambiguous if one cannot distinguish between actual Doppler shifts and aliases that are spaced in frequency by the pulse repetition frequency. The range–velocity product

$$r_a v_a = c\lambda/8 \tag{7.1}$$

typifies the ambiguity resolution capabilities of a Doppler radar with uniformly spaced pulses. If both I and Q samples are processed to resolve the sign of the Doppler shift, the unambiguous velocities span the interval $\pm v_a$. The equation shows the advantage of longer wavelengths, but other factors control this choice. There are radar waveform designs (Deley, 1970) to remove ambiguities for scatterers that are discrete and finite in number (i.e., a squadron of aircraft or missiles). These designs do not work well with weather scatterers that are distributed quasicontinuously over large spatial regions (tens to hundreds of kilometers).

Figure 7.4 illustrates the velocity distributions that can be found in severe storms. Some 20,000 sample points from resolution volumes near ground to about 10 km in altitude are plotted. The centers of the velocity distributions (Fig. 7.4) are displaced relative to one another and to zero, in part because of

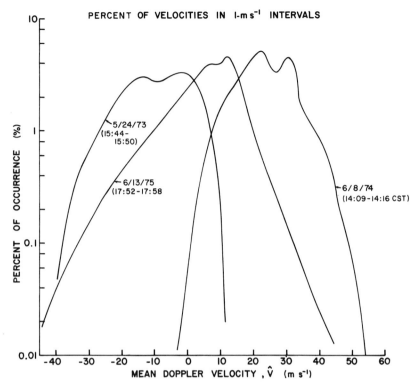

PERCENT OF VELOCITIES IN 1-m s⁻¹ INTERVALS

Fig. 7.4 Relative frequency of occurrences of the mean Doppler velocity estimates for three tornadic storms. Data samples are uniformly spaced throughout most of each convective cell. Note the large spread of radial Doppler velocities, which needs to be measured unambiguously.

storm motion. More important than the mean motion or peak radial speeds is the >50 m s⁻¹ spread in velocities for any of these storms.

7.3 Signal Coherency

In principle one can choose T_s large enough that no second- or higher-order trip echoes (from storms) will ever be received, but increasing T_s is limited in that weather signal samples spaced T_s apart must be correlated for accurate Doppler measurements. Correlation exists if

$$\lambda/2T_s \gg \sigma_v, \tag{7.2a}$$

where σ_v is the velocity spectrum width. Condition (7.2a) merely states that the Doppler width should be much smaller than the unambiguous velocity interval.

When T_s is increased, signal sample correlation (6.5) decreases exponentially, causing the variance in mean Doppler velocity estimates $\hat{v}$ and Doppler width estimates $\hat{\sigma}_v$ to increase exponentially, as can be seen from Eqs. (6.22a or b) and (6.30a or b). This leads one to consider the inequality

$$\sigma_v \leq \frac{v_a}{\pi} \qquad (7.2b)$$

as necessary for accurate estimation of Doppler spectral moments. Equality occurs if $\rho^2(T_s) = e^{-1}$, and this condition is chosen as a convenient correlation threshold. If the correlation decreases below this threshold there is an exponential increase in the variance of the estimate, as shown in Figs. 6.5 and 6.6. In terms of the unambiguous range,

$$r_a \leq \frac{c\lambda}{8\pi\sigma_v} \qquad (7.2c)$$

is the condition to maintain signal sample correlation. Requirement (7.2c) places a limit on r_a, for a given σ_v and wavelength, whereas Eq. (7.1) restricts r_a only if ambiguities due to velocity aliases need to be suppressed by choosing a large v_a. Methods (Section 7.4) to resolve velocity aliases work provided signal samples

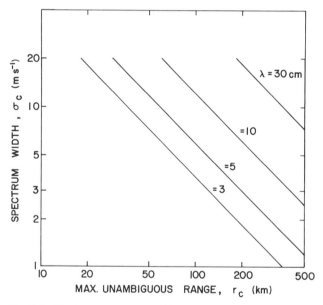

Fig. 7.5 Maximum spectrum width σ_c versus maximum unambiguous range r_c for which contiguous signals samples are correlated; λ is the radar wavelength.

are correlated. Thus relation (7.2c) is a necessary condition to maintain signal sample correlation, and it is a more basic constraint on radar parameters λ, T_s than Eq. (7.1). Note that the spectrum width σ_v is primarily determined by the relative velocities of hydrometeors within the resolution volume, as well as by the antenna rotation rate.

To be definite, let equality in (7.2c) specify the maximum width σ_c to which coherency is maintained for given r_a, or, given a spectral width, the maximum $r_a \equiv r_c$ (i.e., $cT_s/2$). Figure 7.5 relates σ_c to r_c with λ as a parameter. It is apparent that, unless spectrum widths are less than a few meters per second, 10-cm or shorter wavelength radars cannot eliminate range-ambiguous echoes (i.e., by having $r_a \geq 500$ km). Data from severe storms (Fig. 10.11) show a median width of 4 m s^{-1}, and about 10% of measured widths are larger than 8 m s^{-1}. Therefore if velocity estimates are to be accurate in 90% of the storm volume, 10-cm weather radars should have an $r_a \leq 150$ km. Because storm cells are often spread over larger areas (e.g., Fig. 7.1), it becomes apparent that these radars will be plagued by range ambiguities.

7.4 Techniques to Mitigate the Effects of Ambiguities

In the absence of practical methods to eliminate simultaneously range and velocity aliases, schemes have been devised that separate range and velocity measurements or minimize the deleterious effects of overlaid echoes. Some promising techniques will now be discussed.

7.4.1 Phase Diversity

Given a radar of wavelength λ, suppose that a suitable v_a is chosen to provide acceptable velocity aliasing (i.e., correctable using data processing of Doppler moment fields). Then r_a is determined [Eq. (7.1)] but, in most cases, is so small that there can be second- and third-trip weather signals overlaying one another. Let the initial phase ψ_t [Eq. (3.26)] of the transmitted pulse be random from pulse to pulse, as occurs with transmitters using magnetrons. This random phase can be estimated, for instance, by letting a very small fraction of the transmitter pulse leak into the receiver so that its I and Q signals can be measured (Nutten *et al.*, 1979). Thus

$$\psi_k = \tan^{-1}[Q_k(\tau_s)/I_k(\tau_s)], \qquad (7.3)$$

where k denotes the kth transmitted pulse, and τ_s is set to zero indicating that the I and Q are not weather signal samples but samples of the transmitted pulse. It suffices to store the $Q_k(0)$ and $I_k(0)$ for as many PRTs as needed for coherent measurement of velocities in the selected number of $cT_s/2$ range intervals. The signals received must then have their phases corrected to account for the arbitrary phase of each transmitted pulse.

If one wants to make measurements in the first- and second-trip region, ψ_k must be stored for a $2T_s$ period, and phase corrections must be made (Fig. 7.6) in each of two channels, one for the first, the other for the second trip. But, the transmitter reference phase is updated in the first-trip channel immediately after transmission, whereas in the second-trip channel the update of the phase is lagged by one T_s period.

Weather signals outside the selected trip are incoherent (i.e., like white noise because of random phases) and thus appear as a spatially dependent increase in noise level. Although the variances of the spectral moment estimates increase because there is a decrease in the effective signal-to-noise ratio (e.g., Figs. 6.5 and 6.6), mean velocity estimates are not biased. To avoid this increase in noise one can adaptively filter out the unwanted signals in a two-step process: (1) In one of two channels (assuming only first- and second-trip echoes are present) the first-trip echoes are coherently summed, whereas in the other the second-trip echoes are coherently summed; (2) the coherent signal in the first (second) channel is then filtered and the residuals coherently summed, from which moment estimates are made for the second- (first-) trip echoes. The effects of this procedure on signal spectra are shown in Fig. 7.7, and performance characteristics are given by Zrnić and Mahapatra (1985). Because the signal properties change from resolution volume to resolution volume, the filter characteristics should be adaptively adjusted using, for example, maximum-likelihood techniques similar to those described by Waldteufel (1976). Although the procedure would probably give the best overall performance, the complexity of signal processing is much greater than with other techniques.

Deterministic phase coding of transmitted signal can also be employed to reduce the effects of range overlaid echoes. Sachidananda and Zrnić (1986) proposed several phase-coded sequences for this purpose. The coding affects the correlations at lag one so that in the sum (6.18) for $R(T_s)$ the parts from two consecutive estimates due to the second trip echo almost cancel; but the parts contributed by the first trip echo are not altered. For example consider transmit-

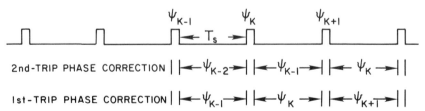

Fig. 7.6 Processing of a random phase pulse train. Transmitted phases ψ_k are random, and in this example the phase correction on echoes is made at different times in two separate receivers, which allows coherent measurement to a range of cT_s.

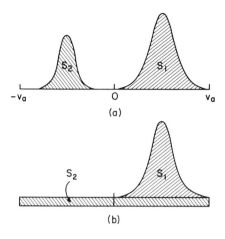

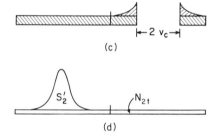

Fig. 7.7 Rejection of overlaid signals using the random phase method. (a) Power spectra for overlaid first- and second-trip signals in a fully coherent receiver. (b) Spectra of echoes for receiver that is coherent for first trip. (c) Spectra of (b) after bandpass filtering. (d) Spectra of (c) after recohering second-trip echoes. $2v_c$ is the width of the filter's notch.

ted pulses with a phase sequence $0, \pi/4, 0, \pi/4, \ldots$, etc. The received first-trip echo samples would also be phase shifted by the same amount (i.e., phase sequence $0, \pi/4, 0, \pi/4 \ldots$), but the phase shift of the second trip signal would be shifted by one pulse repetition interval (i.e., phase sequence $\pi/4, 0, \pi/4, 0, \ldots$). Now, the first or second trip signal phases can be restored if the received samples are phase shifted by $(0, -\pi/4, 0, -\pi/4 \ldots)$ or $(-\pi/4, 0, -\pi/4, 0 \ldots)$, respectively. If the first-trip signal phase is restored the second-trip signal would be phase modulated by the sequence $(\pi/4, -\pi/4, \pi/4, -\pi/4, \ldots)$. Correction for the second-trip signal phases would modify the first-trip signal in the same manner.

It is simple to visualize how phase coding removes the bias error due to second-trip signal in autocovariance processing. Consider the product of the kth and the $(k + 1)$th sample (spaced by T_s)

$$V^*(k)V(k + 1) = A^*(k)A(k + 1) + B^*(k)B(k + 1)e^{j\pi/2} + A^*(k)B(k + 1)e^{j\pi/4}$$
$$+ B^*(k)A(k + 1)e^{j\pi/4}, \tag{7.4a}$$

where $V(k) = A(k) + B(k)$ and $A(k)$ is the first-trip signal and $B(k)$ the second-trip signal. Likewise, the $(k + 1)$ and $(k + 2)$ sample product is

$$V^*(k + 1)V(k + 2) = A^*(k + 1)A(k + 2) + B^*(k + 1)B(k + 2)e^{-j\pi/2}$$
$$+ A^*(k + 1)B(k + 2)e^{-j\pi/4}$$
$$+ B^*(k + 1)A(k + 2)e^{-j\pi/4}. \tag{7.4b}$$

In both Eqs. (7.4a) and (7.4b) the phases of the first-trip signal have been compensated. The additional phase terms are due to the phase coding of the second-trip signal. An estimate of the autocovariance is obtained by averaging M consecutive pair products such as (7.4a) and (7.4b). Because of the phase modulation, the second terms [the autocorrelations of $B(k)$] nearly cancel pairwise if there is high correlation between samples $B(k)$, $B(k + 1)$, and $B(k + 2)$. But, more important, even sums of the second terms have zero mean. It is necessary to average an even number of sample pair products (M even) because the second term cancels pairwise. Furthermore, because $A(k)$ and $B(k)$ are uncorrelated, the third and fourth terms have also zero mean values; they contribute to the variance of mean velocity estimates.

7.4.2 Spaced Pairs with Polarization Coding

Spaced-pulse pairs of orthogonally polarized samples (Fig. 7.8) have several advantages for reducing the occurrence of those echoes that overlay the desired

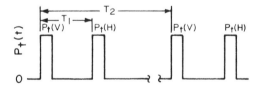

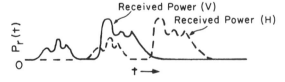

Fig. 7.8 Two orthogonally polarized (V, H) transmitted pulses and received powers; P_t is the transmitter power and P_r the received mean power. Weather-type scatterers are assumed to produce the pattern of P_r.

signal (Doviak and Sirmans, 1973). The $P_t(V)$ and $P_t(H)$ are transmitted powers with vertical and horizontal polarization. The first advantage of the method is that, with T_s sufficiently large, overlay is limited to first- and second-trip storms. Furthermore, overlaid echoes are incoherent, so they do not bias velocity estimates but only decrease the effective signal-to-noise ratio.We can achieve a third advantage if the pulses of a pair are orthogonally polarized because then the overlaid signal power is decreased, possibly by as much as 25 dB (Keeler and Carbone, 1986). This technique has been successfully adapted to radars measuring ionospheric motions (Woodman and Hagfors, 1969). Also, it has been implemented on weather radars but without orthogonal polarization diversity (Campbell and Strauch, 1976).

A factor that needs to be considered is the presence of differential phase shift between the two orthogonally polarized signals caused by (1) propagation through rain and (2) differential phase shift on scattering caused by non-sphericity of the drops within the resolution volume (Section 8.5.2). Because drops are flattened (Fig. 8.1), signals with horizontal electric fields induce stronger dipole moments and have different phase shifts than those with vertically polarized fields. Differential phase shift can bias velocity estimates but, if covariance estimates are made on pairs of pulses in which the sequence of polarization is reversed on each subsequent pair, this bias can be eliminated (Section 6.8.2.2).

A disadvantage of spaced pairs is that a longer time is required to collect a sufficient number of sample pairs to reduce the velocity estimate variance to acceptable limits. Fewer sample pairs but longer dwell times are needed to achieve the same measurement accuracy with spaced pairs as with uniformly spaced pulses because spaced sample pairs are less correlated [Eq. (6.21) and Fig. 6.5). By changing the carrier frequency between pairs of pulses, one can make pairs completely independent without the necessity of having T_2 larger than T_1, and hence the time to acquire the number of sample pairs needed to achieve a desired precision is reduced appreciably. In this case, the spaced pair technique is similar to the frequency diversity method described in Section 7.5.1, but here we have the advantage of polarization coding to reduce overlaid echo power between first- and second-trip echoes. A disadvantage of this technique is that ground clutter canceling is less effective (Banjanin and Zrnić, 1991).

7.4.3 Staggering the PRT to Increase the Unambiguous Velocity

Staggered PRT belongs to a general class of techniques whereby autocovariance or velocity estimates from two PRTs are suitably combined to effectively increase the composite unambiguous velocity. We shall briefly describe the philosophy behind methods that use two PRTs that may or may not be staggered. In a two-PRT technique, a velocity estimate $\hat{v}_1$ is obtained from echo samples

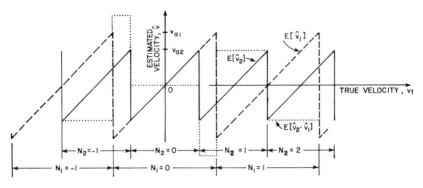

Fig. 7.9 Expected estimated velocities versus the true velocity for $T_{s2}/T_{s1} = 1.5$. N_1, N_2 define ambiguity intervals for samples spaced T_{s1}, T_{s2} apart.

spaced by T_{s1}, whereas a second velocity estimate $\hat{v}_2$ is derived from samples spaced by T_{s2}. For example, uniform spacing T_{s1} can be used during one scan, and the scan can be repeated immediately with the different but uniform sample time spacing T_{s2}. Because $\hat{v}_1$, $\hat{v}_2$ are associated with different unambiguous intervals, velocity aliasing can cause them to be significantly different (Fig. 7.9), and these differences can be used to resolve the true velocity. But mean velocity aliases can only be resolved as long as the expected difference $E(\hat{v}_2 - \hat{v}_1)$ remains unambiguous.

Strictly, $E(\hat{v}_2 - \hat{v}_1)$ becomes ambiguous only if the "waveforms" represented by the dotted lines (Fig. 7.9) repeat. This occurs if

$$v_t = m v_{a1} = n v_{a2}, \tag{7.5}$$

where m and n are integers, v_t is the true velocity, and v_a is the Nyquist velocity. Errors in resolving aliases (dealiasing) may occur because one cannot estimate $E(\hat{v}_2 - \hat{v}_1)$ with zero variance. Sirmans *et al.* (1976) examine the probability of error in dealiasing and give a comprehensive discussion of the statistical precision of the estimates $\hat{v}_1$, $\hat{v}_2$ and $\hat{v}_2 - \hat{v}_1$.

In the staggered PRT technique (Fig. 7.10) autocovariance estimates $\hat{R}_1$ at lag T_{s1} and $\hat{R}_2$ at lag T_{s2} are combined so that the velocity is obtained from the phase difference of the two.

$$\hat{v} = \frac{\lambda}{4\pi(T_{s2} - T_{s1})} \arg\left(\frac{\hat{R}_1}{\hat{R}_2}\right). \tag{7.6a}$$

To decrease the velocity estimate variance, the covariance estimates $\hat{R}_1$ and $\hat{R}_2$ are an average of M sample-pair covariance estimates $\hat{R}_{1i}$ and $\hat{R}_{2i}$. It can be seen from Eq. (7.6a) that the velocity estimate becomes ambiguous if the phase difference, $\arg \hat{R}_1 - \arg \hat{R}_2$, is outside the $-\pi, \pi$ interval. Thus, the "unam-

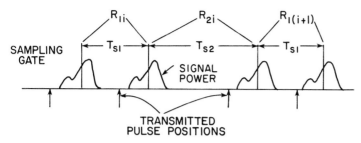

Fig. 7.10 A possible spacing of sampling gates for a staggered PRF system to obtain covariance $\hat{R}_1$, $\hat{R}_2$ estimates at two different lags.

biguous" velocity v_m for this staggered scheme is

$$v_m = \pm \frac{\lambda}{4(T_{s2} - T_{s1})}. \qquad (7.6b)$$

Equation (7.6b) shows that the smaller the difference between T_{s2} and T_{s1}, the larger is the unambiguous velocity. The difference cannot be made too small, because the errors in $\hat{v}$, due to statistical fluctuations in estimates $\hat{R}_1$ and $\hat{R}_2$, are inversely proportional to this difference (Zrnić and Mahapatra, 1985). These errors have been verified in tests on a meteorological research radar (Gray *et al.*, 1989).

The spectrum width can be estimated from the ratio of the magnitudes.

$$\hat{\sigma}_v = \frac{\lambda}{2\pi\sqrt{2(T_{s2}^2 - T_{s1}^2)}} \left[\ln \left| \frac{\hat{R}_1}{\hat{R}_2} \right| \right]^{1/2}. \qquad (7.7)$$

So we see that staggered PRT allows simple mean-velocity and spectrum-width estimation while extending the unambiguous velocity interval.

Before concluding, it is important to review the philosophy employed in selecting staggered or two-PRT parameters. We want v_{a1} and v_{a2} as small (T_{s1} and T_{s2} as large) as possible in order to obtain a large unambiguous range. One must bear in mind that the smallest value of v_{a2} (largest r_{a2}) is dictated by the requirement for coherency [relation (7.2c)]. Spectrum-width data (Fig. 10.11) suggest that for a 10-cm radar, v_{a2} probably cannot be much smaller than 16 m s^{-1} if the coherency condition for a large percentage (75%) of severe storm data is to be maintained. Next we need to decide on a value of T_{s1}, which we shall obtain from the desired unambiguous velocity Eq. (7.6b). An umambiguous velocity $v_m = \pm 48$ m s^{-1} should resolve all but the most extreme velocity aliases (Fig. 7.4). Thus for a 10-cm radar with the smaller unambiguous velocity $v_{a2} = 16$ m s^{-1}, we find $v_{a1} = 24$ m s^{-1}.

What have we gained with the staggered PRT? A single PRT radar having a Nyquist velocity v_a of 48 m s^{-1} has an unambiguous range given by Eq. (7.1), which for a 10-cm radar is about 78 km. A staggered PRT radar would have an unambiguous range r_{a1} [Eq. (3.40a)] equal to 156 km, which gives a fourfold increase in the unambiguous area over the nonstaggered PRT radar. Although this can be an important improvement, an issue is whether obscuration is significantly decreased. Studies with the simplified storm model (Doviak *et al.*, 1978) suggest that the obscuration probability of first-trip echoes is small if $r_a > 130$ km.

There is an additional advantage of the staggered PRT. Assume that there are no storms beyond $c(T_{s1} + T_{s2})/2$. Then an unwanted overlaid signal U is present in only one sample of the T_{s1} or T_{s2} spaced sample pair [e.g., 2, 3 in Fig. 7.11 is an unwanted overlaid signal on the covariance estimate $R(T_{s1})$ of the storm 2]. Thus U will not contribute coherently to the covariance estimate. To demonstrate this, consider the covariance estimate at lag T_{s1}.

$$\hat{R}(T_{s1}) = [U^*(t) + V^*(t)]V(t + T_{s1}). \tag{7.8a}$$

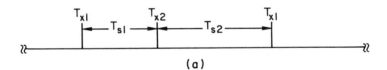

(a)

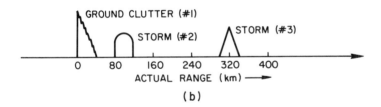

(b)

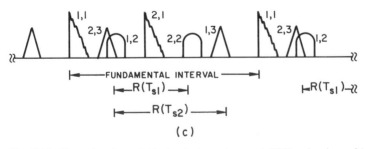

(c)

Fig. 7.11 Example of overlaid signals in a staggered PRT radar ($v_{a1} = 24$, $v_{a2} = 16$ m s^{-1}). A number pair (e.g., 1, 2) identifies the transmitted pulse and its return signal from a cluster of scatterers (i.e., storm). (a) Transmitted pulse sequence; (b) actual ranges to the clusters; (c) distribution of clusters as seen by the radar.

Its expected value is

$$E[\hat{R}(T_{s1})] = E\{[U^*(it) + V^*(it)]V[i(t + T_{s1})]\}$$
$$= E\{V^*(t)V(t + T_{s1})\} \qquad (7.8b)$$

because signal U is uncorrelated with V. Thus overlaid signal will not bias velocity estimates but will only increase the standard error of the covariance estimates if its power is larger than about one tenth of the desired signal power.

In summary, a staggered PRT technique increases the unambiguous range, causes overlaid echoes to be incoherent, and increases the unambiguous velocity to v_m so that velocity aliasing can be reduced significantly.

7.4.4 Interlaced Sampling

Even though Doppler signatures may be obscured by overlaid multiple-trip signals, observers should be given a velocity data field in which the range to a datum is unambiguous and displayed velocity values are credible (i.e., are not in error owing to multiple-trip overlaid signals). This can be accomplished by taking reflectivity samples during an interval T_2 sufficiently long to remove, for practical purposes, all overlaid signals and by having this sampling period interlaced with another whose PRT is short enough to allow coherent measurements for velocity estimates (Fig. 7.12). By interlacing the velocity estimation periods MT_1 with one for reflectivity (T_2), we can have nearly collocated resolution volumes for velocity and reflectivity measurements. Figure 7.12 shows one block of samples that contain $M = 3$ covariance estimates and one reflectivity estimate. To reduce the velocity and reflectivity estimate variance, we need to average covariance and reflectivity estimates from several (K) blocks.

To sample all n trip signals ($n = 2$ in Fig. 7.12) in one T_1 period, sampling should start in the interval nT_1 because the nth multiple-trip echo will not appear

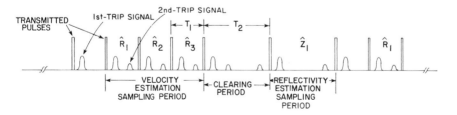

Fig. 7.12 Interlaced sampling technique, where $\hat{R}_1$, $\hat{R}_2$, $\hat{R}_3$ ... are covariance measurements (at equal lags) whose average is used to derive mean Doppler velocity estimates. We depict only first- and second-trip echoes and assume $T_2 = 2T_1$. The clearing period T_2 removes multiple-trip signals from reflectivity estimation in a contiguous T_2 interval.

until then. Interlaced sampling provides reflectivity data without range ambiguities .This allows determination, through comparison of signal powers at range locations separated by $cT_1/2$, of those velocity data that are significantly contaminated (obscured) by scrambled multiple-trip signals, and elimination of them from the display. Furthermore, we can assign correct ranges to the surviving valid velocity data. Such an interlaced sampling system has been in operation at NSSL ($M = 7$, $K = 8$, $T_2 = 4T_1$). Velocity fields displayed in real time are not range ambiguous (Color Plate 1b).

A variant of the interlaced sampling is implemented in the WSR-88D. This variant uses a batch type of PTR consisting of 41 to 80 short T_1 intervals ($0.77 \leq T_1 \leq 1$ ms) for velocity estimates followed by 6 to 12 long T_2 intervals ($1.5 \leq T_2 \leq 3.1$ ms) for reflectivity estimates. The batch processing is implemented at intermediate elevations (Table 7.1) where uniform PRT is not required for good ground-clutter cancellation, but where range-overlaid echoes continue to be troublesome. At the highest elevation angles, range ambiguities are not a problem, hence pulses are transmitted at uniform PRTs (Table 7.1).

The most persistent obscuration to plague the interlaced sampling radar is caused by ground-clutter echoes overlaid onto the second trip (i.e., ground clutter seen just beyond r_a) as well as ground clutter within the first trip. Ground clutter obscuration can be lessened with cancelers and by displacing (through changes in T_1) the second-trip ground clutter ring from the storm of interest; otherwise, at low elevation angles there could be a several-kilometer range interval wherein the second-trip signal is contaminated by clutter (Color Plates 1a and 1b).

Table 7.1
Volumetric Data Acquisition (WSR-88D)

Storm data acquisition mode

16 scans in 5 minutes

El. = 0.50°   2 scans, one at long PRT, (3.1 ms), one at short PRT (0.77 to 1 ms)

El. = 1.45°   2 scans, one at long PRT, one at short PRT

El. = 2.40 to 6.2°   BATCH TRANSMISSIONS, 5 scans

El. = 7.5 to 19.5°   UNIFORM PRT (0.77 to 1 ms), 7 scans

Clear air data acquisition mode
Long pulse ($\tau = 4.7$ μs)

El. = 0.5°, 1.5°, 2.5°   2 scans at each elevation angle; one scan with PRT = 3.1 ms, the other with PRT = 2.2 ms

El. = 3.5°, 4.5°   scans with PRT = 2.2 ms

Adaptive selection of the PRT can also be used to minimize range obscuration by undesirable storms and it need not be part of an interlaced scheme. It is very effective in situations where radar surveillance is localized to a specific area such as an airport (Crocker, 1988).

The interlaced sampling mode can accommodate a staggered PRT during the velocity estimation period to allow an increase in both r_a and v_a at the expense of an increased data acquisition time for a given velocity estimate accuracy.

A closely related method of increasing the range to which reflectivity can be resolved unambiguously is to transmit signals at two different frequencies ω_2, ω_1 (each at different PRTs) so that simultaneous reception is possible. A long PRT yields a large unambiguous range for reflectivity estimates, whereas a short PRT is used for velocity estimation. This technique and its signal processing are analogous to the interlaced sampling technique. Its advantage is a reduction of the acquisition time, and it also offers the possibility of better clutter-canceler design.

7.4.5 Correcting Aliased Velocities

Velocity aliases can usually be identified, because true velocity fields must be continuous whereas aliasing causes unrealistic gradients (discontinuities) in the measured Doppler field. But, reflectivity-free regions disrupt spatially continuous velocity measurements, and furthermore, naturally occurring large shears and poor radar resolution make it difficult to dealias velocities in all situations.

Nevertheless, there is a simple technique that works quite well and has been applied with real-time data. It requires only knowledge of the environmental wind v_e as a function of height. Storms are assumed to perturb the environmental flow, and if $|v_{max} - v_{min}| < 2v_a$ (where v_{max} and v_{min} are the true radial velocity maximum and minimum at that altitude) and wind perturbations are centered on v_e, then velocity aliases can, in principle, be completely resolved.

The estimate of the true Doppler velocity $\hat{v}_t$ is always given by

$$\hat{v}_t = \hat{v} + 2lv_a, \tag{7.9a}$$

where l is a positive or negative integer and $\hat{v}$ is the estimated mean Doppler velocity. Velocity ambiguities are resolved if l can be found for each $\hat{v}$. The parameter l can be estimated for each $\hat{v}$ at altitude h by substituting for $\hat{v}_t$ the radial component v_{er} of the environmental wind and solving for l.

$$\hat{l} = (v_{er} - \hat{v})/2v_a. \tag{7.9b}$$

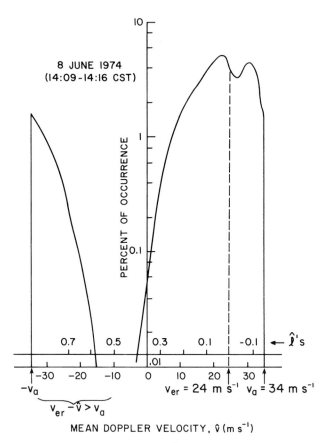

Fig. 7.13 Distribution of $\hat{v}$'s and $\hat{l}$'s from which decisions for dealiasing are made. Frequency of occurrence is given as a percent of velocities in 1 m s^{-1} intervals. All aliased velocity estimates between -34 and -10 m s^{-1} can be corrected.

The distribution of $\hat{l}$'s is a mirror image of the distribution of $\hat{v}$'s shifted by v_{er}. The $\hat{l}$'s are clustered near integers that would correctly dealias the v's when perturbations are small compared to $2v_a$. Whenever the true velocity distribution straddles v_a or its aliases, $\hat{l}$'s form into two clusters, each group near an integer. If all data in each cluster are assigned the integer nearest the group, the $\hat{v}$'s will be correctly dealiased. Figure 7.13 is a sample distribution of $\hat{v}$'s and $\hat{l}$'s for a tornadic thunderstorm. On this day the environmental wind, averaged from the surface to the tropopause, was $230°/30$ m s^{-1}, giving a 24-m s^{-1} radial component v_{er} of averaged environmental wind for this storm that had about a 15° bearing from the radar. Researchers have found that with $\lambda = 10$ cm and $r_a < 160$ km it is not necessary to search for $|l|$ larger than unity (Hennington, 1981). That is, rarely do true speeds exceed 70 m s^{-1}. Therefore, they simply make the following assignments:

$$l = \begin{cases} 1 & \text{if } \hat{l} > 0.5, \\ -1 & \text{if } \hat{l} \le -0.5, \\ 0 & \text{otherwise.} \end{cases} \tag{7.9c}$$

The technique we have described is easily implemented and works well whenever $2v_a$ is large (e.g., >50 m s^{-1}) compared to the width of the velocity distribution. It fails most often in regions of strong divergence (e.g., at storm tops) and circulation, where true velocity spreads can be as large as 100 m s^{-1} (Fig. 7.4 and Color Plate 2a). More sophisticated techniques have been developed to deal with such situations (Merrit, 1984, Borhen *et al.*, 1986, Bergen and Albers, 1988); these are discussed next.

The problem is solved with artificial intelligence methods which mimics an expert meteorologist whose success in velocity dealiasing is rooted in three fundamental principles: (a) spatial continuity of the velocity field; (b) temporal continuity of the velocity field; and (c) use of a crude but reliable model of the wind field. The first step is to identify areas of continuous velocities on surfaces of constant θ_e; then the discontinuities at the boundaries between areas are minimized by offsetting velocities (i.e., choosing a proper l). An independent estimate of the true velocity in some areas is needed initially to start the procedure; this can be obtained from environmental soundings or by inclusion of winds obtained with the velocity azimuth display (VAD, Section 9.3.3) technique. Inclusion of a monitor of the wind field to check for compliance with the model further improves dealiasing. Large computing resources are needed to fully implement this algorithm in real time, therefore a simpler approach that uses only local continuity constraints is implemented in the WSR-88D system (Eilts and Smith, 1990).

In the algorithms with local continuity constraints, velocity difference is checked first along range and then, if needed, between two adjacent radials. The algorithm is adaptive in that it uses simple, efficient checks first, and cascades to more elaborate and time-consuming procedures only if they are needed; therefore, efficiency is optimized, while sophistication and accuracy are maintained. Aliased and successfully dealiased velocity fields in Color Plates 1c and 1d illustrate the potency of this robust algorithm; 99% of aliased velocities in this example were dealiased using only range continuity.

7.5 Methods to Decrease the Acquisition Time

The averaging time for reflectivity and velocity estimation is dictated by the desired accuracy of the estimates. Considerable savings in averaging time can be achieved by judiciously choosing transmitting schemes that increase the equivalent number of independent samples. This can be done either by transmitting several frequencies (simultaneously or consecutively) or by using broadband noiselike signals.

7.5.1 Frequency Diversity

Signals with nonoverlapping spectra are uncorrelated, and such signals, when transmitted, generate uncorrelated and thus independent weather signals. Because an rf pulse with carrier frequency f has its energy concentrated mainly in the band τ^{-1} centered on f, a simple way to create uncorrelated weather signals consists of offsetting the frequencies by more than the reciprocal of the pulse width τ in successive transmitted pulses. A train of M such pulses, each with different frequency (Fig. 7.14a) will produce M independent signals, which, after averaging, will give improved reflectivity estimates for the same dwell time.

Velocity estimates can be obtained if pairs of pulses are transmitted at the same carrier frequency (Fig. 7.14b). Autocovariance-type processing can be used to retrieve the mean velocity or spectrum width, because Fourier analysis is not suitable. A disadvantage of this scheme is that, for all practical purposes, it does not allow ground-clutter canceling. Note that, if implemented as in Fig. 7.14b, measurement of velocities in the second trip involves a more complicated receiver because, in processing the weather signals of f_6 from scatterers in the first trip, one must also process the f_5 signals from scatterers in the second trip. This necessitates two receiving channels following the mixer (more if there are more trip regions to be examined simultaneously). Filters in the two receivers can eliminate the overlay of echoes from the pair f_5 on the pair f_6, etc., but the second-trip echo from the first pulse of f_5 can be overlaid on the first trip from the second pulse of f_5. The effect is the same as with spaced pairs; that is, the overlaid echo behaves like noise and thus does not bias the velocity or spectrum width estimates. An alternative frequency diversity scheme consists of a pulse that is divided into subpulses having different frequencies. A pair of such pulses contains a number of independent samples for autocovariance processing equal to the number of frequencies (Hildebrand and Moore, 1990; Gerardin *et al.*, 1991).

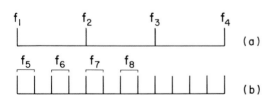

Fig. 7.14 Frequency diversity for (a) reflectivity and (b) velocity measurements. Note that (a) and (b) can be interlaced or transmitted simultaneously.

7.5.2 Random Signal Transmission

A noiselike signal with bandwidth B, transmitted in a pulse of duration τ, consists of a continuum of frequencies, and weather signal samples are essentially uncorrelated along range time if taken at spacings larger than B^{-1} (Section 4.6). If the bandwidth B is much larger than τ^{-1}, integration along range time τ_s for the duration τ will reduce significantly the uncertainties in power estimates without compromising range resolution $c\tau/2$. For example, with $B = 40$ MHz the decorrelation range is about 3.75 m, and if $\tau = 1\ \mu$s ($c\tau/2 = 150$ m), there are 40 independent range samples available for averaging (Section 6.3.1.2). The random signal radar (Fig. 7.15) takes advantage of this concept. It uses a noise source transmitter (a resistor and amplifier with bandwidth B) and a wideband receiver followed by a square-law detector and integrator.

We now briefly demonstrate the reduction of variance achievable with the random-signal radar and compare it with the variance obtained with the pulse-modulated single-frequency radar. The variance of the power estimate $\hat{P}_r$ is the expected value of the square of the second term in Eq. (4.2).

$$\text{var}_\rho(P_r) = \frac{1}{4} \sum_{i \neq k} \sum_{m \neq n} E(A_i A_k^* A_m A_n^*) E(W_i W_k^* W_m W_n^*)$$
$$\times E\{\exp[j4\pi(r_i - r_k + r_m - r_n)/\lambda]\} \tag{7.10}$$

$$= \frac{1}{4} \sum_{i \neq k} E(|A_i^2| \|A_k^2|) E(|W_i W_k|^2).$$

Each summation in Eq. (7.10) is over two indices, and the quadruple sum reduces to a double sum because the expected value of the third term is zero except when $i = n$ and $k = m$. The subscript ρ indicates that the variance is for the random-signal radar, and the weighting function W consists of amplitude and phase fluctuations of the transmitted signal (created by the noise source). If the transmitted pulse contains a single frequency (i.e., there is no amplitude or phase modulation during τ) and receiver bandwidth $B_6 \gg \tau^{-1}$, $|W_i|^2 = |W_k|^2 = 1$ for all i, k in the range interval $c\tau/2$. Furthermore, A_i and A_k are uncorrelated if $i \neq k$, so $E(|A_i|^2|A_k|^2) = E(|A_i|^2)E(|A_k|^2)$. Because we deal with a large number of scatterers, the sum of diagonal terms $i = k$ is considerably smaller than the rest. Then the variance of the echo powers from single-frequency transmitted pulses can be approximated by

$$\text{var}(P_r) = \frac{1}{4}\left[\sum_i E(|A_i|^2)\right]^2, \tag{7.11}$$

which is the square of the expected value of the first term in Eq. (4.2). Thus the variance of P_r is equal to the square of its mean.

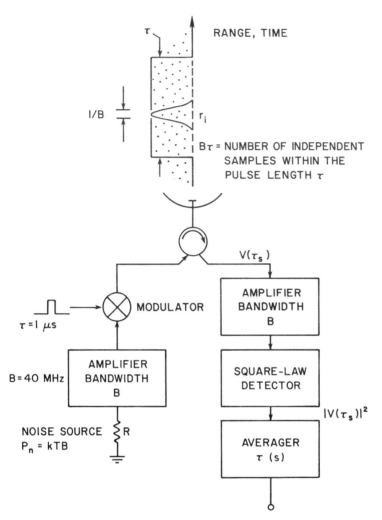

Fig. 7.15 Block diagram of the random signal radar for reflectivity estimation. The noise source is a resistor at a temperature T. (From Krehbiel and Brook, 1979; © 1979 IEEE).

The ratio of Eq. (7.10) to Eq. (7.11) describes the decrease in estimate variance that can be achieved with a random-signal radar over a radar transmitting pulsed sinusoids. A quantitative comparison is made easier if we assume that the scattering medium is homogeneous. Then $E(|A_i|^2|A_k|^2)$ is constant, so the ratio of variances becomes

$$\mathrm{var}_\rho(P_\mathrm{r})/\mathrm{var}(P_\mathrm{r}) = \sum_{i \neq k} |W_i W_k|^2/N_\mathrm{s}^2 = \sum_{l=-N_\mathrm{s}}^{N_\mathrm{s}} \sum_{i=1}^{N_\mathrm{s}-|l|} |W_i W_{i+1}|^2/N_\mathrm{s}^2, \quad (7.12)$$

where N_s is the number of scatterers, and the reasoning leading to Eq. (5.24) was used to obtain the inner sum in Eq. (7.12). N_s is proportional to the pulse length τ, and the term $|W_i W_k|^2$ is the correlation magnitude squared, $|R_\rho(\delta\tau_s)|^2$, at $\delta\tau_s = 2(r_k - r_i)/c = 2r_l/c$. Evaluation of the inner sum in Eq. (7.12) leads to the following formula:

$$\frac{\mathrm{var}_\rho(P_r)}{\mathrm{var}(P_r)} = \sum_{l=-N_s}^{N_s} \frac{(N_s - |l|)|R_\rho(r_l)|^2}{N_s^2} \approx \frac{1}{\tau} \int_{-\tau}^{\tau} \left(1 - \frac{|\delta\tau_s|}{\tau}\right) |R_\rho(\delta\tau_s)|^2 d(\delta\tau_s). \quad (7.13a)$$

The approximation by the integral in Eq. (7.13a) can be safely made because the distance between scatterers is small compared to the spatial pulse length $c\tau$.

As an example, consider a white-noise signal of bandwidth $B \gg \tau^{-1}$ for which the power spectrum is

$$S_\rho(f) = \begin{cases} B^{-1} & \text{for } |f| \le B/2, \\ 0 & \text{otherwise.} \end{cases} \quad (7.13b)$$

$R_\rho(\delta\tau_s)$ differs significantly from zero only for $\delta\tau_s < B^{-1}$. Hence the term $|\delta\tau_s|/\tau$ in Eq. (7.13a) can be neglected, and the integral is approximately

$$\int_{-\tau}^{\tau} |R_\rho(\delta\tau_s)|^2 d(\delta\tau_s) = \int_{-B/2}^{B/2} |S_\rho(f)|^2 \, df, \quad (7.14a)$$

where equality follows from Parseval's theorem (Gabel and Roberts, 1973). With this the ratio of variances simplifies to

$$\mathrm{var}_\rho(P_r)/\mathrm{var}(P_r) = 1/B\tau, \quad (7.14b)$$

and therefore the ratio of the mean value $\bar{P}_r$ to rms fluctuations increases by $\sqrt{B\tau}$. This is significant because it means that a random-signal radar with a bandwidth 10 times larger than τ^{-1} can achieve the same precision of reflectivity estimates using one transmitted pulse as can a monochromatic radar from 10 independent weather signals. (Remember also that the single-frequency radar may have to transmit many more than 10 pulses to obtain 10 independent weather signals.)

The decrease in acquisition time is achieved at the expense of radar sensitivity if the transmitter average power is not changed. Because the random-signal radar has statistically fluctuating transmitter power, one must either make a measurement of the power from pulse to pulse or else τ must be long enough that the average power over τ seconds estimates well the mean power of the noise tube. This is needed to avoid errors in reflectivity estimates. A detailed description of such a radar has been given by Krehbiel and Brook (1979), and a comparison of power fluctuations using a pulse-modulated single-frequency radar and the noise radar, for which power is averaged over a $c\tau/2$ interval, is shown in Fig. 7.16. Note large pulse-to-pulse fluctuations in Fig. 16a and reduced fluctuations in Fig. 16b.

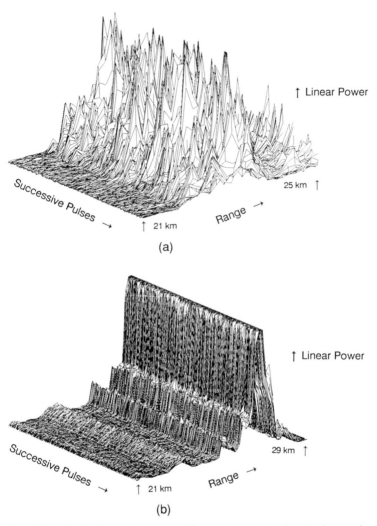

↑ Linear Power

25 km ↑

Successive Pulses →

↑ 21 km

Range →

(a)

↑ Linear Power

29 km ↑

Successive Pulses →

↑ 21 km

Range →

(b)

Fig. 7.16 (a) Weather signal power at the receiver output of a pulsed monochromatic radar (pulse length is 1 μs; 200 successive pulses). (b) Weather signal power for a pulsed wideband noise radar (500 successive pulses). Power is averaged over 1 μs and the bandwidth of the transmitted noise is 300 MHz. (From Krehbiel *et al.*, 1991.)

7.6 Pulse Compression

Pulsed-radar transmitters have limited peak and average power and thus the detection of scatterers is also limited. As discussed in Chapters 2 and 3, the peak transmitter power P_t of a pulsed-Doppler radar is the average power over that cycle of the rf that gives maximum value. The average transmitter power P_{av} is an average of the power over the pulse repetition period. Consider rectangular pulses for which P_t is constant over the pulse length τ. The average power P_{av} then is

$$P_{av} = P_t \tau / T_s, \tag{7.15}$$

and the ratio $\tau/T_s = d$ is referred to as the duty factor. High peak power transmitters (WSR-88D, Table 3.1) have relatively small duty factors (e.g., 0.001), whereas transmitters with lower peak powers may have longer duty factors. Because the minimum detectable cross section is proportional to the average power, one way to improve detection (with low P_t) is by increasing the pulse width [Eq. (4.36)].

Average power cannot be increased by increasing τ without compromising resolution or by decreasing T_s because this may cause additional multiple-trip echoes. But, proper encoding of transmitted signals can be used to increase the average power and improve detection while neither increasing peak power or PRF nor degrading range resolution. Of course, other compromises must be made, and the purpose of this section is to discuss the trade-offs between transmitted power and resolution (i.e., range-weighting function).

To increase the SNR without exceeding the peak permissible power and to improve the range resolution for a given pulse width, designers of radars have invented pulse compression. Here we shall describe this concept by way of an example and discuss its implications for weather radars. To begin with, consider an rf pulse of length τ_p that consists of segments τ seconds long. Suppose that the rf phase of each segment can be independently controlled. A method often used is to change the phases by 180° in a prescribed manner. In Fig. 7.17a such a phase-change sequence is denoted $c(n\tau)$, where $+1$ signifies a certain transmitter phase and -1 a 180° phase shift. A replica of the transmitted signal is stored, and a cross correlation with the returned signal is performed as part of the detection process. The output of a correlation receiver in Fig. 7.17b is for an input signal from a point scatterer. We note that the signal correlation is concentrated near the peak, which in this example is four times larger than the sidelobes. It is

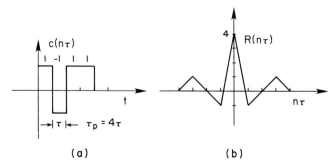

Fig. 7.17   (a) A four-element two-phase code and (b) its autocorrelation function.

important to realize that the output peak $R(0)$ is not the power of the incoming signal, because the correlation was performed with a stored replica of the transmitted signal rather than with the returned signal. Incidentally, if a pulse of length τ with the same peak power is transmitted and a matched correlator is used, we would obtain exactly a symmetrical triangular response having amplitude 1 and base width 2τ. Instead, with a pulse four times longer we have obtained a return that is four times larger at $\tau = 0$, and the resolution (which we define as the width of the main peak) has remained as good as with the short pulse. In this example, the compression ratio $(\tau_p/\tau = \tau_p B)$, also called the time–bandwidth product, characterizes the gain and improvement in resolution achievable with pulse compression. Pulse compression can be applied to increase the number of independent samples by averaging the high-resolution estimates. This may be advantageous in rapid-scan radars contemplated for airborne applications (Strauch, 1988; Gerardin *et al.*, 1991).

The effective increase in signal power over the single pulse of length τ is proportional to the code length squared, or M^2 (the signal voltage increases as M), which for the example of Fig. 7.17 is 16. Although this is an impressive gain, what really matters is the signal-to-noise ratio. The total noise power is

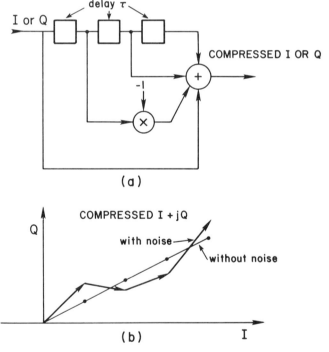

Fig. 7.18 (a) A correlator (decoder) for the code in Fig. 7.17. Each delay circuit introduces a delay of τ. (b) With the indicated delays and a reversal of phase of the second element, an in-phase addition of the complex I,Q signal occurs.

M times larger because it is obtained by summing that many uncorrelated noise samples. Thus the SNR is improved by a factor of M.

A schematic of the correlator for the code of Fig. 7.17 consists of delay circuits, a sign reversal, and an adder (Fig. 7.18a). The device operates in a pipeline mode so that at every τ interval it generates an output that corresponds to an input that occurred 4τ earlier. Two correlators are needed—one for I and the other for Q. Possible output phasors with noise and without noise are plotted in Fig. 7.18b. The price paid for the increased SNR and good resolution is a more complicated system and undesirable range sidelobes.

Why are the sidelobes undesirable? If we recall the explanation of the range-weighting function (Section 4.4.2), we immediately recognize that the weighting function $|W(r)|^2$ is precisely the output of the correlator $|R(n\tau)|^2$, which here has the role of a receiver filter. Thus, in the presence of strong reflectivity gradients, weak scatterers weighted with the main peak can be completely masked by powerful scatterers at the sidelobes. For this reason and because storms generally provide ample return power, most ground-based weather radars do not use pulse compression. The picture changes, however, if one considers airborne and satellite-borne transmitters, which have severe peak power limitations.

7.7 Artifacts

Imperfection in the receiver chain and signal processing affect, to various degrees, the output signal spectrum. First, let us examine receiver nonlinearities that distort the in-phase and quadrature components for the case in which receiver noise N can be neglected. Assume that a nonlinear transfer function g distorts both components equally so that the output signal is

$$V_0(t) = g[I(t)] + jg[Q(t)]. \tag{7.18}$$

If $g(V)$ is an odd function, then $V_0(t)$ has an autocorrelation (Davenport and Root, 1958, p. 292)

$$R_{V0}(\tau) = 2 \sum_{k=1}^{\infty} \frac{h_{0k}^2}{k!} \{\mathrm{Re}^k[\tfrac{1}{2}R(\tau)] + j\,\mathrm{Im}^k[\tfrac{1}{2}R(\tau)]\}. \tag{7.19}$$

$R(\tau)$ is the correlation of the input signal, and k takes odd values. The coefficients h_{0k} are

$$h_{0k} = \frac{1}{\pi j} \int_c f_+(s) \exp(\tfrac{1}{2}\sigma^2 s^2) s^k \, ds. \tag{7.20}$$

where $f_+(s)$ is the Laplace transform of $g(x)$ for $x > 0$, σ^2 is the variance of I or Q, and the integration contour is along the $s = j\omega$ axis. Because $R_{v0}(\tau)$ is odd, the power spectrum of the output [i.e., the Fourier transform of $R_{v0}(\tau)$] will contain odd harmonics. Furthermore, it is less obvious but nevertheless true that the harmonics generated by the odd transfer function alternate between positive and negative frequencies so that at $k = 5, 9, \ldots$ they have one sign, at $k = 3, 7, \ldots$ the other. With sampled signals the higher-frequency components introduced by nonlinearity are aliased into the unambiguous interval unless some kind of filtering is used. From Eqs. (7.19) and (7.20) it may be seen that the power in the undistorted portion of the output spectrum is $h_{01}^2 R(0)$. The noise power (distortion) N_g generated by the nonlinearity g can be calculated from the total power at the output.

$$N_g = \frac{2}{\sqrt{2\pi}\sigma} \int_{-\infty}^{\infty} g^2(I)\exp\left(-\frac{I^2}{2\sigma^2}\right) dI - 2h_{01}^2\sigma^2. \qquad (7.21)$$

The exponential is the probability density of the input I component [Eq. (4.5)], so the first term is the expected value of the output power. In Eq. (7.21) only the I component is used because its statistics equal those of Q. The factor of 2 accounts for the power in both components, and the exponent is due to the Gaussian distribution of I. The ratio of the signal S to distortion noise N_g at the output of the nonlinear device can therefore be defined as

$$S/N_g = h_{01}^2\sigma^2/\{E[g^2(I)] - h_{01}^2\sigma^2\}. \qquad (7.22)$$

7.7.1 Quantization and Saturation Noises

Saturation may occur in the rf or IF portion of the receiver, in the video amplifiers, or in the analog-to-digital (A/D) converter. Saturation and quantization noise of the A/D converter will now be analyzed. In the following discussion $b\sigma$ is the A/D converter's clipping (i.e., saturation) level (Fig. 7.19), n is the number of bits, including sign, $q = 2b\sigma/(2^n - 1)$ is the quantization step size, and 2^n is the number of converter levels. Because the A/D converter has an odd transfer function, the preceding method is suitable to calculate the coefficients h_{0k} and thus the signal-to-noise ratio $S/N_{A/D}$ at the A/D output (Zrnić, 1975c). The plot (Fig. 7.20) shows that the $S/N_{A/D}$ is maximized at a unique value of b. At larger clipping levels, quantization noise is dominant whereas at low levels it is the distortion noise produced by saturation. Normally an automatic gain control circuit controls the gain setting prior to the A/D converter (Fig. 6.1). If it is set so that $b \approx 4$ (with $n = 10$ bits), an optimum results (Fig. 7.20). It is apparent that at least 10 bits are needed to achieve 50 dB of $S/N_{A/D}$.

If a narrowband Gaussian signal is passed through an A/D converter, odd "harmonics" are generated that have alternately positive and negative frequencies. Owing to sampling, the harmonics alias into the unambiguous interval. In

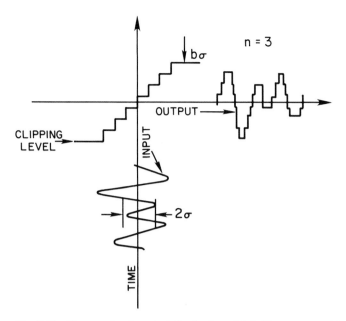

Fig. 7.19 The transfer characteristic of a three-bit A/D converter, with the input signal and its rms value σ indicated.

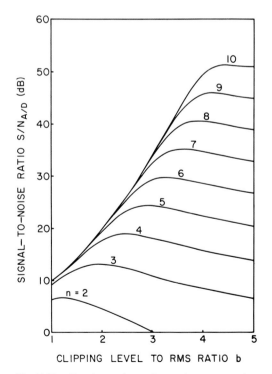

Fig. 7.20 Signal-to-noise ratio at the output of an A/D converter. The number of bits n is indicated.

the case of saturation by receiver nonlinearity (e.g., in the rf or IF amplifiers), this aliasing can be reduced with filtering prior to A/D conversion (Zeoli, 1971). But in the A/D converter, aliasing happens during conversion, and hence the saturation effects cannot be filtered.

Figure 7.21 illustrates how saturation of the A/D converter affects a narrow-band Gaussian spectrum. Figure 7.21a is a periodogram obtained from a weighted (von Hann) weather signal. The mean velocity of the scatterers is 11 m s^{-1}, the spectrum width is 2.6 m s^{-1}, and no significant distortions are present (i.e., harmonics have powers less than 30 dB below the peak). Next, this time series was clipped to cause saturation at the level $b = 1$. The periodogram presented in Fig. 7.21b contains odd alternating harmonics: The third (i.e., the lowest harmonic generated by saturation) is at -33 m s^{-1}, the fifth is at 55 m s^{-1}, and the seventh is buried in the noise.

7.7.2   Amplitude and Phase Imbalances

Imbalances in the gain and nonquadrature phase shift between the I and Q signal channels create an image spectrum symmetric to the actual spectrum. This effect can be seen by considering the Fourier expansion of an arbitrary signal. If the channels are perfectly matched, each term from the Fourier series expansion contributes to the time series an amount

$$A_i e^{j(\omega_i t + \theta_i)}, \tag{7.23a}$$

where A_i is the magnitude and θ_i the phase of the ith coefficient. An imbalance $K = G_I/G_Q$ in the gains G_I, G_Q of the I, Q channels, and a differential phase shift Δ in the Q channel gives

$$A_i K \cos(\omega_i t + \theta_i) + jA_i \sin(\omega_i t + \theta_i \pm \Delta)$$
$$= \tfrac{1}{2} A_i (K + e^{\pm j\Delta}) e^{j(\omega_i t + \theta_i)} + \tfrac{1}{2} A_i (K - e^{\mp j\Delta}) e^{-j(\omega_i t + \theta_i)}. \tag{7.23b}$$

The first term on the right side of Eq. (7.23b) at frequency ω_i is Eq (7.23a) modified in amplitude and phase, while the second term at $-\omega_i$ is the image.

A phase imbalance of 60° is responsible for the image spectrum in Fig. 7.21c. The image peak at -30 m s^{-1} is 5 dB below the signal spectrum peak. A similar result can be obtained by imbalancing the amplitudes (Doviak and Zrnić, 1984a).

Image suppression L (Fig. 7.22), in decibels, is defined as the power ratio of the two terms in Eq. (7.23b).

$$L = 10 \log |(K + e^{\pm j\Delta})/(K - e^{\mp j\Delta})|^2. \tag{7.24}$$

Usually the synchronous detector (Fig. 6.1) introduces imbalances both in phase and amplitude, whereas the video amplifiers may have different gains but normally do not introduce differential phase shifts. To maintain 40 dB of image

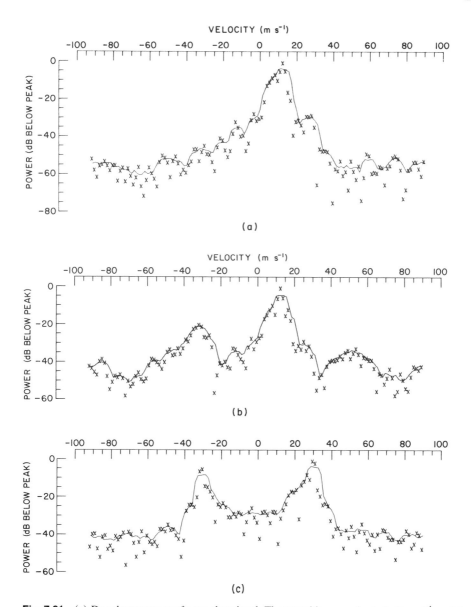

Fig. 7.21 (a) Doppler spectrum of a weather signal. The unambiguous velocity is 91 m s⁻¹, and the time series (von Hann weighted) contains 128 samples; $n = 10$, $b \approx 4$, SNR = 39. (b) The time series that produced (a) is clipped, von Hann weighted, and then Fourier analyzed; $n = 8$, $b = 1$, SNR = 26. (c) Imbalance of 60° in phase created this spectrum, but $K = 1$. This imbalance is for spectral components at 30 m s⁻¹ and was accomplished by shifting the Q component of the signal [spectrum shown in (a)] one T_s interval with respect to I.

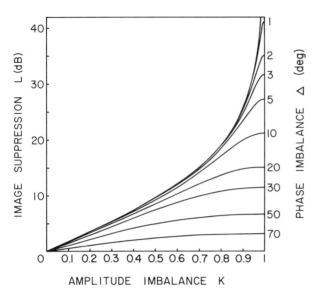

Fig. 7.22 Image suppression L versus amplitude and phase imbalance.

suppression, the amplitude imbalance must be less than 2% and phase imbalance less than 1.2° (Fig. 7.22).

7.7.3 Phase Jitter

The relative phase between transmitted pulses, as well as those phase shifts between echoes sequentially traversing the receiver, must be kept stable (coherent, fixed, or known) to make precise Doppler measurements. We must accept the phase jitter produced by hydrometeor motion or propagation path changes over which we have no control, but additional phase jitter may originate in the transmitter chain, the local oscillator, or in the COHO (Fig. 6.1). This phase jitter also causes phase deviations in the weather signal. For example, ψ_t in Eq. (3.26) fluctuates in an unknown and, most probably, random manner. Aside from phase shifts produced by hydrometeor velocity (i.e., by changes in range) and drop shape change (i.e., ψ_s), the phase deviation at the mth sample is $\psi'_m = \psi_m - \psi_0$ where ψ_0 is the mean phase-shift generated in the radar system. Assuming a joint probability density $p(\psi'_m, \psi'_{m+k})$ of composite phases at times mT_s, $(m + k)T_s$ the following equation for the autocorrelation $R_j(kT_s)$ of echoes with radar-induced jitter is obtained:

$$R_j(kT_s) = E[V^*(mT_s)e^{-j\psi'_m}V(mT_s + kT_s)e^{j\psi'_{m+k}}]$$
$$= R(kT_s)E[e^{j(\psi'_{m+k} - \psi'_m)}]$$
$$= R(kT_s)\int_{-\pi}^{\pi}\int_{-\pi}^{\pi} p(\psi'_m, \psi'_{m+k})e^{j(\psi'_{m+k} - \psi'_m)}\, d\psi'_m\, d\psi'_{m+k}. \quad (7.25)$$

The integrals multiplying the signal autocorrelation $R(kT_s)$ are Fourier transforms of the jitter probability density function. For example, this probability density can be the normalized power spectrum of the jittery transmitted signal. Typically, the power spectral density, for most transmitters, has a narrow peak on a pedestal composed of harmonics. In this case, the peak's width determines the frequency (velocity) resolution, and the pedestal's height reduces proportionately the dynamic range over which the signal spectrum can be observed. Only short-term phase instabilities (with variations that occur during the dwell time) affect the spectral moment estimates. Phase changes due to slow frequency drifts do not significantly broaden the spectrum.

If the phase deviations ψ'_m are independent from PRT to PRT (as is the case for magnetron transmitters), the double integral in Eq. (7.25) reduces to a square of a single integral. Furthermore, assuming that ψ'_m has a Gaussian probability density with a standard deviation σ_ψ, integration of Eq. (7.25) produces

$$R_j(kT_s) = R(kT_s)\exp(-\sigma_\psi^2) \qquad (7.26)$$

valid for $k \neq 0$. In this instance it is instructive to determine the signal-to-jitter-noise ratio (Passarelli and Zrnić, 1989). Consider a pure sinusoid with unit amplitude ($|R(kT_s)| = 1$) and note that jitter has reduced the coherent part to $\exp(-\sigma_\psi^2)$. Therefore the signal-to-jitter ratio becomes

$$\mathrm{SNR}_j = \exp(-\sigma_\psi^2)/[1 - \exp(-\sigma_\psi^2)] \qquad (7.27)$$

Measurements on a magnetron transmitter and simulations (Passarelli and Zrnić, 1989) verify Eq. (7.27).

7.8 Effective Pattern of a Scanning Antenna

As explained in Chapter 6, a number of signal samples need to be processed to reduce the uncertainty in estimates of reflectivity, velocity, and spectrum width. When the antenna is stationary, the resolution in the cross beam dimensions is dictated by the antenna beamwidth. But antenna motion (usually azimuthal rotation) combined with pulse-to-pulse processing creates an effective broadened beamwidth. In the following pages the *effective* antenna pattern is investigated, and a design procedure is presented for selecting the rotation rate and the number of samples processed to achieve a desired azimuthal resolution.

The effective antenna pattern will be derived for the output of a square-law detector, but the results also apply to derived products such as spectral moments calculated from the power spectrum or from the autocorrelation function, Eq. (6.18). For the sake of clarity, we consider the one-dimensional problem and assume time integration (the summation over M) to be continuous. This is a good approximation if the pulse repetition period–antenna rotation rate product ($T_s\alpha$) is much less than the antenna beamwidth. (It should be understood that time-continuous integration sums echoes from a single range bin or resolution volume that changes its azimuthal position in space as the antenna rotates.)

We shall base our development on the procedure followed in Appendix C.2. Consider the antenna positioned at $\theta_e = 0$, ϕ_0 and viewing a distribution of scatterers. Furthermore, assume that the power $2\sigma_\Omega^2(\phi)$ returned from a unit azimuthal angle at a delay τ_s is a function of azimuth and that the antenna pattern function is product separable, $f(\theta, \phi) = f(\theta)f(\phi)$. The equation relating the pattern and the expected echo power $\bar{P}(\phi_0)$ analogous to (C.13) becomes

$$\bar{P}(\phi_0) = I \int_{-\pi}^{+\pi} 2\sigma_\Omega^2(\phi) f^4(\phi - \phi_0)\, d\phi, \tag{7.28}$$

where the integral $I = \int f^4(\theta)\, d\theta$ is a constant that we shall ignore. $\bar{P}(\phi_0)$ is available at the output of the detector (Fig. 7.23).

Let us assume that the antenna turns at a uniform rotation rate α, so that the azimuth position of the beam axis is $\phi_0 = \alpha t$. Hence the expected power $\bar{P}(\phi_0)$ is time dependent if the reflectivity is azimuthally nonuniform. Suppose that an integrator with impulse response $h(t)$ acts on the power samples $\bar{P}(\phi_0)$. Strictly speaking, the integrator is acting on instantaneous power samples that are statistically fluctuating [such as that given by Eq. (C.11)], not on the expected time-continuous function $\bar{P}_0(\phi_0)$. In this discussion, however, we are interested in the mean power $\bar{P}(\phi_0)$ and need to have the expected value after the integrator. Integration and expectation are commutative, and therefore the expected power $\bar{P}(\phi_0)$ at the integrator output is obtained by convolving Eq. (7.28) with the integrator's impulse response $h(\tau)$.

$$\bar{P}_0(t) \sim \int_{-\infty}^{t} h(t - \tau) \int_{-\infty}^{\infty} 2\sigma_\Omega^2(\phi) f^4(\phi - \alpha\tau)\, d\phi\, d\tau. \tag{7.29}$$

Note that the infinite limits of integration in azimuth simplify calculations and can be safely used if the two-way pattern $f^4(\phi)$ has significant values over an angular span, very small compared to 2π. Interchanging the order of integration produces the following formula:

$$\bar{P}_0(t) = \int_{-\infty}^{+\infty} 2\sigma_\Omega^2(\phi) \left[\int_{-\infty}^{t} h(t - \tau) f^4(\phi - \alpha\tau)\, d\tau \right] d\phi. \tag{7.30}$$

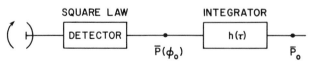

Fig. 7.23 Schematic of the power estimation that affects the azimuth position and width of the effective radiation pattern.

Because Eq. (7.30) is similar in form to Eq. (7.28), we are prompted to define a two-way effective antenna pattern,

$$f_a^4(\phi - \phi_0) = \int_{-\infty}^{t=\phi_0/\alpha} h(t - \tau) f^4(\phi - \alpha\tau) \, d\tau, \qquad (7.31)$$

for the scanning antenna.

It can be shown that $f_a(\phi - \phi_0) \to f(\phi - \phi_0)$ if $\alpha \to 0$. The equivalence in the mean signifies that a motionless antenna with pattern f_a^4 sees, on the average, the same reflectivity factor as an antenna that has a pattern f^4 but scans at a rate α. We caution the reader that the maximum of $f_a(\phi - \phi_0)$ does not occur necessarily at $\phi = \phi_0$, as we shall demonstrate by the example to be given shortly. A discussion concerning the effective pattern for logarithmic or linear detectors is provided by Zrnić and Doviak (1976).

Two parameters of interest for the effective pattern are (1) the displacement of the resolution volume center from the antenna axis position at the end of integration, and (2) the effective beamwidth. Let us examine the finite-time block integration because it can be used for reflectivity estimation and because both the autocovariance and spectral processing employ it. Its impulse response is given by

$$h(\tau) = \begin{cases} 1/MT_s & \text{for } 0 \leq \tau \leq MT_s, \\ 0 & \text{otherwise.} \end{cases} \qquad (7.32)$$

Some thought (Fig. 7.24) reveals that the maximum of Eq. (7.30), i.e., the resolution volume center, occurs if Eq. (7.32) is centered with respect to f^4 in the integral (7.31). This means that the effective pattern maximum occurs at

$$\phi_a(\text{max}) = \phi_0 - \tfrac{1}{2}\alpha MT_s \qquad (7.33)$$

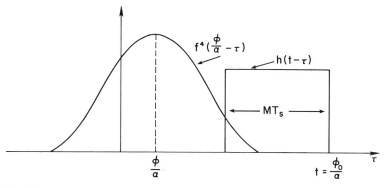

Fig. 7.24 The evaluation of the integral in Eq. (7.31) leading to the effective pattern. For symmetric patterns, the integration is convolution, and the effective pattern maximum occurs when the two curves are centered with respect to each other.

Equation (7.33) shows that the azimuth of scatterers that contribute most (assuming spatially uniform reflectivity) to the integrated output lags behind the beam position at the time the integration is complete.

The one-way half-power effective beamwidth ϕ_a is found by solving for the one-quarter-power points of the two-way effective pattern, Eq. (7.31). Assuming a Gaussian radiation pattern [Eq. (C.16)] and substituting it and Eq. (7.32) into Eq. (7.31), we obtain

$$\text{erf}(2\sqrt{\ln 4}\,\phi/\theta_1) - \text{erf}[2\sqrt{\ln 4}(\phi - \alpha MT_s)/\theta_1] - \tfrac{1}{2}\,\text{erf}(\sqrt{\ln 4}\,\alpha MT_s/\theta_1) = 0,$$

(7.34)

which has two solutions ϕ_1 and ϕ_2 that determine the effective pattern width $\phi_a = \phi_2 - \phi_1$.

A graph of the normalized effective beamwidth versus the product of rotation rate and integration time (Fig. 7.25) demonstrates that there is appreciable increase in ϕ_a for $\alpha MT_s/\theta_1 > 1$. We can use Fig. 7.25 to advantage in choosing values for radar data acquisition parameters (α, M, etc.). For example, given θ_1, the desired resolution ϕ_a determines the ordinate value in Fig. 7.25 and hence a

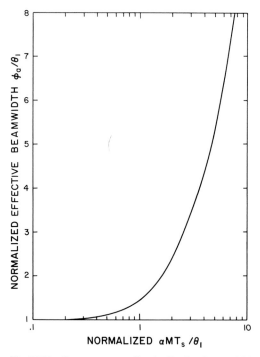

Fig. 7.25 One-way normalized effective beamwidth versus the normalized rotation rate for an assumed Gaussian antenna pattern with a one-way 3-dB beamwidth equal to θ_1.

value for $\alpha M T_s / \theta_1$. Now for a chosen T_s (which is determined by echo coherency and the desired unambiguous range and velocity), the product αM is found. Note here that M is the number of pulses in a contiguous equispaced train. If the transmission is more complicated (such as interlaced pulses), $M T_s$ needs to be replaced with the appropriate dwell time. The desired accuracy of spectral moments dictates M, which then leads to the rotation rate α. If this rate is too slow, either the accuracy requirement must be relaxed or a different signal design and processing scheme should be attempted.

7.9 Antenna Sidelobes

Scatterers seen though antenna sidelobes are detrimental because they can bias the reflectivity, velocity, and spectrum width estimates. Sometimes they may totally obscure weather echoes. Ground clutter is an example of echoes that usually enter the receiver through sidelobes, but this echo power can be reduced (Section 7.10) because its mean velocity is always zero and its spectrum width is usually small (i.e., ≤ 1 m s^{-1}). This is not the case with moving objects such as aircraft, birds, and automobiles. Also, in the presence of intense precipitation cores, the radar measurements several degrees away from cores may be contaminated with echoes received through sidelobes.

The mainlobe shape and width, as well as the near sidelobe levels, are determined by the wavelength, the antenna reflector size, and the tapering of its illumination. More distant sidelobes are caused by imperfection in the antenna and blockage of radiation by the feed horn, its supports (Fig. 3.3), radome, and so on.

Figure 7.26 shows a portion of a measured antenna gain pattern extending in azimuth from $-16.6°$ to $+13.6°$ and in elevation angle from $-11.6°$ to $0.6°$ with

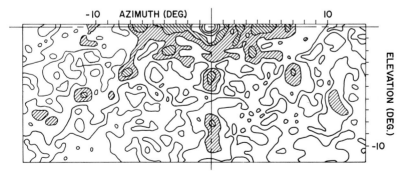

Fig. 7.26 Antenna gain pattern (two-way) of NSSL's Norman Doppler radar. The data were smoothed using a Shuman (1957) nine-point formula. Lines of equal power below maximum are space by 12 dB. ▨, 60–72 dB; ▧, 48–60 dB [from Waldteufel (1976)].

respect to the mainlobe axis. The diagram accuracy is thought to be, on average, ± 1.8 dB in magnitude and $\pm 0.1°$ in position, and the overall picture seems satisfactory in that there is excellent continuity throughout the pattern, with a consistent decrease away from the mainlobe. The diagram is far from annular, as would be expected for a circular symmetric parabolic reflector. The highest sidelobes (i.e., those with two-way gains 48–72 dB below the mainlobe peak gain) have an angular extent much broader than the mainlobe and are distributed along the principal axes (azimuth and elevation) and at $\pm 30°$ from the line $\theta_e = 0°$. The four-legged structure supporting the horn that illuminates the reflector and scattering from radome ribs are instrumental in causing these prominent sidelobes. From Fig. 7.26 it seems reasonable to admit a circular mainlobe with an angular diameter of $2°$, at roughly 40 dB below the two-way pattern peak. Therefore contamination through near sidelobes becomes significant only if reflectivity gradients exceed 40 dB deg^{-1}. Thus, if the main beam is $4°$ away from a strong reflectivity core and pointing at a location with 40 dB less reflectivity (which is not so uncommon), considerable error in the data fields may result. The reflectivity field in a range-height indicator (RHI) display (Fig. 9.22) and the corresponding velocity field in Color Pate 2b illustrate these sidelobe effects. Only reflectivity data above 10 dBZ is presented in Fig. 9.22 whereas all velocity data with SNR ≥ 5 dB are displayed in Color Plate 2B. Velocity data above the storm top (i.e., from about 11 to 15 km AGL) are from echoes received through sidelobes. In this case the main beam is in clear air above the storm at about 2–$4°$ away from the 40-dBZ reflectivity factor region. Similar effects due to sidelobes can be seen on PPI displays (e.g., Fig. 7.27).

Because power enters through all sidelobes, the integrated contribution must be considered. For the pattern in Fig. 7.26 the ratio of power in the mainlobe to power in the sidelobes was found to be 31.5 dB. This simply means that, with uniform reflectivity, the effective signal-to-noise ratio (i.e., mainlobe to sidelobe signals) can be as low as 31.5 dB. If the shear of radial velocities were uniform, there would be no bias in the mean velocity, but there would be an increased spectrum width; otherwise biases would be created. The exact assessment of sidelobe influence on spectral moments requires knowledge of the reflectivity and velocity fields (Waldteufel, 1976; Istok, 1983).

There are no simple solutions to the sidelobe problem. Shrouding the antenna helps, and so does the elimination of the radome (or the use of better ones). The measured pattern of the WSR-88D without the radome (Fig. 7.28) has a first sidelobe at about -29 dB. On this radar the radome is made of randomly shaped panels so that there should be no preferential directions for sidelobes; with the radome the first sidelobe level increases by about 2 dB.

Polarization diversity introduces one more requirement on the antenna pattern. Namely the mainlobes at two orthogonal polarizations must be well matched; otherwise erroneous estimates of polarimetric variables would be produced (Pointin *et al.*, 1988). Antenna patterns are altered by ground obsta-

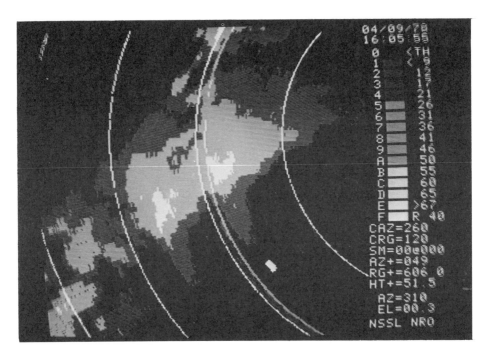

Fig. 7.27 Examples of echoes received through sidelobes. The reflectivity factor spans a range from −5 to nearly 65 dBZ. The faintest gray areas are from −5 to 21 dBZ. The two-way antenna gain at angles as far as 12° away from the mainlobe axis can be as high as −60 dB below the maximum antenna gain for the antenna used here. Thus the reflectivity levels that stretch southward from the reflectivity maximum can be echoes received through antenna sidelobes (i.e., Fig. 7.26) so reflectivity and velocity fields in these regions are questionable. Range marks are 40 km apart, and the inner range arc is at 80 km.

cles that block radiation. The pattern $f^4(\theta, \phi)$ illuminating hydrometeors beyond the blockage will have a different shape (Doviak and Zrnić, 1985). If blockage is not accounted for in the weighting functions, reflectivity estimates will be underestimated (Harju and Puhakka, 1980). Furthermore, in the presence of vertical shear, estimates of Doppler velocity and spectrum width would also be biased (Smith and Doviak, 1984).

7.10 Clutter

Clutter refers to echoes that might interfere with desired signals. The name is descriptive because such echoes "clutter" the radar display and impede recognition of wanted images. In that context weather signals could also be considered as clutter if the radar is not intended for atmospheric measurements. In this

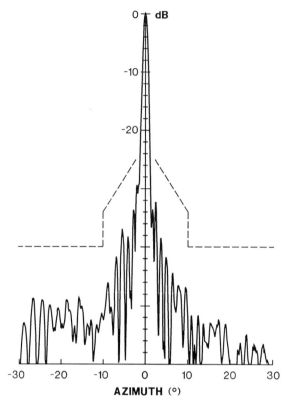

Fig. 7.28 Antenna pattern of the WSR-88D without radome. Sidelobes are specified to be below the dashed envelope.

book we use the term clutter to indicate all echoes or interferences that impede recognition of weather phenomena.

7.10.1 Ground Clutter and Its Suppression

Echoes from the ground occur whenever transmitted energy is incident on the ground by way of main beam or sidelobes. At lowest elevation angles, when a portion of the main beam intersects the ground, ground-clutter echoes are strongest. The beam can also intersect the ground if a strong ground-based temperature inversion causes it to bend (Section 2.2.3.2 and Fig. 2.11b); such echoes are ascribed to anomalous propagation. On the PPI scope of a radar with poor range resolution, (i.e., ≥ 200 m) clutter appears as a large bright area centered in the middle (e.g., Figs. 7.1 and 7.2). If reflectivity is mapped (Fig. 7.29) with a radar having a very narrow transmitted pulse (0.3 μs), a wideband receiver (10 MHz), and an antenna with a narrow beamwidth (0.8°), the

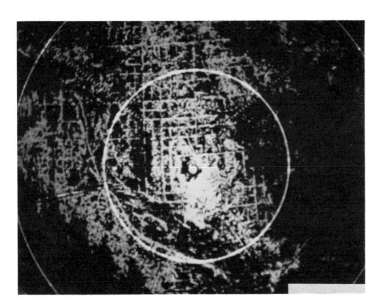

Fig. 7.29   Ground clutter. The transmitter pulse is 0.3 μs long, the elevation angle is 0.6°, the beamwidth is 0.8°, and the range marks are 5 nautical miles (about 9 km) apart. The receiver bandwidth is 10 MHz. The radar is located in Norman, Oklahoma.

clutter pattern will show recognizable structures. The data in Fig. 7.29 were collected at the NSSL radar site in Norma, Oklahoma with a narrow range-weighting function (about 50 m). The echoes depicting roads are most probably caused by trees and posts along the roads. Other diffused returns are from clusters of homes and forested areas. The thin elongated arc (from the south to the west at range 9 km) is from a ridge that borders the southern boundary of the Canadian River valley. The bright round dot in the center represents transmitter pulse leakage and returns from the immediate vicinity of radar. Strong returns from the southeast are from buildings in downtown Norman and at the University of Oklahoma.

In dealing with ground clutter, it is customary to define the ground reflectivity η_c as the cross section per unit area [Eq. (4.10) in Section 4.4 defines volume reflectivity].

$$\eta_c = (\Delta A)^{-1} \sum_{\Delta A} \sigma_{bi}, \qquad (7.35)$$

where, analogous to Eq. (4.10), σ_{bi} is a cross section of scatterers located in the area ΔA. Because the scatterers are spread over an area rather than a volume, the radar equation takes a slightly different form.

$$\bar{P}(r_0) = \frac{P_t g^2 \lambda^2 \eta_c}{(4\pi r_0)^3 l^2} \int_0^\infty \frac{|W(r_0 - r)|^2 \, dr}{\sin \gamma} \int_0^\pi f^4(\theta, \phi) \, d\phi', \qquad (7.36)$$

where all the terms are as in Eq. (4.12) with exception of η_c and γ, which is the angle between the normal to the earth and the range to scatterers on the ground (Fig. 7.30). Note the r^{-3} dependence of the received power in Eq. (7.36). This is a result of the linear increase in area with r and the r^4 decrease in power echoed by ground scatterers. To integrate the pattern, θ and ϕ must be expressed in terms of ϕ', h, and r.

Details about ground clutter, in particular cross section per unit area of various terrains, can be found in books by Long (1975) and Nathanson (1969). Here we stress that echo power has a strong dependence on terrain type and the incident angle of electromagnetic waves. Moreover, because the foliage and water on the ground change, considerable variation with seasons is not uncommon.

As a rule, ground clutter has a long correlation time, which means that its spectrum width is very narrow (i.e., ≤ 1 m s^{-1}). In addition, its mean velocity is zero and thus can easily be recognized in the Doppler spectrum (Fig. 7.31). Therefore, when spectral processing of Doppler signals is employed, removal of the dc line and one or two adjacent spectral coefficients prior to Doppler moment estimation usually restores weather spectrum moment estimates if saturation has not occurred somewhere in the receiver chain.

Recursive filters operating on the digitized I, Q video signals with a notch at dc have been used to suppress ground returns. Groginsky and Glover (1980) present an elliptic filter with three poles that has a 50-dB notch, a 1-dB ripple in the pass band, and a notch width between 2% and 4% of the unambiguous interval. The WSR-88D has a five-pole filter (Table 7.2); stored values of ground-clutter reflectivities (i.e., clutter map) are used to make a decision whether filtered or nonfiltered samples will be further processed.

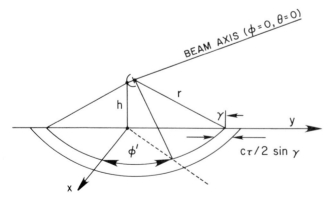

Fig. 7.30 A radar antenna at a height h above ground and pointing at ϕ_0, θ_0 receives echoes from the ground. The annular patch of width $c\tau/2 \sin \gamma$ is illuminated by various portions of the sidelobes.

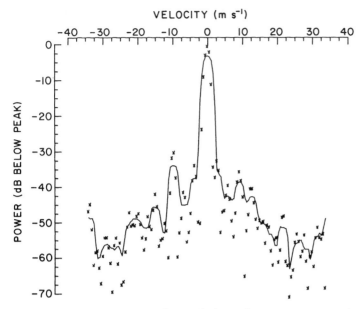

Fig. 7.31 Doppler spectrum of ground clutter for an antenna scanning in azimuth at a rate $\alpha = 10°\,\mathrm{s}^{-1}$. The width is $0.45\,\mathrm{m\,s}^{-1}$; x's are spectral powers, and the solid line is a five-point running average. Elevation $= 0.5°$; range $= 5.315$ km; SNR $= 100$. Data were weighted with the von Hann window.

Table 7.2
Ground Clutter Canceler (WSR-88D)

Type	Infinite impulse response Range selectable
Reflectivity sampling	
Suppression	20–40 dB
Notch 1/2 width at 1/2 power	0.8; 1.2; 1.7 m s^{-1}
Reflectivity bias	≤ 1 dB (for $\sigma_v \geq 4$ m s^{-1})
Velocity sampling	
Suppression	30–50 dB
Notch 1/2 width at 1/2 power	1.3; 1.9; 2.7 m s^{-1}
Velocity and spectral width bias	≤ 1 m s^{-1} at $v = v_a/2$ ($\sigma_v = 4$ m s^{-1})

To effectively use a ground-clutter canceler, signals must not saturate the receiver chain; furthermore one must have an estimate of the clutter spectrum width because the dwell time (over which the filtering is done) should be about equal to the reciprocal of the clutter spectrum width. The wind vibrates foliage and other objects, and if the displacements are a significant fraction of the rf wavelength, they cause fluctuations in the returned signal that are very similar to those produced by the reshuffling of hydrometeors. An intuitively satisfying hypothesis is to model the power spectrum for a stationary antenna as a zero frequency line (i.e., delta function) plus a narrowband fluctuation spectrum centered on this line. The strong line is due to rigid objects (buildings, posts, tree trunks, etc.), while the fluctuation spectrum (of width σ_{cw}) is due to the wind-induced motion of foliage, wires, etc. The spectrum width contribution $\sigma_{c\alpha}$ due to antenna rotation increases with rotation rate according to Eq. (5.69). Because

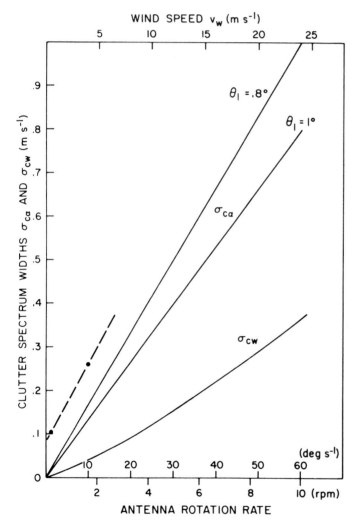

Fig. 7.32   Clutter spectrum widths $\sigma_{c\alpha}$ due to antenna rotation (scale on lower axis) and σ_{cw} due to wind speed (upper axis). Dashed line is through two estimated data points ($\theta_1 = 0.8°$).

the ground scatterers receive changing illumination, even totally rigid objects produce a time-varying signal and thus contribute to spectrum width.

Groginski and Glover (1980) have fitted the data of Nathanson (1969) for spectrum width σ_{cw}, caused by foliage moving in the wind, with the following regression line:

$$\sigma_{cw} = 0.008\ v_w^{1.2} \qquad (\text{m s}^{-1}) \qquad\qquad (7.37)$$

where wind speed v_w is in m s^{-1}. The curves for σ_{cw} and $\sigma_{c\alpha}$ are plotted in Fig. 7.32 where one can observe that a 1° beamwidth antenna rotating at 3 rpm produces a spectrum width of 0.23 m s^{-1}. At wind speeds of 10 m s^{-1} the spectrum width is broadened by 0.12 m s^{-1}. Thus the total clutter spectrum width would be 0.26 m s^{-1}.

Histograms of spectrum widths (Fig. 7.33) obtained over a prairie (Zrnić and Hamidi 1981) show that at an antenna rotation rate of 10° s^{-1} the mean widths are about four times larger than at 1° s^{-1}, rather than 10 times as suggested by Eq. (5.61). This is due to bias by the spectral window and the effects of wind speeds; both can dominate at slower scan rates. The unbiased mean width for $\alpha = 10°$ s^{-1} would be about 0.26 m s^{-1} and is plotted in Fig. 7.32. An accurate unbiasing of the width at 1° s^{-1} rotation rate is difficult because the predicted window bias is often larger than the total measured width; furthermore, wind also could have contributed. Thus the measured value of 0.1 m s^{-1} is included in Fig. 7.32.

A display of ground clutter before and after filtering (Fig. 7.34) demonstrates significant improvement; reflectivities in excess of 35 dBZ are reduced to less than 15 dBZ. Similar results are expected for other sites in the plains. An infinite impulse response filter with a dwell time of 50 ms was used in this example; the stop band (defined as positive velocities where clutter is attenuated by more than 50 dB) was 1 m s^{-1} and the pass band (defined as velocities for

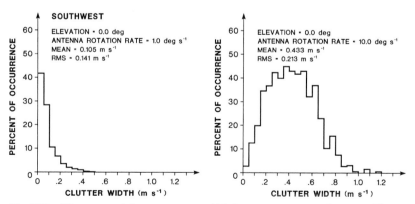

Fig. 7.33 Histograms of clutter spectrum width for two antenna rotation rates. The mean and rms values for each histogram are indicated; returns are from prairies southwest of the radar. The spectrum window bias has not been removed.

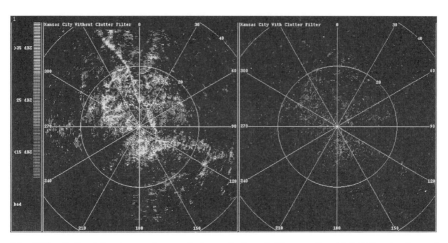

Fig. 7.34 Reflectivity factor of ground clutter before (left) and after (right) filtering by a clutter filter. These data were obtained by a 10-cm wavelength weather radar of the MIT Lincoln Laboratory from a location in Kansas City. The antenna rotation rate is $20° \text{ s}^{-1}$, the elevation angle is $0.4°$, the beamwidth is $1°$, and range rings are 20 km apart. (Courtesy Sen Lee, Lincoln Laboratory.)

which the filter characteristic changes by less than ± 1 dB) extended from 3 m s^{-1} to the unambiguous velocity.

Even after a good clutter canceler (e.g., 50 dB) there may remain substantial residue that could contaminate meteorological data. Automatic algorithms for recognition of low reflectivity phenomena such as gust fronts and dry microbursts are susceptible to errors from residues. This has prompted designers of weather radars to store the field of residual echo powers from ground clutter in a clutter map (Hynek, 1990). A two dimensional spatial filtering is then used to eliminate signals that have powers comparable to the stored powers.

7.10.2 Other Clutter

Other echoes that impede recognition of weather phenomena are from the sea, birds (Larkin, 1991), and airborne objects (balloons, aircraft etc). Insects also interfere with the weather signal but, if they are tracers of wind, they would not be considered clutter. To ornithologists and entomologists echoes from birds and insects are valuable sources of information about migratory patterns, insect density, and the roosting locations of birds (Riley, 1989). Backscattering cross sections of insects and birds are discussed in the review article by Vaughn (1985).

A reflectivity field of a flock of birds departing a roosting location is displayed in Fig. 7.35 (at the cursor). These data were collected by the WSR-88D radar at 5:40 CST when the birds had already left their roost; thus the annular ring of reflectivity about the cursor. At takeoff the signature was filled and looked like a small storm cell. Other features on this figure are echoes from

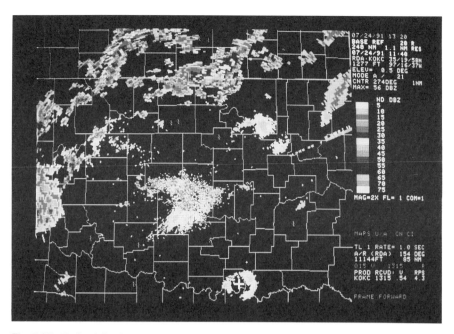

Fig. 7.35 Reflectivity factor of storms, a flock of birds (at the cursor), individual birds or airplanes (points), and microwave radiation from the sun (line at 66°). Data were obtained on 24 July 1991 at 5:40 CST by the WSR-88D radar at Twin Lakes, OK. (Courtesy of the staff of the NWS Operational Support Facility, Norman, Oklahoma.)

storms, and airplanes or individuals birds (points on the display); the line image at 66° is caused by microwave radiation from the sun.

Problems

7.1 Consider a circular storm with a diameter of 10 km. The radar has an unambiguous range of 100 km. (a) The storm center is at 80 km from the radar. Where would it appear to be in the second trip? How big (approximately) is the obscured area in the second trip? (b) The storm center is at 180 km. How big is the obscured area in the first trip?

7.2 Compare the acquisition times needed to reduce the standard deviation of mean velocity estimates to 1 m s^{-1}. Radar parameters are $T_s = 1$ ms; $\lambda = 0.1$ m; SNR = 20 dB; spectrum width is 4 m s^{-1}. In one case, frequency diversity betwen pairs is used to obtain independent sample pairs, while in the other, the frequency is not changed. If $\tau = 1 \ \mu$s, how large a change in frequency is needed to create independent pulses? How much does this change affect the wavelength λ?

7.3 A scanning weather radar has an antenna beamwidth of 1°. The required resolution due to the effective beamwidth should not be worse than 1.5°. Assume that SNR $\geq$ 20 dB and that velocity estimates are obtained using contiguous pulse pairs with $T_s = 1$ ms. Suppose that errors (SD) for velocities with spectral width of 4 m s^{-1} must be less than 1 m s^{-1}. Determine the number of samples M and the rotation rate that satisfy the stated conditions.

7.4 Suppose that a storm at a range of 60 km spans a 20° azimuthal angle and is moving toward the radar with a velocity of 50 km h^{-1}. Bear in mind that the internal storm motions will generate a considerable spread of radar-measured radial velocities about the velocity of the storm. If the radar unambiguous velocity is ±30 m s^{-1}, which velocities would you correct and how?

7.5 Consider interlaced sampling. If the powers of two range-overlaid signals differ by more than 10 dB, it is possible to obtain a good velocity estimate of the stronger echo. Let the pulse repetition time for velocity measurements be 1 ms and consider a situation in which a second-trip signal is overlaid on the first-trip signal. Assume autocovariance processing, equal spectrum widths, and 10-dB power difference; plot the measured velocity as a function of actual velocity difference.

7.6 A staggered PRT with a 10-cm Doppler radar is employed with $v_{a1} = 15$ m s^{-1} and $v_{a2} = 20$ m s^{-1}. (a) Find the extended unambiguous velocity v_m and the unambiguous range. (b) If the v_{a1} channel measures 8 m s^{-1} and v_{a2} channel estimates -5 m s^{-1}, what is your estimate of true velocity?

7.7 A storm with a 20-km diameter is moving at 20 m s^{-1} from 200°. The spread of velocities around the mean is ±30 m s^{-1}. The radar's unambiguous velocity interval is ±25 m s^{-1}. Assume that the storm is 80 km from the radar and that it may be at any azimuth. Find the azimuthal limits where perfect velocity dealiasing (i.e., without use of continuity) is possible.

8

Precipitation Measurements

A long-standing problem in meteorology is distinguishing, by radar, ice and liquid phases of precipitation. This is especially challenging in convective storms, where liquid water can exist at temperatures colder than 0°C and ice can be found at temperatures warmer than 0°C. Virtually all multiparameter radar techniques conceived so far have, at one time or another, been aimed at this problem (e.g., *Special Issue of Radio Science*, Vol. 19, No. 1, Jan.–Feb. 1984). Equally important is the problem of quantifying rain-, snow-, or hail-fall rates. Accurate estimates from radar measurements of these rates or of the liquid water content M require detailed knowledge of the precipitation size distribution $N(D)$. Different measured distributions giving the same Z can cause rainfall rates to differ by as much as a factor of 4 (Richards and Crozier, 1981).

Although radar techniques have practical limitations and their accuracy in rainfall rate estimation is highly suspect, they have a decided advantage because radars can survey vast areas and make millions of measurements in minutes. Radars (e.g., dual polarization and/or Doppler) capable of measuring more than one parameter (e.g., vertical and horizontal reflectivities and/or a spectrum of terminal velocities) in each resolution volume offer improved estimates of critically important parameters of the hydrometeor size distributions so that high-resolution measurement of the spatial distribution of water can be made. Multiple-parameter radars in combination with satellites (Atlas *et al.*, 1982), rain gauges, and other instruments may give the sought-for accuracy in rainfall estimates. It has been demonstrated that these radars can discriminate hail from other hydrometeor types (Bringi *et al.*, 1984) and distinguish rain from frozen precipitation (Hall *et al.*, 1980).

To characterize accurately the relations between water density and cloud dynamics with good spatial resolution and to sense reliably the threat of unusual but significant events such as flash floods, large hail, or heavy snowfall, efforts have been and still are under way to improve the accuracy with which radars can estimate precipitation.

8.1   Drop Size Distributions

The distribution of hydrometeor sizes is of central importance in determining the reflectivity factor Z, water content M, and precipitation rate R. The size distribution is the volume density of hydrometeors per unit diameter and therefore has units of m^{-4}. Distributions of some precipitation types such as rain drops, cloud droplets, and hail stones have been measured and theoretically computed. Cloud droplets are usually formed by water vapor condensing on particulates that serve as condensation nuclei. Super saturation (humidity of $>100\%$) is required to condense water vapor in pure air, but condensation is too slow to produce precipitating drops (rain) within the lifetime of clouds. The coalescence of colliding drops is required for rapid growth of raindrops.

Drops of diameter $D < 0.35$ mm are essentially spherical, and drops up to 1 mm in diameter have a shape well approximated by an oblate spheroid (Fig. 8.1; Pruppacher and Pitter, 1971). Larger drops have a progressively flattened and then concave base ($D \geq 4$ mm); the largest drops have a high probability of breaking up into smaller fragments. The processes of coalescence and breakup determine the size distribution of raindrops. Srivastava (1971) gives a review of the theoretical models used to determine size distributions and compares theory with observations. We present here a brief description of drop formation and give some results from drop growth models to assess the size distribution of cloud particles and raindrops. For a thorough treatment, the reader is referred to texts on cloud physics, such as that by Mason (1971). We also discuss the distribution sizes and shapes of hailstones.

8.1.1   Cloud Drop Size Distributions

Consider a droplet of water in an environment of water vapor at a pressure such that equilibrium is achieved between the rate of evaporation of the water molecules in the drop and rate of condensation of the molecules surrounding the drop. The vapor pressure required to achieve this equilibrium depends exponentially on the inverse diameter D^{-1} of the drop—smaller droplets have much higher vapor pressure. Air above a plane surface ($D \to \infty$) of water has a humidity $H = 100\%$ if equilibrium is established. Thus a droplet introduced into a barely saturated ($H = 100\%$) atmosphere will evaporate; it cannot persist, much

Fig. 8.1   Typical shape of large drops falling at terminal velocity w_t. The diameter D of the drops, from left to right, is 8.00, 7.35, 5.80, 5.30, 3.45, and 2.70 mm, corresponding to the following values of w_t: 9.2, 9.2, 9.17, 9.13, 8.46, and 7.70 m s^{-1}. (From Pruppacher and Beard, 1970.)

less grow, unless the environment is supersaturated ($H > 100\%$) by the amount equal to the vapor pressure of the droplet.

The droplet diameter at which equilibrium is achieved is a critical diameter because larger drops continue to grow and smaller drops evaporate as long as the vapor pressure is maintained. In pure air, water vapor will not condense until the air becomes supersaturated by several hundred percent. In this highly saturated state, condensation does occur on ions produced by passing cosmic particles. The Wilson cloud chamber, used to map the trajectories of ionizing nuclear particles, is based on this fact.

A pure environment rarely exists in the atmosphere because of the myriad aerosols, both natural and manufactured, that serve as centers of condensation (i.e., condensation nuclei). These condensation nuclei require considerably lower vapor pressure to achieve equilibrium than pure water drops of the same diameter, and hence they prevent great supersaturation from occurring in the natural atmosphere.

Not all condensation nuclei are solid particles. Some are hygroscopic liquids, such as nitric and sulfuric acid. Some have the property of condensing water vapor at pressure less than saturation (i.e., $H = 100\%$). As the droplet grows on the aerosol, be it solid or liquid, the increased drop volume shields the nuclei's influence, and the drop looks more like pure water. The vapor pressure required for continued growth then increases and the drop reaches an equilibrium size. If $H < 100\%$, droplets never achieve sufficient sizes to form a cloud (i.e., drop diameters $\geq 5~\mu m$, where scattering of light is practically nonselective with respect to color), although they may contribute to haze, which causes a moderate diminution of visibility. If drops reach the critical diameter and $H \geq 100\%$ then clouds form, and cloud droplets may continue to grow to precipitation size. Nevertheless many drops compete in the reservoir of vapor and will decrease saturation, so the growth of droplets is restricted.

Now let us consider a rising parcel of moist air. Drops immediately begin to form about condensation nuclei when the air parcel, lifted by convection, is brought below its dew-point temperature (i.e., the temperature at which water vapor begins to condense). More efficient nuclei allow drops to grow more rapidly, and when moisture is made available by the updraft at the rate equal to the rate at which it is being condensed on the nuclei, the concentration of cloud drops equals the concentration of nuclei activated. The air is supersaturated at this point, but subsequently supersaturation decreases rather rapidly to a quasi-steady state in which the rate of condensation is very nearly balanced by the rate at which moisture in excess of saturation is made available by the updraft. Because the rate of drop diameter increase is inversely proportional to D (Mason, 1971, p. 122), the size distribution narrows as drops grow. Figure 8.2 shows a computed size distribution for droplets near the base of a cloud and a comparative observation.

The depicted cloud drop distribution can be used to estimate the reflectivity factor Z and the cloud liquid water, as discussed in Section 8.3. For example the

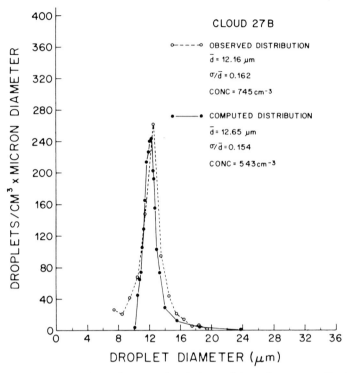

Fig. 8.2 Comparison of computed and observed drop size spectra. Condensation alone is considered. (After Fitzgerald, 1972.)

reflectivity factor of some common clouds has the following approximate values (Gossard and Strauch, 1983): 0 to -18 dBZ for developing cumulus, -15 dBZ for stratocumulus, 5 dBZ for altocumulus, -8 dBZ for altostratus, and -5 dBZ for cirrus. The reflectivity factor for the drop sizes illustrated in Fig. 8.2 is about -27 dBZ.

8.1.2 Raindrop Size Distributions

Water drops commonly do not coalesce on collision but rebound. But, when slightly electrified, drops do unite on collision. Thus, small drops are lost and bigger ones created while the number density decreases. The theoretical framework for drop size growth by stochastic coalescence is given by Telford (1955). The growth of big drops and the loss of small ones are balanced by drops breaking up when they reach large sizes. The probability of breakup has been observed to increase exponentially with drop diameter (Komabayasi *et al.*, 1964). For example, the lifetime of a 7.5-mm drop is observed to be 10 s. An exponential distribution of raindrop sizes had been observed by Marshall and

Palmer (1948), who used filter paper to measure directly the density of drop diameters at the surface, and Sekhon and Srivastava (1971) inferred from vertically pointed Doppler radar data that spectra were exponential at several altitudes below the melting layer.

It is interesting to note that for the Marshall–Palmer (M–P) data (Fig. 8.3a), the drop size distribution $N(D)$ has a tendency to flatten at small drop diameters. This is consistent with the observations of Laws and Parsons (1943), whose data (Fig. 8.3b) show, in addition, a rapid increase in $N(D)$ for even smaller drops, as also suggested by the theoretical steady-state distributions (Srivastava, 1971).

The parameters of the exponential size distribution that fit the Marshall–Palmer (M–P) data are[1]

$$N(D) = N_0 \exp(-\Lambda D), \tag{8.1a}$$

$$\Lambda = 4.1R^{-0.21} \quad \text{mm}^{-1}, \tag{8.1b}$$

$$N_0 = 8 \times 10^3 \quad \text{m}^{-3}\,\text{mm}^{-1}, \tag{8.1c}$$

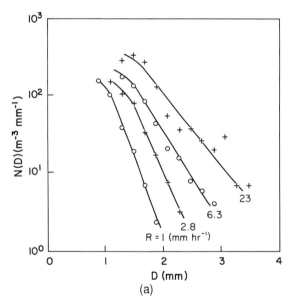

(a)

Fig. 8.3 (a) Raindrop size distribution versus drop diameter (data from Marshall and Palmer, 1948). (b) Measured drop-size distributions by Laws and Parsons (————) with a few corresponding Marshall–Palmer distributions (–––). (*Figure continues.*)

1. Unless otherwise stated, formulas in this chapter, $N(D)$, and other quantities are in mks units, but in the figures and in discussion conventional units are often used.

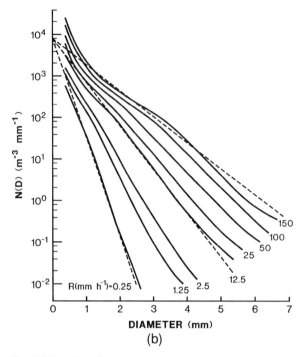

Fig. 8.3 (*Continued*)

where R is the rainfall rate in millimeters per hour. This distribution is widely used to compute rainfall rates from reflectivity factor measurements, although actual drop-size spectra differ vastly depending on geographical location, type of rainstorm, season, and region within the storm. To better represent the drop size distributions Ulbrich (1983) proposed the general gamma distribution

$$N(D) = N_0 D^\mu \exp(-\Lambda D) \tag{8.1d}$$

in which the parameter μ may have values between -3 and 8.

Comparison of the M–P raindrop spectra with the theoretical stationary spectra (Srivastava, 1971) for the middle drop-size range shows that (1) the stationary distributions are flatter than the M–P ones and (2) the stationary distributions are roughly parallel to each other. Although the stationary spectra appear quite different from the M–P exponential spectra, they could be fitted with truncated exponential functions in the drop diameter range (1–7 mm) that contributes most to R and Z (Section 8.2) so that simple analytic formula may still be used. In general N_0 and Λ, as well as the maximum drop diameter, depend on the type of rain. Indeed, many observations suggest that both N_0 and Λ are highly variable (see, for example, Waldvogel, 1974). For many practical purposes, $N(D)$ can be represented by an untruncated exponential function.

There are other functions containing two parameters that might better fit the $N(D)$ observed (Cataneo and Stout, 1968; Austin and Geotis, 1979). Although functions containing three or more parameters can better fit a wider variety of observed $N(D)$, more remotely sensed measurands are required to specify these parameters (Section 8.5.2).

The theoretical distribution, derived from experimental data for the model parameters, demonstrates that raindrops have an upper diameter limit at about 8 mm. But, larger drops are effectively created by partial melting of giant snowflakes or graupel, and actual distributions can be markedly different than the theoretical.

8.1.3   Hailstone Size Distributions

Theoretical studies and observations (Morgan and Summers, 1986) have shown that hail can grow at many different locations and altitudes within thunderstorms. Thus it appears that there are several distinct classes of hailstones caused by the different interaction of microphysics and storm dynamics. Hail growth begins at altitudes between 5 and 10 km (Browning and Foot, 1976) in a cloud of supercooled droplets (temperature $< 0°C$), and a broad region of moderate updraft is necessary to suspend hail aloft in the prime growth layer (Nelson, 1983). The major ice particles involved in the growth of hailstones are graupel, small hail, and ice pellets (List, 1986). Graupel consists of a central ice crystal that has accreted supercooled droplets that freeze after impact. The graupel often has a conical shape, a diameter of about 5 mm, and a density of 800 kg m^{-3}. Because graupel is composed of a myriad small frozen particles it appears white. Small hail represents an intermediate stage between graupel and a hailstone. Small hail forms from graupel by intake of liquid water into the air capillaries of its ice structure wherein liquid water is produced either by accretion of warm cloud droplets or by partial melting of the graupel. The size can be similar to graupel but the density would be larger because of water accretion. Ice pellets (sleet) are produced by freezing of raindrops or refreezing of partly melted snowflakes. They have diameters less than 5 mm and densities nearly equal to ice (917 kg m^{-3}), or higher if liquid water is present in their interior. According to List (1986) hailstones are lumps of ice or ice and water, with air inclusions and diameters >5 mm. Shapes depend somewhat on size and are roughly spherical or conical (diameters 5–10 mm), ellipsoidal or conical (10–20 mm), ellipsoidal, often with small lobes or knobs and indentations along the shortest axis (10–50 mm), and approximately spherical (40–100 mm) with small or large protuberances.

First measurements of hailstone size distributions were made by Douglas (1964) who collected samples in wire-mesh baskets. He found an exponential form with $N_0 = 10$ m^{-3} mm^{-1} and $\Lambda = 0.31$ mm^{-1} (Hitchfeld and Stauder, 1965,

p. 47). Values of the exponent close to 4 mm^{-1} were obtained for the mean distributions with a hailstone disdrometer (Federer and Waldvogel, 1975) and also with an airborne laser hail spectrometer (Spahn and Smith, 1976). Hailstone samples collected on the ground from seven storms in Alberta led Cheng and English (1983) to propose a single parameter exponential size distribution for hailstones in which

$$N_0 = 115\ \Lambda^{3.63}\quad (\text{in m}^{-3}\ \text{mm}^{-1}\ \text{and } \Lambda \text{ in mm}^{-1}). \tag{8.2}$$

This affords the convenience of using a single parameter to describe hailfall rate (Section 8.4.1.3) Although the three-parameter gamma distribution fits better data samples (Ziegler *et al.*, 1983), the simplicity of the Cheng–English distribution has made it attractive in several modeling studies.

8.2 Terminal Velocities

Gunn and Kinzer (1949) made precise measurements of the terminal velocity w_t of water droplets in stagnant air at sea level. These data are commonly used in calculating the rainfall rate, on the ground, from $N(D)$ and in deriving the Doppler velocity power spectra for vertically pointed radars. Figure 8.4 shows the terminal velocity for drops in the diameter range 0.27–5.76 mm. Drops of diameter larger than about 6 mm are unstable and break into smaller ones. A useful formula for terminal velocity is (Atlas *et al.*, 1973)

$$w_t(D) = 9.65\ -\ 10.3\ \exp(-600D)\quad \text{m s}^{-1}, \tag{8.3}$$

where D is the diameter in meters; if D lies between 6×10^{-4} and 5.8×10^{-3} m, there is less than 2% error from the measured values of Gunn and Kinzer; we shall find it useful to have a power law fit to $w_t(D)$ data. Atlas and Ulbrich (1977) show that

$$w_t(D) \approx 386.6D^{0.67} \tag{8.4}$$

fits the data of Gunn and Kinzer in the diameter range $5 \times 10^{-4}\,\text{m} < D < 5 \times 10^{-3}$ m. Equation (8.4) is also plotted in Fig. 8.4 for comparison. At low terminal velocities (low Reynolds number corresponding to drop diameter <0.1 mm), such that pressure gradients and inertial forces are negligible compared to viscous forces, theory predicts an exponent of 2 (Rogers, 1976). At higher Reynolds numbers inertial forces are balanced by drag, and theory [Eq. (8.6b)] predicts an exponent of 0.5. At the larger drop diameters found in moderate-to-heavy rainfall, drops are distorted and a theoretical solution is difficult, so we resort to experimental data and use empirical formulas.

Drops fall faster in rarified air, so $w_t(D)$ needs to be adjusted at higher altitudes. Foote and duToit (1969) have examined data on terminal velocities of drops falling in a partially evacuated tube. They deduce that dependence of the

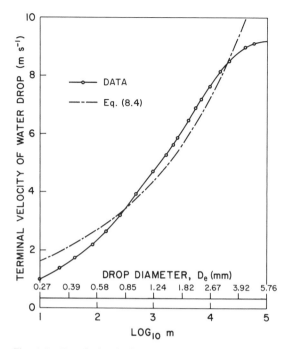

Fig. 8.4 Terminal velocity (solid line) of distilled water droplets (in stagnant air at 76-cm-Hg pressure, 20°C, and 50% relative humidity) as a function of the mass m (in micrograms) or the equivalent spherical diameter D_e. (Data from Gunn and Kinzer, 1949.)

terminal velocity $w_t(D, \rho)$ on the air density can be approximated by

$$w_t(D, \rho) = w_t(D)(\rho_0/\rho)^{0.4}, \tag{8.5}$$

where ρ_0 is the density of air at a pressure of 760 mm Hg and a temperature of 20°C.

Theoretical terminal velocity of hailstones can be obtained through Newton's third law by equating the total drag force to the weight. This produces the expression

$$w_t = (2mg/\rho A C_d)^{1/2}, \tag{8.6a}$$

where m is the mass of the stone, g acceleration of gravity, A the cross-sectional area of the stone normal to the airflow, and C_d the drag coefficient. For spherical hail this equation becomes

$$w_t = (4\rho_h g/3\rho C_d)^{1/2} D^{1/2}, \tag{8.6b}$$

where ρ_h is the hailstone density and D its diameter. The most uncertain parameter for determining the terminal velocity is the drag coefficient because hailstones have irregular shapes. In a thorough experimental study, Matson and Huggins (1980) determined the median values of the drag coefficient to lie in the range between 0.82 and 0.96 for hailstones with diameters less than 25 mm. Their best-fit terminal velocity-versus-diameter relationship produced

$$w_t = 3.62D^{1/2} \quad \text{(in m s}^{-1} \text{ for } D \text{ in mm)}, \tag{8.6c}$$

which corresponds to the drag coefficient of 0.85. This drag coefficient is higher than the one for smooth (0.45) or rough (0.6) spheres and is not unexpected considering that hailstones resemble rough ellipsoids.

In a series of tests to evaluate the utility of chaff, Battan (1958) found a nominal terminal speed of about 0.8 m s^{-1}, but some types have terminal speeds below 0.3 m s^{-1}.

Gunn and Marshall (1958) used the formula

$$w_t = 0.98D^{0.31} \quad \text{(m s}^{-1}) \tag{8.6d}$$

to estimate terminal velocities of aggregate snowflakes where D (in mm) is the diameter of a water sphere with the same mass as the snowflake.

8.3   Rainfall Rate, Reflectivity, and Liquid Water Content

In Chapter 4 we showed that the expected power of weather signal samples is proportional to the reflectivity factor Z if drop diameters are much smaller than wavelength. In this section we present the assumptions that need to be made to relate the measurands Z, specific attenuation K, etc., to meteorological variables, such as the liquid water content M of the cloud and the rainfall rate R.

8.3.1   Liquid Water Content

An important parameter in rain production is the efficiency with which the water vapor is converted into cloud water. Updraft speeds and storm intensities are related to the amount of latent heat released by production of cloud water. If the liquid water concentration is large, jet aircraft engines ingesting water may "flame out." Thus radar measurements of the distribution of M throughout the cloud are often desired.

The liquid water mass density contributed by drops having diameters between D and $D + dD$ is

$$dM(D) = (\pi/6)D^3 \rho_w N(D) \, dD, \tag{8.7}$$

where ρ_w is water density of the drop (10^3 kg m^{-3}). The cloud's water density is then

$$M = \int dM(D) = (\pi\rho_w/6) \int_0^\infty D^3 N(D) \, dD. \tag{8.8}$$

We shall find the following formula quite useful for exponential drop size distributions:

$$\int_0^\infty x^{\nu-1} e^{-\mu x} \, dx = (1/\mu^\nu)\Gamma(\nu), \qquad \mathrm{Re}\ \mu > 0, \qquad \mathrm{Re}\ \nu > 0, \qquad (8.9)$$

where $\Gamma(\nu)$ is the gamma function [if ν is an integer, n, $\Gamma(\nu) = (n-1)!$]. Values of $\Gamma(\nu)$ for noninteger ν are given in mathematical tables (e.g., Abramowitz and Stegun, 1964). If we assume an untruncated exponential drop diameter distribution [Eq. (8.1a)] we can easily solve the integral (8.8) using Eq. (8.9) to obtain a cloud water density of

$$M = \pi \rho_w N_0 / \Lambda^4. \qquad (8.10)$$

The diameter of a drop such that half the water is contained in larger drops is defined as the median volume diameter D_0. Thus, D_0 is the solution of the equation

$$(\pi \rho_w / 6) \int_0^{D_0} D^3 N(D) \, dD = M/2. \qquad (8.11)$$

This can be solved for D_0, giving

$$D_0 = 3.67/\Lambda. \qquad (8.12)$$

The median drop diameter is used frequently instead of Λ because D_0 is an easily identified physical attribute.

   Equation (8.1a) may be a good approximation in many situations even if N_0 and Λ are not limited to those values specified by Marshall and Palmer. In general, N_0 and Λ are independent and unknown, and therefore cannot be determined from the measurement of the single parameter Z. Section 8.5 discusses methods whereby radar can be used to measure N_0 and Λ and hence M.

8.3.2   Reflectivity Factor Z

Z is defined as the sixth power of the hydrometeor diameter summed over all hydrometeors in a unit volume [Eq. (4.32)]; in terms of the hydrometeor size distribution,

$$Z = \int_0^\infty N(D) D^6 \, dD. \qquad (8.13)$$

It is instructive to plot, the integrand of Eq. (8.13) to show the relative weights that different diameters give to Z. For the sake of illustration, we used the exponential drop diameter distribution with parameters deduced by Marshall and Palmer (Fig. 8.5a).

   Even though the smallest drops are most numerous, the sixth power of D causes the fewer larger diameter drops to be the most important contributors to Z. Therefore, even if the exponential distribution poorly estimates the actual

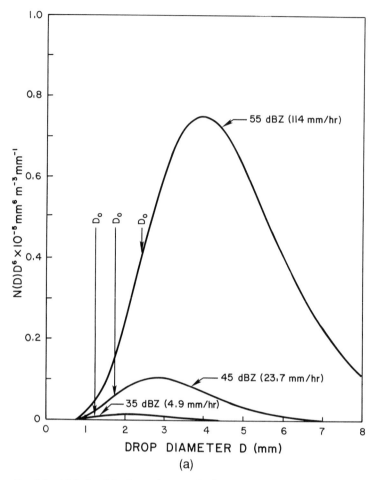

Fig. 8.5 (a) Reflectivity factor integrand $D^6N(D)$ and (b) the integrand of rainfall rate [Eq. (8.18)] versus drop diameter showing which drops contribute most. The M–P drop size distribution is assumed, so R is computed from Eq. (8.1b). The rainfall rate integrand is $(\pi N_0/6)f(D)\ 10^{-9}$; $f(D) = (9.65 - 10.3e^{-0.6D})\,D^3e^{-\Lambda D}$. (*Figure continues.*)

drop size density at small drop diameters, this should not produce much error in Z provided that Λ and N_0 accurately describe the distribution for diameters that strongly contribute to Z. Also shown in Fig. 8.5a is the median volume drop diameter for each distribution. Note that the reflectivity factor increases by two orders of magnitude but diameters that contribute most to Z increase by no more than a factor of 2. For diameter distributions that can be expressed by Eq. (8.1a), we derive the relation

$$Z = N_0(6!)(D_0/3.67)^7 = N_0(6!)\Lambda^{-7} \tag{8.14}$$

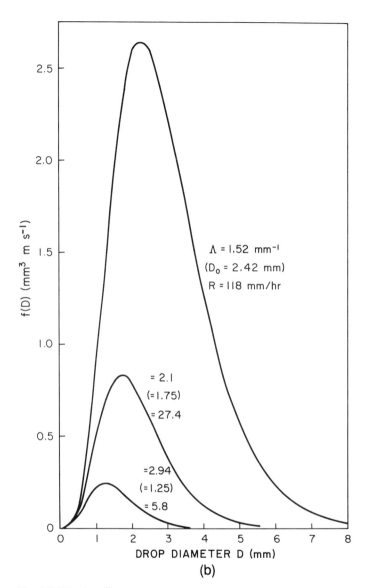

Fig. 8.5 (*Continued*)

between the reflectivity factor and the two parameters defining the distribution.

Equation (8.14) was obtained under the assumption that drop diameters extend to infinitely large values, which is unrealistic. The estimate of Z is not significantly in error if the actual drop size spectrum is well approximated by the M–P exponential function and integration is made assuming all sizes are present, but large errors can result for spectra that are flatter (i.e., have a smaller Λ) and not truncated at large diameters.

Consider the exponential drop size distribution truncated at $D = D_{max}$. Then it can be shown that the reflectivity factor is given by

$$Z = N_0 \Lambda^{-7} \gamma(7, a), \tag{8.15}$$

where $\gamma(7, a)$ is the incomplete gamma function (Gradshteyn and Ryzhik, 1965) and

$$a = \Lambda D_{max}. \tag{8.16}$$

A truncated exponential drop size distribution with $R = 125$ mm h^{-1}, $N_0 = 4 \times 10^2$ m^{-3} mm^{-1}, $\Lambda = 0.644$ mm^{-1}, and $D_{max} = 6.5$ mm fits well a stationary distribution theoretically deduced by Srivastava (1971). Using these values to compute $\gamma(7, a)$, we can show that Eq. (8.14) overestimates Z by almost an order of magnitude relative to the more accurate Eq. (8.15).

8.3.3 Rainfall Rate

When we know the drop diameter spectrum, we can directly compute the rainfall rate in still air using $w_t(D)$ from Eq. (8.3). The number of drops $N_A(D)$ of diameter D between D and $D + dD$ impacting on a differential area dA in time dt is

$$N_A(D)/(dA\ dt) = N(D)w_t(D)\ dD. \tag{8.17}$$

Each drop has a liquid water mass $m(D)$.

$$m(D) = (\pi/6)D^3 \rho_w$$

Therefore the total mass of water per unit area and time is

$$\int_0^\infty N_A(D)\ m(D)\ dD \left/ dA\ dt = (\rho_w \pi/6) \int_0^\infty D^3 N(D) w_t(D)\ dD. \right. \tag{8.18}$$

The rainfall rate R is usually measured as depth of water per unit time, so

$$R = (\pi/6) \int_0^\infty D^3 N(D) w_t(D)\ dD. \tag{8.19}$$

To derive a relation between R and the two parameters of the exponential distribution, substitute Eqs. (8.1a) and (8.3) into Eq. (8.19) and integrate; this gives the formula

$$R = \frac{\pi N_0}{\Lambda^4} \left[9.65 - \frac{10.3}{(1 + 600/\Lambda)^4} \right] \text{ m s}^{-1}, \tag{8.20}$$

where mks units are implied throughout. To convert to the more commonly used units of millimeters per hour, multiply Eq. (8.20) by the factor 3.6×10^6.

It is instructive to find the drop diameters that contribute most to the rain. We have plotted in Fig. 8.5b the rainfall rate integrand of Eq. (8.19) assuming an M–P drop-size distribution. We can see that, even for the highest rain rate shown, most of the contribution to rain comes from drops having diameters less than 5 mm. If the spectra are flatter (i.e., Λ is smaller), as some observations show, then larger-diameter drops contribute significantly to R and Eq. (8.20) is then not accurate. In this case one needs to integrate a truncated size spectra to derive a formula for R that depends on three parameters, N_0, Λ, and D_{max}.

In deriving Eq. (8.20) we have assumed Eq. (8.3) to be valid for all drops. Obviously if $D \to 0$ we obtain unrealistic values for w_t. But, Fig. 8.5b shows that drops smaller than 0.5 mm contribute very little to the rainfall rates $R \geq 5$ mm h^{-1}, and thus the error has little practical effect on our results. Of course, it would be best to integrate Eq. (8.19) numerically using the exact data for $w_t(D)$.

We note that if the Λ and N_0 from the Marshall–Palmer relations (8.1b) and (8.1c) are substituted into Eq. (8.20) we obtain a value of R somewhat higher than that implied by Eq. (8.1b). This difference arises because Eq. (8.1b) is an empirical relation based on measured R for each observed $N(D)$ distribution, and the Marshall–Palmer measurements extend over only a limited drop-diameter interval (1 mm $< D <$ 3.5 mm), within which $N(D)$ is nearly exponential. The actual distribution can significantly depart from the assumed exponential distribution for both large and small drops. This departure from the actual $N(D)$ could account for the small differences in R.

8.4 Single-Parameter Measurement of Precipitation

In this section we discuss remote measurements of one parameter, such as the reflectivity factor Z or specific attenuation K, to estimate the rate of precipitation, the liquid water content, cumulative precipitation amount, or to identify hail.

8.4.1 Reflectivity Factor Method

R, Z relationships for rain and snow are commonly used to estimate the rate of precipitation. Alternatively, the reflectivity factor can be used to delineate areas where rain fall rate exceeds a specified threshold; from the area and duration of rainfall, the cumulative volume of water can be estimated.

8.4.1.1 R, Z Relations for Rain

Remote measurement of R has considerable practical interest. For many years radar meteorologists have attempted to find a useful formula that relates R to the reflectivity factor Z. Unfortunately, there is no universal relationship connecting these parameters, although it is common experience that larger rainfall rates produce more intense echoes. Examining Eq. (8.20), we see that, for the exponential drop-size distribution, we need to measure two parameters in order to obtain R. Moreover, a real drop-size distribution requires an indefinite number of parameters to characterize it, and thus the radar-determined value of Z alone cannot provide a unique measurement of R.

There has been considerable effort to establish whether some of the drop-size distribution parameters might be known for a given type of rain (stratiform, thunderstorm, etc.). Measurements of drop-size distributions around the globe under different climatic conditions have been made, and Battan (1973) lists no fewer than 69 different R, Z relations (Fig. 8.6). Even for rain conditions that were supposedly the same (stratiform), Atlas and Chmela (1957) report considerable variability in the R, Z relations. For example, their data show that the same reflectivity factor Z could be associated with either $R = 33$ mm h^{-1} or $R = 11$ mm h^{-1}, a possible 300% error depending on which measured drop-size distribution is used. Because these R, Z relations were obtained from actual drop-size measurement, they should be accurate for each of the rain events. Even though both rains were classified as stratiform, they differed significantly in size distribution.

It is quite difficult to calibrate radars to within a decibel, and there could be a systematic bias in the radar-measured reflectivity. Some of this error can be compensated by choosing an appropriate R, Z relation. According to Cain and Smith (1976), the relation $Z = 155R^{1.88}$ removes any pervasive bias in the radar-estimated rainfall (RER) in North Dakota, whereas Woodley *et al.* (1975) report that the relation $Z = 300R^{1.4}$ worked better in Miami, Florida. We should recognize that even if the actual drop-size distribution were on the

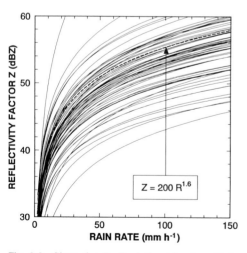

$$Z = 200 R^{1.6}$$

Fig. 8.6 Sixty-nine R, Z relationships from Battan (1973).

average the same at two different locations, errors in radar calibration could lead meteorologists to develop different R, Z relations appropriate to each region. Because neither R, Z relation holds in an absolute sense—another radar at the same location could find a different R, Z relation—the radars need to be reliably calibrated.

Other sources of error are (1) horizontal winds that cause ground-level droplet deposition to be displaced from the elevated radar resolution cell location (radars rarely detect precipitation close to the ground except for very near ranges in flat country); (2) attenuation from atmospheric gases, rain, and wetted radome; (3) reflectivity enhancement in the melting layer (Hill *et al.*, 1981); (4) incomplete beam filling; (5) evaporation (Hardy, 1963); (6) beam blockage (Harju and Puhakka, 1980); (7) rain rate gradients (Zawadzki, 1981); (8) polarization effects (Section 8.5.3.1); and (9) vertical air motion, which can induce R, Z errors by changing the vertical rainwater flux and by changing the drop-size distribution in vertically inhomogeneous situations (Carbone and Nelson, 1978).

If the vertical air velocity w is subtracted from w_t [Eq. (8.3), for example], the result is substituted into Eq. (8.19), and this equation is then integrated for an assumed exponential drop-size distribution, the rainfall rate equation obtained is similar to Eq. (8.20) except that w must be subtracted from 9.65 m s^{-1}. From this result it is easy to determine that a 2-m s^{-1} updraft causes a 50% overestimate and a 2-m s^{-1} downdraft a 25% underestimate, both relatively independent of R. But these errors occur at the height of the measurement. The following discussion based on physical models of Kessler (1987) and Lee (1988) shows that the rain rate measurement at the ground below the radar is not necessarily in error. This is because (1) an R, Z relationship aloft assumes zero vertical air velocity; therefore it underestimates (overestimates) the rain rate in downdrafts (updrafts) aloft. (2) As a volume of air descends (rises) it spreads (Fig. 9.33) because of divergence (convergence) so that the lower surface becomes larger (smaller) than the upper. (3) For steady-state conditions the area integral of rain rate through the top surface of the volume equals the integral through the bottom surface, if we ignore evaporation, because there is no accumulation between the surfaces. Hence the rain rate measured in downdrafts (updrafts) aloft can be equal to the point rain rate at the ground, but the area will be underestimated (overestimated). This areal error leads to errors in total accumulated water over storm complexes. Furthermore, even if the spatial average of w should vanish, the error in the rate of total waterfall (and hence total accumulated water) may not, as can be seen from the following general formula for the error in water mass per unit time:

$$\text{error} = \iint w(\mathbf{r})M(\mathbf{r})\, dA, \qquad (8.21)$$

where A is the area over which accumulated water is to be computed (this is usually a catchment area). In general, the integral does not vanish, and usually significant rains occur in regions of downdrafts (negative w) so that total water mass rates are underestimated.

In spite of the superfluity of R, Z relations, many of them do not differ greatly at rainfall rates between 2 and 200 mm h^{-1}. For stratiform rain the relation

$$Z = 200R^{1.6}, \qquad (8.22a)$$

referred to as the Marshall–Palmer formula (Marshall *et al.*, 1955) with R in mm h^{-1} and Z in mm^6 m^{-3}, has proven quite useful, although there are notable exceptions (e.g., Jorgensen and Willis, 1982; Cataneo, 1969). Laws and Parsons drop-size spectral measurements give

$$Z = 400R^{1.4}. \qquad (8.22b)$$

More recently Joss and Waldvogel (1970) have used

$$Z = 300R^{1.5} \qquad (8.22c)$$

and showed a 42% standard deviation between radar- and disdrometer-measured daily rainfall accumulations for 47 days of rain events throughout 1967. For 25 days in which rain accumulation was larger than 10 mm, the standard deviation was reduced to 28%. The use of three different R, Z relations (one for drizzle, one for widespread rain, and one for thunderstorms) doubled the accuracy of radar-measured amounts of precipitation (standard deviation of $\approx 13\%$); but these accurate estimates are valid for daily accumulations, not the rain rate. On the other hand, Richards and Crozier (1981) conclude that for southern Ontario, Canada, the improvement in accuracy of radar measurements of precipitation from separating the observations into different rainfall types is insignificant.

8.4.1.2   The Area–Time Integral

Although the use of Eqs. (8.22a), (8.22b), or (8.22c) may produce large dispersion from the actual rain rate measured by a gauge over short periods (e.g., ≤ 5 min.), the accuracy of rainfall measurements can be greatly improved by averaging in space or time. Leber *et al.*(1961) used Eq. (8.22a) to obtain hourly averaged rainfalls for each radar resolution cell during extremely heavy rain and integrated these hourly accumulations to produce a 24-h isohyet map that compared very well with the accumulated distribution of rainfall obtained from a dense network of rain gauges. The good agreement between accumulated rainfall estimated by radar and that measured by gauges suggests that radar estimates can be considered a "standard" against which other rainfall estimators can be compared.

One such estimator that does not require use of any R, Z relation but only a well-calibrated radar, is the area–time integral (ATI) algorithm (Doneaud *et al.*, 1984). The hypothesis is that rainfall accumulated over large areas and time is independent of how rain intensity is distributed within the storm. This is related to the observation that the probability density of rainfall is a statistically stable log-normal function (Kedem *et al.*, 1990). Byers (1948) was apparently the first author to emphasize the close relationship between the amount of rain falling from a shower and the shower's size and duration. The rainy areas have been defined by Doneaud *et al.* (1984) as those regions within which the reflectivity factor is larger than some threshold value, but other measurands (e.g., area of convective cloud cover) might be considered. The relationship proposed by Doneaud is

$$\text{Rain volume (km}^2 \text{ mm)} = 3.68 \, \text{ATI}^{1.01} \qquad (8.23)$$

where ATI is measured in units of km^2 h. An example of the data from North Dakota in Fig. 8.7 demonstrates the close relationship between rainfall and ATI and shows a correlation coefficient of 0.98 and a standard error of 0.16 (a linear factor of 1.45) in logarithmic units of rain volume. In percentages, the error is between -31% and 45%, which is comparable to the uncertainty in rain volume estimates obtained from radar data using a R, Z relationship. Atlas *et al.* (1990)

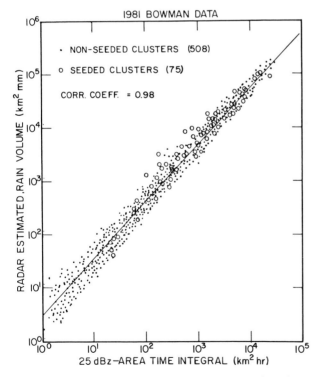

Fig. 8.7 Scatter plot and regression line of the rain volumes for storm clusters versus ATI (threshold = 25 dBZ); for June–August 1981, Bowman data (from Doneaud *et al.*, 1984).

showed how rainfall estimates depend on the reflectivity thresholds that define the rain area, and Rosenfeld *et al.* (1990) added the measurement of storm height to further improve the estimates. It is very likely that techniques based on area–time integrals will be most useful for global estimation of rainfall using satellites (Doneaud *et al.*, 1987).

8.4.1.3   R, Z Relations for Snow and Hail

Both accepted conventions and physical factors complicate the determination of the R, Z relationships for ice hydrometeors (Smith, 1984). First, radars are calibrated to measure the reflectivity factor Z_w of water drops [Eq. (4.35)]. Thus for ice particles Z_i is related to Z_w [Eq. (4.32)] by

$$Z_i = (|K_i|^2/|K_w|^2)Z_w. \tag{8.24}$$

Snowflakes and ice crystals have shapes that are quite different from spherical. Yet, as suggested by Marshall and Gunn (1952), if the sizes are in the Rayleigh scattering region the backscattering cross section [Eq. (3.6)] of an irregular particle composed of a weak dielectric like ice is the same as that of a sphere with the same mass. Thus the exact shape of the particle is immaterial and if the equivalent sphere diameter of the ice hydrometeor is known, the value for $|K_i|^2$ should be 0.176. But to obtain the equivalent diameter researchers (Sekhon and Srivastava, 1970) often use melted drop diameters that are smaller by a factor of $0.92^{1/3}$ (0.92 is the specific gravity of solid ice). In such a case either K_i should be changed to 0.208 or the melted drop diameter must be increased by the factor $0.92^{-1/3} = 1.028$ (Smith, 1984) so that the appropriate relationship (for $\lambda = 3$–10 cm) between Z_w and Z_i is $Z_i = 0.224Z_w$.

The Sekhon–Srivastava (1970) R_s, Z_w relationship for snow is

$$Z_w = 1780R_s^{2.21} \tag{8.25a}$$

and it corresponds to

$$Z_i = 399R_s^{2.21} \tag{8.25b}$$

This is an appropriate expression for estimating snowfall in terms of an equivalent rainfall rate R_s in mm h^{-1}.

A widely applicable R_H, Z relationship for hail does not exist because it is very difficult to detect hail with a single-parameter measurement and because there are few simultaneous measurements of radar reflectivity factor and hail rates on the ground. Using Cheng and English (1983) data, Torlaschi *et al.* (1984) have related the equivalent rainfall rate of R_H (mm h^{-1}) to Λ (mm^{-1}) by

$$\Lambda = \ln(88/R_H)/3.45. \tag{8.26a}$$

Substitution of Eq. (8.26a) in Eq. (8.14) and use of N_0 from Eq. (8.2) produces the following R_H, Z_H relationship:

$$Z_H = 5.38 \ 10^6 \ [\ln(88/R_H)]^{-3.37} \tag{8.26b}$$

The simplest method to detect hail is to choose a high reflectivity threshold and classify as hail those data that exceed this threshold. For hail to form, the storm cell must contain a region of temperatures below $-10°C$ with at least one large particle (diameter ≈ 5 mm). This led Waldvogel *et al.* (1979) to propose that if the height of a 45-dBZ contour is 1.4 km above the freezing level the cell is very likely to produce hail. Information about the storm's velocity structure may further improve hail discrimination. Yet there is no reliable single-parameter hail detection method.

8.4.1.4 Hail Signatures in the Reflectivity Field

In some instances large hail poduces radially elongated reflectivity signatures (Fig. 8.8) behind high reflectivity cores (Zrnić, 1987). The signature is caused by (a) unusually strong scatter from the hydrometeors to the ground, (2) scattering from the ground back to the hydrometeors, and (3) scattering by the hydrometeors back to the radar. This "three-body" signature in Fig. 8.8 is over 10 km long with a reflectivity factor between 10 and 25 dBZ.

A pictorial representation of the process that leads to the signature is given in Fig. 8.9. Indices i and j in the figure indicate two of the scatterers that interact

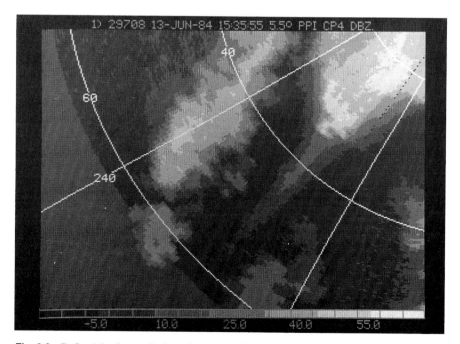

Fig. 8.8 Reflectivity factor display of a storm with a pronounced three-body scattering signature. The reflectivity factor in dBZ_e is given in seven shades of gray by the scale at the bottom. This storm produced extensive hail damage in Denver, Colorado on 13 June 1984 at 1535. Range marks are at 20-km intervals and the antenna elevation angle is 5.5° (from Wilson and Reum, 1988).

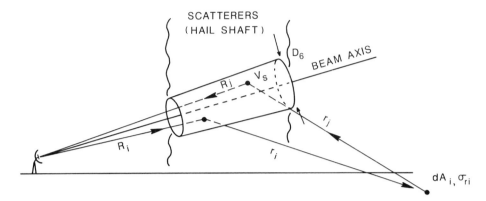

Fig. 8.9 Illustration of the interaction between ith and jth scatterer via the ground patch.

via ground reflection to contribute to the "three-body" scattering. Note that any two scatterers for which the total path length is the same contribute to the signal at the same time. Therefore the scattering volume length is determined by the region containing pairs of hydrometeors that contribute significantly to this interaction, and this volume can be much longer than $c\tau/2$. Because each scatterer interacts with all other (M) scatterers there are M^2 possible interactions. Thus if scatterers are uniformly distributed in the scattering volume, M is proportional to the square of range, and hence the returned signal is independent of range. Signal power depends only on the scattering characteristics of hydrometeors, and the backscattering cross section of the ground. Signal power is inversely proportional to the cube of the range from the center of the scattering volume to the reflecting ground (Section 7.10, Eq. 7.36). The signature is stronger for horizontally polarized electric fields because scattering toward ground (in the vertical plane) is enhanced by oblate hydrometeors and because scattering of vertically polarized waves in the direction directly under the hydrometeors is minimal.

Strong scattering capable to produce three-body signatures is attributed to the large number of scattering pairs and the large cross sections. The latter is the reason why it has been observed more often with the 5-cm wavelength radar than with the 10-cm one (Wilson and Reum, 1988). Whereas three-body signatures in Colorado and Oklahoma were observed at the lowest elevation and were associated with hail on the ground, signatures in Alabama (Wilson and Reum, 1988) were observed above the height of 1 km and did not associate with significant hail at the surface.

Doppler velocities within the signature are usually toward the radar and have been observed to be as strong as -20 to -40 m s^{-1}. Both the vertical and radial velocity components determine the Doppler velocity. But the vertical component contributes more significantly and, because large hydrometeors are usually associated with strong downdrafts, the Doppler velocity is predominantly negative.

8.4.2 Attenuation Method

Communication engineers and radar meteorologists have observed, for a wide range of rainfall rates and rain types, a consistent relationship between specific attenuation of microwaves and the rainfall rate measured with rain gauges along the propagation path. Furthermore, the observations show that K and R are related by a power law for a wide range of rainfall rates and rain types. The relation is nearly linear at 1 cm wavelength; we shall illuminate the underlying cause of this unique property of the specific attenuation that could make its measurement attractive for remote estimation of R.

It can be shown that if $\sigma_e(D)$ is well approximated by a power law dependence on D in the range of diameters that contribute significantly to K, then the liquid water content M and K are also related by a power law, a result that is consistent with many experiments. Moreover, we shall now demonstrate that the power law approximation,

$$\sigma_e(D) \approx CD^n, \qquad (8.27)$$

for the normalized extinction cross sections shown in Fig. 3.4, leads to a rainfall rate as a function of specific attenuation that can be independent of the drop-size distribution! Atlas and Ulbrich (1974) first illustrated that the power law dependence of the rainfall rate on the microwave attenuation implies the effective power law dependence given in relation (8.27).

Now let us examine again the rainfall rate formula (8.20). For R to be expressed as an integral of a power law function of D, we are motivated to use the approximation (8.4) for the terminal velocity w_t. Using (8.4) in (8.19) yields

$$R = [\pi(386.6)/6] \int_0^{\infty} D^{3.67} N(D)\, dD. \qquad (8.28)$$

Substitution of relation (8.27) into Eq. (3.15) gives

$$K = 4.34 \times 10^3 C \int_0^{\infty} N(D)D^n\, dD \quad \text{dB km}^{-1}. \qquad (8.29)$$

Comparison of Eqs. (8.28) and (8.29) immediately shows that if the power exponent n were equal to 3.67, then R and K would be linearly related and, moreover, independent of drop-size distribution.

Atlas and Ulbrich (1974) fitted Waldteufel's (1973) numerically computed K, R values with a power law in the R interval of 1–100 mm h^{-1}. From these fitted curves they have obtained values of C and n for a wide range of wave-lengths (0.1–10 cm) and temperature (0.1–40°C). These data show that $n = 3.67$ for $\lambda = 0.86$ cm. Furthermore, at this wavelength, C and n are essentially inde-pendent of temperature over the range 0–18°C and are altered only negligibly as the temperature rises to 40°C. For comparison, we have plotted in Fig. 3.4 a line (thinly dashed) having the slope $n = 3.67$ (1.67 for the normalized extinction cross section) to show how well it fits the $\sigma_e(D)$ data at $\lambda = 0.86$ cm in the diameter range (0.5–5 mm) of drops that contribute significantly to R (Fig. 8.5). *In situ* measurements of the rain rate with rain gauges having excellent spatial (45 m) and temporal (20 s) resolution, and made close (2 m) to a microwave path $\lambda = 0.84$ cm, confirmed the linear relation between K and R for both thun-derstorm and frontal rain (Norbury and White, 1972).

Any wavelength near 1 cm would similarly show that attenuation and rain-fall rate relations are relatively independent of the drop-size distribution. This conclusion is quickly accepted if one refers to Fig. 8.10, which contains specific attenuation versus rainfall rate for four wavelengths. The circles are K and R values numerically computed by Atlas and Ulbrich (1977) using measured drop-size distributions (204 of them for three days) and the exact attenuation cross sections for each drop diameter at $T = 10°C$. The numerical values of K versus R are from the best fit to the data of a regression equation $\log K = \alpha + \beta \log R$. The regression equations and the average percentage deviation of R are shown in the inset table of Fig. 8.10. We note how closely packed the data are for $\lambda = 0.86$ cm in spite of the large number of different drop spectra used. We do notice at $\lambda = 3.22$ cm a larger scatter, attributed to differences in drop-size spectra.

Even though accurate measurement of K can be related to R if short wavelengths are used, there are practical difficulties in acquiring the data at high rain rates over large areas. For example, the strong attenuation at these short wavelengths imposes limitations on the magnitude and range over which atte-nuation can be measured. Atlas and Ulbrich (1977) show (at $\lambda = 1.25$ cm) that the maximum R that can be measured using the two-way radar method is 20 mm h^{-1} if this rate extends over a range of 30 km. One can trade off range for higher rainfall rate measurements, but in any case the areal coverage is severely limited. There would be a significant increase in areal coverage if one-way measurements could be made; but this implies separated transmitters and receiv-ers, increasing the cost. Another practical obstacle to radar measurements of attenuation is a need for a reflector of known cross section at the end of the path where measurements are being made.

The attenuation method requires the beam to be uniformly filled, although the pulse can experience different levels of attenuation (due to cells along the path having different rain rates) as it propagates along the beam. In this case the measured attenuation is path averaged, and the deduced rainfall rate is the average along the path if R and K are linearly related (as at $\lambda \approx 0.88$ cm).

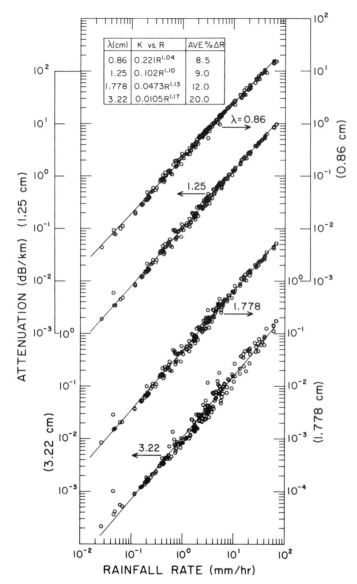

Fig. 8.10   Scattergrams and regression line of K and R at indicated wavelengths. ($T = 10°C$.) The inset table gives the regression equations and average errors in R. (Courtesy of C. W. Ulbrich, Clemson University.)

8.4.3 Differential Phase Method

In addition to attenuation, electromagnetic waves experience phase shift as they propagate through precipitation-filled media. This phase shift adds to the free-space phase shift, and in anisotropic media like rain it depends on polarization. Because rain drops are oriented so that their larger dimension is horizontal (Fig. 8.1) the horizontally polarized electric fields experience larger phase shift than the vertically polarized fields. As explained in Section 6.8.3 the specific differential phase K_{DP} along a path can be measured if a sequence of alternately polarized waves is transmitted and received. In terms of the drop-size distribution K_{DP} is given by (Oguchi, 1983)

$$K_{DP} = \frac{180\lambda}{\pi} \text{Re} \left\{ \int_0^{D_m} [f_h(D_e) - f_v(D_e)] N(D_e)\, dD_e \right\} (\text{deg m}^{-1}), \quad (8.30)$$

where D_e is the equivalent volume diameter of a drop, λ is the free-space wavelength and f_h, f_v are the forward scatter coefficients of the drop for horizontal and vertical polarizations.

Fitting aD_e^b to $f_h - f_v$ produces the exponent $b = 4.24$ at $\lambda = 10$ cm (Sachidananda and Zrnić, 1986). This is significant because it shows the relative insensitivity of the rain rate R or liquid water content M to variations in $N(D_e)$. Note that M depends on D_e^3 in the integrand (8.7), and R has a $D_e^{3.67}$ in the integrand (8.28). If the best-fit power law were $b = 3.67$ or 3 then the relationship between K_{DP} and R or M would be linear and insensitive to the variations in the drop-size distribution. Nevertheless, both of these exponents are much closer to 4.24 than to 6, which is the exponent in the expression (8.13) for Z; this suggests that K_{DP} should provide estimates of R and M, which are almost independent of the drop-size distribution. Jameson (1985) has related the liquid water content to K_{DP}. Sachidananda and Zrnić (1987) proposed the following single-parameter relationship between K_{DP} and rain rate that is valid for Rayleigh scattering.[2]

$$R = 5.1(K_{DP}\lambda)^{0.866} \ (\text{mm h}^{-1}) \qquad (8.31a)$$

in which K_{DP} is in deg km^{-1} and λ is in cm. A similar relationship between M and K_{DP} is

$$M = 0.34(K_{DP}\lambda)^{0.702} \ (\text{g m}^{-3}). \qquad (8.31b)$$

There are additional reasons to use the K_{DP} in radar meteorology. Simulations by Chandrasekar et al. (1990) demonstrate that at high rain rates (>70 mm h^{-1}) the estimator (8.31a) is more accurate than the estimator that uses reflectivity or differential reflectivity (Section 8.5.3.1). Furthermore, hail is often mixed with intense rain, and we show in Section 8.5.3.3 how K_{DP} senses mainly the rain portion of the mixture. Comparison of rain rates in a convective

---

2. Equations 8.31a and b were obtained from a fit at $\lambda = 10$ cm and are approximately valid to wavelengths of 3 cm.

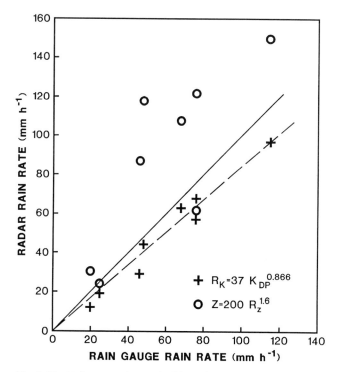

Fig. 8.11   Rain rates observed with a 10-cm radar versus rain gauge from a storm in Alberta. (Courtesy A. R. Holt, University of Essex.)

storm (Fig. 8.11) demonstrates a much tighter clustering of estimates obtained with Eq. (8.31a) than obtained with Eq. (8.22a).

Another very important advantage of differential phase measurement is its independence from system gain calibration. This suggests a method to derive a R, Z relationship based on Eq. (8.31a) at high rain rates. The derived relationship could then be extended to lower rain rates where Eq. (8.31a) does not provide as good an estimate for R as does radar measuring Z alone (Chandrasekar et al., 1990). Other robust characteristics of K_{DP} are (a) it is not affected by rain attenuation, (b) it is almost immune to beam blockage, (c) it is less sensitive to beam filling, (d) it is very well suited for estimation of the average rain rate along the radar beam, and (e) the bias in Eq. (8.31a) due to presence of hail is much smaller than the bias that hail causes to the R, Z relation.

8.5   Multiple-Parameter Measurements of Precipitation

We now turn our attention to methods whereby measurement of two or more parameters of the weather signal can be used to deduce two or more parameters of $N(D)$. For example, measurement of specific attenuation and reflectivity factor can in principle be used to better infer the rainfall rate, liquid water content, or anything else derived from a two-parameter drop-size distribution.

We focus our attention on three techniques whereby several variables can be measured: (1) the dual wavelength method, in which the reflectivity factor Z and specific attenuation K are remotely measured; (2) the polarization method, in which signals at two different polarizations are processed to derive several measurands that can quantify precipitation; and (3) the radar–rain gauge method.

8.5.1   Dual Wavelength

In Section 8.4.2 we showed that, at wavelengths around 1 cm, attenuation alone can be used to estimate R. But, because of the practical limitations that direct attenuation measurements impose, investigators have turned to dual wavelength radars in an attempt to measure K remotely. The dual wavelength technique does not limit attenuation measurements to λ near 1 cm; rather it allows remote measurement of Z and K, from which one can determine two parameters of the drop-size spectra.

Alternatively, we can measure the equivalent reflectivity factor at two wavelengths. One involves Rayleigh scattering, the other a wavelength sufficiently short that substantial Mie scattering (i.e., $D > \lambda/16$) occurs for the larger drops that comprise the scatter domain. From the differences in measured reflectivity factors, one can deduce drop-size parameters (Wexler and Atlas, 1963).

Although there are several dual wavelength techniques to estimate rainfall rates, we shall confine our attention to one promising method—the independent estimate of Z and K. Eccles (1979) used attenuation measured remotely with a dual wavelength radar and computed R from a modified R, K relation of Atlas and Ulbrich (1974). The dual wavelength radar routinely observed severe convective phenomena in northeastern Colorado in 1972–1974, and the precipitation rate computed from the attenuation was found to be more accurate than that computed from Z.

Although the dual wavelength technique applies to any two-parameter $N(D)$ we shall assume an exponential drop-size spectrum to show how the parameters N_0 and Λ can be determined from measurement of the reflectivity factor Z and specific attenuation K. We shall arrive at two simultaneous equations, containing N_0 and Λ, derived from the backscattered power at two wavelengths, one (λ_l) long and nonattenuating, the other (λ_s) short and attenuating. One equation gives the specific attenuation at λ_s; the other, the reflectivity factor at λ_l. We select a nonattenuating wavelength (λ_l) to simplify the problem; the specific attenuation values measured at two attenuating wavelengths can also be used to determine N_0 and Λ, as can Z and K (Goldhirsh and Katz, 1974). Although the measurement of specific attenuation at two wavelengths eliminates the need to calibrate the radar for absolute power (thus disposing of a major source of error in reflectivity measurements), uniform reflectivity must exist along the path over which K is measured.

We assume the two radar beams are matched in resolution volume size. The mean power backscattered at the two wavelengths are, at range $r = r_1$ [Chapters 3 and 4, Eqs. (3.13b) and (4.34)].

$$P_l = \frac{C_l Z}{r_1^2}, \tag{8.32a}$$

$$P_s = (C_s Z_s / r_1^2)\exp\left(-2 \int_0^{r_1} k_s \, dr\right), \tag{8.32b}$$

where Z is the reflectivity factor and Z_s is the equivalent reflectivity factor at λ_s (Section 4.4.5). Because λ_s is small, drop diameters may be larger than the Rayleigh limit $\lambda/16$, so Z_s might not equal Z. The short-wavelength specific attenuation k_s [Eq. (3.15)] is

$$k_s = \int N(D)\sigma_e(D) \, dD \quad \text{m}^{-1}. \tag{8.33}$$

At range r_2 we have two equations of identical form except that Z and Z_s may be different. Taking the natural logarithm of the ratio of Eq. (8.32b) at two ranges r_1 and r_2, we obtain

$$\ln\left[\frac{P_s(r_1)Z_s(r_2)r_1^2}{P_s(r_2)Z_s(r_1)r_2^2}\right] = 2 \int_{r_1}^{r_2} k_s \, dr, \tag{8.34}$$

in which both $Z_s(r_2)$ and $Z_s(r_1)$ are unknown. Nevertheless, from measurements of Z at λ_l we might deduce Z_s. For example, it can be shown that for R between 1 and 100 mm h^{-1} the reflectivity factor Z_s approximates Z well for wavelengths as short as 3 cm (Wexler and Atlas, 1963). Assume $\lambda \geqslant 3$ cm and use Eq. (8.32a) to arrive at the following expression for the path-averaged specific attenuation rate:

$$2\bar{k}_s = \frac{2}{r_2 - r_1} \int_{r_1}^{r_2} k_s \, dr = \frac{1}{r_2 - r_1} \ln\left[\frac{P_s(r_1)P_l(r_2)}{P_s(r_2)P_l(r_1)}\right]. \tag{8.35}$$

The right side is obtained directly from measurements of the echo power at the two wavelengths and two ranges.

If $N(D)$ is uniform across the beam the path-averaged $\bar{Z}$, $\bar{K}$, and $\bar{R}$ are not dependent on how drops are distributed along the averaging length $(r_2 - r_1)$. At a wavelength of 0.86 cm, the R, K relation is nearly independent of $N(D)$, so that the path average $\bar{K}$ alone determines $\bar{R}$.

At other wavelengths we need to assume a two-parameter $N(D)$ for drops within the beam volume between r_1 and r_2. Then $\overline{R}$ should be obtained from $\overline{Z}$ and $\overline{K}$. Assume $N(D)$ to be an exponential drop-size distribution. With this assumption we can determine the path-averaged k_s,

$$\overline{k}_s = N_0 \int e^{-\Lambda D} \sigma_t(D)\, dD \quad (\text{m}^{-1}). \tag{8.36a}$$

For precision, Eq. (8.36a) needs to be numerically evaluated because $\sigma_t(D)$ is not an analytic function; but in practice it can be approximated by a power law Eq. (8.27). Thus, substituting Eq. (8.27) into Eq. (8.36a) and integrating, we obtain

$$\overline{k}_s = \Gamma(n + 1)C'N_0\Lambda^{-(n+1)}, \tag{8.36b}$$

where $C' = 10^{2n-4}C$, and Λ is in inverse meters. The factor 10^{2n-4} converts data for C [given by Atlas and Ulbrich (1974)] from the centimeter to the meter units used in the equations. Likewise the path-averaged reflectivity factor is obtained from Eq. (8.14).

$$\overline{Z} = N_0(6!)\Lambda^{-7}. \tag{8.37}$$

We thus have two equations and two unknowns. Let us change Eq. (8.36b) to units of decibels per kilometer and divide it by Eq. (8.37) to eliminate N_0.

$$\overline{K}_s/\overline{Z} = [4.34 \times 10^3 \Gamma(n+1)C'/(6!)]\Lambda^{6-n} \quad \text{dB km}^{-1}\, \text{m}^{-3}. \tag{8.38}$$

Thus Λ (in inverse meters) is determined directly from the ratio of $\overline{K}_s$ to $\overline{Z}$.

The difficulties inherent in dual wavelength measurements are pointed out by Eccles and Mueller (1973), who propose a technique to solve the problem caused by the statistical fluctuations of the two signals. It is important to note here that we are focusing our attention on path-averaged values, and these can often be estimated with greater accuracy than point (i.e., resolution volume) estimates of Z and K_s. Still, the radar needs to be well calibrated, and a dual wavelength radar might well be supplemented with a smaller number of rain gauges for purposes of calibration.

A more detailed discussion of error and radar requirements is given by Goldhirsh (1975), who shows that R-estimate errors are significantly less if $\lambda_s = 1$ cm. Note that at $\lambda \approx 1$ cm the measurement of $\overline{K}$ alone estimates $\overline{R}$ as accurately as $\overline{K}_s$ is estimated (Fig. 8.10). Although N_0 and Λ need to be estimated using both $\overline{Z}$ and $\overline{K}_s$, $\overline{Z}$ does not add more information to the estimate of $\overline{R}$. Nevertheless, severe attenuation of 1-cm waves in intense precipitation limits the rain area that can be surveyed.

Hail can have a significant effect on measurements with a dual wavelength radar. The ratio of reflectivities at two wavelengths is independent of $N(D)$ for Rayleigh scattering. But if scatterers are large so that Mie scatter effects become significant, this ratio changes. Eccles and Atlas (1973) have proposed a hail-detection technique that recognizes this change, which becomes apparent when spherical hail, wet or dry, has a diameter larger than 8 mm (the radar wavelengths are 10 and 3 cm). The technique is based on the range dependence of the decibel ratio of the reflectivities. Tuttle and Rinehart (1983) define the attenuation-corrected dual frequency reflectivity ratio as the hail signal if it exceeds 3 dB. Hail signature locations found with this criterion have been in good agreement with locations obtained using reflectivity and differential reflectivity (Bringi *et al.*, 1986; Section 8.5.3.2); furthermore, the ratio can be indicative of hail size.

8.5.2 Polarization Diversity

Radars with polarization diversity have a variable transmitted and/or received wave polarization, or provide for dual channel reception of orthogonally polarized waves. This allows measurement of hydrometeor characteristics such as size, shape, spatial orientation, and discrimination of thermodynamic phase. The purpose of this section is to introduce principles of polarization that are applicable to weather radar and show the meteorological utility.

8.5.2.1 Backscattering Matrix

Polarization characteristics of a single hydrometeor are of fundamental importance. These characteristics are described in terms of the backscattering matrix $[\mathbf{S}]$ (e.g., McCormick and Hendry, 1985) that relates the backscattered electric field $[\mathbf{E}]^b$ at the antenna to the incident electric field $[\mathbf{E}]^i$ by

$$\begin{bmatrix} \mathbf{E}_1 \\ \mathbf{E}_2 \end{bmatrix}^b = \begin{bmatrix} s_{11} & s_{12} \\ s_{21} & s_{22} \end{bmatrix} \begin{bmatrix} \mathbf{E}_1 \\ \mathbf{E}_2 \end{bmatrix}^i \frac{\exp(-jkr)}{r} , \qquad (8.39)$$

where the subscripts 1 and 2 refer to two orthogonal polarizations such as linear horizontal and vertical or right-hand circular and left-hand circular. In this convention the first index of the scattering elements refers to the polarization of the backscattered field and the second to the polarization of the incident electric field. In reciprocal media, such as is a distribution of hydrometeors, $s_{12} = s_{21}$, and this will be assumed throughout the rest of the chapter. Circularly polarized orthogonal fields can be expressed in terms of orthogonal linearly polarized fields via the transformation (Bringi and Hendry, 1990)

$$\begin{bmatrix} \mathbf{E}_r \\ \mathbf{E}_l \end{bmatrix}^i = [\mathbf{G}] \begin{bmatrix} \mathbf{E}_h \\ \mathbf{E}_v \end{bmatrix}^i \qquad (8.40a)$$

in which the matrix

$$[\mathbf{G}] = \frac{1}{\sqrt{2}} \begin{bmatrix} j & 1 \\ -j & 1 \end{bmatrix}$$

and the indices r, l stand for right-hand and left-hand circular polarization; h, v are for horizontal and vertical. A similar relationship holds for the backscattered fields except that a conjugate of the matrix $[G]$ must be used because the "handedness" of the reflected wave is opposite to that of the transmitted wave. Thus

$$[E_c]^b = [G^*][E_+]^b, \tag{8.40b}$$

and subscripts c and $+$ stand for the circularly polarized and linearly polarized orthogonal pairs. Starting with Eq. (8.40b), substitute Eq. (8.39), then premultiply Eq. (8.40a) with G^{-1} and substitute this result into Eq. (8.40b) to obtain the following relationship between the scattering matrices for linear and circular polarization

$$[S_c] = [G^*][S][G]^{-1}. \tag{8.41a}$$

Therefore the individual elements are explicitly related by

$$
\begin{aligned}
s_{rr} &= (s_{vv} - s_{hh} - j2s_{vh})/2 \\
s_{ll} &= (s_{vv} - s_{hh} + j2s_{vh})/2 \\
s_{lr} &= s_{rl} = (s_{vv} + s_{hh})/2
\end{aligned}
\tag{8.41b}
$$

8.5.2.2 Backscattering Covariance Matrix and Polarimetric Measurands

We define the polarimetric measurand or variable as a non-redundant backscattering quantity that depends on polarization. For the sake of clarity and because of practical significance we consider linear orthogonal polarization, but the discussion is valid for any other orthogonal polarization basis. Also propagation effects are neglected at first but will be introduced later.

Consider the linearly polarized electric field backscattered by a hydrometeor located at r_n

$$E_{ij} = P_j^{1/2} \exp(-j2kr_n)s_{ij}(n)\eta_0^{1/2}g^{1/2}f(\theta, \phi)/(2(\pi)^{1/2}r_n^2) \tag{8.42}$$

where s_{ij} is an element of the backscattering matrix (8.39) for the nth hydrometeor, k is the wave number, P_j is the transmitted power that produces linearly polarized incident electric field, η_0 is the free-space impedance, and E_{ij} is the received field. The magnitude of the incident field at the scatterer is given by $P_j^{1/2}\eta_0^{1/2}g^{1/2}f(\theta, \phi)/(2(\pi)^{1/2}r_n)$ so that the convention regarding the scattering coefficients in Eq. (8.42) agrees with that of McCormick and Hendry (1975) in which $|s_{hh}|^2 = \sigma_b/4\pi$.

In weather radars, signal voltages v_{ij} are processed to retrieve properties of hydrometeors. The voltage v_{ij} for the nth hydrometeor is proportional to the scattering coefficient and can be written as

$$v_{ij}(\mathbf{r}_n) = s_{ij}(n)F(\mathbf{r}_n) \exp(-j2kr_n). \tag{8.43a}$$

The proportionality factor $F(\mathbf{r}_n)$ contains range dependence, attenuation, the weighting function, and other system parameters (Section 5.2.2). For an ensemble of scattering hydrometeors the composite voltage V_{ij} is a superposition of voltages from each individual scatterer (Section 4.2, Eq. 4.1)

$$V_{ij} = \sum_n s_{ij}(n) \exp(-j2kr_n) F(\mathbf{r}_n). \qquad (8.43b)$$

The mean value of V_{ij} is zero because contributions by the phase terms in the summation over n cancel each other. Thus radar meteorologists use various second-order moments, $\langle V_{ij} V_{kl}^* \rangle$ to characterize the polarized signals (brackets denote expectations) and relate these to properties of the hydrometeors. Detailed derivation of the relationship between some second-order moments and the scattering coefficients can be found elsewhere (e.g., Jameson, 1985). Starting from Eq. (8.43b) the expected value of the general term is

$$\langle V_{ij} V_{kl}^* \rangle = \left\langle \sum_n \sum_m [s_{ij}(n)s_{kl}^*(m) \exp\{-j2k(r_n - r_m)\}F(\mathbf{r}_n)F^*(\mathbf{r}_m)] \right\rangle$$

$$= \sum_n \langle [s_{ij}(n)s_{kl}^*(n)] \rangle |F(\mathbf{r}_n)|^2$$

$$= \langle s_{ij} s_{kl}^* \rangle \int |F(\mathbf{r}_n)|^2 dV. \qquad (8.44a)$$

In the last equality the summation over n is replaced with the integral over the resolution volume weighting function, and it is assumed that the scatterer distribution is homogeneous.

For the most general case the second-order moments [Eq. (8.44a)] can be grouped in a four-by-four covariance matrix but, because of the reciprocity, the term $V_{ij} = V_{ji}$ so that the covariance matrix reduces to a three-by-three dimension (Borgeaud *et al.*, 1987). It is evident from Eq. (8.44a) that the voltage covariance matrix is a scalar multiple of the backscattering covariance matrix defined as

$$
\begin{matrix}
\langle |s_{hh}|^2 \rangle & \langle s_{hv}s_{hh}^* \rangle & \langle s_{vv}s_{hh}^* \rangle \\
\langle s_{hh}s_{hv}^* \rangle & \langle |s_{hv}|^2 \rangle & \langle s_{vv}s_{hv}^* \rangle \\
\langle s_{hh}s_{vv}^* \rangle & \langle s_{hv}s_{vv}^* \rangle & \langle |s_{vv}|^2 \rangle.
\end{matrix}
\qquad (8.44b)
$$

There are other ways to define this matrix, notable is the one due to Tragl (1990) who scales the elements so that powers are conserved when the matrix is multiplied with orthogonal vector components of the electric field. The sole purpose here is to illustrate the number of measurands and to relate these to commonly observed variables; thus the simple form (8.44b) is chosen.

The expectations in (8.44) are expressed in terms of the distribution of the hydrometeor's properties (i.e., equivalent volume diameter, shape, canting angle, etc.). Thus the general term is

$$\langle s_{ij} s_{kl}^* \rangle = \int N(\mathbf{X}) s_{ij} s_{kl}^* \, d\mathbf{X} \qquad (8.45)$$

$N(\mathbf{X})$ is the probability density of the scatterer's properties. These properties are represented by a vector $\mathbf{X}$.

The off-diagonal symmetric terms in the backscattering covariance matrix are conjugates of each other; therefore there are nine real quantities (three real on the main diagonal and the remaining six real from the off-diagonal terms) that a polarimetric radar can measure (Ioannidis and Hammers, 1979).

Most terms of the backscattering covariance matrix have been used by themselves or in combination with others to infer properties of hydrometeors. A number of attributes describing the properties of an ensemble of hydrometeors is at least six; two for the distribution of sizes, two for the thermodynamic phase, one for shape and one for orientation. Thus one would expect that some bulk properties of the ensemble could be estimated from the nine polarimetric measurands. This is indeed the case in special situations as when precipitation is pure rain (Section 8.5.3.1). But hydrometeors are often mixed, do not have well-defined shapes, and their polarimetric signatures are ambiguous. Furthermore the relationship between the bulk parameters and the measurands is nonlinear, obfuscated by the expectation integrals in Eq. (8.45). Therefore researchers have used clever combinations of measurands that eliminate some parameters of the hydrometeor properties and isolate others. Also physical considerations including location of measurements with respect to the melting layer and models of precipitation are used to unravel polarmetric signatures. Some quantities that are derived from the covariances are

1. Reflectivity factor at horizontal polarization

$$Z_{\mathrm{h}} = (4\lambda^4/\pi^4|K_{\mathrm{w}}|^2) \langle|s_{\mathrm{hh}}|^2\rangle, \tag{8.46a}$$

2. Reflectivity factor at vertical polarization

$$Z_{\mathrm{v}} = (4\lambda^4/\pi^4|K_{\mathrm{w}}|^2) \langle|s_{\mathrm{vv}}|^2\rangle, \tag{8.46b}$$

3. Differential reflectivity

$$Z_{\mathrm{DR}} = 10 \, \log(\langle|s_{\mathrm{hh}}|^2\rangle/\langle|s_{\mathrm{vv}}|^2\rangle), \tag{8.46c}$$

4. Linear depolarization ratio

$$\mathrm{LDR}_{\mathrm{hv}} = 10 \, \log(\langle|s_{\mathrm{hv}}|^2\rangle/\langle|s_{\mathrm{vv}}|^2\rangle), \tag{8.46d}$$

or

$$\mathrm{LDR}_{\mathrm{vh}} = 10 \, \log(\langle|s_{\mathrm{hv}}|^2\rangle/\langle|s_{\mathrm{hh}}|^2\rangle), \tag{8.46e}$$

5. Correlation coefficient at zero lag

$$\rho_{\mathrm{hv}}(0) = \langle s_{\mathrm{vv}}s_{\mathrm{hh}}^*\rangle/[\langle|s_{\mathrm{hh}}|^2\rangle^{1/2}\langle|s_{\mathrm{vv}}|^2\rangle^{1/2}]. \tag{8.46f}$$

This list contains five variables, three real from the diagonal terms and one complex off-diagonal term. The other two complex measurands have been less utilized, although Jameson (1985) has shown their explicit dependence on the scattering coefficients. The three complex terms of the backscattering covariance matrix can be expressed as functions of amplitudes and phases

$$\langle s_{vv}s_{hh}^* \rangle = \langle |s_{vv}s_{hh}^*| \exp[j(\delta_{vv} - \delta_{hh})]\rangle \tag{8.47a}$$

$$\langle s_{hv}s_{vv}^* \rangle = \langle |s_{hv}s_{vv}^*| \exp[j(\delta_{hv} - \delta_{vv})]\rangle \tag{8.47b}$$

$$\langle s_{hv}s_{hh}^* \rangle = \langle |s_{hv}s_{hh}^*| \exp[j(\delta_{hv} - \delta_{hh})]\rangle, \tag{8.47c}$$

where the phase δ_{ij} of the scattering coefficient s_{ij} corresponds to the lag or lead angle of the backscattered field (polarization i) with respect to the incident field (polarization j) referenced to the location of the scatterer.

Hydrometeors for which the Rayleigh–Gans theory is applicable (usually at 10-cm wavelengths) have very small differences of backscattering phase angles (Jameson and Mueller, 1985) so the three terms (8.47a,b, and c) are nearly real. Thus the total number of useful measurands is reduced to six. But at shorter wavelengths this reduction may not be possible; on the other hand the backscattering phase differences could have meteorological significance.

Analogous to Eq. (8.46f) two more correlation coefficients can be defined as

$$\rho_v = \langle s_{hv}s_{vv}^* \rangle / [\langle |s_{hv}|^2 \rangle^{1/2} \langle |s_{vv}|^2 \rangle^{1/2}] \tag{8.48a}$$

$$\rho_h = \langle s_{hv}s_{hh}^* \rangle / [\langle |s_{hv}|^2 \rangle^{1/2} \langle |s_{hh}|^2 \rangle^{1/2}]. \tag{8.48b}$$

The symbols ρ_h, ρ_v are chosen to designate the correlation coefficients between the co-polar (indicated by the subscript) and cross-polar weather signals, and distinguish these from the correlation ρ_{hv} between the co-polar signals (where both indices stand for the received polarization).

For hydrometeors, small compared to radar wavelength, the two correlations (8.48a) and (8.48b) are about equal, and that would further reduce the number of useful measurands to five. Even for non-Rayleigh scatterers with regular shapes, such as spheroids, the correlations (ρ_v and ρ_h) should be similar. Moreover it remains to be demonstrated that ρ_v and/or ρ_h could add significantly different information to what other measurands already carry. If there is no added information the number of useful measurands would be four; this is a realistic possibility in view of the observations by Illingworth and Caylor (1989) who claim that ρ_h through the bright (Section 8.5.3.2) band did not contain useful information. Methods to obtain all these polarimetric measurands with a single-receiver radar are proposed by Zrnić (1991).

8.5.2.3 Propagation Effects

The discussion so far deliberately ignored propagation effects in order to clearly identify the backscattering properties of the hydrometeors in the radar resolution volume. Both attenuation and phase shift along the propagation path affect the received signals and an account of these need to be made (Bebbington et al., 1987). At a 10-cm wavelength, the differential phase along propagation paths can be measured and related to attenuation; differential attenuation is rather small and is also linearly related to differential phase shifts (Bringi *et al.*, 1990). Therefore we will consider only the differential phase shifts. The shifts can be introduced directly into the off-diagonal terms of the back-scattering covariance matrix. Thus they also affect the correlation coefficients (8.46f), (8.48a), and (8.48b). Because radar measurements contain both the differential phase shift upon backscatter and the differential phase shift along the propagation path, it is instructive to write the correlation coefficient in which the two are separated. Thus the correlation of the co-polar signals is given by

$$\rho_{hv}(0)\,\exp[j(\phi_{hh} - \phi_{vv})] \tag{8.49a}$$

and the correlations of the co- and cross-polar echoes are

$$\rho_v\,\exp[j(\phi_{vv} - \phi_{hv})] = \rho_v\,\exp[j(\phi_{vv} - \phi_{hh})/2] \tag{8.49b}$$

$$\rho_h\,\exp[j(\phi_{hh} - \phi_{hv})] = \rho_h\,\exp[j(\phi_{hh} - \phi_{vv})/2]. \tag{8.49c}$$

The phase shifts ϕ_{ij} are cumulative for the total round trip between the radar and the resolution volume. Reciprocity is again invoked by setting $\phi_{hv} = \phi_{vh}$, and the right sides of Eqs. (8.49b) and (8.49c) are obtained by inserting the equality

$$\phi_{vh} = (\phi_{hh} + \phi_{vv})/2. \tag{8.50}$$

Note that the differential phase

$$\phi_{DP} = \phi_{hh} - \phi_{vv} \tag{8.51}$$

can be obtained directly from any one of the equations (8.49a, b, c) if the scattering is Rayleigh because ρ_{hv}, ρ_h, or ρ_v would all be real. Otherwise there would be differential phase shifts on scattering that cannot be separated from ϕ_{DP} without accounting for propagation effects.

Differential phase shifts due to propagation can have significant effects on circularly polarized waves. Before discussing these we briefly point out some practical uses of circular polarization. Radars designed to track aircraft in precipitation use circular polarization because it is possible to easily remove a large portion (e.g., 99%) of precipitation echo. For simplicity consider a back-scattering matrix of a single spheroidal hydrometeor whose minor axis is vertical. Symmetry considerations require the backscattering matrix for linear horizontal and vertical polarization to contain only the diagonal terms s_{hh} and s_{vv}. Use of Eq. (8.41b) establishes that the s_{rr} and s_{lr} are related to s_{hh} and s_{vv} as follows

$$s_{rr} = (s_{vv} - s_{hh})/2 \tag{8.52a}$$

$$s_{lr} = (s_{vv} + s_{hh})/2. \tag{8.52b}$$

In circularly polarized radars the power received by the antenna is separated into the right-hand channel and the left-hand channel. For a rain composed of small spherical drops $s_{hh} = s_{vv}$, and the ensemble average of Eq. (8.52a) is zero;

this signifies that the transmitted right-hand circular polarization has, on reflection, become left-hand polarized. In this idealized case all the power is in the "main" left-hand channel. In the presence of larger drops s_{vv} is slightly smaller than s_{hh} (Section 8.5.3.1) and there would always be some power in the "orthogonal" channel. Circularly polarized echoes from large irregular objects such as aircraft would have on the average about the same amount of power in either polarization and would appear in both channels. Therefore the signal, in the orthogonal channel (i.e., where s_{lr} is processed), from an aircraft could be much stronger than the weather signal.

The scenario just described applies to situations where rain between the antenna and the resolution volume of the measurement is low to moderate. If the path is filled with heavy rain, depolarization of circularly polarized echo can be so severe that most echo power shifts back into the orthogonal channel. The amount of power in either channel can be quantified by considering the transmission matrix $[\mathbf{T}]$ for linear orthogonal polarization given by

$$[\mathbf{T}] = \begin{bmatrix} \exp(-j\phi_{DP}/2) & 0 \\ 0 & 1 \end{bmatrix}. \tag{8.53}$$

In this simplified model it is assumed that the differential phase shift is the dominant propagation effect and that there is no depolarization of linear components through propagation as would occur if drops were canted or wobbling; hence the off-diagonal terms in $\mathbf{T}$ are zero. Note also that $\phi_{DP}/2$ is the one-way differential phase shift. The effective backscattering matrix for linearly polarized waves that includes propagation is $[\mathbf{T}][\mathbf{S}][\mathbf{T}]$ and after it is substituted in Eq. (8.41a) in place of $[\mathbf{S}]$ the elements of the corresponding $[\mathbf{S}_c]$ (for circular polarization) are found. For example,

$$s_{rr} = [s_{vv} - s_{hh} \exp(-j\phi_{DP})]/2 \tag{8.54a}$$

$$s_{lr} = [s_{vv} + s_{hh} \exp(-j\phi_{DP})]/2. \tag{8.54b}$$

Now powers in the two orthogonal channels are proportional to the ensemble average of $|s_{rr}|^2$ and $|s_{lr}|^2$. For Rayleigh scatter $I_m(s_{vv}s_{hh}^*) = 0$ and the averages are

$$\langle |s_{rr}|^2 \rangle = (\langle |s_{vv}|^2 \rangle - 2 \operatorname{Re}\langle s_{vv}s_{hh}^* \rangle \cos(\phi_{DP}) + \langle |s_{hh}|^2 \rangle)/4 \tag{8.55a}$$

$$\langle |s_{lr}|^2 \rangle = (\langle |s_{vv}|^2 \rangle + 2 \operatorname{Re}\langle s_{vv}s_{hh}^* \rangle \cos(\phi_{DP}) + \langle |s_{hh}|^2 \rangle)/4. \tag{8.55b}$$

Because $\langle |s_{vv}|^2 \rangle$ is proportional to the reflectivity factor Z_v of vertically polarized waves, $\langle |s_{hh}|^2 \rangle$ to the Z_h, and ρ_{hv} is given by Eq. (8.46f), we can write the ratio of the powers in the "main" and "orthogonal" channel as

$$\frac{P_{lr}}{P_{rr}} = \frac{Z_h + Z_v + 2 \operatorname{Re}[\rho_{hv}(0)](Z_h Z_v)^{1/2} \cos \phi_{DP}}{Z_h + Z_v - 2 \operatorname{Re}[\rho_{hv}(0)](Z_h Z_v)^{1/2} \cos \phi_{DP}}, \tag{8.56a}$$

where the reflectivity factors are in units of $mm^6\ m^{-3}$. The arithmetic mean and geometric mean of the reflectivity factors in Eq. (8.56a) are very close because

the two reflectivities are not much different (Section 8.5.3.1). Furthermore because differential phase shift on scattering is zero, ρ_{hv} is real. Consequently a
very good approximation of Eq. (8.56a) is

$$\frac{P_{lr}}{P_{rr}} = \frac{1 + \rho_{hv} \cos \phi_{DP}}{1 - \rho_{hv} \cos \phi_{DP}} \ . \tag{8.56b}$$

In rain media the average value of the correlation coefficient was observed
(Balakrishnan and Zrnić, 1990b) to be 0.98, and introduction of this value
further simplifies Eq. (8.56b).

A look at either Eq. (8.56a) or Eq. (8.56b) reveals that the ratio of powers
at $\phi_{DP} = 180°$ is the reciprocal of the ratio at $\phi_{DP} = 0°$. At phase shifts of
$180° + 360°n$ (n is an integer) the depolarization is maximal, and therefore most
of the echo power ends up in the orthogonal channel. This is a realistic possibility even at S-band frequencies in heavy rain; for example, a rain rate of
50 mm h^{-1} over a 60-km distance (not uncommon in squall lines) could generate

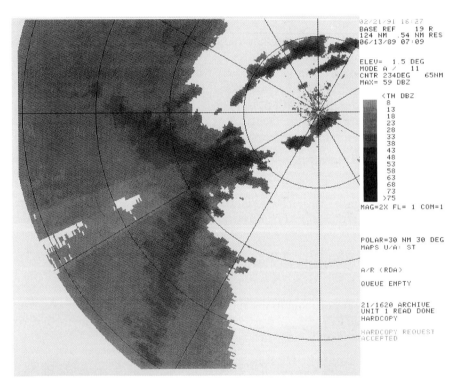

Fig. 8.12 Reflectivity factor field of a storm complex obtained with a prototype WSR-88D radar
using circular polarization. Range marks are spaced by 30 nm and the elevation is 1.5°. Significant
depolarization loss is seen behind the high reflectivity core between azimuths of 240° and 250°.

180° of cumulative phase shift. This large phase shift has occurred along some radials in Fig. 8.12, causing the streak-like reduction in the reflectivity behind strong cores. An example of a cumulative phase shift along a radial in Fig. 8.13a indicates that 55° accumulates over a 10-km path if the reflectivity is over 50 dBZ. Note the 1.5–2.5-dB values of the differential reflectivity (Fig. 8.13b) that also typifies rain (Section 8.5.3.1); the dip in Z_{DR} (dashed in Fig. 8.13b) is caused by the presence of hail, and it coincides with the decrease of the correlation coefficient (Fig. 8.13b and Section 8.5.3.4) and the maximum of reflectivity.

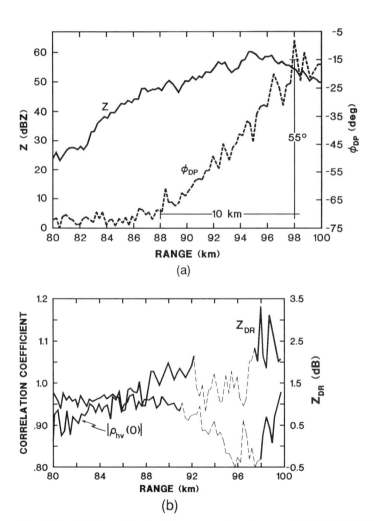

Fig. 8.13 Radial profiles of multiparameter data taken on 29 July 1987. (a) Profiles of Z_h and ϕ_{DP}. (b) Profiles of Z_{DR} and $|\rho_{hv}(0)|$. The dashed portions of the curves between 92 km and 98 km indicate possible regions of rain/hail mixture (from Golestani *et al.*, 1989).

8.5.2.4 The Backscattering Matrix Coefficients and Reflectivities for Oblate Spheroids

The basis for the dual polarization rainfall measurement is the observation that drops are not spherical but have an oblate spheroidal shape (Fig. 8.1). In still air the drops fall with their minor axis vertical and have eccentricities that depend only on the diameter of the equivalent sphere. Although falling drops generate a spectrum of shapes, consider the effect that a single drop falling in still air has on the backscattering cross sections.

The axis ratio b/a is related to the diameter D_e of an equivalent-volume spherical raindrop as (Green, 1975)

$$D_e = 2\{(T_s/g\rho_w)[(b/a)^2 - 2(b/a)^{1/3} + 1](b/a^{1/3}\}^{1/2}, \qquad (8.57)$$

where $T_s = 72.75 \times 10^{-3}$ J m^{-2} is the surface tension of water, g is the accelera-tion due to gravity, and ρ_w is the water density. The plot of Eq. (8.57) in Fig. 8.14 shows that if $D_e > 1$ mm, drops become flattened and a nearly 2:1 ratio of width to thickness is attained for drops approaching the breakup limit of 8 mm. The diameter D_e of the sphere having a volume $\pi D_e^3/6$ that equals the volume $4\pi b^2 a/3$ of the spheroid is chosen as a convenient parameter not only because it uniquely specifies the water drop shape but because it is also the parameter used in relating the terminal velocity to drop size (Fig. 8.4).

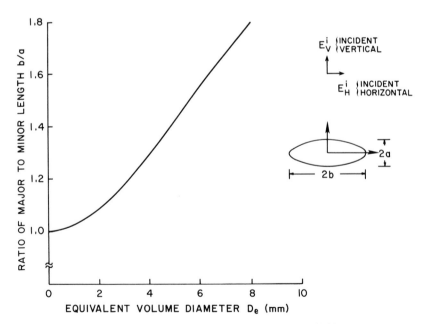

Fig. 8.14 The ratio of the major axis length to minor axis length (b/a) as a function of D_e given by Eq. (8.57).

We shall now use a simplified theory attributed to Gans (1912) and employed by Seliga and Bringi (1976) to determine the backscattering matrix coefficients of oblate spheroids for incident waves of either horizontal or vertical polarization. Gans's work is essentially an extension of the Rayleigh theory for spheres applied to the case of oblate and prolate spheroids. According to Gans's theory, the components of dipole moments aligned along the axes of the ellipsoid are independently excited by the components of the incident electrical field along each axis. As with scattering from a sphere, the dipoles that produce the radiation or scattering have moments not only proportional to the incident electric field strength and to the volume and dielectric properties of the scatterer, but also to factors that depend on the shape and orientation.

The backscattering matrix elements for an oblate drop are given by (Stapor and Prat, 1984)

$$s_{hh} = k_0^2[(p_v - p_h) \sin^2 \delta \sin^2 \psi + p_h] \tag{8.58a}$$

$$s_{vv} = k_0^2[(p_v - p_h) \cos^2 \delta \cos^2 \psi + p_h] \tag{8.58b}$$

$$s_{hv} = 0.5 k_0^2 (p_v - p_h) \sin^2 \delta \sin 2\psi, \tag{8.58c}$$

where

$$p_{h,v} = \frac{ab^2}{3} \left\{ \frac{m^2 - 1}{A_{h,v}(m^2 - 1) + 1} \right\} \tag{8.59a}$$

$$A_v = \frac{1}{e^2} \left\{ 1 - \left(\frac{1 - e^2}{e^2} \right)^{1/2} \sin^{-1}(e) \right\} = 1 - 2A_h \tag{8.59b}$$

and e is the eccentricity of the ellipsoid,

$$e = \{1 - (a/b)^2\}^{1/2}. \tag{8.59c}$$

The refractive index of water is m (Section 3.2), δ is the angle between the propagation direction of the incident field and the symmetry axis, and ψ is the canting angle, that is, the angle between the incident electric field and the projection of the axis of symmetry on the plane of polarization (Fig. 8.15). The forward scattering amplitudes needed for the determination of differential phase [Eq. (8.30)] for a drop that does not appear canted (i.e., $\psi = 0$) are given by $f_{h,v} = k_0^2 p_{h,v}$ (Oguchi, 1983; Sachidananda and Zrnić, 1986).

Starting with Eq. (8.42) it follows that the backscattering cross section of a scatterer is related to the scattering coefficients by

$$|s_{ij}|^2 = \sigma_{ij}/4\pi. \tag{8.60}$$

Hence from Eq. (8.58a) an oblate spheroidal drop has the cross sections for horizontal and vertical incident fields given by

$$\sigma_{h,v} = \frac{\pi^5 D_e^6}{9\lambda^4} \left| \frac{m^2 - 1}{1 + (m^2 - 1)A_{h,v}} \right|^2. \tag{8.61}$$

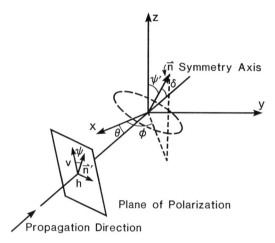

Fig. 8.15   Scattering geometry where **n** is the symmetry axis of the scatter, **n′** is the projection of **n** onto the constant phase plane, θ is the radar elevation angle, **h** and **v** are the linear polarization base vectors, and ψ is the canting angle of the scatterer. The vector **v** denoting vertical polarization is in the x, z plane.

It can be shown that when $a/b \rightarrow 1$, $A_h = A_v \rightarrow 1/3$ and $\sigma_h \rightarrow \sigma_v$, both of which then equal the backscattering cross section given by Eq. (3.6).

The ratio of σ_h/σ_v for oblate spheroids (Fig. 8.16) indicates marked differences in the σ_h, σ_v backscattering cross sections for water in the range of b/a expected for liquid drops. This ratio would also be the ratio of reflectivities η_h/η_v for drops of uniform size. Note that the dashed portion of the line for liquid is at ratios where drops are unstable; however, it can be shown (Battan, 1973) that if 25% (10% of the radius) of a spherical ice mass melts, it appears to the radar as a liquid hydrometeor of the same diameter. Curves labeled graupel and snow are for oblate spheroids with densities of graupel and snow. These were obtained from the Debye theory, which postulates that $|K_w|/\rho$ [K_w is the factor containing refractive index (Section 3.2) and ρ is density] is constant. For snow $\rho = 0.1$ g cm^{-3} and $m = 1.069$; and for graupel $\rho = 0.5$ g cm^{-3} and $m = 1.372$ (Illingworth *et al.*, 1987).

The logarithm of the reflectivity ratio defines the differential reflectivity.

$$Z_{DR} = 10 \log(\eta_h/\eta_v). \tag{8.62}$$

The reflectivity, a function of the drop-size distribution, can be expressed in terms of D_e.

$$\eta_{h,v} = \int \sigma_{h,v}(D_e) \, N(D_e) \, dD_e. \tag{8.63a}$$

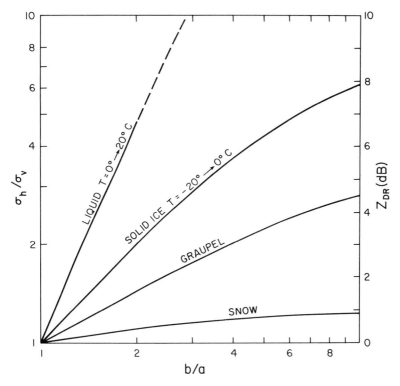

Fig. 8.16 The ratio of backscattering cross sections σ_{h}, σ_{v} for an oblate spheroidal scatterer as a function of its axis ratio and composition. The curves for graupel and snow have been adapted from Frost *et al.*, 1989.

We shall assume a truncated exponential drop-size distribution, so

$$\eta_{\mathrm{h,v}} = \frac{N_0 \pi^5}{9\lambda^4} \int_0^{D_{\max}} D_{\mathrm{e}}^6 \left| \frac{m^2 - 1}{1 + (m^2 - 1)A_{\mathrm{h,v}}(D_{\mathrm{e}})} \right|^2 e^{-\Lambda D_{\mathrm{e}}} \, dD_{\mathrm{e}}. \quad (8.63b)$$

We substitute Eq. (8.63b) into Eq. (8.62) to obtain

$$Z_{\mathrm{DR}} = 10 \log \left[\frac{\displaystyle\int_0^{D_{\max}} D_{\mathrm{e}}^6 S_{\mathrm{h}}(m, b/a)(e^{-\Lambda D_{\mathrm{e}}}) \, dD_{\mathrm{e}}}{\displaystyle\int_0^{D_{\max}} D_{\mathrm{e}}^6 S_{\mathrm{v}}(m, b/a)(e^{-\Lambda D_{\mathrm{e}}}) \, dD_{\mathrm{e}}} \right], \quad (8.64)$$

where

$$S_{\mathrm{h,v}}(m, b/a) = \left| \frac{m^2 - 1}{1 + (m^2 - 1)A_{\mathrm{h,v}}(D_{\mathrm{e}})} \right|^2$$

gives the dependence of Z_{DR} on the shape as well as the temperature and wavelength (through the complex refractive index m) provided $D_e \ll \lambda$. The parameter $A_{h,v}(D_e)$ is a function of D_e through Eqs. (8.57) and (8.59a,b,c). We note that Eq. (8.64) is independent of N_0, and thus we can directly determine Λ of the truncated drop size distribution if D_{max} is known.

8.5.3 Application of Dual Polarization

The potential of polarization diversity radar to measure rainfall and discriminate hydrometeor types is examined in this section. We consider linear orthogonal polarization and use some of the measurands discussed previously.

8.5.3.1 Rainfall Rate Estimates

Seliga and Bringi (1976) have evaluated Eq. (8.64) for $D_{max} = 10$ mm and $T = 20°C$. Their results are plotted in Fig. 8.17, where for comparison the Z_{DR} values using an exact theoretical formulation evaluated by Al-Khatib $et\ al.$ (1979) are also shown. We see that the simplified theory agrees well with the more exact formulation for $\Lambda \geq 2$ mm^{-1}. The large difference for $\Lambda < 2.0$ mm^{-1} is mostly caused by the difference in D_{max}. When both limits are equal, there is relatively little difference in the entire indicated range of Λ (Seliga and Bringi, 1978).

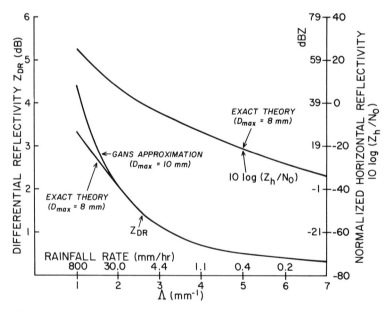

Fig. 8.17 Variations of Z_{DR} and $10 \log(Z_h/N_0)$ with Λ. The rainfall rate and dBZ scales are for $N_0 = 8 \times 10^3$ m^{-3} mm^{-1}, and Z_h is in mm^6 m^{-3}.

Once we obtain Λ from Z_{DR} measurements, we need to determine N_0 to completely specify the drop-size distribution under the assumption that it is exponential. N_0 can be determined from measurements of η_h or η_v. The reflectivity factors $Z_{h,v}$ in terms of $\eta_{h,v}$,

$$Z_{h,v} = (\lambda^4/\pi^5|K_w|^2)\eta_{h,v}, \tag{8.65}$$

have been evaluated by Al-Khatib *et al.* (1979) using the exact formulation for scattering. We also plot Z_h in Fig. (8.17); for reference, we have shown in Fig. 8.17 the rainfall rate along the abscissa and reflectivity factor (in dBZ) along the ordinate using the Marshall–Palmer N_0 value. Both parameters of the drop-size distribution, and hence the rainfall rate, can now be computed from Z_h and Z_{DR} measurements.

If D_{max} is an independent variable, then D_{max} needs to be estimated; otherwise, we can have different R for the same measured Z_h and Z_{DR}. Ulbrich and Atlas (1984) point out that, if $\Lambda D_{max} > 9.2$, Z_{DR} is not sensitive to the truncation of an exponential distribution. But scientists at the Air Force Geophysics Laboratory using disdrometer data have found that for rain the maximum drop diameters generally adhered to the relationship (Plank, 1977)

$$D_{max}\Lambda = 7.5. \tag{8.66}$$

Sachidananda and Zrnić (1987) varied parameters of a gamma drop-size distribution to obtain a best-fit relationship between rain rate and Z_h, Z_v that can be expressed as

$$R = 6.84 \times 10^{-3} Z_h^{-3.86} Z_v^{4.86} \quad (\text{mm h}^{-1}), \tag{8.67}$$

where the reflectivity factors ($\text{mm}^6 \text{ m}^{-3}$) are for the indicated polarizations. A similar relationship was proposed by Ulbrich and Atlas (1984) and a slightly modified one by Seliga *et al.* (1986).

It should be evident that the single parameter R, Z relationship must depend on polarization. If the relationship is valid at one polarization it would have to be transformed through formula (8.67) for use at the other polarization (Zrnić and Balakrishnan, 1990).

An example of the difference that polarization makes to the R, Z relation follows. Crane (1975) cites that $270R^{1.3}$ was well suited for rain in Virginia; the polarization was vertical. Crozier *et al.* (1989) claim that $295R^{1.43}$ fits best data in the southern Ontario region; the polarization for these measurements was horizontal. Substitution of the relation for the horizontal polarization in Eq. (8.67) produces $Z_v = 255R^{1.34}$, which is very close (Fig. 8.18) to the Crane relation. It is clear that the two relations are nearly equivalent, tied by the polarization transformation; application of a valid relation to an inappropriate polarization could result in 30% errors. The familiar Marshall–Palmer relationship (8.22a) seems to be appropriate for horizontal polarization.

To illustrate the complexity of measurement and the difficulty in comparing the rain-gauge and radar estimates we have plotted in Fig. 8.19 the rain rates obtained with various radar techniques and a rain gauge. The estimates from

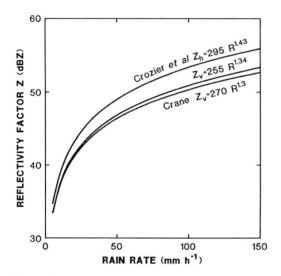

Fig. 8.18 Z_h, R relationship from Crozier *et al.* (1989); Z_v, R relation from Crane (1975); and a Z_v, R relationship obtained from the Crozier *et al.* relationship using Eq. (8.67).

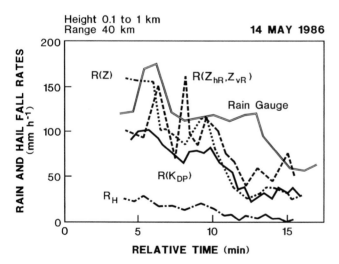

Fig. 8.19 Rain and hail rates versus time. Radar data were obtained from vertical cross sections at heights between 0.1 and 1 km above ground and are smoothed over half-minute intervals.

radar data are averages over 2.25 km in range and 1 km in height. Note that $R(Z)$ from formula (8.22a) and $R(Z_{hR}, Z_{vR})$ from formula (8.67) produce larger rain rates (second subscript R stands for contribution by rain, Section 8.5.3.2) in the region of higher reflectivities than does $R(K_{DP})$ [i.e., Eq. (8.31a)]. That is because small hail (<1 cm in diameter), present until about 10 minutes into a data collection, biases the Z, R relation, but K_{DP} is sensitive only to liquid water (Section 8.5.3.2). The reflectivity factor Z_{hR} contains contributions from hail and so does the differential reflectivity. Hail rate R_h is discussed in Section 8.5.3.3.

Under ideal conditions the drop-size distribution variability produces, at most, errors of the order of 30–35% in rain rates if the $R(Z)$ method is used. This has been shown by simulating estimates from disdrometer measurements (Balakrishnan et al., 1989). Use of differential reflectivity reduces the error to 20–25% and the differential propagation constant has the lowest error of 10–15%. But the disdrometer simulation does not consider radar errors and the needed dwell times to make the measurements. When these effects are introduced (Chandrasekar et al., 1990) the estimators' accuracies depend on the rain rate. Thus at a wavelength of 10 cm the R, Z relationship is superior for $R \leq 20$ mm h^{-1}; it outperforms $R(Z_h, Z_v)$ because random errors in Z_{DR} mask any improvement obtained by the extra information. $R(K_{DP})$ has a fractional error that is inversely proportional to the rain rate. Therefore it is superior for heavy rainfall ($R \geq 70$ mm h^{-1}). This would leave the range 20–70 mm where $R(Z_h, Z_v)$ appears to be the best.

There is at least one distinct advantage of the dual polarization technique over the dual wavelength method previously discussed. The dual polarization technique measures the rainfall rate averaged over the resolution volume and does not require the rainfall to be uniformly distributed within this volume. This is not the case for the dual wavelength method, in which the measured value of the attenuation depends strongly on the distribution of the rain across the beam.

8.5.3.2 *Discrimination between Ice and Liquid Hydrometeors — Use of Reflectivity Factors*

Another benefit of the dual polarization technique is its capability to differentiate at times between the ice phase and the liquid phase of water. For instance, in the high-reflectivity column (55 dBZ) at 35 km (Fig. 8.20b), water drops below 2 km are responsible for the large (>2 dB) Z_{DR} (Fig. 8.20a). Above 2 km, Z_{DR} is only 0.13 dB, most likely because of randomly oriented ice hydrometeors. The sharp change in Z_{DR} occurs near the 0°C isotherm and marks the transition between ice particles and water. As mentioned earlier (Section 8.5.2.4), water-coated ice can backscatter as a liquid hydrometeor of the same diameter. Assuming this holds for flattened ice-crystal structures, a large increase in Z_{DR} due to change in phase occurs when the ice is partially melted (Fig. 8.16). The reflectivity factor Z often increases (as ice melts) many times more (~ 10) than could be caused by phase change alone (4.7 times). These large reflectivity increases are attributed to the coalescence of ice crystals, which begins above the melting layer. Humphries and Barge (1979) have simultaneously observed with dual wavelength (5 and 10 cm) and polarization diversity radar (10 cm) the reflectivity and circular depolarization ratio (CDR) in stratiform precipitation. Their measurements suggest the presence of water-coated non-Rayleigh scatterers from about 200 m above the 0°C level to 500 m below it,

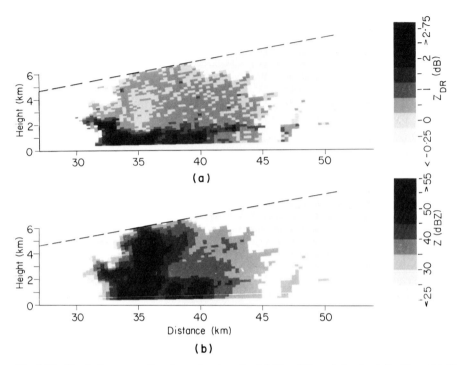

Fig. 8.20 Vertical cross section through rain cells. (a) The differential reflectivity Z_{DR}. (b) The horizontal reflectivity Z_h. (Courtesy of Appleton Laboratory, Slough, England.)

and they attribute this to snowflakes coalescing in the presence of water drops at temperatures below freezing just above the 0°C height. Because Z depends on the sixth power of hydrometeor size (for Rayleigh conditions), it is easily accepted that explosive increases in Z occur as crystals coalesce and melt. The increase in fall speed as ice melts and the inevitable breakup of large melting crystals lead to decreases in reflectivity below the melting layer, which forms an enhanced region of high reflectivity, called the bright band. In Fig. 8.20b the bright band is seen between 41 and 44 km. In the bright-band region Z_{DR} is also enhanced, as seen in Fig. 8.20a, because it is there that wet hydrometeors are the most oblate.

It is evident in Fig. 8.20 that the peak of Z_{DR} in the bright band occurs a little below the peak of reflectivity, a result consistent with that of Humphries and Barge (1979), who noted that for circularly polarized radars the maximum CDR is located below the peak of reflectivity. These observations suggest that hydrometeors falling through a melting layer are flattest below the height of maximum size. This results because large quasispherical ice hydrometeors begin to melt and flatten (producing enhanced Z) before they are broken by the airstream.

Using linear-polarized radar data, Leitao and Watson (1984) established a

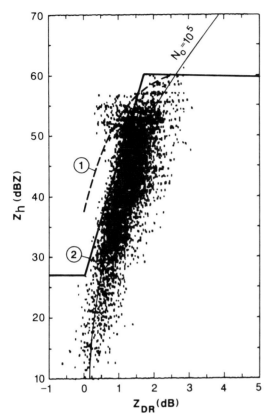

Fig. 8.21   Scattergram of Z_h versus Z_{DR} for a rainfall on 10 June 1986, from the elevation of 1°. The curve for an $N_0 = 10^5$ mm^{-1} m^{-3} is plotted (Steinhorn and Zrnić, 1988). (1) is the rain–hail boundary proposed by Leitao and Watson (1984), and (2) is the boundary from Aydin *et al.* (1986).

boundary in the $Z_h - Z_{DR}$ space that separates rain from hail (Fig. 8.21). Bringi *et al.* (1984) presented convincing evidence of hail detection with the Z_h, Z_{DR} pair measurement. The analysis of simulated radar data from disdrometer measurements led Aydin *et al.* (1986) to propose a hail detection signal H_{DR}. H_{DR} is the departure of the observed Z_h from the hail–rain boundary in the $Z_h - Z_{DR}$ space, Fig. (8.21), and is defined as

$$H_{DR} = Z_h - F(Z_{DR}) \qquad \text{dB}$$

$$F(Z_{DR}) = \begin{cases} 60; & Z_{DR} > 1.74 \\ 19\,Z_{DR} + 27; & 0 < Z_{DR} \le 1.74 \\ 27 & Z_{DR} \le \quad 0 \end{cases} \qquad (8.68)$$

These studies, prima facie, established that the polarization radars are aptly suited for detection of, and discrimination between, ice and liquid phases of precipitation.

A hail signature in the Z, Z_{DR} data is seen in Fig. 8.22. At a location between 18 and 19 km the minimum in the Z_{DR} coincides with the peak Z_h (50 dBZ) at a height of 2.5 km (above mean sea level). Note that the shape of the Z_{DR} contour suggests descent of hail toward ground.

Several recent observations suggest that minima of differential reflectivity are collocated or very close to the maxima of reflectivity. Husson and Pointin (1989) recorded negative Z_{DR} and have measured, with hail pads, maximum sizes of 2.3 cm. Their data show definite negative correlation between Z and Z_{DR}. A similar finding was reported by Illingworth et al. (1987). Bringi et al. (1986) also report negative differential reflectivity and attribute it to elongated and vertically oriented hailstones. These observations and their own measurements lead Balakrishnan and Zrnić (1990b) to speculate that oblate spheroidal hailstones with diameters larger than about 2 cm would produce Z_{DR} values in the neighborhood of -1 dB. Measurements by Aydin et al. (1990) from a Colorado hailstorm imply that vertically elongated hail with diameters between 1.2 and 4 cm produced Z_{DR} values of -0.5 dB or less. The evidence points toward a common cause but there are no in situ measurements that could relate large reflectivity and negative differential reflectivity to the shape and fall behavior of hail. Crude models of vertically oriented hail qualitatively reproduce some

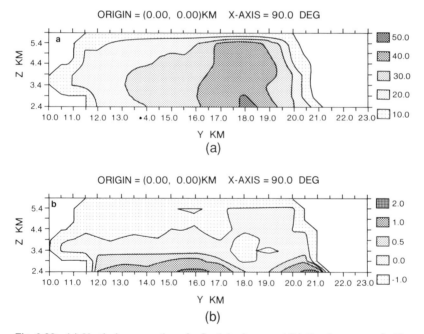

Fig. 8.22 (a) Vertical cross section of reflectivity factor and (b) Z_{DR} in contours/halftones through the core of a convective cell on 28 May 1983. Contours for reflectivity start at 10 dBZ in steps of 10 dBZ; differential reflectivity contours start at -1 dB. Note the depression of the Z_{DR} field in the region of high reflectivity ($18 \leq Y \leq 19$ km) indicating hail (from Bringi et al., 1986).

of the reported results, but sophisticated aero dynamic modeling is required to understand the physical reasons that lead to the observed backscattering properties.

Oblateness of hail is not simply related to size as is oblateness of raindrops. Yet there are indications that larger hail is more oblate and has axis ratios between 1.25 and 1.66, as reported by Knight (1986). Hailstone oblateness and the observations that larger stones seem to be vertically oriented while falling may be related. If this relationship is true there would be significant practical and scientific implications. One obvious outcome would be better warnings of large hail.

The hail–rain boundary, in the Z–Z_{DR} space can be effective in discriminating between pure rain and hail. This approach becomes ambiguous and less effective when the radar beam encounters mixed-phase precipitation. For instance, the large spread of Z_{DR} around a mean value, due to drop-size distribution variations, makes it difficult to establish the boundary precisely. In reality, the boundary between the hail and rain phases is diffuse, being characterized by hydrometeor phases where rain and hail coexist. The presence of large (i.e., >2 cm) nonrain hydrometeors, even in small amounts, if not inferred and accounted for, leads to large errors in the rain rates predicted using Z_h and Z_{DR}.

It is desirable that the boundaries for discrimination between rain and hail be based on appropriate polarization measurands such that the estimated rain-and-hailfall rates are within acceptable accuracies even in mixed-phase precipitation region. An attempt to handle mixed-phase precipitation has been made by Golestani et al. (1989). They propose the logarithm of the difference between the reflectivity factors Z_{DP} defined as

$$Z_{DP} = 10 \log(Z_h - Z_v) \qquad (8.69)$$

(where $Z_h > Z_v$ in mm^6 m^{-3}) be used to estimate the fraction of rain in mixed-phase precipitation. This is valid for a model in which hail contributions to the reflectivity factors at orthogonal polarizations are equal (i.e., hailstones are assumed to be spherical on the average). Thus Z_{DP} would be due solely to rain; an example seen in Fig. 8.23a demonstrates the unique relationship between Z_{DP} and Z_h for pure rain (i.e., Z_{hR}). The solid line was obtained from simulated gamma drop-size distributions. Lateral displacement of the data from the line is indicative of mixed-phase precipitation (Fig. 8.23b). From a data point Z_h, Z_{DP} and the rain line (Fig. 8.23) we can infer the contribution of rain Z_{hR} to the total reflectivity and thus its fraction f, which is

$$f = Z_{hR}/(Z_{hR} + Z_{hH}). \qquad (8.70)$$

Consequently the reflectivity ratio of hail to rain is

$$Z_{hH}/Z_{hR} = (1 - f)/f. \qquad (8.71)$$

Substantial deviation from a rain line is seen (Fig. 8.23b) in a scatter plot of Z_{DP} versus Z_h for a region of mixed-phase precipitation. The horizontal depar-

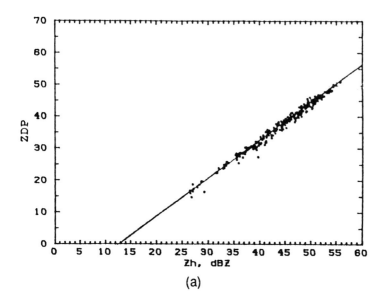

(a)

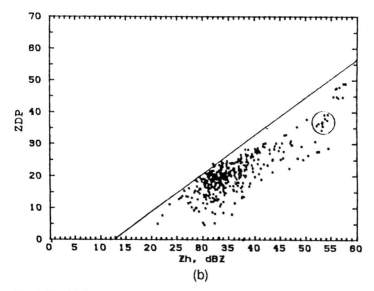

(b)

Fig. 8.23 (a) Scatter of $10 \log(Z_h - Z_v)$ versus dBZ for data from rainfall. (b) Same as in (a) except that data is from regions of rain/ice mixture (from Golestani *et al.*, 1989).

ture for the cluster of points where Z_{DP} is between 35 and 40 dB is 10 dB, therefore the fraction $f = 0.1$. To relate Z_h, Z_R to the masses of hail and rain, one would need to assume a size distribution for hail and rain.

8.5.3.3 Discrimination between Ice and Liquid Hydrometeors—Use of Reflectivity and Specific Differential Phase

For the purpose of discriminating between rain and hail it is useful to examine a scatter plot of Z and K_{DP} pairs in rain (Fig. 8.24). Z is averaged over 2.25 km, and Eq. (6.62) is used to calculate K_{DP} over the same distance. A curve depicting the boundary of $Z–K_{DP}$ scatter is included in Fig. 8.24. The empirical relationship for this boundary that distinguishes pure rain from mixed-phase and hail is given by

$$Z = 8 \log(2K_{DP}) + 49 \qquad (8.72)$$

The lower the measured K_{DP} from that given by Eq. (8.72), the higher the probability that the precipitation contains hail.

In addition to the discrimination between rain and frozen precipitation, K_{DP} and Z may provide a quantitative estimate of liquid water and ice in the mixture. This is because K_{DP} is affected by anisotropic hydrometeors, as described next.

Consider a homogeneous mixture of isotropic and anisotropic hydrometeors along a propagation path. Let the ensemble average of the specific phase be $\langle k_v \rangle$ for vertical polarization and $\langle k_h \rangle$ for horizontal. Without loss of substance,

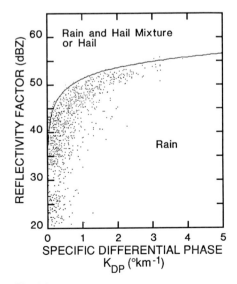

Fig. 8.24 Scatter plot of Z, K_{DP} from about 1400 radar measurements in a storm on 10 June 1986. Z is an average of reflectivity factors Z_h (horizontal and in dBZ) and Z_v (vertical polarization).

attenuation can be neglected and these phase constants can be written as (Oguchi, 1983)

$$\langle k_v \rangle = k_0 + \langle k_{vR} + k_{vH} \rangle, \tag{8.73a}$$

$$\langle k_h \rangle = k_0 + \langle k_{hR} + k_{hH} \rangle, \tag{8.73b}$$

where k_0 is the free-space propagation constant, k_{vR} and k_{vH} are contributions by rain and hail to the constant for vertically polarized waves, and k_{hR} and k_{hH} are similar contributions for horizontally polarized waves. Now for isotropic hail $\langle k_{vH} \rangle = \langle k_{hH} \rangle$, so after Eq. (8.73a) is subtracted from Eq. (8.73b) the specific differential phase becomes

$$K_{DP} = (\langle k_h \rangle - \langle k_v \rangle) = (\langle k_{hR} \rangle - \langle k_{vR} \rangle). \tag{8.74}$$

This remarkable yet simple result states that the specific differential phase is affected by anisotropic hydrometeors only (in this instance, rain). The physical explanation behind this fact is trivial. Isotropic hydrometeors produce equal phase shifts for either polarization, and the difference is due only to the anisotropic constituents of the medium. For example, if statistically isotropic hail is mixed with rain, the K_{DP} will be mainly affected by rain drops.

Not only can the rain rate be sensed but the method allows separation of hail and rain contributions to the reflectivity factor (Balakrishnan and Zrnić, 1990a). To obtain the portion of reflectivity factor due to hail one needs to subtract from the measured Z the part produced by rain, Z_R. A direct estimate of Z_R is not available, but an indirect one may be obtained from the K_{DP}, R and Z_R, R relationships. For a 10-cm radar as an example, combine Eq. (8.22a) with Eq. (8.31a) to obtain the following Z_R, K_{DP} relationship:

$$Z_R = 65,800(K_{DP})^{1.386} \quad (\text{mm}^6 \text{ m}^{-3}). \tag{8.75}$$

Now the reflectivity factor of hail is estimated as $Z_H = Z - Z_R$ (mm^6 m^{-3}). The hail rate in Fig. 8.19 has been obtained using this procedure and the Cheng–English Z, R_H relation (8.26b). A similar procedure can be used to obtain the ice water content [with use of Eq. (8.8) and appropriate density of a hailstone] whereas the liquid water content is related to K_{DP} by Eq. (8.31b). Three limitations are associated with estimates of hail reflectivity. First, the Z_R, R relationship (8.22a) might not be appropriate for a particular rain–hail mixture, in which case an unknown bias will be present. Second, errors in Z and Z_R will cause errors in Z_H. Third, the Cheng–English model (8.2) might not be applicable.

A set of curves in the K_{DP}, Z plane (Fig. 8.25a) illustrate the effects of variation in the relative amount of hail (hail rate). These curves were computed from the Cheng–English distribution of hailstone sizes and the Marshall–Palmer drop-size distribution (Balakrishnan and Zrnić, 1990a). Hailstones are assumed to be spherical and either very wet (refractive index $9.0585 + j1.3421$) or dry (refractive index $1.78 + j0.007$). Because the computations used a specific size distribution [Eq. (8.1a)], the resulting curves in Fig. 8.25a determine the

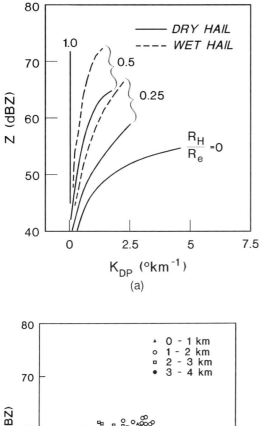

(a)

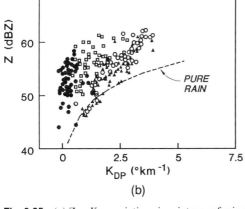

(b)

Fig. 8.25   (a) $Z - K_{DP}$ variations in mixtures of rain and wet spherical hail; R_e is the sum of rainfall and hailfall rates. The curve for wet hail and $R_h/R_e = 1$ coincides with the curve for dry hail. (b) Scatterplots Z and K_{DP} measured from 1559:00 CST to 1605:00 CST on 14 May 1986, $Z = (Z_h + Z_v)/2$ in dBZ.

relative amounts of hail and rain in a mixture to the extent that Eq. (8.2) is valid for the hail size distribution. A scattergram of measured Z and K_{DP} stratified according to altitude is given in Fig. 8.25b for a storm at a time it was producing small hail and rain on the ground. The theoretical Z and K_{DP} curve bounding pure rain is shown by a dotted line. Remarkable similarity with theoretical curves (Fig. 8.25a) suggests a progressively larger amount of hail in the mixture with height, with the result that between 3 and 4 km most of the hydrometeors are frozen.

8.5.3.4   Use of the Correlation Coefficient

The correlation coefficient [Eq. (8.46f)] between horizontally polarized weather signals and vertically polarized signals is mainly affected by the variability in the ratio of the vertical-to-horizontal size of individual hydrometeors. This is because the intensity of backscattering for the Rayleigh condition depends monotonically on the dimension of the hydrometeor in the direction of the electric field; the relation is more complicated for the Mie condition because it involves the differential phase shift on scattering. We shall discuss the following effects that influence the correlation: distributions of eccentricities, differential phase shifts on scattering, canting angles, irregular shapes of hydrometeors, and mixture of two types of hydrometeors.

In rain the eccentricity distribution is tied to the distribution of equivalent volume diameter D_e. The physical reason why $\rho_{hv}(0)$ depends on eccentricity is the continual, but small, change in the drop-size distribution within the resolution volume. Because increments in reflectivities at horizontal and vertical polarizations are not equal for the same increment in D_e, variations in drop-size distributions are sufficient to reduce $|\rho_{hv}(0)|$; this is also the reason why drop oscillations reduce $|\rho_{hv}(0)|$. Thus $|\rho_{hv}(0)|$ depends on the breadth of the axis ratio distribution (Jameson, 1987; Jameson and Davé, 1988). Sachidananda and Zrnić (1985) indicate that the shape effects in rain are small, and compute theoretical values larger than 0.99. Measurements in rain (Fig. 8.26) reveal an average $|\rho_{hv}(0)|$ of 0.98, close to, but less than the theoretical value because canting angle variations and noise, not accounted for in the theory, act to further decrease the correlation; instead of the expected increase in $|\rho_{hv}(0)|$ for small reflectivity (i.e., small drops are almost spherical; Fig. 8.1), there is a decrease possibly due to sidelobes and receiver noise.

Changes in the differential phase shift on scattering (at a specific large diameter for a given dielectric constant) cause the composite horizontally and vertically polarized signals to fluctuate differently [the ensemble average in the numerator of Eq. (8.46f) is less coherent]. Similar reduction in correlation occurs if canting angles have a probability distribution of finite width (Balakrishnan and Zrnić, 1990b). This dependence on canting angles may offer possibilities to discriminate (with a vertically pointed beam) between liquid drops and ice needles if the latter are randomly oriented in the horizontal plane. For dry

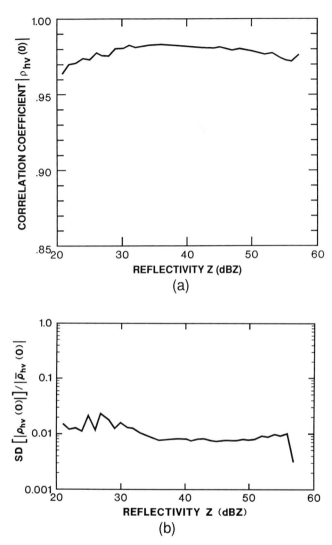

Fig. 8.26   Data from a rain storms on 6 June 1986, in Oklahoma. (a) Mean values of the correlation coefficient. (b) Normalized standard deviations of the estimates in (a); $Z = (Z_h + Z_v)/2$ in dBZ.

needles (modeled as prolate spheroids) the correlation reduces to about 0.9; $|\rho_{hv}(0)|$ for wet needles is much less (Fig. 8.27).

Irregular shape of hydrometeors is another factor that reduces the correlation. Larger hailstones (4–10 cm diameter) are roughly irregular with small and large protuberances (List, 1986). Hailstones do not obey fractal laws, that is

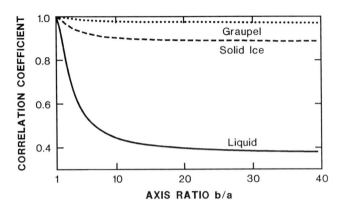

Fig. 8.27 Correlation coefficient for randomly distributed prolate spheroids, versus major-to-minor-axis ratio (b/a). Dotted curve is for graupel (refractive index $m = 1.372$); dashed curve is for dry spheroids $(m = 1.78 + j0.007)$; and full curve is for predominately wet spheroids $(m = 9.0585 + j1.3421)$. The frequency is 2.88 GHz.

the protuberance-to-diameter ratio is not constant, but may increase with hailstone size. Although this is not always the case, when true it will result in a noticeable decrease of $|\rho_{hv}(0)|$. For distortions independent in horizontal and vertical directions and small compared to the diameter of the hailstone, $|\rho_{hv}(0)|$ of monodispersed hailstones is given in the Rayleigh limit as (Balakrishnan and Zrnić, 1990b)

$$|\rho_{hv}(0)| = (1 + 3\sigma_D^2/D_e^2)^2/(1 + 15\sigma_D^2/D_e^2 \\ + 45\sigma_D^4/D_e^4 + 15\sigma_D^6/D_e^6), \tag{8.76}$$

where σ_D is the rms value of the protuberance and D_e is the equivalent diameter of the hailstone. Plot (Fig. 8.28) of Eq. (8.76) indicates that σ_D/D_e of 0.1 may reduce the correlation to 0.92. We expect similar or larger effects to be produced by snowflakes.

In a mixture of precipitation types the reduction of the correlation coefficient is due to the broader spread in the composite distribution of eccentricities and sizes compared to a distribution of single precipitation type. The drop in correlation would be largest if reflectivity weighted distributions of the two hydrometeor types are comparable. If either one of the populations prevails, the $|\rho_{hv}(0)|$ will tend to the intrinsic value for that hydrometeor type. When the precipitation medium is rain, the shape effect is possibly the dominant cause for values of $|\rho_{hv}(0)| < 1$. However, when the precipitation habitats include hailstones or snowflakes, further reductions in $|\rho_{hv}(0)|$ are due to the cumulative effects of all the physical factors described so far.

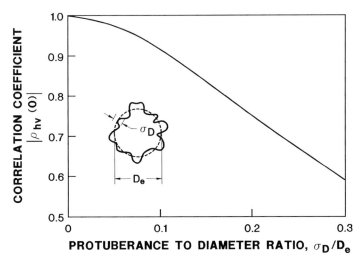

Fig. 8.28 Correlation coefficient for Rayleigh scatterers with protuberances.

8.5.3.5 *Use of the Linear Depolarization Ratio*

Linear depolarization ratio LDR_{hv} (or LDR_{vh}) is defined as a ratio of the cross-polar signal power to the co-polar power [i.e., Eq. (8.46d or e)]. By inspection of Eq. (8.58c) it follows that a cross-polar signal occurs only if spheroidal hydrometeors fall with their major or minor axis not aligned nor orthogonal to the electric field (i.e., when $\psi \neq 0$ or $\neq 90°$, Fig. 8.15). This result comes from symmetry considerations, which imply that if a vertically (horizontally) polarized incident electric field is aligned with one axis there can be no horizontally (vertically) polarized backscatter, hence $LDR \rightarrow -\infty$ dB. During fall, oblate spheroidal particles wobble and thus there would be a distribution of canting angles, which increases LDR. Also, irregularly shaped hydrometeors can cause increases in LDR.

Computations of LDR_{hv} for randomly tumbling oblate spheroids (Fig. 8.29) are in qualitative agreement with observations (Frost *et al.*, 1989). The values of LDR_{hv} rise as the particles either become more oblate or their refractive index increases. Even if snowflakes tumble they have such a low refractive index that their LDR_{hv} is about -32 dB. Oblate dry hail or graupel could have values up to -20 dB if the axes ratios were as high as 2.5 or 6. Larger depolarizations are expected only for wet, tumbling ice particles. Depolarization by rain is very small, in the range of -30 dB (Jameson, 1987; Hendry *et al.*, 1987).

Observations with a narrow beam (0.25°, 10-cm wavelength) radar (Frost *et al.*, 1991) suggest a possibility to distinguish between wet ice particles and graupel. When the particles start to melt and become coated with water, peak

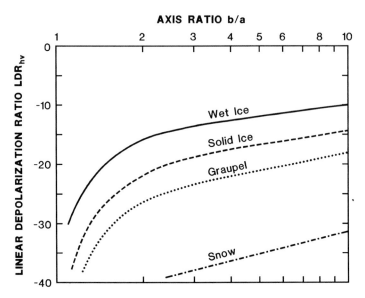

Fig. 8.29   Linear depolarization ratios for randomly tumbling oblate spheroids as a function of axial ratio (adapted from Frost *et al.*, 1989).

LDR_{hv} can reach -16 dB. This occurs in stratiform clouds and some showers (Fig. 8.30a) and could be due to frozen hydrometeors having a rocking or spinning motion and an axial ratio larger than about 2. In contrast, some vigorous showers (Fig. 8.30b) produce LDR_{hv} in the bright band of -26 dB or less, which is perhaps due to water-coated graupel pellets with b/a of about 1.25. The depression in the bright band caused by graupel and a difference of over 6 dB in LDR between graupel and aggregate regions is dramatically evident in Fig. 8.30b.

Linear depolarization ratio is more susceptible to noise contamination than Z_{DR} or $|\rho_{hv}(0)|$ because the cross-polar power is more than two orders of magnitude below the co-polar signal. Furthermore, at short wavelength (<10 cm) depolarization caused by propagation through precipitation mars the measurement's utility.

8.5.3.6   Combined Measurements

The multifaceted information obtained by combining polarization measurements should provide more insight regarding precipitation types and their evolution than is possible from examination of each measurand separately. This is illustrated in the vertical profiles of Z, Z_{DR}, K_{DP}, and $|\rho_{hv}(0)|$ (Fig. 8.31). Large reflectivities through the depth of 8 km suggest intense precipitation with presence of hail, at least near and above the freezing level. The monotonic increase of K_{DP} below the freezing level signifies continuing melting and a rain rate of 82 mm h^{-1} at the ground. Substantial negative Z_{DR} below the freezing

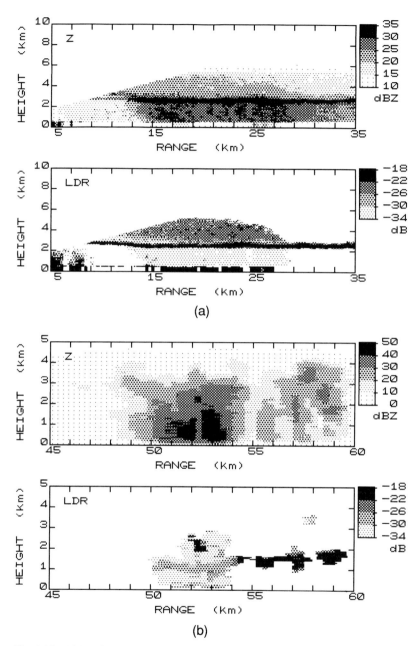

Fig. 8.30 (a) Vertical section (RHI) of Z and LDR_{hv} through stratiform precipitation. LDR in the bright band is high (> -22 dB), it is moderate (-26 to -22 dB) in the dry ice region, and low in rain (-34 to -30 dB). (b) Vertical section of Z and LDR_{hv} through a convective cloud. Values at the melting layer (1 to 2 km above ground) suggest graupel at 50–55 km and snow at 55–60 km (from Frost *et al.*, 1991).

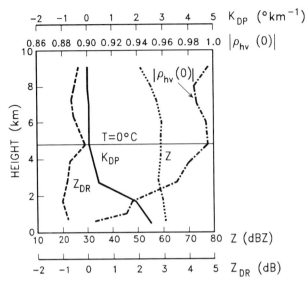

Fig. 8.31 Vertical profiles of reflectivity factor [$Z = (Z_h + Z_v)/2$], differential reflectivity (Z_{DR}), specific differential phase (K_{DP}), and the correlation coefficient $|\rho_{hv}(0)|$ measured on 2 June 1985, at 1859 CST. The range is 82.5 km; azimuth is 242°. Data are averaged over a range interval of 2.1 km. The 0°C isotherm is at 4.8 km.

level indicates presence of large (>2 cm in diameter) wet hailstones (Balakrishnan and Zrnić, 1990b); Z_{DR} first decreases probably due to wetting of hailstones, but near ground it increases because some hailstones have melted. Below the freezing level the correlation coefficient continually decreases because hail is melting but the reflectivity weighted distribution of hailstones is still larger than the one of raindrops. Judging from the negative Z_{DR} immediately above the freezing level, vertically oriented large hydrometeors dominate the population. A small dip in $|\rho_{hv}(0)|$ near 8 km could mean that the reflectivity weighted distribution of graupel is close to the one of hail; note also that the reflectivity factor decreases. There are no other supporting data to corroborate the described speculation, but the self consistency among the polarimetric measurands adds credibility to the interpretation.

Although there are other multiparameter techniques such as dual wavelength and radar–radiometer combination, the polarimetric technique allows a single instrument to be used and is thus a likely practical candidate for operational applications. The basic problem facing polarimetric researchers consists of deciding what variables to measure and how to relate them to the properties of hydrometeors. Relating these properties to polarimetric parameters, especially quantitatively, is difficult. Put in simple terms a multiparameter decision rule is sought that will partition the five-dimensional space of Z_h, Z_{DR}, K_{DP}, LDR_{hv}, and $|\rho_{hv}(0)|$ so that each partition corresponds to a distinct hydrometeor

Table 8.1
Values of Polarimetric Measurands for Various Precipitation Types

	Z_h (dBZ)	Z_{DR} (dB)	$|\rho_{hv}(0)|$	K_{DP} (deg km^{-1})	LDR$_{hv}$ (dB)
Drizzle	<25	0	>0.99	0	< −34
Rain	25 to 60	0.5 to 4	>0.97	0 to 10	−27 to −34
Snow, dry, low density	<35	0 to 0.5	>0.99	0 to 0.5	< −34
Crystals, dry, high density	<25	0 to 5	>0.95	0 to 1	−25 to −34
Snow, wet melting	<45	0 to 3	0.8 to 0.95	0 to 2	−13 to −18
Graupel, dry	40 to 50	−0.5 to 1	>0.99	−0.5 to .5	< −30
Graupel, wet	40 to 55	−0.5 to 3	>0.99	−0.5 to 2	−20 to −25
Hail, small <2 cm wet	50 to 60	−0.5 to 0.5	>0.95	−0.5 to 0.5	< −20
Hail, large >2 cm wet	55 to 70	< −0.5	>0.96	−1 to 1	−10 to −15
Rain & hail	50 to 70	−1 to 1	>0.9	0 to 10	−20 to −10

type. Positive results that rely on partitions of two-dimensional subspaces, as in Figs. 8.21, 8.24, and 8.25, have been achieved. In Table 8.1 we list a range of values that polarimetric measurands are likely to have from various forms of precipitation. These come from modeling, measurements, and experience of various investigators. Much more modeling, measurements, and verification remains to be done before the full polarization potential can be realized.

8.5.4 Rain Gauge and Radar

The most direct way to measure the rainfall rate is to use a rain gauge—a catchment that measures the depth of water per unit time. Although tipping buckets and weighing gauges are commonly used, they are subject to significant errors caused by wind (Neff, 1977). Rain gauges measure rainfall only at a point. More often, interest lies in accurate estimation of rainfall in a unit time averaged over large catchment areas, and these estimates are usually expressed in millimeters of water depth. The areal averages find application in hydrology and flash flood forecasting.

Because there may be large errors in the rain depth at any one gauge representing the areal average, hydrologists have resorted to a network of rain gauges and to radar to improve areal average rainfall estimates. There is no doubt that a sufficiently dense network of gauges can measure rainfall better than a radar. In fact, gauge measurements are accepted as the standard against which other measurement techniques are compared. Yet no matter how accurate gauges may be for point measurements (errors are typically 5–10%), their accuracy for areal averages is a function of the gauge density and the spatial variability of rainfall (Huff, 1970; Doviak, 1983).

Although the accuracy of the radar-measured rainfall rate is highly suspect, radar has the decided advantage of being able to survy remotely vast areas and to make millions of measurements in minutes. The cost of a gauge network to match these capabilities in spatial continuity and to send data to a central location would be prohibitive. Therefore, meteorologists have combined radar and rain gauge data to take advantage of the best of each—the accuracy of gauge data and the spatial coverage of radar data. The combination adjusts the error-prone radar measurements.

Before we can confidently accept radar estimates of rainfall, we should be aware of the phenomena that can cause variance from gauge estimates. If only reflectivity factor measurements are available, one needs to choose an appropriate R, Z relation. Because the parameters of an assumed drop-size distribution vary considerably from point to point, one may also expect the R, Z relation to vary. On the other hand, for areal averages of rainfall, there is suggestive evidence that these parameters may be appropriately chosen to produce an R, Z relation that in the mean predicts the average rainfall measured by a network of gauges (see Fig. 8.32 and the discussion in Section 8.4.1.1).

Figure 8.33 is a comparison of gauge- and radar-estimated water depths ($\bar{G}/\bar{R}$) at 65 gauge locations for a shower measured by a WSR-57 radar located in Central Oklahoma. Atmospheric gas attenuation accounted for most of the correction seen in Fig. 8.33b. Although for this day the gauge-estimated rainfall

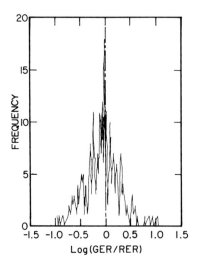

Fig. 8.32 Distribution of the logarithmic ratio of gauge-estimated rainfall to radar-estimated rainfall (from Cain and Smith, 1976), for 300 gauge-hour events in North Dakota during 1972.

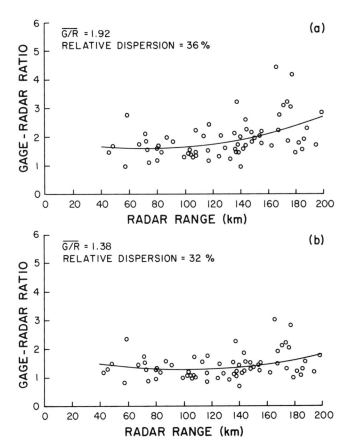

Fig. 8.33   Range distribution of G/R ratios using radar measurements that were (a) uncorrected and (b) corrected for atmospheric absorption, rainfall attenuation, and biases due to reflectivity gradients. Each point is a radar–gauge comparison, at a gauge location, of the estimated rainfall rate integrated in time (i.e., the depth of water) over 24 hours. The relative dispersion is the standard deviation expressed as a percentage of the mean. (From Brandes and Sirmans, 1976.)

(GER) was 1.38 times higher than that estimated by radar (RER), the average ratio $(\overline{G}/\overline{R})$ for 14 rain events is 1.05 when the relation $Z = 200R^{1.6}$ is used (Brandes and Sirmans, 1976). Because daily radar calibrations can be in error, and because different R, Z relations might be appropriate for different days as well as for different locations surveyed by the same radar, it becomes mandatory to adjust radar rain estimates in accordance with the *in situ* gauge measurements.

Brandes (1975) has suggested a technique whereby gauges can be used to adjust the RER. This methodology minimizes the impact of choosing an inappropriate R, Z relation and the need for a nearly perfect radar calibration.

The major steps in Brandes's technique are as follows:

1. Radar-estimated rainfall rates are integrated in time for the selected period for each resolution volume, using Eq. (8.22a).

2. G/R ratios are calculated at each gauge having at least 2.5 mm of rain, using the radar data from within a fixed radius about the gauge. The radius is chosen to be small with respect to the gauge spacing and the scale of the precipitation.

3. The G/R data are then used to determine field-of-adjustment factors, which are applied to the RER field to generate a first-guess corrected radar rainfall field.

4. Differences between corrected radar-measured rainfall and actual gauge estimates are then used to determine a refinement in the adjustment factors so that the final corrected RER will agree with the gauge data at gauge locations.

The radar–gauge technique would work better if a single-parameter radar measured the spatial variability in R rather than the variability in Z or K_{DP}. It is expected, however, that the K_{DP} parameter would provide a better estimate of the spatial variability of R because it is better correlated with R, at high rain rate, than is Z (Section 8.4.3). Furthermore, the Z, R-gauge technique assumes that the gauge measures an R that fills the resolution volume at the individual gauge location and that rain alone is responsible for Z. The capability of a single-parameter (especially K_{DP}) radar to measure variations in R has not been adequately verified. A multiparameter radar might map the spatial variations in R better than R from K_{DP} or Z. A few sparsely sprinkled rain gauges would then be helpful to remove pervasive biases. In the studies by Wilson (1970) and Brandes (1975), the gauge–radar mean rainfall estimates were more accurate than the estimates using only gauges for large-area (29,000 km^2), low-gauge density (no more than one gauge per 700 km^2) and long-duration rainfall cases. Brandes (1975) shows that radar-measured rainfall corrected by gauge data improved the accuracy from 24% for measurements by gauge alone to 14% for combined radar–gauge measurements with a gauge density of one gauge per 1600 km^2. A summary of improvement in radar-measured rainfall, corrected by gauge data for denser spacings and smaller areas, is given by Wilson and Brandes (1979, Table 3).

8.6 Distribution of Hydrometeors from Doppler Spectra

Measurements of the Doppler spectrum or its moments with a vertically pointed beam may be employed to make inferences concerning the drop-size distributions. In a sense, this can be considered a multiparameter method, since several simultaneous measurements (i.e., reflectivity for each resolved Doppler velocity) are nearly independent. As a matter of fact, in the first Doppler radar applications the entire Doppler spectrum (as opposed to its moments) was used to determine the drop-size distribution (Barratt and Browne, 1953; Boyenval, 1960).

In essence, the method relies on the existence of a unique relationship between the terminal speed and diameter of drops falling in still air (Fig. 8.4). In the presence of up- or downdrafts of velocity w, uniform within the resolution volume, the relationship between the normalized Doppler spectrum and the drop-size distribution is

$$S_n(w - w_t) \, dw_t = \sigma_b(D)N(D) \, dD/\eta, \qquad (8.77)$$

where it is assumed that the resolution volume is much larger than the average spacing between drops for all diameters. Furthermore, S_n must be free from all other spectrum broadening contributions (Section 5.3). The relation $\sigma_b(D)/\eta = D^6/Z$, from Eqs. (3.6a) and (4.32), can be inserted in Eq. (8.77) to obtain

$$S_n(w - w_t) \, dw_t = D^6 N(D) \, dD/Z. \qquad (8.78)$$

Thus, if the vertical speed of air is known, measurement of the reflectivity factor and the Doppler spectrum suffices to determine $N(D)$ with no need for assumptions regarding its functional form.

The most serious obstacle to drop-size measurements from the Doppler spectra is the need for precise determination of w. Rogers (1964) was the first to use a relationship between the mean terminal velocity and reflectivity for an assumed exponential drop-size distribution. Battan (1973) proposed a lower-bound method that involves assumptions on the minimum size of the detectable hydrometeors and their terminal velocity.

Atlas et al. (1973) used Eq. (8.78) as the starting point for a detailed analysis of the properties of Doppler spectra for vertical viewing. They calculated theoretical spectra and spectral moments for an exponential $N(D)$ and $w_t(D)$ approximated by various analytical expressions. For an assumed power law dependence of terminal velocity, $w_t = CD^\alpha$, the reflectivity-weighted mean terminal speed is

$$\bar{w}_t = 720^{(7-\alpha)/7} \Gamma(7 + \alpha) C(Z/N_0)^{\alpha/7}, \qquad (8.79)$$

which must be added to the mean Doppler velocity (i.e., the fall speed with respect to the ground) to determine the airspeed w.

Hauser and Amayenc (1981) proposed a least squares fit of the three parameters N_0, Λ, and w of a model Doppler spectrum to the observed spectra. Their results exhibit a fine structure in the parameters deduced (the vertical air velocity, liquid water content, rainfall rate, N_0, and Λ) on scales of the order of 200 m. The general coherence of the results leads to the conclusion that the method is well suited to the study of stratiform precipitation when the hypothesis of exponential drop-size distribution holds and when spectral broadening due to turbulence is negligible.

If an independent estimate of w were possible, the drop-size distribution could be readily obtained from the Doppler spectrum. Such a possibility is offered when lightning occurs in the resolution volume (Zrnić *et al.*, 1982), because the ionized channel acts as a perfect tracer. This is seen in the Doppler spectrum as a distinct peak (Fig. 8.34) located at the vertical speed of air. The speed of the updraft was estimated to be 11 m s^{-1}, and a terminal velocity for spherical hail at 7 km above ground is assumed to follow the Matson–Huggins (1980) relation $w_t = 3.62 D^{1/2}$ (in m s^{-1} for D in mm). An estimate of the turbulent broadening was made from the positions of the lightning peak in several consecutive spectra. The peak had an excursion of 4 m s^{-1}, and we therefore assume the turbulent broadening to be $(4/3)^{1/2}$ m s^{-1}. Spectral skirts have been reduced accordingly, and Eq. (8.78) has been applied over a 25-dB dynamic range of the spectrum. $N(D)$ obtained from the Doppler spectrum of Fig. 8.34 is plotted in Fig. 8.35. The equivalent reflectivity factor (here 55 dBZ) was scaled with the ratio of the refractive indexes of hail to water. Evident in Fig. 8.35 is the huge (10^{13}) span of the distribution and the remarkably good fit to an exponential $N(D) = 4 \times 10^5 \exp(-2D)$ m^{-3} mm^{-1}, for sizes larger than 1 mm. With these values, the calculated equivalent reflectivity factor [Eq. (8.14)] is 55.5 dBZ. In this particular experiment the resolution volume was about 10^8 m^3, and we conclude from Fig. 8.35 that, even though the number of hailstones between 9.5 and 10.5 mm is 10^{-4} m^{-3}, there are on the average 10^4 such stones in the resolution volume.

It is also possible to make an independent estimate of w with vertically pointing VHF and UHF radars that sense refractive index irregularities as well

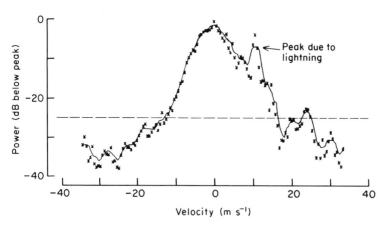

Fig. 8.34 Reflectivity spectral density for precipitation and lightning at 7.1 km versus vertical velocity obtained with a vertically pointing antenna. The solid curve is a five-point running average of the data (x), which were obtained by averaging 12 spectra. Powers below the dashed line and the peak at 24 m s^{-1} are considered to be unreliable (i.e., because of window effects, noise sources, etc.).

Color Plates

Plates 1a,b   Correction of range-ambiguous Doppler velocity fields of storms. (a) Doppler field for the radar operating with uniform PRT and no range corrections applied. Red (green) colored areas are radial velocities away from (toward) the radar. The shades of color from dark to light correspond to median velocities (in m s^{-1}) indicated to the right of the color bar. Range circles are at 100-km intervals. The light-blue color tags echoes from ground objects and airborne scatterers having radial velocities between -3 and 3 m s^{-1}. (b) Same as (a) but the velocity field is obtained 80 s later, when an interlaced PRT was used and range dealiased adjustments had been applied. The purple color (labeled "RF" on the color bar) denotes regions where range overlaid echoes had powers within 10 dB.

Plates 1c,d   Dealiasing of velocity ambiguities in a strong density current on 27 May 1982 in central Oklahoma. (From Elits and Smith, 1990.) (c) Doppler field with no corrections for velocity ambiguities. Color bar shows the velocity categories (in m s^{-1}). Range-arc labels are in kilometers. All color categories (i.e., measured Doppler velocities) must lie between $v_a = \pm 24.5$ m s^{-1}. (d) Same as (c) but velocity dealiasing corrections have been applied.

Plate 2   Doppler velocity fields. Color categories are as in Plate 1(a). (a) The Doppler velocity field for divergent flow near the top of storms in northwest Oklahoma, 2 May 1979. (PPI displays of reflectivity and spectrum width are shown in Fig. 9.20.) Range and height of the data field are about 150 and 12 km, respectively; range arcs are 40 km apart. (b) The vertical cross section of the Doppler velocity field for a storm on 16 June 1980 in central Oklahoma. Range arcs are 20 km apart; left arc is at 60-km range. Evident are the boundary layer flow toward the storm and the outflow in the anvil cloud ($\approx$10 km AGL). The circular cursor marks the 14.7-km altitude. (The reflectivity in the same cross section is in Fig. 9.22.) (c) A PPI display of Doppler velocities from converging flow a few hundred meters AGL in the Edmond, Okla., tornadic storm of 8 May 1986; range arcs are 20 km apart. (d) As in (c) but from an almost purely rotating flow at 2.3 km AGL in the Edmond storm's cyclone.

Plate 3   (a) The Doppler velocity field of a density current from a cluster of storm cells that passed through central Oklahoma on 1 June 1985. The elevation is 0.2°. Velocity/color scale is as in Plate 1(a), except the white color tags areas where velocity is near zero. (b) Doppler velocity field of a microburst near Denver, Colo. The elevation is 0°, range arcs are at 20 and 30 km, and the velocity scale (in m s^{-1}) is on the bottom. (Courtesy of J. Wilson, NCAR.) (c) The Doppler velocity field for one of a pair of solitary waves launched by the density current seen in (a), but 34 min later. The wave appears as a strip of negative velocities (band of green) at a range of about 20 km (at its nearest point). (d) The Doppler velocity field of an undular bore that propagated over central Oklahoma on 22 June 1987. (A satellite image of clouds produced by this wave is in Fig. 9.43.) Range circles are in kilometers and the velocity/color bar scale is at the base of the figure. The elevation angle is 0.5°.

Plate 4   (a) The Doppler velocity field of a nearly two-dimensional squall line observed with the WSR-88D Doppler radar in Norman, Okla., on 4 May 1989 (the reflectivity field of this storm is in Fig. 9.46). The radar is located at the circle near the left border of the figure. County lines and villages in Oklahoma are presented. The velocity (m s^{-1}) categories are presented next to the color bar. (Courtesy of Dan Purcell, OSF/NWS.) (b) The Doppler velocity field of Hurricane Bob at 1902 UT, 19 August 1991, observed with the TDWR radar in eastern Massachusetts. Range circles are 10 nautical miles apart, elevation is 1.2°, and velocity (knots) is on the color bar. (Courtesy of T. McDonagh, Raytheon.) (c) The Doppler velocity field of a cold front passing a stationary dry front observed with the Oklahoma City WSR-88D Doppler radar on 30 April 1991, 2220 UT. (Thin lines of reflectivity marking the intersecting fronts are in Fig. 9.50.) Color bars categorize velocity in knots. Color labeled "RF" tags areas where range folded echoes interfere (Chapter 7). (Courtesy of Dan Purcell, OSF/NWS.) (d) Doppler velocity display of three vertically stacked jets observed with the WSR-88D in Norman, Okla. (Courtesy of R. Murnan, OSF/NWS.)

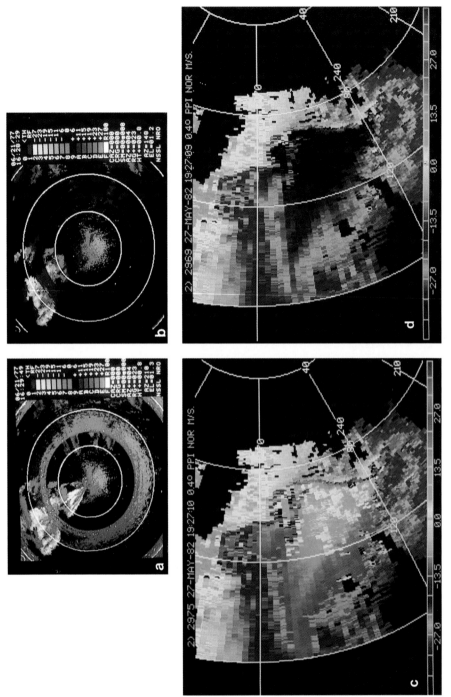

Plate 1

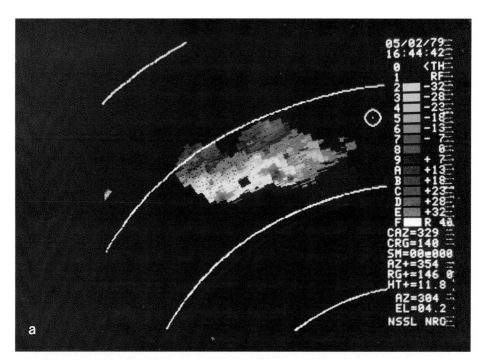

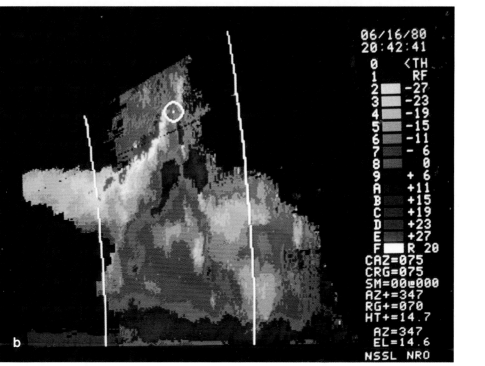

Plate 2

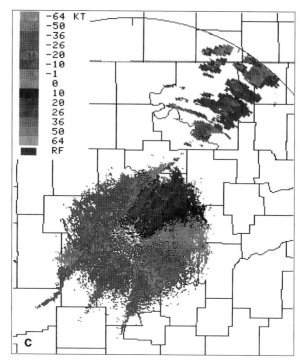

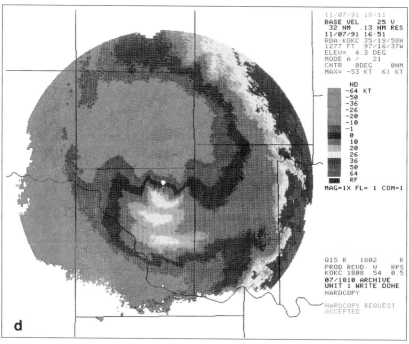

Plate 4 *(continued)*

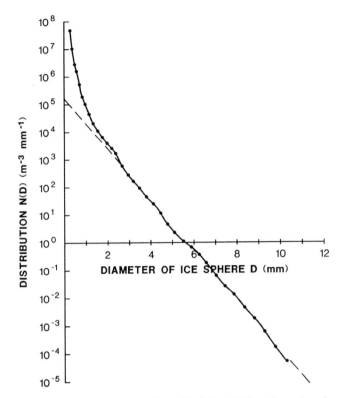

Fig. 8.35 The size distribution of hail (modeled as ice spheres), obtained from the Doppler spectrum in Fig. 8.34.

as hydrometeors (Wakasugi *et al.*, 1986, Gossard *et al.*, 1990). These radars are designed to measure the vertical profiles of wind. As in the case of lightning echoes the spectrum of "clear air" echoes from refractive index irregularities is distinct from the spectrum of precipitation, and hence the two can be separated. The spectrum in Fig. 8.36a was obtained with a vertically pointing 50-MHz wind-profiling radar (Wakasugi *et al.*, 1987). At this wavelength cross sections of drops are significantly reduced (by λ^4) compared to cross sections at shorter wavelength [Eq. (3.6a)]. No such reduction happens for echoes from refractive index irregularities. That is why the peak corresponing to mean air motion, at 3 m s^{-1}, is comparable to the peak from precipitation at -5 m s^{-1}, which in this case corresponds to 42 dBZ. The dashed curve is a parametric fit that assumes an exponential drop-size distribution and a Gaussian spectrum for clear-air signal. The deduced drop-size distributions are in Fig. 8.36b, where the two curves are obtained by subtracting the clear-air spectra from the spectra in Fig. 8.36a and transforming to $N(D)$ as in Eq. (8.78). The least square fitted line is also included from which $N_0 = 23000 \text{ m}^{-3} \text{ mm}^{-1}$ and $\Lambda = 2.73 \text{ mm}^{-1}$; the line does

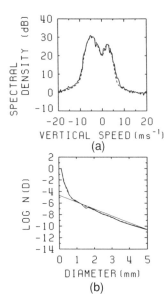

Fig. 8.36 (a) A one-minute average Doppler spectrum observed with a vertically pointing beam (solid curve) and the fitted spectrum (dashed curve). (b) The drop-size distribution is given by the solid curve; the dashed curve is the least squares fit, and the straight line is a fitted exponential (from Wakasugi *et al.* 1987).

not coincide with the curves because the latter contains effects of turbulence and finite length of the data window.

Problems

8.1   A drop-size distribution is uniform for drops with diameters between 0 and 5 mm, and equals $10 \ \mathrm{m}^{-3} \ \mathrm{mm}^{-1}$. (a) Which drops contribute most to the reflectivity factor? (b) Determine the reflectivity factor in dBZ units. (c) Assume that the terminal velocity of drops is given by Eq. (8.4). Calculate the rain rate (in mm h^{-1}) that corresponds to this drop-size distribution. (d) Obtain the Z, R relationship for this drop-size distribution, assuming that only N_0 varies.

8.2   The electric field at a point in space is described by

(a) $Ae^{j\omega(t-r/c)}\mathbf{a}_z$;

or

(b) $Ae^{j\omega(t-r/c)}\mathbf{a}_z + Be^{j\omega[(t-r/c) + \psi]}\mathbf{a}_x$.

$\mathbf{a}_x$ and $\mathbf{a}_z$ are unit vectors. A and B are real numbers in V m^{-1}. Determine the polarization (i.e., linear, RH circular, etc.) in (a) and in (b) for $\psi = 0$, $\pi/2$, $A = B$, $A > B$.

8.3   Assuming an exponential drop-size distribution with an N_0 given by Eq. (8.1c), use the formula (8.20) to derive an approximate expression that relates Z and R. How does this compare with the relation $Z = 200R^{1.6}$? Discuss the reasons for deviations.

8.4 Suppose that a +3-dB bias error exists in the calibration of a radar. Plot the error in rain rate obtained from the formula $Z = 200R^{1.6}$. Find a different Z, R relation that can be used with the faulty calibration so that the error for the range of rain rates between 5 mm h^{-1} and 150 mm h^{-1} is less than 10%.

8.5 Calculations of liquid water content M and attenuation rate K using exact formula for extinction cross sections and measured drop-size distributions show that M and K are reasonably well-related by a power law

$$M = aK^b. \tag{P.1}$$

Assuming exponential drop-size spectra, and the power law dependence between σ_e and D,

$$\sigma_e = CD^n, \tag{P.2}$$

prove Eq. (P.1), for a constant N_0. Which value must n be for Eq. (P.1) to be independent of drop-size spectra? What can be said about the drop-size spectra parameters knowing that a and b are nearly invariant for different rain types (i.e., $N(D)$)?

8.6 Derive Eq. (8.57). For a $b/a = 1.4$, find the equivalent drop diameter D_e. Then determine the differential reflectivity Z_{DR} for a hypothetical rain in which all the drops have equal diameters corresponding to $b/a = 1.4$.

8.7 The radar measures an effective reflectivity factor of 30 dBZ at 50 km. It is known that rain has produced this value. Furthermore, an observer who was in this shower reported that all drops were of equal size and that there were about 1000 drops per m^3. (a) Which of the two observations is physically less plausible? (b) Assuming the observations are correct, calculate the drop's diameter. (c) Calculate the rain rate and reflectivity factor produced by the drops and compare it to the rain rate that would be derived from Eq. (8.22a); give reasons for the differences. (d) The rain that produced 30 dBZ at 50 km is observed at 100 km from the radar, and because of beam blockage only half of the resolution volume is illuminated. How much less power than at 50 km, without blockage, does the radar receive? What would be the measured effective reflectivity factor?

9

Observations of Winds, Storms, and Related Phenomena

The great utility of centimeter-wavelength pulsed-Doppler radar for storm observation derives from its capability to estimate the reflectivity η and radial velocity v fields inside the storm's shield of clouds. If single-beam radars are used, a three-dimensional picture of a storm takes about 2–5 minutes of data collection time, not only because of antenna rotation limitations, but also because a large number of echoes from each resolution volume needs to be processed to reduce the statistical uncertainty in the η and v estimates (Chapter 6). Although the storm can change significantly during this period, leading to distortion of the radar image of the reflectivity and velocity fields, highly significant achievements have been made in depicting its structure and evolution.

The meteorologically interesting variables, however, are not η nor v but parameters such as the rainfall rate (on the ground) and wind. Doppler radar most often measures the radial speed of hydrometeors, not air, and in certain situations, such as vertically directed beams, these speeds can differ significantly from the radial component of wind. Nevertheless, scatterers such as water drops have small mass and quickly respond to wind forces. Stackpole (1961) has shown that, for the energy spectrum following a power of $-5/3$ law (Section 10.1.3) to at least 500 m, more than 90% of the rms wind fluctuations are acquired by the drops if their diameter is less than 3 mm. For radar beams at low elevation angles, hydrometeor terminal velocities (Section 8.2) give negligible bias error in the radial wind component. At high elevation angles terminal velocities need to be estimated.

Incoherent radars map η and not wind but, if the radar's resolution volume is sufficiently small, these radars can resolve small-scale reflectivity structures, and if reflectivity is conserved (i.e., if scatterers following the flow don't change cross section), they can map the vector wind (Crane, 1979; Rinehart, 1979). Flow in the unstable boundary layer is quite turbulent and the associated velocity structures have been found to have longer lifetimes than reflectivity (Section 10.1.4). Symthe and Zrnić (1983) and Tuttle and Foote (1990) have applied tracking techniques to radial velocity structures to map boundary layer flows.

9.1 Thunderstorm Structure

Although there are authors who comprehensively examine the structure and characteristics of thunderstorms (e.g., Kessler, 1982), a brief description will be given to help interpret the weather radar observations shown in this and other chapters. We focus our attention on the supercell thunderstorm not only because it produces damaging wind and hail, but also because it is often part of a larger weather system such as a squall line. The less devastating and short-lived ordinary thunderstorm is discussed in Section 9.2.3.

Clouds give a good picture of the thunderstorm, and even though most of the storm's violent interior is obscured from visual observation, the cloud patterns delineate much of the storm's internal motion. The cloud structure of a supercell thunderstorm is diagrammed in Fig. 9.1a as it might appear to an observer. Even when clouds, precipitation, trees, and buildings do not obstruct one's view, a composite of phenomena such as that shown could not be observed at one time because the scales are so vastly different. Even a thunderstorm photograph (Fig. 9.1b) that shows nearly the entire cloud envelope does not display all the phenomena present in this storm. The diagrammatic view sketched in Fig. 9.1a is toward the northwest. The main storm tower contains the largest and strongest updrafts, although there is a multitude of smaller-scale turbulent vertical motions. Broad zones of warm, dry downdrafts of weak (<5 m s^{-1}) intensity have been observed in the clear air surrounding the upper parts of a severe storm (Sinclair, 1973; Fritsch, 1975), and sometimes the small-scale turrets that form the envelope of the cauliflower cloud structure can be seen descending on the upwind side of the storm. The strongest downdrafts, however, are usually encountered within the regions where precipitation exerts drag and evaporatively cools the air, locally increasing its density.

When there is strong shear of the environmental wind and convective instability is considerable, fast updrafts are created (Hess, 1959). The shear tilts the updrafts so that the precipitation falls out rather than through the updraft, with a resultant increase in the net buoyancy. The interaction of the updrafts with the sheared environmental wind organizes the flow so that dry middle-level air is continually cooled by evaporating precipitation, and it descends rapidly (because cooled air is more dense). Dense low-level outflow also initiates new convection because it lifts the less dense but potentially unstable boundary-layer air. In that case the storm may be long-lived (several hours) and have a quasi-continuous flow of up- and downdrafts. These are the characteristics of supercell storms.

Ordinary thunderstorms (Section 9.2.3) can be considered closed systems in the sense that the storm air does not sufficiently interact with the environment to maintain a continuous quasisteady flow of ascending and descending air. The supercell storm, sketched in Fig. 9.1a, is an open system hypothesized to organize a steady flow and to propagate either to the left or right of the mean

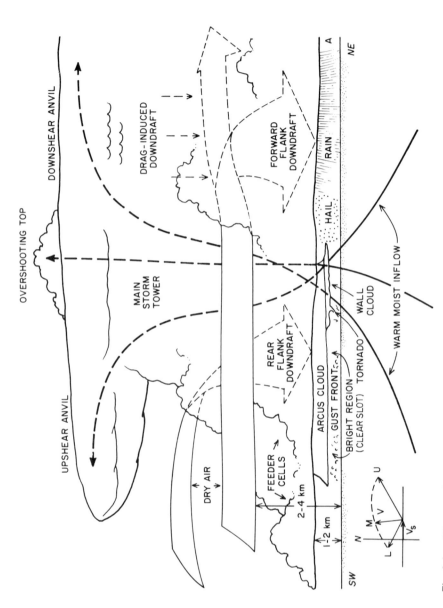

Fig. 9.1a Diagram of a tornadic supercell storm as viewed from the southeast. The sketch is not to scale. Dashed lines are air trajectories inside the cloud. Environmental wind vectors at low (L), middle (M), and upper (U) levels of the troposphere are drawn at the lower left. The storm motion vector is $\mathbf{v}_s$.

Fig. 9.1b A severe thunderstorm as seen from an aircraft. The view is toward the upshear anvil cloud. Three distinct cloud columns associated with updrafts are seen in this multicell severe storm. (Courtesy of J. Lee, NSSL.)

Fig. 9.1c Photograph toward the rear flank downdraft location, showing the wall and arcus clouds, the clear slot and a tornado. (Courtesy of H. Bluestein, University of Oklahoma.)

Fig. 9.1d Southeast view of a bubble of updraft air overshooting the anvil of clouds. A feeder cell is to the right (southwest) of the storm. (Courtesy of H. Bluestein, University of Oklahoma.)

tropospheric wind with a velocity that allows it a steady interaction with the large-scale environment of the storm (Browning, 1982). In either of the two storm types, air rising and descending does not follow a closed circuit, because the air that goes up does not return and the air that comes down does not immediately go up again; the thunderstorm produces an interchange between the surface and the upper air.

In the middle latitudes of the North American Great Plains, the updraft is fed by boundary layer air, usually entering the storm from the east to southeast in a coordinate system translating at a velocity v_s equal to that of the storm (Fig. 9.1a). Storm motion is usually northeast, although supercells move more easterly and even southeast. Within the updraft, condensation takes place, and above the 0°C altitude freezing of rain droplets can commence, releasing more latent heat, which further increase the buoyancy of the ascending air. At levels where temperatures decrease to below -30°C, the proportion of ice increases (Anthes et al., 1982). The rapidly ascending air is associated with a low-pressure region beneath the main tower, where a wall cloud (Fig. 9.1c) is formed from saturated subcloud air that has been cooled by expansion. It is here where the most intense updraft below the cloud base is located and where the most intense tornados are found. Usually a wall cloud rotates cyclonically and is part of a larger mesocyclone, which extends up through the main storm tower. Increasing rotational velocity is accompanied by lowering pressure.

The intensity of the updraft can be so strong that its momentum can carry air into regions of large negative buoyancy in the lower stratophsere, and over-shooting cloud tops can be observed (Fig. 9.1d). Above the tropopause (Fig. 2.4) the air is stably stratified, and a strong negative buoyancy force acts on the updraft, forcing it back down to the tropopause, where it spreads into a relatively thin layer (which may be more than 3000 m thick, however; see Color Plate 2b), known as the anvil cloud. The outflow speed near the storm top is so fast that air can move upstream against the strong upper level winds and form the upshear portion of the anvil cloud. Although the upwind anvil may extend only several kilometers from the main storm tower (Fig. 9.1b), the downwind portion can be visible for many hundreds of kilometers as a plume of cloud carried off by the environmental winds. The air in the updraft column is usually found to be rotating cyclonically.

The downdraft contains large amounts of dry mid-level air that is turbulently mixed with the updraft air containing precipitation. As dry air approaches the storm from the southwest to west, some of it is deflected around the updraft tower, which acts partially like an obstacle. Two distinct zones of downdrafts are depicted on the surface map in Fig. 9.2a: (1) the forward-flank downdraft (FFD) and (2) the rear-flank downdraft (RFD) (Lemon and Doswell, 1979).

Precipitation falling downstream from the updraft column may drag air to contribute to the FFD. This downdraft, abetted by evaporation, is depicted in Fig. 9.1a by downward-directed dashed lines that emanate high in the storm

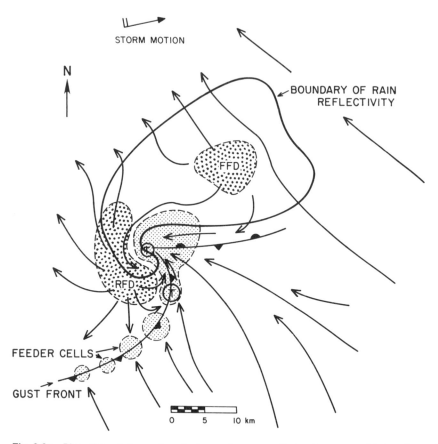

Fig. 9.2a Plan view of the surface features of a supercell thunderstorm, with locations of the forward-flank (FFD) and rear-flank (RFD) downdrafts. The solid line outlines the rainy area usually mapped by radar (note the hooklike feature). Downdraft and updraft regions are delineated by coarse and fine stippling. T designates the most likely location where tornados form. (Adapted from Lemon and Doswell, 1979; courtesy of R. Davies-Jones.)

beneath the downwind anvil. Because the updraft acts as an obstacle to the environmental flow, a pressure deficit can be created near the downstream side of the main updraft tower. Thus pressure-gradient forces may divert dry air into the precipitation zone, where it mixes with the forward flank downdraft. An example of a downdraft embedded in rainy air is shown in Fig. 9.2b, where precipitation near the surface is deflected outward as depicted in Fig. 9.1a (region A).

The rear flank downdraft (RFD) is hypothesized by Lemon and Doswell (1979) to begin at the upper levels of the storm, where an upwind stagnation point and a high-pressure region develop that may initiate the downward flow.

Fig. 9.2b View of a rain shaft; deflection of precipitation is due to downdraft air being turned outward at the ground. (Courtesy of H. Bluestein, University of Oklahoma.)

Fig. 9.2c Photograph toward the rear-flank downdraft, which evaporates cloud particles and thus allows sunlight to penetrate to the surface. Strong outflow of the RFD lifts the moist boundary-layer air that forms the arcus cloud. (Courtesy of H. Bluestein, University of Oklahoma.)

Air trajectory analysis for numerically modeled storms as well as early Doppler data did not exhibit environmental air descending much more than a kilometer or two (Klemp *et al.*, 1981). More recent observations with multiple Doppler radar show that air from as high as 5 km descends to the ground (Knupp and Cotton, 1982; Brown, 1989), in agreement with deductions based on the moisture and temperature characteristics of downdraft (Barnes, 1978).

If the relative humidity of the downdraft remains less than 100%, cloud particles and the smallest raindrops evaporate quickly, leaving behind a nearly cloud-free column through which sunlight can be seen as a bright region at the surface of the rear flank of the storm (Figs. 9.1a and 9.2c). The downdraft builds a pool of air on the ground that is more dense than the surrounding air and often carries with it the horizontal momentum of the lower- to middle-level wind. The air then quickly spreads outward as a density current and creates at its leading edge a gust front whose presence may be marked by dust clouds and, extending above and along the front, an arcus cloud (Figs. 9.1a and 9.2c). When water vapor condenses in the moist environmental air (lifted by the density current) it creates a shelflike cloud (the arcus cloud) that can extend several kilometers beyond the surface location of the density current front. New feeder cells (Figs. 9.1a and 9.1b) develop along the density current front, and these drift with the southwesterly winds and carry moisture into the main tower to enrich the updrafts.

Many flow characteristics implicit in this broad description are elaborated in the Doppler radar estimated wind and reflectivity fields presented throughout this chapter.

9.2 Wind Measurement with Two Doppler Radars

A single Doppler radar maps a field of velocities that are directed toward (or away from) the radar. A second Doppler radar spaced far from the first produces a field of different radial velocities because the true velocities are projected on different radials. The two radial velocity fields can be vectorially combined to retrieve the two-dimensional velocities in the planes containing the radials. It is customary to accomplish the synthesis on common grid points to which radar data are interpolated. Mean radial velocities (assigned a location at the center in each of the radar's resolution volumes surrounding a grid point) are not measured simultaneously but are separated in time up to the few minutes required for each radar to scan the storm volume. Furthermore, the respective resolution volumes are usually quite different in size and orientation.

Nevertheless, useful estimates of wind can be made on scales of air motion large compared to the largest resolution volume dimension if during the period required to scan a storm the velocity field is nearly stationary *in the coordinate frame translating with the storm*. Then data collected at different times can be displaced spatially (displacement for each datum depends on advection velocity and differences in datum and reference times) to obtain an adjusted "radial" velocity field with a reference time common for data from each radar. Note that the adjusted "radial" velocities are no longer directed toward the radars and if displacements are significant, this change needs to be taken into account (Brown, 1989). Advection velocities are usually obtained by tracking the position of the storm, but Gal-Chen (1982) formulated a more general procedure using either the reflectivity or Doppler velocity field. In this approach, one finds a moving coordinate frame wherein the observations are as stationary as possible (in the least squares sense). Stationarity is probably the most severely violated of the many assumptions required to derive vector wind fields from dual Doppler data. If the wind field is not stationary, then interpolation of a time series data for each resolution volume could be used to estimate the radial velocity at a common reference time (Schneider, 1991; Gal-Chen, 1978).

9.2.1 Wind Field Synthesis

Wind field synthesis using two-Doppler radar data is greatly simplified if the synthesis is performed in cylindrical coordinates with axis chosen to be the line connecting the two radars. That is, radial velocities at data points (centers of resolution volumes) are acquired or interpolated to nearby grid points on planes having a common axis (the COPLAN technique; Lhermitte, 1970). Cartesian wind components can be derived from these synthesized cylindrical components. Although one could solve directly for Cartesian wind components, this necessitates solving an inhomogeneous hyperbolic partial differential equation to derive the vertical wind (Armijo, 1969).

9.2.1.1 COPLAN

The cylindrical coordinate system is illustrated in Fig. 9.3 where effects of the earth's curvature are ignored. The measured Doppler velocity needs to be corrected for the scatterers' terminal velocity w_t. In this chapter overbars, used in Section 5.2, are dropped; nevertheless, it is important to be aware of the reflectivity and $I(\mathbf{r}_0, \mathbf{r}_1)$ weights. Thus the estimate of the radial component of air motion is

$$v_{1,2} = v_{1,2}^h + w_t \sin \theta_{e1,2}, \tag{9.1}$$

where $v_{1,2}^h$ are the weighted mean hydrometeor velocities measured by radars 1 and 2, w_t is positive, and θ_e is the elevation angle. To estimate w_t for each resolution volume, we could use Eq. (8.79) or the empirical expression (Atlas et al., 1973)

$$w_t = 2.65 Z^{0.1114} \left(\frac{\gamma_0}{\gamma} \right)^{0.4} \quad \text{m s}^{-1}, \tag{9.2}$$

where the parenthetical term is a correction [Eq. (8.4)] to account for the height-dependent air density γ. This relation represents, to within a standard

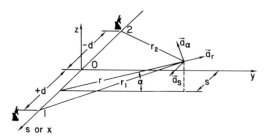

Fig. 9.3 Cylindrical coordinate system used for dual radar data analysis. The radars are located at the points 1 and 2, and $\mathbf{a}_r$, $\mathbf{a}_s$, $\mathbf{a}_\alpha$ are the unit normals defining the direction of the three orthogonal velocity components. The cylinder axis is along the line connecting the radars, and r is the range from the axis to the data point.

error of 1 m s^{-1}, the experimental data of Joss and Waldvogel (1970) over a large range of Z (i.e., $1 \leq Z \leq 10^5 \text{ mm}^6 \text{ m}^{-3}$) and drop-size distributions for regions of liquid water; but large errors (up to several meters per second) in w_t estimates can be caused by using Eq. (9.2) for regions of hail. Terminal velocity relations for hail can be found using, for example, Eq. (8.6c). Usually there is little or no information to identify hail regions, and errors in vertical wind w can result. It has been shown, however, for typical arrangements of storms relative to the two-radar placement, that these errors are significantly smaller than errors in the estimate w_t (Doviak *et al.*, 1976).

The estimated radial velocities $v_{1,2}$ of the air can be interpolated to uniformly spaced grid points in planes at an angle α to the horizontal surface containing the baseline. Interpolation filters the data and reduces the variance. Several interpolation schemes are possible, and a particularly simple and effective one employs the Cressman function (Cressman, 1959) W_i to weight data at points inside a volume centered on a grid point from which the distance to the ith datum is D_i.

$$W_i = \begin{cases} (R_i^2 - D_i^2)/(R_i^2 + D_i^2) & \text{for} \quad D_i \leq R_i, \\ 0 & \text{for} \quad D_i > R_i. \end{cases} \tag{9.3}$$

R_i is the influence radius that determines the size of the interpolation volume. The shape of this volume, usually selected to be a sphere, is dictated by the functional dependence of R_i on the direction of a datum from the grid point. The cylindrical wind components w_r, w_s in the r, s plane (Fig. 9.3) are related to $\bar{v}_1$, $\bar{v}_2$ as

$$w_r = [(s + d)r_1\bar{v}_1 - (s - d)r_2\bar{v}_2]/2dr, \tag{9.4a}$$
$$w_s = (r_2\bar{v}_2 - r_1\bar{v}_1)/2d, \tag{9.4b}$$

where $\bar{v}_{1,2}$ are the interpolated Doppler velocities of air. Here the overbar signifies two averaging processes: (1) an $I(\mathbf{r}_0, \mathbf{r}_1)$ and η weighted spatial average, and (2) an interpolated average of $v_{1,2}$ over several resolution volumes.

The wind component w_α normal to the plane is obtained by introducing the continuity equation for air density γ,

$$\partial\gamma/\partial t + \nabla \cdot (\gamma\mathbf{v}) = 0, \qquad (9.5a)$$

where $\mathbf{v}$ is the vector wind. Unfortunately, this introduces another unknown. But since perturbations in air density are much smaller than the ambient density, it can be shown by using scale analysis for deep convection (Pielke, 1989) that the first-order velocity field is produced by the first-order density and pressure perturbations, and to this order the continuity equation is

$$\nabla \cdot (\gamma_0\mathbf{v}) = 0, \qquad (9.5b)$$

where γ_0 is the ambient air density, which is assumed to depend only on height. This form of the continuity equation, first introduced by Ogura and Phillips (1962), is called anelastic because the energy conservation equation no longer has contribution from elastic energy (i.e., there are no sound waves; Ogura, 1963). In the cylindrical coordinate system this equation takes the form

$$\frac{1}{r}\frac{\partial}{\partial r}(r\gamma_0 w_r) + \frac{1}{r}\frac{\partial}{\partial\alpha}(\gamma_0 w_\alpha) + \gamma_0\frac{\partial}{\partial s}(w_s) = 0. \qquad (9.5c)$$

If the ambient atmosphere has constant potential temperature θ_0, γ_0 is given by Hess, 1959,

$$\gamma_0(r, \alpha) = \gamma_s \left[\frac{\theta_0 - \dfrac{g}{c_p} r \sin\alpha}{\theta_0} \right]^{C_p M/R - 1}, \qquad (9.5d)$$

where γ_s is air density at $\alpha = 0$ (e.g., at the earth's surface), $C_p = 1000$ J kg^{-1} K^{-1} is the specific heat at constant pressure, g is the gravitational constant (9.8 m s^{-2}), M is the mean molecular weight of air (0.029 kg mol^{-1}), and R is the universal gas constant (8.314 J mol^{-1} K^{-1}). Estimates of θ_0 can be obtained from soundings of the environment within which the storm develops. Several techniques used to compute the vertical velocity are compared by Ray *et al.* (1980), and Ziegler *et al.* (1983) solve the anelastic form of the continuity equation using a variational formulation with downward integration to reduce errors in vertical velocity.

9.2.1.2 Errors

The wind fields derived from dual Doppler radar measurement contain errors from several sources. Some of these are (1) variance in the mean Doppler velocity and Z estimates caused by the statistical nature of the weather echo; (2) nonuniform shear and asymmetric reflectivity factor within resolution volumes (Section 5.2); (3) the use of an incorrect w_t, Z relationship; (4) inaccuracies in the location of the resolution volume; (5) increases in the vertical velocity variance with each integration step owing to error in the divergence used in the continuity equation; (6) nonstationarity of the storm during a data collection scan; and (7) echoes received through sidelobes that contaminate signals associated with the resolution volume.

The accuracy with which wind vectors can be synthesized from the two radial velocity fields depends principally on the angle β subtended by $\mathbf{r}_1$ and $\mathbf{r}_2$. Formulas for wind estimation accuracy as a function of the cylindrical coordinates r, s, and α (Fig. 9.3) are given by Doviak et al. (1976). For example, the sum of the variances of w_r and w_s is

$$\sigma_r^2 + \sigma_s^2 = (\sigma_1^2 + \sigma_2^2)/\sin^2 \beta, \tag{9.6a}$$

where σ_1^2, σ_2^2 are the variances of $\bar{v}_1$, $\bar{v}_2$. Thus, for angles β less than about 30° (or more than 150°) error variance in w_r and/or w_s increases rapidly. The volume within which β lies between 30° and 150° is one that is swept out by rotating, about the baseline, a crescent-shaped area with vertices at the radar sites (see Fig. 11.30 for an illustration of the area).

Large areas for vector wind estimation can only be obtained by increasing radar separation, which of course, worsens spatial resolution, and consequently compromises need to be made (Davies-Jones, 1979). Furthermore, because of the earth's curvature, a larger separation reduces the area of common surveillance at low altitudes where weather phenomena is hazardous to human life. These considerations, and the need to observe the wind and reflectivity structure of storm cells near the surface as well as at their tops, have led NSSL to select a radar separation of about 40 km.

Of the three components of wind, the vertical component w is the most difficult to reliably estimate from observations. Assuming $\alpha \ll 1$, the variance of w_α (roughly w for small α) is (Doviak et al., 1976)

$$\sigma_\alpha^2 = \frac{\alpha \, \Delta \alpha}{2(\Delta r)^2 \gamma_0^2(\alpha)} \left[\frac{r_1^2 r_2^2}{4d^2} C_1 + C_2 \right], \tag{9.6b}$$

where

$$C_1 = \frac{1}{l} \int_0^l (\sigma_1^2 + \sigma_2^2) \gamma_0^2(l) \, dl; \qquad C_2 = \frac{1}{l} \int_0^l \sigma_t^2 \gamma_0^2(l) l^2 \, dl,$$

α is in radian units; $r \, \Delta \alpha$ and $\Delta r = \Delta s$ are grid point spacings; $l = r\alpha$ is the arc length; and σ_t^2 is the variance in estimates of w_t. Doviak et al. (1976) estimate σ_t^2 to be approximately $1 \, \text{m}^2 \, \text{s}^{-2}$. In deriving Eq. (9.6b) it has been assumed that errors at grid points are uncorrelated, and that centered differences approximate derivatives. It has been our experience, from theory and observations of wind in the turbulent air of the fair weather boundary layer, that σ_1^2, σ_2^2 have a lower bound of about $1 \, \text{m}^2 \, \text{s}^{-2}$ (see Rabin and Zrnić, 1980, Fig. 6; and Doviak and Berger, 1980, Table 2a). In thunderstorms, where turbulence is stronger, σ_1^2, σ_2^2 values could be larger. Istok and Doviak (1986) separated Doppler velocity fields in a supercell thunderstorm into turbulent and ordered flows and showed that turbulent velocity fluctuations (at scales smaller than about 3 km) are larger than 3 m s^{-1} throughout much of the storm; especially in regions of updrafts (Section 10.4).

Assuming $\sigma_1 \approx \sigma_2 \approx 1\,\mathrm{m\,s}^{-1}$ and upward integration of Eq. (9.5c), (i.e., from $\alpha = 0$ to α), Doviak et al. (1976) demonstrate that σ_z, the standard deviation of w, at an altitude of 14 km can be as large as 30 m s^{-1}. This large uncertainty can be reduced, however, if a boundary condition on w is known at the top of the storm and integration of Eq. (9.5c) is performed downward (Bohne and Srivastava, 1975). The ratio of σ_α for upward integration $\sigma_{\alpha u}$ to that for downward integration $\sigma_{\alpha d}$, obtained from Eq. (9.6b), is

$$\frac{\sigma_{\alpha u}}{\sigma_{\alpha d}} = \frac{\gamma_s}{\gamma(\alpha)}, \qquad (9.6c)$$

where γ_s is the air density at the surface and $\gamma(\alpha)$ the density at the upper boundary. For storm tops near 14 km the ratio $\gamma_s/\gamma(\alpha)$ is about five and hence downward integration should result in a $\sigma_{\alpha d}$, at the earth's surface, of about 6–7 m s^{-1}. Distributions of estimated $w(0)$ at the ground (where w should be zero), computed from downward integrations of actual storm data assuming $w = 0$ at the storm top (i.e., where $Z \approx 10\,\mathrm{dBZ}$), show for one storm a measured σ_z of about 9 m s^{-1} (Nelson and Brown, 1982; 1987). Likewise, Vasiloff et al., 1986, have, for another storm, observed a ± 5 to $\pm 10\,\mathrm{m\,s}^{-1}$ error level. Thus, both observations are in close agreement with theory. The measured values, which are slightly larger than the 6–7 m s^{-1} theoretically estimated, suggest that σ_1, σ_2 are larger than 1 m s^{-1}, a reasonable deduction in view of the fact that the 1 m s^{-1} value is appropriate for fair weather.

Vertical wind errors can be reduced by imposing both top and bottom boundary conditions on w (e.g., by using a variational adjustment technique to produce a wind field that satisfies both boundary conditions as well as the continuity equation; Ziegler, 1978). Comparison of vertical velocities, derived from multiple Doppler radar data and those measured in situ with an instrumented sail plane, is reported by Ziegler et al. (1991). The two independent measurements agreed within $\pm 4\,\mathrm{m\,s}^{-1}$ over a 20-minute period as the sail plane spiraled from an altitude of 5 to 8 km (MSL) in updrafts ranging from 2 to 14 m s^{-1}.

9.2.2 Supercell Thunderstorms

Doppler radar observations of two tornadic storms will be briefly discussed to illustrate some of the general storm features depicted in Figs. 9.1 and 9.2.

Figure 9.4a shows a horizontal cross section of a tornadic storm that formed in central Oklahoma on 2 May 1979. The three components of wind were obtained from radial velocity measurements made by two radars spaced 70 km apart, a reflectivity–terminal velocity relation, and integration of the mass continuity equation using suitable boundary conditions at the ground and the

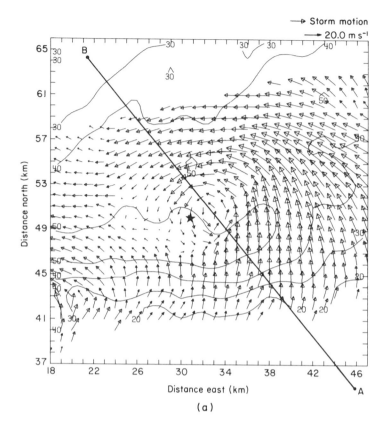

(a)

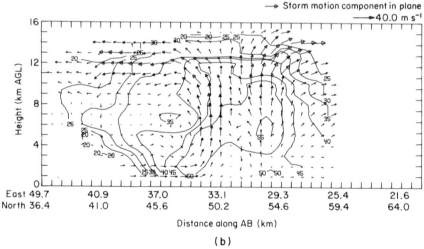

(b)

Fig. 9.4a,b Contours of reflectivity factor (in dBZ) and the storm-relative wind field in a horizontal cross section at 2 km AGL (a), and a vertical cross section (b), of a tornadic thunderstorm on 2 May 1979 at 1658 C.S.T. Line AB in (a) is the location of the vertical cross section. Distances are from a radar at Roman Nose State Park, Oklahoma (Alberty *et al.*, 1979). The star pinpoints the tornado location in (a). The arrow at the top right of each plot gives the speed scale, which is proportional to the arrow's length. (Courtesy of E. Brandes and B. Johnson, NSSL.)

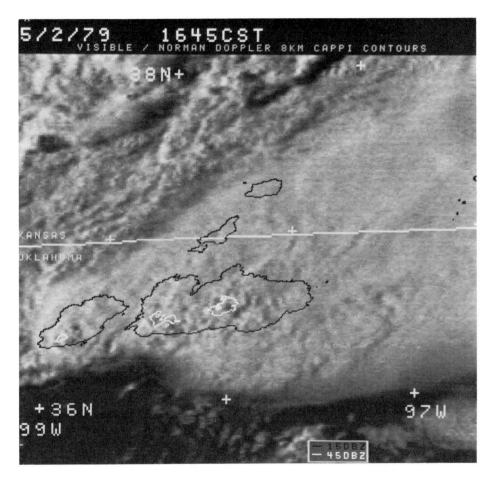

Fig. 9.4c The anvil of clouds from the 2 May 1979 storm at 1645 C.S.T. The Norman Doppler radar reflectivity factor contours of 45 (white) and 15 (black) dBZ at the 8-km altitude are superimposed. The fields of reflectivity, spectrum width, and Doppler velocity for this storm are presented in Figs. 9.20, 9.21, and Color Plate 2a. (Courtesy of Dr. G. Heymsfield, GLAS, NASA.)

cloud top. Satellite photographs with superimposed reflectivity contours at 8 km and horizontal wind mapped at 11.9 km by the dual Doppler radars are in Figs. 9.4c and 9.4d.

The large circulation of air seen in Fig. 9.4a is a thunderstorm cyclone. Because thunderstorm cyclones are much smaller than extratropical or tropical cyclones, including hurricanes, these circulations are called *mesocyclones*. In all the cases observed, large tornados have been preceded by mesocyclones. Usually pressure decreases significantly within the interior of a mesocyclone, and when cyclonic circulation reaches the surface, moist boundary-layer air entering

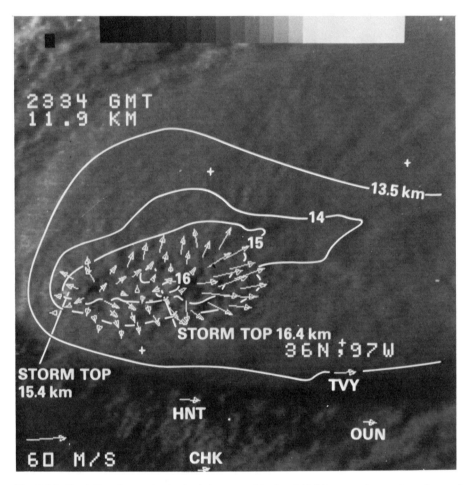

Fig. 9.4d Dual Doppler synthesized winds at an altitude of 11.9 km superimposed on the anvil cloud (2 May 1979, 1734 C.S.T.). The height contours of the cloud tops are obtained from a stereo pair of satellite images. The vector at 60 m s^{-1} scales the wind vectors. Winds at 11.9 km, but measured at four rawinsonde stations [Norman (OUN), Hinton (HNT), Chickasha (CHS), and Oklahoma City (TVY)], are also displayed. (Courtesy of G. Heymsfield, GLAS, NASA.)

the cyclone condenses and forms a wall cloud, making the circulation visible (Fig. 9.1c). Mesocyclones have diameters ranging from a few kilometers to ~10 km and can extend in height from near the ground to near the tropopause.

Figure 9.4e shows a tornado embedded in the larger mesocyclone, which is also visible. Tornadoes are smaller-scale vortices usually embedded within the mesocyclone, and if their centers do not coincide (as in Fig. 9.4a) the tornado circulates around the mesocyclone center, forming a trochoidal path on the ground as the mesocyclone and tornado are translated by the moving storm. Note that the small-scale tornado vortex is not resolved in Fig. 9.4a.

Fig. 9.4e A tornado embedded in the larger-scale mesocyclone (Norman, Oklahoma 1973). The mesocyclone cloud base was seen to be rotating counterclockwise. (Courtesy of Don Burgess, NSSL.)

The inflow of air from the southeast in the first 2 km above ground level (AGL) and updrafts as large as 40 m s^{-1} can be seen in the vertical cross section in Fig. 9.4b. Large gradients of reflectivity mark the region of inbound air, and strong outflow occurs in the anvil (Fig. 9.4d). Although the cross section in Fig. 9.4b contains mostly updrafts, other regions of the storm have downdrafts.

The second example of Doppler-synthesized wind fields is from a storm that occurred on 20 May 1977 (in Section 9.5.3 we present the Doppler spectra of winds around the tornado that developed in this storm). On the morning of this day a pool of cold air, resulting from the outflow of nocturnal thunderstorms in northwest Oklahoma, formed a boundary that extended southwestward from northeast Oklahoma. A morning rawinsonde released from Oklahoma City revealed a potentially very unstable air mass with strong vertical shear. Southeasterly boundary layer flow brought warm moist air above the denser cold air. Thunderstorms in a wide area were initiated over this cold air boundary. Individual cells propagated northeastward, whereas the band of storms progressed slowly eastward. Figure 9.5a shows the reflectivity field observed with the Norman Doppler radar at 1840 C.S.T. The Del City storm, about 35 km north-northeast of the radar at Norman, was scanned by four Doppler radars. At 1847 C.S.T. the Del City tornadic storm was in an excellent position (Fig. 9.5a) to be sampled by NSSL's Norman and Cimarron 10-cm Doppler radars.

Data from all four radars entered into the synthesis of the storm's vector wind fields portrayed in Fig. 9.5b–e, but most weight was given to the radial

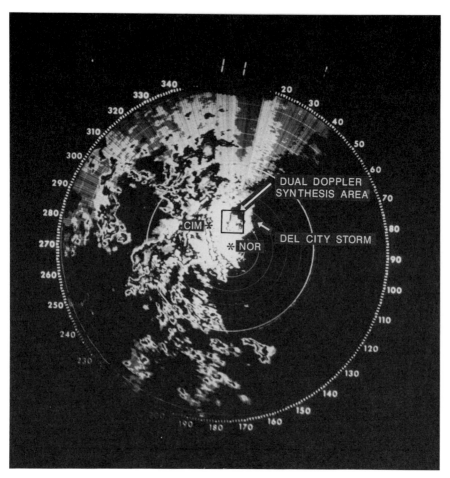

Fig. 9.5a PPI display of reflectivity for storms at 1840 C.S.T., 20 May 1977. The locations of NSSL's Cimarron and Norman Doppler radars are shown with asterisks. The box outlines the area within which winds of the Del City storm were synthesized. The elevation is 1.1° and range circles (faintly bright) are spaced 20 km apart.

velocities measured by the Norman and Cimarron radars, which had the best aspect angles and resolution. The method of synthesis is given by Ray *et al.* (1981). The strong downdraft to the north of the strongest updraft, (Fig. 9.5b) is identified as a forward flank downdraft (Fig. 9.1a). There is an indication (Fig. 9.5b) of descending air being drawn around the western portion of the updraft. Earlier the updraft was confined to a single cyclonically rotating region, whereas at 1847 two major updraft branches have formed. The western branch (A; coordinates 3, 31) contained a vorticity maximum of $2.5 \times 10^{-2}\,\text{s}^{-1}$ at an altitude of 2 km, while the eastern branch (B; 8, 29) had a $1.5 \times 10^{-2}\,\text{s}^{-1}$ vorticity maximum.

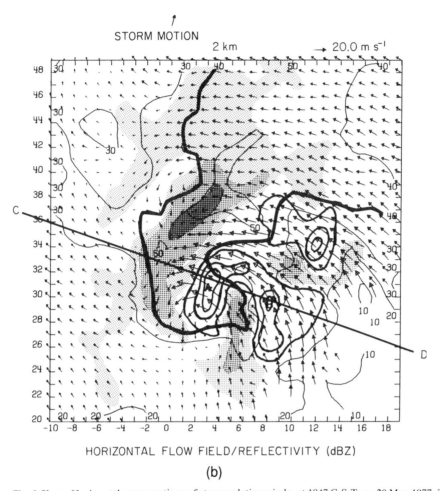

Fig. 9.5b,c Horizontal cross sections of storm-relative winds, at 1847 C.S.T. on 20 May 1977. The origin is at the Norman Doppler radar. (b) Height above ground $h = 2$ km; (c) $h \approx 0$ km. The velocity scale is the arrow labeled 20 m s^{-1} in the upper-right corner. Reflectivity (dBZ) is contoured with thin lines except the 40-dBZ contour in (b) is a bold line showing a hooklike feature. The lines of medium thickness in (b) are contours of updrafts in steps of 10 m s^{-1} from the first contour at 5 m s^{-1}. Areas of downdrafts faster than 1 m s^{-1} are lightly stippled, and darker stipples mark areas where downdrafts are faster than 5 and 15 m s^{-1}. In (c) the tornado is at 4 km, 29.5 km, but its damage path is the elongated north–south lightly stippled area; the outflow front is the curved line. Line CD in (b) locates the cross section shown in Fig. 9.5d,e. (From Ray *et al.*, 1981.)

Another updraft (coordinates 12, 34) formed northeast of the updraft associated with the Del City storm. This updraft was caused by the strong convergence along the line indicated in the 0-km height wind fields (Fig. 9.5c). Updraft B in Fig. 9.5b appears to have been caused by the convergence along the outward propagating density current (delineated in Fig. 9.5c), where feeder cells form

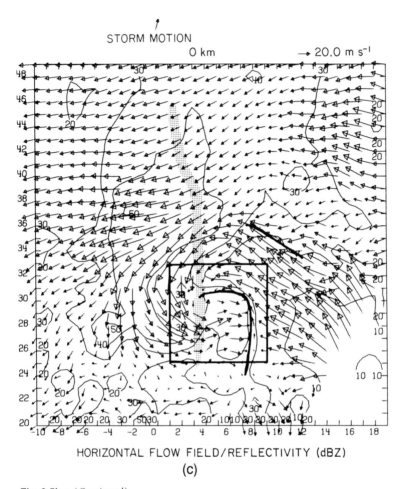

HORIZONTAL FLOW FIELD/REFLECTIVITY (dBZ)

(c)

Fig. 9.5b,c (*Continued*)

(Fig. 9.2a). The tornado's location is between the updraft air of branch A and the downdraft forming the density current front in the region of strong horizontal gradient of vertical wind. The strongest inflow (9.5c) approaches from the southeast, although environmental air from the east and northeast enters into this storm.

The wind in the vertical cross section (CD in Fig. 9.5b) is shown in Figs. 9.5d and 9.5e. Clearly (Fig. 9.5d), there are two branches of updraft with high-reflectivity regions within each. The contours in Fig. 9.5e show the airflow perpendicular to the plot, whereas the in-plane wind is the same as in Fig. 9.5d. The strong surface wind from the north is quite evident in the lowest levels of Fig. 9.5e.

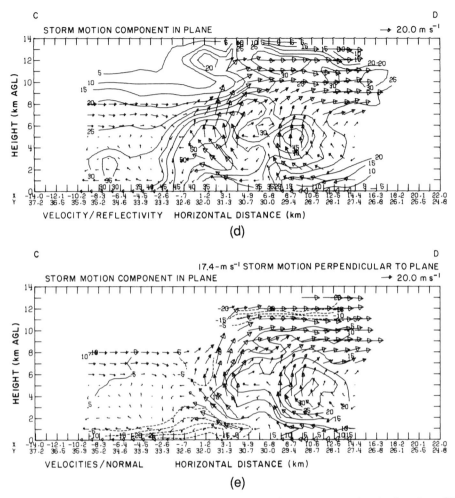

Fig. 9.5d,e Vertical cross section of storm winds relative to storm motion in the plane CD. Reflectivity contours in (d) are in dBZ. Velocity contours in (e) are wind speeds into (solid lines) and out of (dashed lines) the cross section CD. Speed contours are in steps of 5 m s^{-1} starting from ±5 m s^{-1}. (From Ray *et al.*, 1981.)

9.2.3   Ordinary Thunderstorms

At the other end of the spectrum of thunderstorms, opposite the quasisteady-state supercell type, are the ordinary variety, sometimes referred to as "air mass" storms; these were the first to be intensively studied (Byers and Braham, 1949). If the environmental shear is weak, only showers and up- and downdrafts are present, but if convective instability is strong, hail and damaging winds may also occur. Updrafts are usually short-lived (i.e., less than an hour), because

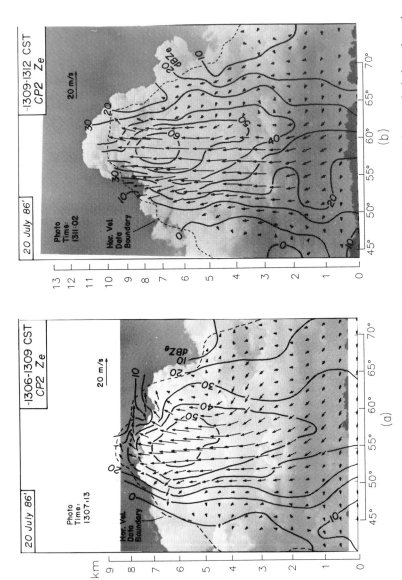

Fig. 9.6 Wind vectors derived from Doppler radar analysis and reflectivity factor contours in a vertical plane through the storm center and orthogonal to the line of sight from the photographer. The solid contours of reflectivity factor are in steps of 10 dBZ and were obtained by the 10-cm CP2 radar of the National Center for Atmospheric Research (Courtesy of D. Kingsmill, UCLA). (*Figure continues.*)

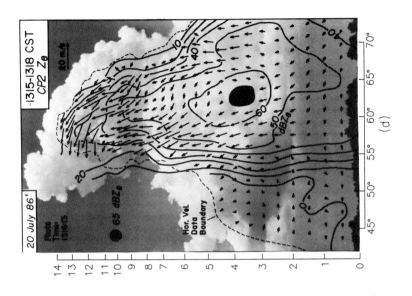

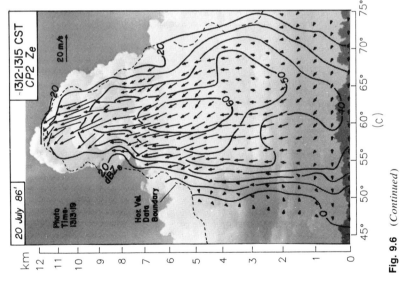

Fig. 9.6 (Continued)

303

intensifying precipitation causes drag that slows the upward flow, eventually turning it downward. In this case, precipitation falls to the ground through the air column in which it was formed. These storms occur seemingly randomly within large areas of an unstable boundary-layer air of high humidity and are most often observed in tropical climates.

A comprehensive study of an ordinary thunderstorm has been made by Kingsmill and Wakimoto (1991) using data from three-Doppler radars in an over-determined dual-Doppler analysis (Kessinger *et al.*, 1987). Illustrative combinations of cloud photos of this ordinary thunderstorm and its internal wind and reflectivity field for the storm's growing (cumulus) and mature stages are in Fig. 9.6a–d. The Doppler fields are in cross sections through the center of the storm and perpendicular to the line of sight. In these images the storm is about 13 km away from the camera and the angle scale at the base of Fig. 9.6a indicates the azimuth, which gives a measure of the storm's lateral dimension (about 4 km across the 20 dBZ contour at mid-levels).

The cumulus stage of the storm (Fig. 9.6a,b) contains updrafts with speeds as large as 18 m s^{-1}, and reflectivity factors growing to 60 dBZ at about 7 km AGL, near cloud top (≈ 10 km). Buoyancy forces the cloud top to ascend 1.5 km in about 4 minutes, and upper-level divergence is evident as faster air in the center of the buoyant bubble overtakes the slower-ascending upper boundary of the cloud turret.

The mature stage of this storm is illustrated in Fig. 9.6c–d. This is the stage where the core of high reflectivity (in this case the volume of 60 dBZ or greater reflectivity factors) begins its downward descent. The prominent cloud turret is evidence of a vigorous updraft, which attains velocities as large as 25 m s^{-1}. The reflectivity core descends in about 3 minutes from a height of 5 to 3 km. The upper altitude wind shear is causing the cloud turret to tilt, and the Doppler-derived wind field above 11 km shows this effect.

In the dissipating stage (not shown here), downdrafts dominate the vertical velocity field; it is these downdrafts that can create strong outflows near the surface (Section 9.6). The entire life of this storm, from the beginning of the rapid vertical development to the time of peak divergence of the outflow at the ground, was about 20 minutes. Thus, it is apparent that observations of similar storms require rapid coordinated scans of the Doppler radars to avoid errors in wind retrievals caused by the evolution of the storm.

9.3   Wind Measurement with One Doppler Radar

Weather systems of sizes or scales larger than a few hundred kilometers, such as those associated with the area of high and low pressures as seen on global or continental sized weather maps, do not produce much change in the wind field over the horizontal range surveyed by the radar. Lhermitte and Atlas (1961) described how a single Doppler radar could be used to determine the vertical profile of *horizontally* uniform wind. They proposed a data collection mode in which the radar beam is directed at a constant elevation angle and the radial velocity is continuously recorded at several ranges (to obtain soundings at different heights) as the radar beam sweeps through a full circle; this produces on an oscilloscope a series of traces called a *velocity azimuth display* (VAD).

Although analyses of data collected in this way have increased in sophistication (Caton, 1963; Browning and Wexler, 1968), the data collection method is still often referred to as VAD. Because the wind is rarely uniform over the large volumes surveyed by the radar we are led to consider the analysis of Doppler velocities for wind that is linear within analyses volumes (Fig. 9.7) or on horizontal areas that do not necessarily lie over the radar. For example the wind at constant height along a frontal band might be well represented by a linear wind model. This analysis technique was first studied by Waldteufel and Corbin (1979), who introduced the term *volume velocity processing* (VVP).

One might question how well the linear model represents the actual wind's spatial dependence. For example, measurements of the spatial variation

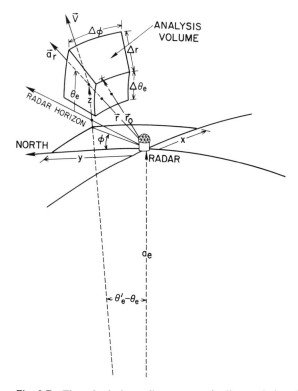

Fig. 9.7 The spherical coordinate system for linear wind analysis. Analysis volume size is typically $\Delta r \approx 20\text{--}30$ km; $\Delta \theta_e \approx 1\text{--}2°$; $\Delta \phi \approx 30\text{--}40°$.

of wind made with radar (e.g., Figs. 10.7 and 10.8) and instrumented towers, aircraft, and rawinsondes (e.g., Vinnichenko and Dutton, 1969) show that wind usually exhibits variability simultaneously at all horizontal wavelengths Λ (i.e., scales) extending from hundreds of kilometers to centimeters. Nevertheless, the linear wind model can account for most of the observed wind variance. But wind often has a persistently strong nonlinear dependence on height; the analysis presented here, however, has general applicability and allows the wind to be horizontally linear and to have any dependence on height.

9.3.1   Linear Wind Fields

In this section we examine the linear wind model in three dimensions. The theory of multivariate regression analysis will be used to determine the accuracy (bias and variance) of our wind estimates. We shall show that the accuracy of estimates of the kinematic properties (divergence, deformation, etc.) of the wind depends on the size and shape of the analysis volume, the measurement error in the radial velocity, the number of parameters in the model, and the linearity of the actual wind.

We begin by deriving an expression relating the radial velocity, measured by the radar, to the wind at the measurement point (r, θ_e, ϕ). A spherical earth coordinate system is used with x and y as arc distances from the radar along two orthogonal great circle paths. We choose y to be along a longitude, with north the positive direction, as shown in Fig. 9.7. The earth's radius vector, drawn through the measurement location, defines the vertical axis along which z measures the height above the surface. The earth is assumed to have an effective radius a_e (Chapter 2) due to the mean vertical gradients of the refractive index. The horizontal components u and v of the vector wind $\mathbf{v}$ are tangent to the great circle arcs at x and y and are directed eastward and northward, respectively. The vertical component w of $\mathbf{v}$ is along z.

The vector wind $\mathbf{v}$ at r, θ_e, ϕ is assumed to be well represented by a first-order (linear) Taylor series in a region (i.e., analysis volume) about some point (x_0, y_0, z_0), usually the center of the analysis volume. Therefore,

$$\mathbf{v}(x, y, z) = \mathbf{v}(x_0, y_0, z_0) + \frac{\partial \mathbf{v}}{\partial x}(x - x_0) + \frac{\partial \mathbf{v}}{\partial y}(y - y_0) + \frac{\partial \mathbf{v}}{\partial z}(z - z_0). \quad (9.7)$$

If $r \ll a_e$, the coordinates x, y, and z are related to the radar coordinates r, θ_e, and ϕ by

$$x \approx r \cos \theta_e' \sin \phi.$$
$$y \approx r \cos \theta_e' \cos \phi.$$
$$z = (a_e^2 + r^2 + 2ra_e \sin \theta_e)^{1/2} - a_e. \quad (9.8)$$

The angle θ'_e is the sum of the beam's elevation angle to the data point and the angle subtended by the verticals at the radar and at the measurement (i.e., data) point.

$$\theta'_e = \theta_e + \tan^{-1}[r \cos \theta_e / (a_e + r \sin \theta_e)]. \tag{9.9}$$

The radial velocity v_r is the projection of $\mathbf{v}$ onto $\mathbf{r}$, the vector from the radar to the point (r, θ_e, ϕ). Again, if $r \ll a_e$,

$$v_r = \mathbf{v} \cdot (\hat{\mathbf{i}} \cos \theta'_e \sin \phi + \hat{\mathbf{j}} \cos \theta'_e \cos \phi + \hat{\mathbf{k}} \sin \theta'_e), \tag{9.10}$$

where $\hat{\mathbf{i}}, \hat{\mathbf{j}}, \hat{\mathbf{k}}$ are unit vectors at $\mathbf{r}$ in the x, y, z directions. Substitution of the first-order expansion (9.7) and the relations for x and y, from Eqs. (9.8), into Eq. (9.10) and rearranging terms as recommended by Easterbrook (1974), we obtain

$$
\begin{aligned}
v_r ={}& u'_0 \cos \theta'_e \sin \phi + u_x \cos \theta'_e \sin \phi (r \cos \theta'_e \sin \phi - x_0) \\
&+ u_z \cos \theta'_e \sin \phi (z - z_0) + v'_0 \cos \theta'_e \cos \phi \\
&+ v_y \cos \theta'_e \cos \phi (r \cos \theta'_e \cos \phi - y_0) + v_z \cos \theta'_e \cos \phi (z - z_0) \\
&+ (u_y + v_x) \cos \theta'_e [r \cos \theta'_e \sin \phi \cos \phi - \tfrac{1}{2}(x_0 \cos \phi + y_0 \sin \phi)] \\
&+ w_0 \sin \theta'_e + w_x \sin \theta'_e (r \cos \theta'_e \sin \phi - x_0) \\
&+ w_y \sin \theta'_e (r \cos \theta'_e \cos \phi - y_0) + w_z \sin \theta'_e (z - z_0), \tag{9.11}
\end{aligned}
$$

where $u'_0 \equiv [u_0 + \tfrac{1}{2} y_0 (v_x - u_y)]$, $v'_0 \equiv [v_0 - \tfrac{1}{2} x_0 (v_x - u_y)]$ are termed the *modified wind components* and the subscripts x, y, z denote partial derivatives. If the ground below the analysis volume is horizontal, it can be shown (by applying the continuity equation (9.5b) and the boundary condition $w = 0$ at $z = 0$) that $w_x = w_y = 0$ (Problem 9.9). Because each of the 11 unknowns u'_0, v'_0, $u_y + v_x$, u_x, v_y, u_z, w_0, w_x, w_y, w_z is multiplied by a unique function of r, θ_e, ϕ, we can, in principle, estimate each of these unknowns through the different dependencies they predict on the measured radial component $\hat{v}_r$ as $\mathbf{r}$ scans the analysis volume. Because the linear wind field is described by 12 parameters (u_0, u_x, u_y, u_z, v_0, v_x, v_y, v_z, w_0, w_x, w_y, w_z), four of which appear in combined forms multiplied by unique trigonometric functions, additional assumptions must be made to determine all 12 components. For example, one difficulty is that vertical vorticity $v_x - u_y$ of horizontal wind cannot be discriminated from u_0, v_0 because they share a common trigonometric dependency. If x_0, y_0 were zero, then u_0, v_0 could be retrieved but not vorticity. This would restrict the study to linear wind in an analysis volume with its center at the radar site. For many situations it is preferable not to choose such an x_0, y_0 because then some kinematic properties (e.g., the horizontal divergence, $u_x + v_y$) can be mapped. Unless other assumptions are introduced, the wind u_0, v_0 and the vorticity cannot be retrieved.

To facilitate manipulation of Eq. (9.11), we introduce matrix notation. Define the vector $\mathbf{K}$ by

$$\mathbf{K}^T = (u'_0, u_x, u_z, v'_0, v_y, v_z, u_y + v_x, w_0, w_x, w_y, w_z), \tag{9.12}$$

where the T indicates the transpose. For later derivations we shall find it useful to define the vector $\mathbf{K}_m$ containing a subset of m elements of $\mathbf{K}$. For example, in the VAD analysis scheme (Section 9.3.3) only the elements u_0, u_x, v_0, v_y, $u_y + v_x$, w_0, w_x, w_y are required (i.e., $m = 8$) to specify v_r on the measurement circle if the wind is linear. The model (with $m = 8$) is then said to be adequate [i.e., it contains all the data attributes of the radial velocities; Draper and Smith, 1966, p. 59]. Let the predictor vector $\mathbf{P}_i$ contain the following functions of $\mathbf{r}_r$, associated with the ith observation:

$$\mathbf{P}_i^T = \begin{bmatrix} p_{i,1} \\ p_{i,2} \\ p_{i,3} \\ p_{i,4} \\ p_{i,5} \\ p_{i,6} \\ p_{i,7} \\ p_{i,8} \\ p_{i,9} \\ p_{i,10} \\ p_{i,11} \end{bmatrix} = \begin{bmatrix} \cos \theta'_e \sin \phi \\ \cos \theta'_e \sin \phi (r \cos \theta'_e \sin \phi - x_0) \\ \cos \theta'_e \sin \phi (z - z_0) \\ \cos \theta'_e \cos \phi \\ \cos \theta'_e \cos \phi [r \cos \theta'_e \cos \phi - y_0) \\ \cos \theta'_e \cos \phi (z - z_0) \\ \cos \theta'_e [r \cos \theta'_e \sin \phi \cos \phi - \frac{1}{2}(x_0 \cos \phi + y_0 \sin \phi)] \\ \sin \theta'_e \\ \sin \theta'_e (r \cos \theta'_e \sin \phi - x_0) \\ \sin \theta'_e (r \cos \theta'_e \cos \phi - y_0) \\ \sin \theta'_e (z - z_0) \end{bmatrix} \tag{9.13}$$

With this notation we can express the true radial velocity of air as $v_r = \mathbf{PK}$. $\mathbf{P}_{im}$ will denote a subset of the predictor functions of $\mathbf{r}_i$ corresponding to $\mathbf{K}_m$.

The estimates $\hat{v}_r^h$ of the reflectivity-weighted radial velocity of hydrometeors will contain some error $\varepsilon(\hat{v}_r^h)$, and a component $\hat{w}_t \sin \theta'_e$ due to the terminal fall speeds whose estimate will also contain error, $\varepsilon(\hat{w}_t)$. If Eq. (9.2) is used to estimate $\hat{w}_t$, then Z estimate errors will contribute to $\varepsilon(\hat{w}_t)$. Finally, the wind field is very likely to deviate from linearity; let this error be denoted $\varepsilon(\delta v)$. Thus, the relationship between the estimated radial component of wind velocity $\hat{v}_{ri}$ and $\mathbf{P}_i\mathbf{K}$ the linear radial component of the wind at the ith measurement point becomes

$$\hat{v}_{ri} = \hat{v}_{ri}^h + \hat{w}_{ti} \sin \theta'_{ei} = \mathbf{P}_i \mathbf{K} + \varepsilon_i, \tag{9.14}$$

where $\varepsilon_i = \varepsilon(\hat{v}_{ri}^h) - \varepsilon(\hat{w}_{ti}) \sin \theta'_{ei} + \varepsilon(\delta v)$ is the combined error. Doviak *et al.* (1976) show that the Z estimate variance produces much smaller error in w_t than that caused by uncertainties in Eq. (9.2). Unless an accurate measurement of the drop-size distribution can be made, one may reasonably assume that the standard deviation of $\varepsilon(\hat{w}_t)$ is 1 m s^{-1} (Section 9.2.1.).

Equation (9.14) can be used to find linear wind parameters from an estimate of $\mathbf{K}$. There are as many as 11 quantities in $\mathbf{K}$ to be estimated. The number of radial velocity measurements n will generally be at least several hundred. The most common technique for fitting such an overdetermined set of equations is least squares.

The fitting equations are concisely represented with the matrix notation. The n estimates $\hat{v}_r$ are collected into a vector $\hat{\mathbf{V}}_n^{\mathrm{T}} = (\hat{v}_{r1}, \hat{v}_{r2}, \ldots, \hat{v}_{rn})$; the associated predictor functions are collected into an $n \times m$ matrix $\mathbf{P}_{nm}^{\mathrm{T}} = (\mathbf{P}_{1m}^{\mathrm{T}}, \mathbf{P}_{2m}^{\mathrm{T}}, \ldots, \mathbf{P}_{im}^{\mathrm{T}}, \ldots)$, where $\mathbf{P}_{im}^{\mathrm{T}}$ represents Eq. (9.13) for m of the predictors. As shown in Draper and Smith (1966, Ch. 2), least squares estimates of $\mathbf{K}_m$, are computed from

$$\hat{\mathbf{K}}_m = (\mathbf{P}_{nm}^{\mathrm{T}}\mathbf{P}_{nm})^{-1}(\mathbf{P}_{nm}^{\mathrm{T}}\mathbf{V}_n). \tag{9.15}$$

We would like expressions for the bias and variance of the estimates $\hat{\mathbf{K}}_m$. $\hat{\mathbf{K}}_m$ is an unbiased estimate (i.e., $\mathbf{E}(\hat{\mathbf{K}}_m) = \mathbf{K}_m$), if Eq. (9.14) is an adequate model. Two reasons why Eq. (9.14) would not be an adequate model are now considered. First, if the wind field is nonlinear, the radial velocities will have variations associated with terms of higher order than in Eq. (9.7). If the residuals,

$$\hat{\varepsilon}_i = \hat{v}_{ri}^{\mathrm{h}} + \hat{w}_{ti} \sin \theta_{ei}' - \mathbf{P}_{im}\hat{\mathbf{K}}_m, \tag{9.16}$$

are not zero mean random variables, then the linearity assumption is not valid and the estimates $\hat{\mathbf{K}}_m$ may be biased.

The second reason is relevant if the wind field is linear but m is chosen to be less than the number required for an adequate model. Then all of the variations of v_r versus $\mathbf{r}$ are not modeled. For example, if data is analyzed over a volume and $m = 11$, the model is adequate; moreover, if analysis is performed on a surface at constant height and $m = 8$ (Section 9.3.3), the model is adequate. But for some practical cases m less than required for an adequate model reduces the variance of the estimates (Section 9.3.4). With this in mind, the bias is

$$\mathbf{E}(\hat{\mathbf{K}}_m - \mathbf{K}_m) = (\mathbf{P}_{nm}^{\mathrm{T}}\mathbf{P}_{nm})^{-1}(\mathbf{P}_{nm}^{\mathrm{T}}\mathbf{P}_{nl})\mathbf{K}_l^{\mathrm{T}}, \tag{9.17}$$

where $\mathbf{K}_l$ contains the remaining $l = 11 - m$ parameters (i.e., for volume analyses) and $\mathbf{P}_{nl}$ the associated predictors. Evaluation of Eq. (9.17) is difficult in practice because it requires an estimate $\mathbf{K}_l$. One can assign, however, an upper value to the elements of $\mathbf{K}_l$ to estimate the maximum bias.

Even when $\hat{\mathbf{K}}_m$ is unbiased, it will contain random errors, termed the variance errors. Assuming an adequate model, the covariance matrix of $\hat{\mathbf{K}}_m$ is

$$\mathbf{C}_{mm} = \mathbf{E}[(\hat{\mathbf{K}}_m - \mathbf{K}_m)(\hat{\mathbf{K}}_m - \mathbf{K}_m)^{\mathrm{T}}] = (\mathbf{P}_{nm}^{\mathrm{T}}\mathbf{P}_{nm})^{-1}\sigma_\varepsilon^2, \tag{9.18}$$

where σ_ε^2 is the variance of the measurement error. In practice σ_ε^2 can be estimated by the residual variance $s^2 = \hat{\mathbf{E}}_n^{\mathrm{T}}\hat{\mathbf{E}}_n/(n - m)$, where $\hat{\mathbf{E}}_n^{\mathrm{T}} = (\hat{\varepsilon}_1, \hat{\varepsilon}_2, \ldots, \hat{\varepsilon}_n)$.

The $\mathbf{P}_{nm}^{\mathrm{T}}\mathbf{P}_{nm}$ matrix appearing in these equations is the covariance matrix of the predictors. Since the predictors contain sine–cosine products, the off-diagonal terms can be large compared to the diagonal terms, especially for small analysis volumes. Because the variance of the estimates $\hat{\mathbf{K}}_m$ are proportional to the diagonal elements of the inverse matrix, the accuracy of $\hat{\mathbf{K}}_m$ improves with the size of the analysis volumes, as is intuitively expected. Equation (9.18) allows a quantitative measure of the variances. Thus $(\mathbf{P}_{nm}^{\mathrm{T}}\mathbf{P}_{nm})^{-1}$, which depends on the size and shape of the analysis volume and distribution of data therein, determines how well the chosen analysis predicts the components of a linear wind field.

We now have sufficient theoretical development to allow quantitative evaluation of the errors in the estimates of the linear wind components.

9.3.2 Uniform Wind along a Circular Arc

Because of its simplicity, we shall consider in detail the case in which wind is strictly uniform and horizontal over a circular arc. For this case, $m = 2$, $\mathbf{K}_m^{\mathrm{T}} = (u_0, v_0)$, is an adequate model. The n radial velocity estimates are made at the same elevation angle and range along an arc of length $\Delta\phi$ centered at the azimuth ϕ_0. Note that the vertical shear of the horizontal wind does not affect the analysis because the data are analyzed at a constant height. Therefore, in this case, the model is adequate even if wind is not uniform in height. Thus,

$$\mathbf{P}_{n2}^{\mathrm{T}}\mathbf{P}_{n2} = \cos^2\theta_{\mathrm{e}}'\begin{bmatrix} a & b \\ b & c \end{bmatrix},$$

where

$$a = \sum_{i=1}^{n}\sin^2\phi_i \approx \frac{n}{\Delta\phi}\int_{\phi_0-\Delta\phi/2}^{\phi_0+\Delta\phi/2}\sin^2\phi\,d\phi$$

$$= (n/2\,\Delta\phi)(\Delta\phi - \cos 2\phi_0\sin\Delta\phi),$$

$$c \approx (n/2\,\Delta\phi)(\Delta\phi + \cos 2\phi_0\sin\Delta\phi),$$

$$b \approx (n/2\,\Delta\phi)\sin 2\phi_0\sin\Delta\phi.$$

The determinant

$$\det[(\mathbf{P}_{n2}^{\mathrm{T}}\mathbf{P}_{n2})] = (\Delta\phi^2 - \sin^2\Delta\phi)\left(\frac{n}{2\,\Delta\phi}\right)^2\cos^4\theta_{\mathrm{e}}'$$

is independent of the center azimuth ϕ_0. The trace of the covariance matrix (9.18),

$$\mathrm{tr}(\mathbf{C}_{22}) = \frac{(a+c)\sigma_\varepsilon^2\cos^2\theta_{\mathrm{e}}'}{\det(\mathbf{C}_{22})} = \frac{n\sigma_\varepsilon^2\cos^2\theta_{\mathrm{e}}'}{\det[(\mathbf{P}_{n2}^{\mathrm{T}}\mathbf{P}_{n2})]}$$

is equal to the sum of the variances of u_0 and v_0, and provides a measure of the accuracy of the uniform wind analysis. Simplifying somewhat,

$$\mathrm{tr}(\mathbf{C}_{22}) = \frac{n\sigma_\varepsilon^2}{\left(\dfrac{n}{2\,\Delta\phi}\cos\theta_{\mathrm{e}}'\right)^2(\Delta\phi^2 - \sin^2\Delta\phi)} \approx \frac{12\sigma_\varepsilon^2\sec^2\theta_{\mathrm{e}}'}{n\,\Delta\phi^2},$$

where $\sin\Delta\phi$ has been approximated by $\Delta\phi - (\Delta\phi)^3/6$.

It can be seen from this result that the most accurate uniform horizontal wind analysis is for low elevation angles (where $\sec^2 \theta'_e \approx 1$). The accuracy depends inversely on the number of data points n and on the square of the arc length (i.e., azimuthal width). Thus the variance of the estimates of the horizontal wind increases rapidly as $\Delta\phi$ decreases (below 1 rad).

The more complicated and complete models (i.e., $m > 2$) have similar (square) increases in the variance of estimates of linear wind components with decreasing analysis volume size. The algebra becomes more complicated, however, and a numerical evaluation of the inverse matrix by a computer is necessary.

9.3.3 Linear Wind over a Circle (VAD)

Because wind fields are usually characterized by relatively strong changes along the vertical, we are lead to relax our restrictions (i.e., linearity) on the vertical variation of wind and examine the equation for data collected on circles at constant height where the wind is assumed to be horizontally linear. If the antenna beam is scanned in azimuth ϕ while θ_e is fixed, the radial component of wind along a circle of measurement (Fig. 9.8) has the ϕ dependence

$$v_r = w \sin \theta_e + v_h \cos \theta_e \cos(\delta - \phi), \tag{9.19}$$

where $D = \delta + \pi$ is the wind direction (i.e., the direction *from which the wind blows*), and w and v_h respectively are the wind's vertical and horizontal components. On expanding the second term and using the relation between Cartesian

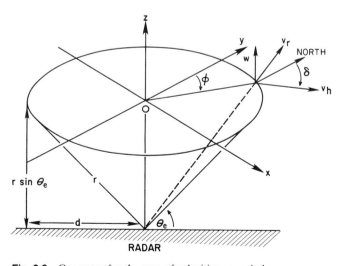

Fig. 9.8 Geometry for the scan of velocities on a circle.

and polar wind components, Eq. (9.19) is reformulated as

$$v_r = w \sin \theta_e + u \sin \phi \cos \theta_e + v \cos \phi \cos \theta_e. \quad (9.20)$$

Combining Eq. (9.7), but in which $x_0 = y_0 = 0$ and $z = z_0$, with Eq. (9.20) and expressing x, y in terms of the radar coordinates, one obtains

$$v_r = w_0 \sin \theta_e + \frac{1}{2}\left(\frac{\partial u}{\partial x} + \frac{\partial v}{\partial y}\right) r \cos^2 \theta_e$$

$$+ u_0 \cos \theta_e \sin \phi + v_0 \cos \theta_e \cos \phi \quad (9.21)$$

$$+ \frac{1}{2}\left(\frac{\partial v}{\partial x} + \frac{\partial u}{\partial y}\right) r \cos^2 \theta_e \sin 2\phi + \frac{1}{2}\left(\frac{\partial v}{\partial y} - \frac{\partial u}{\partial x}\right) r \cos^2 \theta_e \cos 2\phi.$$

This equation also follows from Eq. (9.11) when earth's curvature is ignored and it is recognized that $w_x = w_y = 0$ above a horizontal ground. Examination of Eq. (9.21) reveals that a horizontally linear field contributes only to the first three components of a Fourier expansion of v_r over the fundamental period $0 \le \phi \le 2\pi$. The measured Doppler velocity $\hat{v}_r^h$ contains errors and a bias due to hydrometeor terminal velocity. From Eq. (9.14), the unbiased estimate $\hat{v}_r^h$ is

$$\hat{v}_r^h = v_r - [(\hat{w}_t + \varepsilon(\hat{w}_t)] \sin \theta_e + \varepsilon(\hat{v}_r^h) + \varepsilon(\delta v). \quad (9.22)$$

A series of n uniformly spaced $\hat{v}_r^h$ and $\hat{w}_t$ data along the circle of measurement is, in general, expressible in terms of the discrete Fourier transform (Chapter 5).

$$\hat{v}_r^h(\phi) + \hat{w}_t(\phi) \sin \theta_e = \sum_{k=0}^{n-1} C_k e^{jk\phi}, \quad (9.23)$$

where the C_k are complex Fourier coefficients. The coefficients, symmetrically placed about $n/2$, are complex conjugates because $\hat{v}_r^h$ and $\hat{w}_t$ are real; but their pair-wise sums are real and unambiguously represent harmonics at wave numbers less than $n/2$, the Nyquist wave number (Section 5.1). If wind is strictly linear and measurements are error free, only the first three coefficients differ from zero. Use the properties, introduced in Chappter 5, that relate the complex amplitude of the various spectral coefficients and assume that errors are negligible; then comparison of Eqs. (9.23) and (9.21) reveals the following:

1. Horizontal divergence

$$\text{div } v_h \equiv \left(\frac{\partial u}{\partial x} + \frac{\partial v}{\partial y}\right) = \frac{2}{r \cos^2 \theta_e}(C_0 - w_0 \sin \theta_e); \quad (9.24a)$$

2. Horizontal wind speed

$$v_h = 2|C_1|/\cos \theta_e; \quad (9.24b)$$

3. Horizontal wind direction

$$D = \arg C_1 - \pi; \quad (9.24c)$$

4. Deformations

$$\text{(stretching)} \qquad \frac{\partial u}{\partial x} - \frac{\partial v}{\partial y} = -\frac{4}{r \cos^2 \theta_e} \operatorname{Re}(C_2), \tag{9.24d}$$

$$\text{(shearing)} \qquad \frac{\partial u}{\partial y} + \frac{\partial v}{\partial x} = \frac{4}{r \cos^2 \theta_e} \operatorname{Im}(C_2). \tag{9.24e}$$

The divergence, vorticity, wind speed, and the sum of the squares of each deformation are invariant with respect to rotation of the x, y axes about z and hence are fundamental properties of the wind. Fourier analysis of $\hat{v}_r^h + \hat{w}_t \sin \theta_e$ directly determines the wind speed and direction and the deformation. Other important properties of motion, such as the vorticity, divergence, and vertical velocity, cannot be directly determined using the VAD method. But, by using the anelastic form of the continuity equation (9.5b) expressed in Cartesian coordinates, we obtain the following first-order linear differential equation:

$$\frac{dw_0}{dz} + \left(\frac{1}{\gamma_0} \frac{d\gamma_0}{dz} - \frac{2z}{r^2 - z^2} \right) w_0 = -\frac{2rC_0(z)}{r^2 - z^2}, \tag{9.25}$$

which can be solved analytically (e.g., Doviak and Zrnić 1984a) for w_0 in terms of integrals of known variables and a boundary condition. Usually one has C_0 values from near the surface to levels through the convective boundary layer (1–2 km) in clear air (Section 11.6) and much higher in showery weather; so $w_0(0) = 0$ at the ground is a convenient boundary condition. Thus a single Doppler radar can measure the vertical profiles of the three components of horizontally linear wind over a sampling circle of radius d (Fig. 9.8).

As an example, the velocity trace obtained from the azimuthally scanning Cimarron radar (located 40 km northwest of Norman, Oklahoma) is presented in Fig. 9.9. The mean velocities were calculated with the pulse pair alogrithm (Section 6.4.1) on samples of echoes from clear-air refractivity fluctuations. (For relations between reflectivity and the refractive index fluctuations, see Chapter 11.) It is visually evident that there is no significant power other than in the zeroth and first harmonic. The large data gaps are the results of editing uncertain radial velocities because of low signal-to-noise ratios. Gaps occur, for example, at 20° and 110°, where tall trees interfere with the low-elevation radar beam. Because the data are spaced nonuniformly, the Fourier coefficients were obtained by least squares fitting the zeroth harmonic and the fundamental to the data (Rabin and Zrnić, 1980). Substantial scatter of the accepted velocities exists even though velocities from three consecutive range gates were averaged to give an effective range interval of 450 m. Some of this scatter is attributed to low SNR. Because the VAD method is most useful for determining the magnitude and direction of wind at scales large compared to the diameter of the scan, it is interesting to compare the results obtained from two closely spaced radars. In this comparison, the radius of the circle used for the velocity data was changed in steps of 0.3 km, and the elevation angle θ was fixed at 0.5°. Hence the following discussion will refer to curves of the wind speed and direction (Fig. 9.10a) and

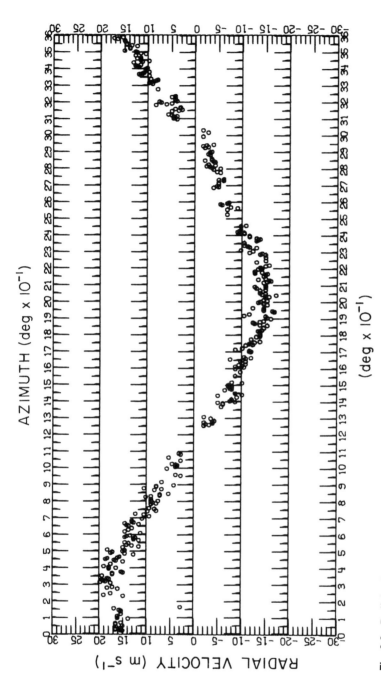

Fig. 9.9 Radial velocity at a range of 40 km versus azimuth of the Cimarron radar, 1411–1414 C.S.T., 27 April 1977. The elevation is 0.5°.

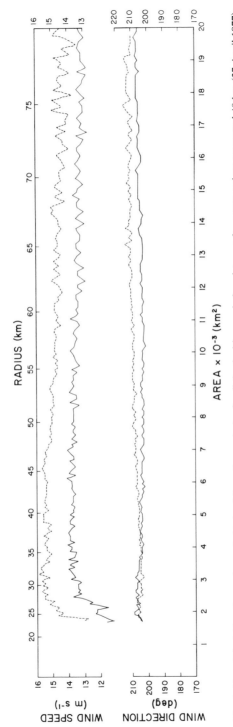

Fig. 9.10a Wind direction and speed obtained from the analysis of radial velocities on circles; data are from two radars separated 40 km (27 April 1977); ———, Norman; ———, Cimarron.

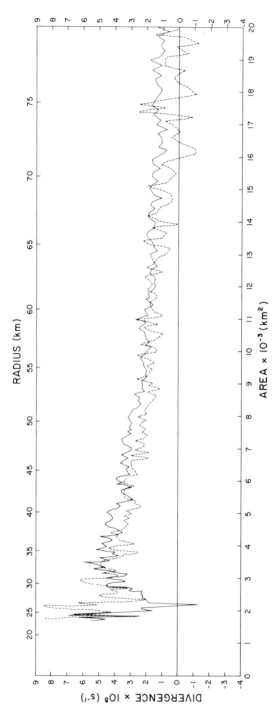

Fig. 9.10b Divergence over the circular areas above the Norman (———) and Cimarron (----) radars at 1525–1529 C.S.T. on 27 April 1977.

divergence (Fig. 9.10b), plotted against the radius and area of the azimuthally scanned circle, for the Cimarron and Norman radars.

The apparent decrease in wind speed at short range is due to ground clutter, which biases the pulse pair velocity estimate toward zero. The slight difference in observed wind speeds between the Norman and Cimarron radar sites (Fig. 9.10a) indicates a wind speed gradient of about $1 \text{ m s}^{-1}/40 \text{ km}$ to the northwest. (The difference in wind speeds was even smaller at an earlier time.) This is of the same order as the synoptic-scale gradient $(0.6 \text{ m s}^{-1}/40 \text{ km})$ observed at that time from a surface network of anemometers. The difference in wind direction between Norman and Cimarron is about $5°$ (Fig. 9.10a) owing to a curved flow.

The horizontal divergence as a function of range for both radars (Fig. 9.10b) shows excellent consistency between 30 and 65 km. At shorter ranges ground clutter and the noncommon area are responsible for the differences.

The gradual decrease of the divergence for areas $\geq 5 \times 10^3 \text{ km}^2$ may be due to an increase in height of the sampling circle rather than an increase in sampling area. (This conclusion is based on the divergence profile obtained from the detailed dual radar analysis for a smaller area contained within the circle of measurement.) The positive divergence measured on 27 April 1977 is consistent with the anticyclonic geostrophic wind flow and subsidence over central Oklahoma.

9.3.4   Prestorm Application

With a volume of data, all 11 parameters can be discriminated. But we are free to select both m and the size and shape of the analysis volume to control the bias and variance errors. One interesting application of linear wind field analysis is estimation of the low-altitude prestorm divergence. Near-ground convergence has been recognized as an important mechanism for storm initiation (Byers and Braham, 1949; Watson and Blanchard, 1984). Often such interactions can be seen on satellite images (Purdom, 1982), and Purdom and Marcus (1982) indicate that 73% of afternoon thunderstorms in the southeast portion of the United States formed on or near convergence boundaries. High-performance Doppler radars are well suited to map winds, convergence zones, and leading edges of density currents in the planetary boundary layer.

If only the low altitudes are of interest, we can neglect the vertical velocity terms (which contain $\sin \theta_e$), but not the vertical shear of u and v, which for many meteorological conditions will be much larger than the horizontal gradients. Ignoring terms such as vertical velocity (which is poorly sampled by quasihorizontal radar beams) can actually improve the estimation of retrieved terms (Koscielny et al., 1982). If the analysis volume includes at least two elevation angles, then the vertical and horizontal gradients can be discriminated. Alternatively, analyses could be performed on data at constant height whereby vertical shear terms are no longer required for an adequate model.

The bias caused by the excluded vertical velocity terms depends on the analysis volume. Before Eq. (9.17) can be applied, we must specify the analysis volume; for convenience we shall choose a sector 30° wide in azimuth, 30 km in range, and containing data at the elevation angles 0.4° and 0.8°. In addition, we assume that the data are spaced 1° in azimuth and 450 m in range. Evaluation of Eq. (9.17) gives

$$
E
\begin{bmatrix}
\hat{u}_0' - u_0' \\
\hat{u}_x - u_x \\
\hat{u}_z - u_z \\
\hat{v}_0' - v_0' \\
\hat{v}_y - v_y \\
\hat{v}_z - v_z \\
(\hat{u}_y + \hat{v}_x) - (u_y + v_x)
\end{bmatrix}
=
\begin{bmatrix}
0.004 & 532.0 & 141.0 & 0.162 \\
4 \times 10^{-8} & 0.0015 & 0.0005 & 1 \times 10^{-5} \\
6 \times 10^{-6} & 0.935 & -0.246 & 0.0043 \\
0.015 & -141.0 & 44.30 & 0.603 \\
-2 \times 10^{-7} & 0.0025 & 0.0142 & 2 \times 10^{-7} \\
2 \times 10^{-5} & -0.2460 & 0.0841 & 0.0160 \\
-2 \times 10^{-7} & 0.0064 & 0.0064 & -6 \times 10^{-6}
\end{bmatrix}
$$

$$
\times
\begin{bmatrix}
w_0 \\
w_x \\
w_y \\
w_z
\end{bmatrix}
\tag{9.26}
$$

Ogura and Chen (1977) found, from a dense network of surface stations and balloon soundings, that divergence preceding storms was about $10^{-4}\,\text{s}^{-1}$. Thus, if divergence can be estimated with an accuracy of about $10^{-5}\,\text{s}^{-1}$, it may be possible to improve short-term (2-hr) forecasts of local storms by combining Doppler data with other remote and *in situ* observations. Although there is bias when vertical velocity predictor terms are excluded, u_x and v_y have a bias less than $10^{-5}\,\text{s}^{-1}$ if $w_0 < 10\ \text{m s}^{-1}$; w_x, $w_y < 10^{-4}\,\text{s}^{-1}$; and $w_z < 1\ \text{s}^{-1}$. (Recall, if wind is linear above a horizontal ground, $w_x = w_y = 0$.) Evaluating Eq. (9.18) for this analysis volume gives the variance of the estimates. For the divergence $u_x + v_y$, the variance is given by $c_{22} + c_{55} + 2c_{25}$, which, with about 4000 data points and $\sigma_\varepsilon = 1\ \text{m s}^{-1}$, is $(1.6 \times 10^{-5}\,\text{s}^{-1})^2$. Thus, if the wind is linear, a VVP analysis over the volume sector (30 km, 30°, 1°) will allow the estimation of low-level mesoscale divergence with an accuracy on the order of $10^{-5}\ \text{s}^{-1}$.

The VVP technique was used to analyze the kinematic wind structure in the clear-air convective boundary layer before the onset of thunderstorms. On 19 June 1980 a nearly stationary front was located just southwest of the Norman radar. With the possibility of thunderstorm development near the front, radar data collection began at 1530 C.S.T.

Figure 9.11 shows the results of an analysis of single-Doppler radar data for an assumed uniform wind over each analysis volume. Then only two parameters (wind speed and direction) are estimated to produce a least squares fit to the observed radial velocity data. The sky was generally free of clouds; however,

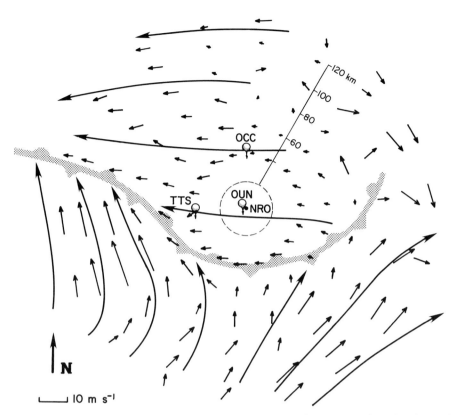

Fig. 9.11 Wind field obtained from a VVP analysis of radar data in a prestorm boundary layer, 19 June 1980, 1530 C.S.T. The winds are assumed uniform within the VVP volume ($\Delta r = 30$ km, $\Delta \phi = 30°$, elevation angles are 0.4° and 0.8°).

there was sufficient backscattering to obtain wind data to a range of 120 km and a height of about 2 km. The arrows indicate the direction and speed of the horizontal wind on a conical surface elevated 0.6° from the radar horizon. The position of the front is clearly evident; winds are easterly to its north, and southerly diffluent flow appears to its south. The three rawinsondes located to the north of the front measured easterly winds below the frontal surface (0.6–0.9 km AGL) above which the wind had a southerly component, in agreement with the VVP results.

The National Weather Service's surface and 850 mb data support the wind pattern shown here. But, due to the slope of the front, its position at the earth's surface appears south of where it is intersected by the conical surface scanned by the radar beam (~500 m AGL).

Estimates of horizontal divergence with the seven-parameter (i.e., the parameters in Eq. 9.26) VVP technique applied to data at 1530 C.S.T. are plotted

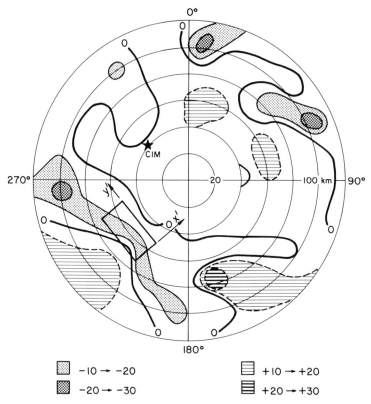

Fig. 9.12 Horizontal divergence $\times 10^5$ s^{-1}, 19 June 1980, 1530 C.S.T., obtained from a VVP analysis as in Fig. 9.11 but the winds are assumed to be linear within each VVP volume. (From Koscielny *et al.*, 1982.)

in Fig. 9.12. There is a band of maximum horizontal convergence just southwest of the frontal boundary in Fig. 9.11. This convergence zone, as well as the one northeast of the radar, are superimposed on a satellite photograph of clouds 2 hours later (1730 C.S.T.) in Fig. 9.13. The band of cumulus to the southwest of NRO persisted during the 2-hour period without any perceptible movement. The displacement of the cloud line to the northeast of maximum convergence in Fig. 9.13 can be accounted for by the trajectory of the air as it is lifted to the height of the condensation level. Satellite photographs showed that the cloud to the northeast developed between 1700 and 1730 in the vicinity of the convergent area. An area of divergence appears in Fig. 9.12 extending from northwest to southeast over the radar. This is consistent with the lack of clouds in central Oklahoma as seen from satellite.

An area of deeper convection (ADC; Harrold and Browning, 1971), a region of increased depth of the convective boundary layer, formed in the area of convergence maximum (Fig. 9.12) to the west of the radar. At 1800 C.S.T.

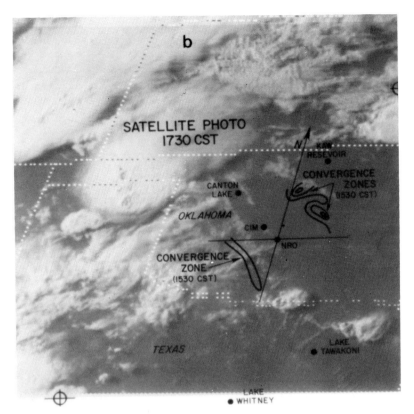

Fig. 9.13 Satellite photograph (1730 C.S.T.) of the cloud cover superimposed on the horizontal convergence pattern of Fig. 9.12.

this ADC did not extend above 3 km, reflectivities were less than −3 dBZ, and no outstanding cloud appeared on satellite photographs at this time; 100 minutes later, 1940 C.S.T., clouds had developed from this ADC to a height of 10 km and attained reflectivities in excess of 25 dBZ.

To evaluate the performance of the VVP method with actual data, the estimated divergence field is compared with one obtained 3 minutes later for data collected before storms developed on 17 May 1981. Figure 9.14 is a comparison of the divergence for the two times for analysis volumes having sector widths $\Delta\phi = 40°$ at ranges $r_0 = 40, 60,$ and 80 km. Let us assume that the actual divergence does not change in 3 minutes over each analysis volume. Also, no system bias is expected between the two sample sets, because they are obtained with the same radar. With these assumptions the data can be compared with a regression line with slope of 1 passing through the origin to obtain the noise N_D (i.e., residual variance) and signal S_D in the horizontal divergence estimates (Smith, 1986).

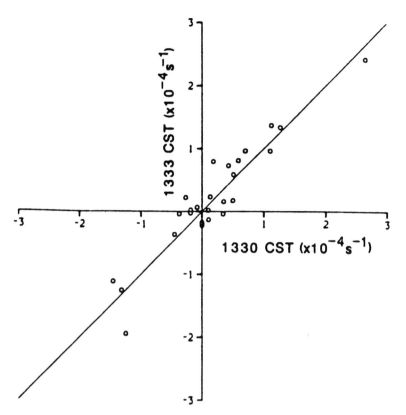

Fig. 9.14 Comparison of the horizontal divergence at 1330 and at 1333 C.S.T. for a prestorm environment on 17 May 1981. VVP analysis volume size is the same as for Fig. 9.11.

$$N_D = \frac{1}{2k} \sum_{i=1}^{k} (y_i - x_i)^2,$$

$$S_D = \frac{1}{2}(\sigma_x^2 + \sigma_y^2 + \bar{x}^2 + \bar{y}^2) - N_D,$$

where x_i and y_i are estimates at times 1 and 2, respectively; σ_x and σ_y are their standard deviations; $\bar{x}$, $\bar{y}$ are their means; and k is the number of estimates.
Using the data of Fig. 9.14 we obtain

$$\sqrt{N_D} = 2.2 \times 10^{-5} \quad s^{-1}, \qquad \sqrt{S_D} = 9.2 \times 10^{-5} \quad s^{-1}.$$

The residual variance in the divergence estimate is about five times larger than expected from the model as discussed in the preceding section. Some of this increased uncertainty of the divergence estimate may be attributed to the inadequacy of the linear wind assumption, some to unexplained variance in the Doppler velocity estimates.

9.3.5 Large-Scale Horizontal Wind

When large-scale weather produces vertically sheared but essentially horizontally uniform wind over the area under radar surveillance, interpretable patterns of Doppler velocity fields appear on plan position indicator (PPI) displays as the radar beam is scanned at a constant θ_e. These patterns can be easily diagnosed by trained personnel to estimate the vertical profiles of wind. PPI displays of Doppler velocity contours corresponding to a variety of vertical profiles of wind speed and direction are shown in Fig. 9.15, where the center of the PPI display represents the radar location.

Winds that have a component away from the radar (termed "outbound") have positive Doppler velocity values indicated by solid contours in Fig. 9.15, whereas those toward the radar (termed "inbound") have negative values indicated by dotted contours. Winds directed normal to the beam produce zero Doppler velocities. If wind speed is constant with height (left column in Fig. 9.15), all contours pass through the center of the display. Also, maximum and minimum Doppler velocities occur along the heavier (nonzero) contour lines, rather than at one point (black dots) as with other wind speed profiles. If the surface speed is zero, only the zero velocity contour (thick, long dashes) passes through the center of the radar display (three rightmost columns of Fig. 9.15.

If the wind speed profile has a peak within the height interval on the display, there will be a pair of closed contours 180° from each other; the azimuth of the inbound (or outbound) maximum is the direction from (or toward) which the velocity jet is approaching (or departing) the radar, and the height of the peak value can be computed from the radar antenna's elevation angle and the slant range to that point. Velocity contours of a modeled double jet are displayed in the right column in Fig. 9.15; a PPI display of an observed triple jet that occurred in an evolving snow storm can be seen in Color Plate 4d.

Whereas the wind speed profile controls the overall pattern including the spacing between contours, the vertical profile of wind direction uniquely specifies the zero velocity contour. Note that the zero contours are identical in each row (a reflection of wind direction profile), although the overall patterns in each row differ significantly (a reflection of wind speed profile).

Since wind direction is perpendicular to a radial line (from display center) at the point where the line intersects the zero velocity contour, wind direction variation with height (range on the PPI display) can be determined by inspection. Air motion is from the negative side of the zero contour toward the positive side. Looking at the middle row of Fig. 9.15, we see that there are southerly winds at the ground (practically, just above the ground) because at $r = 0$ the zero velocity line is oriented east–west, and air is approaching from the south and flowing away toward the north. Halfway between the center (zero height) and edge (height = H) of the display, southwesterly winds are perpendicular to the radial line. At the edge of the display, wind is from the west because the radial line intersecting the zero contour is oriented north–south.

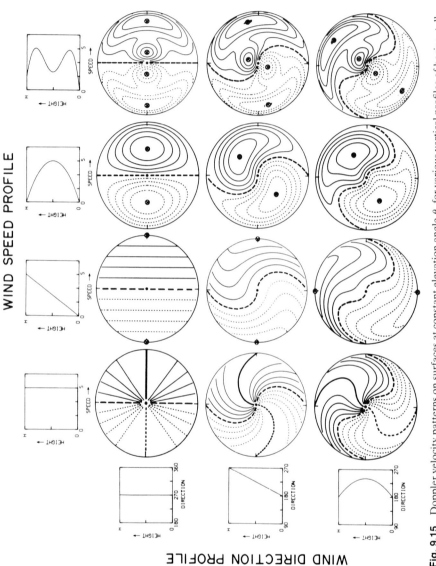

Fig. 9.15 Doppler velocity patterns on surfaces at constant elevation angle θ_e for various vertical profiles of horizontally uniform wind. The outer circle of each display depicts Doppler velocities at height H and range $r = H/\sin \theta_e$. Contour interval is 0.2 of the maximum wind speed. (From Wood and Brown, 1983.)

Veering winds, representative of warm air advection, produce a striking S-shaped pattern of the zero velocity contours, whereas backing winds, representative of cold air advection, produce a backward S. These features are seen in the middle and bottom rows.

Other patterns associated with horizontal variation of wind speed and direction (e.g, fronts, mesocylone, *etc.*) are presented by Wood and Brown (1983).

9.3.6 Large-Scale Vertical Wind

Estimates of the average vertical velocity over the measurement circle do not require any assumption about the spatial distribution of wind. We apply Gauss's theorem,

$$\oiint \mathbf{v} \cdot \mathbf{n} \, dS = \int_V \nabla \cdot \mathbf{v} \, dV, \tag{9.27}$$

to the volume V (Fig. 9.16) enclosed by the area S_1 sampled by the radar and an area S_2 at constant height, over which the vertical velocity is to be averaged. Assuming that the mass density is height dependent only, we obtain from Eq. (9.5a)

$$\nabla \cdot \mathbf{v} = -\frac{w}{\gamma_0} \frac{d\gamma_0}{dz},$$

so that Eq. (9.27) can be written

$$\iint_{S_1} v_r \, dS_1 + \iint_{S_2} w \, dS_2 = -\int_v \frac{w}{\gamma_0} \frac{d\gamma_0}{dz} \, dV. \tag{9.28}$$

The average vertical velocity is defined as

$$\bar{w} = \frac{1}{S_2} \iint_{S_2} w \, dS_2, \tag{9.29}$$

and differentiating Eq. (9.28) with respect to z, but keeping r constant, we obtain

$$\frac{dS_2 \, \bar{w}}{dz} + \frac{S_2 \bar{w}}{\gamma_0} \frac{d\gamma_0}{dz} = -\int_0^{2\pi} v_r r \, d\phi. \tag{9.30}$$

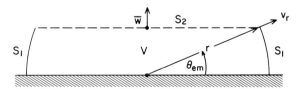

Fig. 9.16   Geometry for estimating the vertical velocity averaged over the circular area S_2.

Expressing S_2 in terms of r and z and using C_0 for the average of v_r around the circle of measurement we finally obtain

$$\frac{d\bar{w}}{dz} + \left(\frac{1}{\gamma_0}\frac{d\gamma_0}{dz} - \frac{2z}{r^2 - z^2}\right)\bar{w} = \frac{-2rC_0(z)}{r^2 - z^2}. \qquad (9.31)$$

This differential equation has exactly the same form as Eq. (9.25) except that in place of w_0 we now have $\bar{w}$. We therefore conclude that vertical velocity, averaged over a circle of measurements, can be obtained from single-Doppler radar data without assumption on the spatial dependence of winds. This is in contrast to the average wind speed and direction which cannot, in general, be obtained from analysis of azimuthally scanned velocities if the wind is not linear (Hondl, 1990).

Errors in estimating C_0 lead to errors in the estimates of $\bar{w}$. Because γ_0 decreases with height with a roughly exponential dependence, it can be shown that these errors increase exponentially with the height to which $\bar{w}(z)$ is estimated (Section 9.2.1.2).

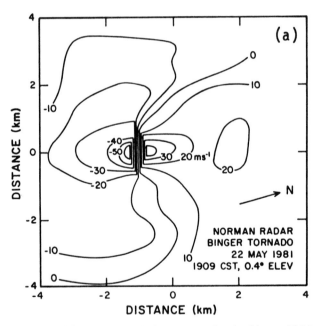

Fig. 9.17 (a) The radial velocity contours for the Binger, Oklahoma tornadic storm on 22 May 1981. The center of the mesocyclone is 70.8 km from the Norman Doppler radar at azimuth 284.4°; the data field has been rotated so the radar is beyond the bottom of the figure. (b) Mesocyclone and tornadic vortex parameters used to simulate the contours in (a); Rs and Vs denote the radii and velocities of the maximum tangential winds. (c) Simulation of radial velocity contours for parameters in (b). Dark dots indicate the centers of the mesocyclone and tornado vortices. (*Figure continues.*)

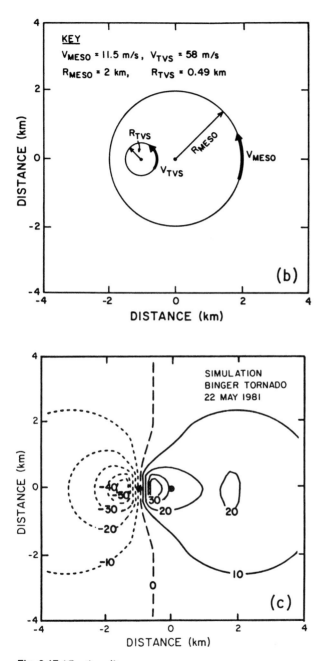

Fig. 9.17 (*Continued*)

9.3.7   Flow Models

Although the Doppler radar measures only the wind component toward and away from the radar, the spatial distribution of Doppler velocities can signify meteorological features such as tornadoes, microbursts (i.e., the divergent flow of strong thunderstorm outflow near the ground), buoyancy waves, etc. Brown and Wood (1991) show that combinations of three relatively simple flow fields [uniform, a Rankine vortex (Section 9.5), and a uniform axisymmetric divergence (or convergence) core surrounded by a region of zero divergence outflow (or inflow)] can replicate a rich variety of actual Doppler velocity patterns. For example, the Doppler velocity pattern of a pair of Rankine vortices agrees well with the observed Doppler pattern for a tornado imbedded in a mesocyclone (Fig. 9.17). The combinations of such simple flow fields has aided many to interpret and quantify single-Doppler data from severe thunderstorms (e.g., Brown and Crawford, 1972; Lemon *et al.*, 1978; Burgess, 1976; Brown and Wood, 1987). Thus, Doppler data, subjected to methods of pattern recognition and artificial intelligence, can be a useful operational tool to detect severe storm phenomena and to provide more accurate and timely warnings than present observing systems (Zrnić *et al.*, 1985).

9.4   Severe Storms

Although the reflectivity field cannot be reliably used for tornado detection, it has proved valuable for hydrological studies and severe weather warnings. Those warnings are primarily based on reflectivity values, storm top heights, and sometimes on circulatory features or reflectivity hooks (Fig. 9.18). The reflectivity indentions or weak echo regions penetrating the southern edge of the cells sometimes indicate the presence of a mesocyclone and hence tornado potential. At best this signature is a poor indicator of circulation because of the large number of false alarms (Table 9.1 and Section 9.5.3).

Fields of large spectrum width are a distinguishing feature of a severe storm. The various signatures of mesocyclones, tornados, and other severe storm hazards may be revealed quickly by displaying simultaneously reflectivity, Doppler velocity, and spectrum width fields.

A tremendous advantage is obtained with Doppler radar because it can sort out among many storms those that have intense circulation and hence the potential for tornado development. Figures 7.1 and 7.2 show a mesoscale convective system composed of many individual storms. Each storm can be systematically interrogated, and for the example in Fig. 7.1 only the storm outlined by the box produced a mesocyclone signature that was identified in a Doppler velocity display. This mesocyclone signature was tracked for almost an hour; the mesocyclone contained a tornado, and the signature relative to the damage path is plotted in Fig. 9.19. It is noticed that damage lies to the right of the path where

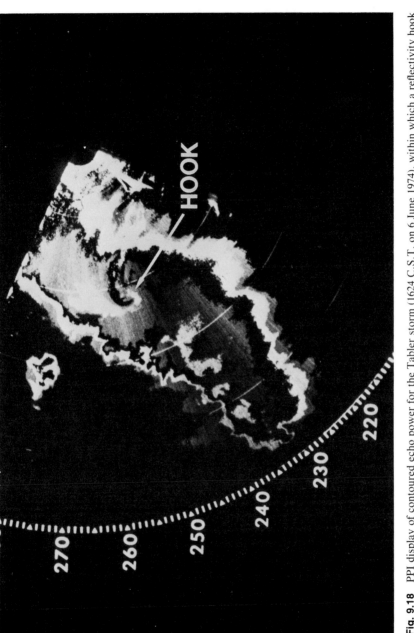

Fig. 9.18 PPI display of contoured echo power for the Tabler storm (1624 C.S.T. on 6 June 1974), within which a reflectivity hook is seen. Arcs at constant range are spaced 20 km apart. Brightness categories (dim, bright, black, dim, etc.) start at 10 dBZ and are incremented in 10-dBZ steps. (From Brandes, 1977.)

Table 9.1

Comparison of Tornado Warning Performance, 1977–1978 [a]

	Non-Doppler system	Doppler radar alone
Probability of detection [b]	0.64	0.69
False alarm rate [c]	0.63	0.25
Critical success index [d]	0.30	0.56
Lead time (min) [e]	−0.8	21.4

[a] Approximately 200 cases were examined. The performance for a WSR-88D radar, in Oklahoma during spring 1991, is given in Section 9.5.3.

[b] $X/(X + Z)$, where X is the number predicted tornados that occurred, Y the number of predicted tornados that did not occur, and Z the number of tornados that occurred without a prediction.

[c] $Y/(X + Y)$.

[d] $X/(X + Y + Z)$.

[e] Between issuing of warning and tornado occurrence.

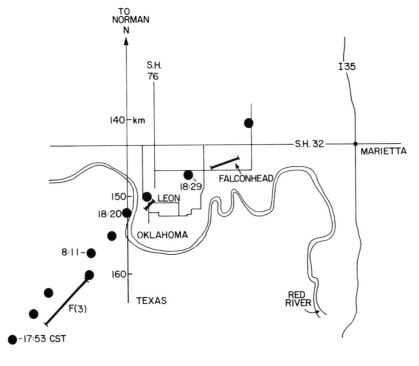

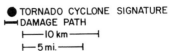

Fig. 9.19 Mesocyclone signature track for the Falconhead tornado with the damage path locations (19 April 1976). The reflectivity field of the storms at 1816 C.S.T. is shown in Figs. 7.1 and 7.2, where the boxed area contains the storm that produced the only tornado on this day.

speeds were faster due to addition of ambient wind ($\approx$15 m s^{-1} to the northeast) and counterclockwise cyclonic wind.

Even though the storm was in the Doppler radar's second trip, the beamwidth was sufficiently small ($\sim$2 km) to track its circulation from a range of 170 km, where the signature was first noticed. The Doppler radar had an unambiguous range of 115 km, but the distribution of storms was such that no storm was range overlaid onto this mesocyclone (Fig. 7.2), thus allowing an unobscured measurement of its velocity signature. The overlaying of storms due to the small unambiguous range (Chapter 7) associated with Doppler radars can result in obscuration of signatures.

Figures 9.20, 9.21, and Color Plate 2a display Doppler radar data for thunderstorms in northwestern Oklahoma during the afternoon of 2 May 1979. These figures depict fields of the three principal moments: reflectivity (Figs. 9.20a, 9.21a), spectrum width (Figs. 9.20b, 9.21b), and Doppler velocity (Color Plate 2a).

Three cells are evident in Fig. 9.20, two of which are closely spaced and appear as a merged reflectivity; all have reflectivities of 46 dBZ or more. The easternmost storm has an annular region within which reflectivity exceeds 55 dBZ, surrounding a weak reflectivity ($\leq$40 dBZ) core. This weak reflectivity region is attributed to a strong updraft that does not allow sufficient time for large hydrometeors to form.

Although the reflectivity display shows no evidence of a mesocyclone, the mean Doppler velocity display (not shown here) depicted it remarkably well. The two eastern cells both produced tornadoes that remained on the ground for over an hour (approximately 1620–1720 C.S.T.). The easternmost tornado produced considerable damage. The display (Fig. 9.20b) of the spectrum width is an indicator of turbulence or large shear within the radar's resolution volume (Section 10.4). High spectrum width values occur in the vicinity of the circulations.

At an altitude of about 12 km, the reflectivity (Fig. 9.21a) of each of the two eastern cells is still greater than 50 dBZ, while the western cell has tops less than 12 km and therefore is not visible in this display. As will be discussed in Section 9.5, the pattern of Doppler velocities for circularly symmetric divergent flow is similar to that for pure circulatory flow, but rotated 90°. At higher altitudes in these storms, the pattern of circulation gave way to one of divergence (Color Plate 2a) centered almost directly above the mesocyclones. The high-speed outflow toward the radar exceeds the negative Nyquist limit of -34 m s^{-1} and aliases into the positive velocities shown by the red areas surrounded by green. The Doppler velocities in the eastern storm range from about -60 to $+40$ m s^{-1}. The velocities toward the radar are higher than those away because the storm's outflow is embedded in an environmental wind from the northwest. Most spectrum widths in the display (Fig. 9.21b) exceed 5 m s^{-1}, indicating strong shear or turbulence (Section 10.4) nearly everywhere at the storm top.

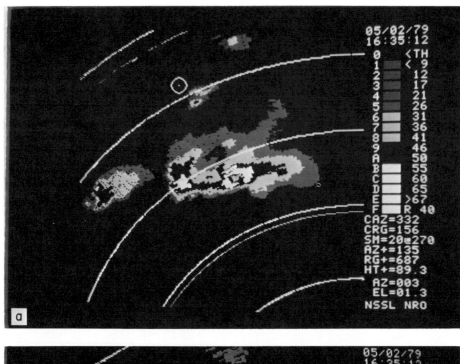

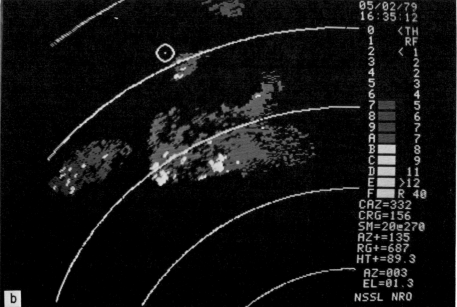

Fig. 9.20 Plan position indicator display of (a) the reflectivity factor in dBZ; the scale is indicated to the right of the bars; and (b) the spectrum width (in m s^{-1}) for an Oklahoma tornadic storm on 2 May 1979, at 1635 C.S.T. Range marks are 40 km apart, and the two closely spaced cells are 140 km from the radar. The height above ground is about 3 km. The elevation angle is 1.3°. Data with SNR < 15 dB are not displayed.

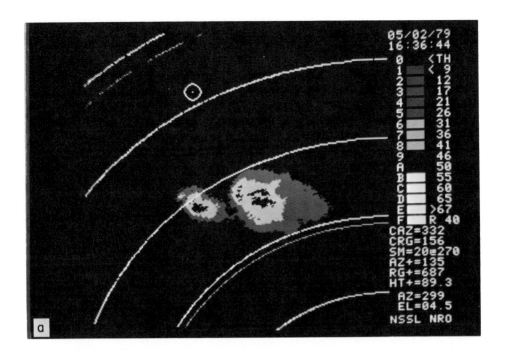

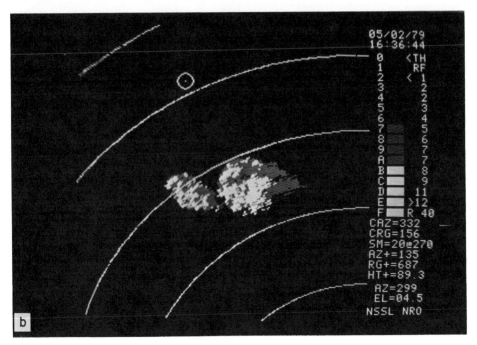

Fig. 9.21 Same as in Fig. 9.20 but at an altitude of about 12 km and at a time 1636 C.S.T. The elevation angle is 4.5°. The Doppler velocity field corresponding to this time is presented in Color Plate 2a.

Storm structure is well illustrated when data from a radar are displayed on a range height indicator (RHI) that gives vertical cross sections of Z, v, and σ_v. Figure 9.22 and Color Plate 2b show Z (dBZ) and v in cross sections along a 347° azimuth, for a severe thunderstorm. Visible in the Z fields is an anvil of cloud particles extending to the left (south–southeast) at ~11 km AGL. More of the anvil is evident in the velocity display, for which the SNR threshold has been reduced to show the motion of the weakly reflective particles, which are probably ice crystals.

Storm overhang and strong reflectivity gradients below 2 km AGL on the storm's south side suggest strong inflow into the storm. Weakly reflecting scatterers drifting with the flow in the clear-air boundary layer (the red region in Color Plate 2b between ranges of 50 and 60 km at low altitude) trace the air moving toward the storm. The northwesterly winds outside the storm carried particles causing the anvil cloud to grow southeastward. In the anvil, air velocities were more negative than -27 m s^{-1}, the Nyquist limit, so they appear as high positive values (but less than 27 m s^{-1}) and are seen as patches of lightest

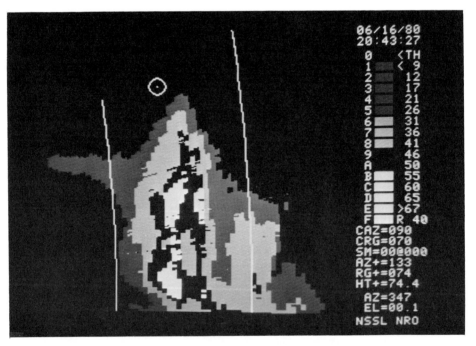

Fig. 9.22 A vertical cross section of the reflectivity factor Z through a severe storm that occurred in central Oklahoma on 16 June 1980. The storm top (near the dot in the white circle) is at 14.7 km above ground level. The grey shades of reflectivities in dBZ are indicated at the right; values above 20 dBZ are displayed. The radar location is to the left, and the vertical arcs are at 60 and 80 km from the radar. The display of Doppler velocities at this time is presented in Color Plate 2b.

red embedded in regions of lightest green. The momentum of updraft air carried condensate into the lower stratosphere, where negative buoyancy forced air downward and outward around the updraft. This latter motion is evident in Color Plate 2b above the anvil on the storm's southern side, where we see the highest radial velocities.

In Color Plate 2b the velocity data above the highest altitudes of the reflectivity field (Fig. 9.22) are due to antenna sidelobes that receive echoes from the high-reflectivity regions when the main beam lobe is pointed above the storm. These falsely mapped velocities do not portray the true wind there.

9.5 Mesocyclones and Tornadoes

Because the radar maps the distribution of Doppler velocity inside a storm, significant meteorological events (unseen by the eye) such as tornado cyclones (e.g., Fig. 9.4a) can produce tell-tale signatures. Donaldson (1970) stipulated criteria whereby a vortex signature can be distinguished from the azimuthal shear (i.e., the Doppler velocity gradient along an arc at constant range) signature of a nonrotating region. Briefly, there must be a localized region of persistently high, 5×10^{-3} s^{-1}, azimuthal shear that has a vertical extent equal to or larger than its diameter.

It can be shown that nontranslating cyclones have isodops forming a symmetric couplet of closed contours with an equal number of isodops encircling positive- and negative-velocity maxima (Fig. 9.23). If the inner portion of an idealized cyclone is a solidly rotating mass of air, its tangential velocity increases linearly with radius to a maximum (i.e., the inner region has uniform vorticity); outside this maximum in a region of zero vorticity the velocity decreases inversely with the radial distance (Kundu, 1990). Such a Rankine vortex has isodop contours that are circular arcs connected with straight lines (Fig. 9.23). There are only minor deviations in this pattern even if the Rankine vortex is as close to the radar as two times the diameter of the mesocyclone (Brown and Wood, 1991). Burgess (1976) found that mesocyclones associated with rotating updrafts within Oklahoma severe thunderstorms had core diameters typically about 5 to 6 km, and peak tangential winds between 20 and 25 m s^{-1}.

If there is outflow from the cyclone as well as rotation, and if the functional dependence of outflow velocity on distance from the vortex center is the same as for the tangential component, the pattern of isodops remains unchanged except for a counterclockwise angular displacement α (Fig. 9.23) proportional to the strength of divergence. If there is no rotation but pure axially symmetric convergence (divergence), then $\alpha = -90°$ ($+90°$).

These patterns have been observed many times, and in Color Plates 2c and 2d are examples of signatures of nearly pure convergence and rotation. The reflectivity patterns do not indicate circulation (Fig. 9.24) nor convergence.

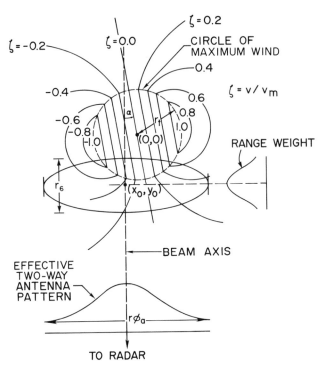

Fig. 9.23 Plan view of the isodop pattern for a stationary Rankine vortex located at a range large compared to vortex diameter. ζ is the Doppler velocity normalized to its peak value. The radar is located beyond the bottom of the figure. A resolution volume and antenna and range weighting functions are depicted. The angle α determines the radial inflow ($\alpha < 0$) or outflow ($\alpha > 0$).

There is a slight clockwise angular displacement of the pattern of rotation (Color Plate 2d) suggesting some convergence within the vortex at an altitude of 2.3 km. The pattern of convergence (Color Plate 2c) is observed at an altitude of about 200 m; but also notice, across the line of zero velocity, separating the closely spaced peak outbound and inbound Doppler velocities, the presence of a strong azimuthal shear region at 53° azimuth and 33 km range. This strong azimuthal gradient of Doppler velocities, which extends over about a 1-degree interval, is a tornado vortex signature (TVS).

Large changes in the first Doppler moment between contiguous azimuth samples and large second-moment (i.e., σ_v) magnitudes have been judged potentially important as tornado signatures. Storms having such signatures have either a tornado or the potential to produce one and are considered dangerous. Sufficiently intense azimuthal shear is the feature used to locate the TVS; this has been correlated with many tornadoes (Brown *et al.*, 1978). A TVS is not seen separately in the mesocyclone pattern (Color Plate 2d) because the tornado's center is coincident with that of the larger mesocyclone. Nevertheless, the

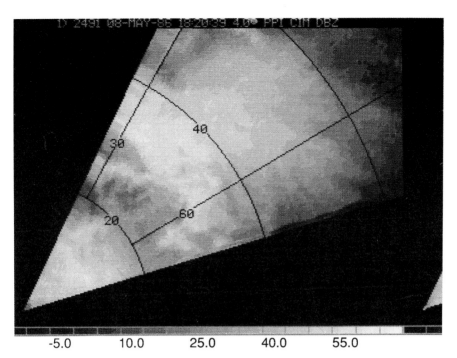

-5.0   10.0   25.0   40.0   55.0

Fig. 9.24 A PPI display of the reflectivity factor field of the Edmond tornadic storm on 8 May 1986; the reflectivity brightness scale (dBZ) is at the figure bottom, and range arcs are in kilometers. The radial velocity fields of this storm are shown in Color Plates 2c, 2d.

large azimuthal shear in the mesocyclone signature suggests the presence of a strongly rotating core; although peak winds appear to be less than 28 m s^{-1}, higher speeds are likely but may have been attenuated by resolution volume filtering (Section 9.5.1).

A pattern of Doppler velocities for another mesocyclone appears in Fig. 9.25. Although a tornado was embedded in the cyclone that produced the data displayed in Fig. 9.25, its velocity pattern is not resolved because its size is small compared to the radar's resolution volume. Positive evidence of tornado wind speeds has been obtained, however, and will be given in Section 9.5.3. The larger-scale mesocyclone (this can also be called a tornado cyclone because a tornado occurred within it) produces radial velocity contours that bear a strong resemblance to the idealized isodop pattern depicted in Fig. 9.23. Notice the large asymmetry, however, in the magnitude of the velocity even though the cyclone's translation speed has been subtracted. This asymmetry is probably due to the additive effects of cyclonic shear associated with the forward flank downdraft air that is west of the tornado cyclone and is flowing southward (see Fig. 9.5c) as well as to the combined effects of the two circulations (Fig. 9.17).

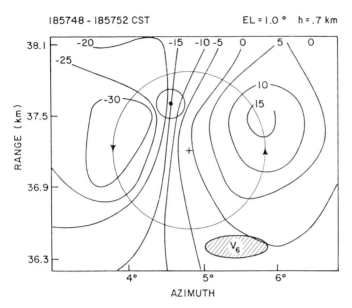

Fig. 9.25 Position of the Del City tornado (small circle drawn to scale) with respect to the mesocyclone on 20 May 1977 at 1857:50 C.S.T. Radial velocity contours are drawn from data spaced 0.6 km in range and 0.2° in azimuth. The mean radial motion of the mesocyclone is removed. The hatched area gives the size of the resolution volume V_6. The elevation angle is 1.0°; height is 0.7 km.

9.5.1 Smoothing of a Vortex Signature

Recognition of atmospheric vortices from the mean Doppler velocities is not feasible if the resolution volume size is larger than the vortex radius r_t (Fig. 9.23). To illustrate this point, we present a simplified example. We assume the beam axis is horizontal, the reflectivity η is constant, the velocity field is independent of z, and the two-way gain pattern is Gaussian and one dimensional with the following normalized azimuthal dependence:

$$f^4(\phi - \phi_0) = \exp[-(4 \ln 4/\phi_a^2)(\phi - \phi_0)^2], \tag{9.32}$$

where ϕ_a is the effective beamwidth (Fig. 7.25). The range weighting function is assumed to be narrow (i.e., $r_6 \ll r_t$) and to cut through the vortex center, and $r \gg r_t$ so that the radar senses only the tangential velocities v_t,

$$v_t = \begin{cases} v_m x/r_t, & x < r_r, \\ v_m r_t/x, & x > r_t, \end{cases} \tag{9.33}$$

of the Rankine vortex where v_m is the peak tangential wind, and $x \simeq r\phi$ is the distance along an azimuthal arc. Using Eq. (5.52) for an antenna pointed at ϕ_0 and a vortex centered at ϕ_v, the normalized velocity spectrum is calculated to be

$$S(v) = \frac{4\sqrt{\ln 2}}{\sqrt{2\pi}} \frac{r_t}{r\phi_a} \left\{ \frac{1}{v_m} \exp\left[-\frac{8r_t^2 \ln 2}{r^2 \phi_a^2} \left(\frac{v}{v_m} - \frac{x_0}{r_t} \right)^2 \right] \right.$$

$$\left. + \frac{v_m}{v^2} \exp\left[-\frac{8\ln 2 r_t^2}{r^2 \phi_a^2} \left(\frac{v_m}{v} - \frac{x_0}{r_t} \right)^2 \right] \right\} \qquad v \le v_m, \qquad (9.34)$$

where $x_0 = (\phi_0 - \phi_v)r$ is the separation between the resolution volume center and the vortex center. The first term in Eq. (9.34) is the spectral power contribution from the region inside r_t, whereas the second term is from outside. If the beam is centered on the tornado, Eq. (9.34) also describes the tornado spectral signature (Section 9.5.3), which is a double-peak spectrum having a total velocity span equal to the maximum Doppler velocity difference across the vortex.

Integration between $-v_m$ and v_m of $vS_n(v)$ produces the mean Doppler velocity (it is the fields of mean Doppler velocities that are displayed in PPIs; e.g., the Color Plates). The mean velocity depends on the radius of maximum winds r_t, the distance r to the vortex, the effective beamwidth ϕ_a, the maximum wind speed v_m, and the azimuthal separation of the centers of the resolution volume and the vortex. With proper normalization it is possible to reduce the number of parameters to three; the *normalized parameters* are plotted in Fig. 9.26. Accepting (somewhat arbitrarily) that the vortex will be recognized if the peak in radial velocity is pronounced, we note that the resolution $r\phi_a$ should be finer than the vortex diameter. Figure 9.26 represents the mean velocities as if they were densely sampled by the radar. In reality, discrete azimuthal measurements are taken, and therefore the peak of the velocity is often missed.

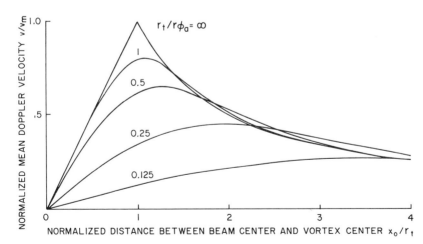

Fig. 9.26 Smoothing of the vortex signature by the weighting function (Section 5.2). (From Brown and Lemon, 1976.)

Also, velocity aliasing may be present, further complicating pattern recognition.

As an example, consider a vortex at a range of 240 km with a radius of 2 km and a radar having an effective beamwidth of 1°. Then the normalized parameter $r_t/r\phi_a$ would be about 0.5 and a significant loss in detecting the peak wind would occur. Furthermore, if samples were collected every beamwidth, the data spacings δx in the azimuthal direction would be 4 km (i.e., $\delta x/r_t = 2$), making the probability even greater that significant wind speeds would be missed! All these effects somewhat reduce one's ability to recognize vortices from mean velocity signatures. To reduce significantly sampling errors, the velocity should be estimated at intervals smaller (e.g., $\phi_a/2$) than the effective beamwidth.

9.5.2 High-Resolution Images of a Tornado

Wakimoto and Martner (1992) combined high-resolution single-Doppler radar measurements with photographs to reveal the detailed structural relationship of a tornado's reflectivity and Doppler velocity fields with its visual features seen from the radar site (Fig. 9.27). This intensifying tornado was observed with NOAA's 3-cm Doppler radar located 22 km from the tornado. The reflectivity factor and Doppler velocity fields are those in a vertical cross section through the center of a reflectivity hook (see Fig. 9.18 for an example of a PPI representation of a reflectivity hook of another tornadic circulation) and the tornado's center of circulation.

The tornado first became visible as a wide swirling dust cloud near the ground at 1421 and it had a diameter of about 450 m at the time of the photograph (Fig. 9.27a,b). No mesocyclone was apparent in the mid-altitude tropospheric wind fields, a feature that usually precedes tornadoes in severe thunderstorms, and the tornado developed along a convergence line separating air masses. It has been hypothesized that shear instabilities along convergence lines are the cause of vortices and hooklike features in reflectivity fields. Such features have been seen along wind shear lines by Carbone (1983) and Wilson (1986), and in association with water spouts by Golden (1974). Thus, as pointed out by Wakimoto and Wilson (1989), and earlier by Burgess and Donaldson (1979), tornadoes can form even in the absence of a parent mesocyclone. These tornadoes (sometimes called gustnadoes), however, are usually much weaker and of shorter duration than those that form within mesocyclones.

Apparent in Fig. 9.27 is the weak echo region coincident with the core of the tornado's circulation, and a descending higher reflectivity region surrounding the core. This core is also the weak echo hole of the reflectivity hook. This hole progressed upward above cloud base as the tornado matured. The descending higher reflectivity region surrounding the tornado is seen to appear in clear air, so the measured reflectivity is attributed to sparse but large raindrops (Wakimoto and Martner, 1992), a deduction made earlier for reflectivity fields in the

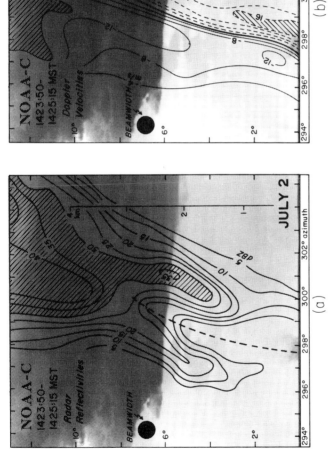

Fig. 9.27 Photographs of the 2 July 1987 tornado near Denver. Contours of reflectivity (left) and Doppler velocity (right) fields, on vertical sections through the reflectivity hook and tornado cyclone, are superimposed on the images; data are from NOAA Wave Propagation Laboratory's 3-cm radar. Z values greater than 30 dBZ and $|v_r| > 16$ m s^{-1} are hatched. Solid and dashed contours represent Doppler velocities toward and away from the photographer located at the radar. Dimensions of the beam are shown in some of the diagrams. (From Wakimoto and Martner, 1992.) *(Figure continues.)*

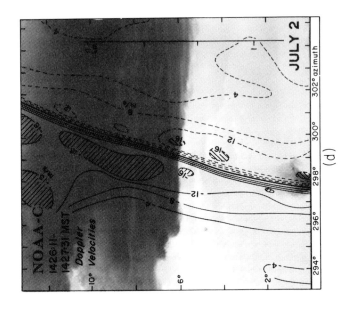

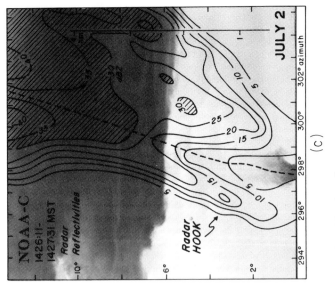

Fig. 9.27 (*Continued*)

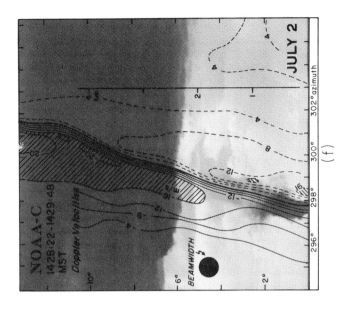

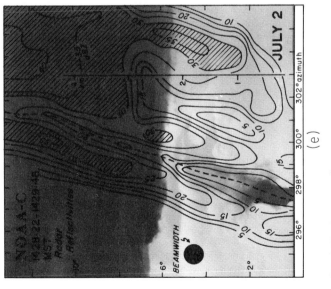

Fig. 9.27 (*Continued*)

343

vicinity of other tornadoes (e.g., Brandes, 1978) and documented by field scientists a few kilometers from a tornado (Zrnić and Doviak, 1975). Thus the descent of higher reflectivity is due to the large hydrometeor's terminal velocity overcoming weaker updrafts that were observed in the wind fields deduced from dual Doppler analysis, as well as from analysis of videotapes of the motion of debris packets (Wakimoto and Martner, 1992).

At the earliest time the maximum tangential velocities were confined to the lowest altitudes where airborne debris was first noticed, but reflectivity there is less than 5 dBZ. The dust that is visible in the photographic images contributes insignificantly to the reflectivity; apparently the debris is comprised of particles so small that the D^6 dependence of its backscattering cross section causes negligible reflectivity. Later (Fig. 9.27e) reflectivity peaks appear near the ground as well as around the tornado's axis at higher altitudes. The lower-altitude reflectivity peak is hypothesized to be associated with larger particles of debris that are eventually centrifuged by the tornado's cyclonic circulation, whereas the larger reflectivity surrounding the tornado at higher altitudes is attributed to centrifuging of water drops.

Although peak tangential velocities depicted in these figures are less than 16 m s^{-1}, larger peak values that are present are not observed because of the relatively large beamwidth of the Doppler radar (Section 9.5.1). Wakimoto and Martner (1992) used the beam-smoothing corrections for a TVS simulated by Brown et al. (1978) to deduce a maximum speed of about 25 m s^{-1} at 1424.

9.5.3 Doppler Spectra of Tornadoes

In 1961 Smith and Holmes reported a tornado spectrum that was observed with a cw Doppler radar; 12 years later, a tornado was first observed by a pulsed-Doppler radar (Zrnić and Doviak, 1975).

To reveal the radial component of motions within a resolution volume, signal samples are Fourier analyzed and the power spectrum is obtained (Section 5.1). Radar principally senses that portion of a circulation which lies within the resolution volume, but all scatterers moving with the same velocity contribute to a spectral coefficient according to their backscattering cross section, their number (i.e., in proportion to isodop spacing), and the weight received from $I(\mathbf{r}_0, \mathbf{r}_1)$ [i.e., Eq. (5.40)]. Only scatterers in those volumes between isodop surfaces which have an $I(\mathbf{r}_0, \mathbf{r}_1)$ weighted reflectivity sufficiently large will return signal power of magnitude to be detected above the spectral noise level. Nevertheless, it has been our experience that tornadoes in a resolution volume offer enough reflectivity, due to debris and hydrometeors, that a large-velocity span can be observed. Currently it is not known whether motions within a few meters per second of the peak tornado wind speed can be reliably resolved.

To relate the expected Doppler spectra of tornadoes to radar measurements, we shall show spectra for a model tornado circulation. The vortex is

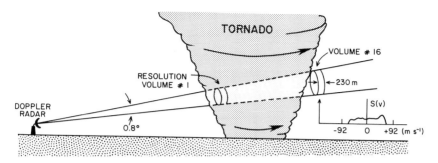

Fig. 9.28 Schematic of a radar beam and resolution volumes intersecting a tornado vortex.

approximated by a Rankine model, with provision made for particle inflow or ejection. Positions of maximum radial and tangential velocities are assumed to be coincident, and the radius of their location is referred to as maximum wind radius r_t. Given the model reflectivity and Doppler velocity fields (Fig. 9.23), one can calculate, using Eq. (5.52), the mean Doppler spectra associated with a specified resolution volume. Figure 9.28 illustrates the locations of 16 resolution volumes along the beam intersecting a tornado vortex.

Although tornadoes have nonuniform reflectivity, we assume (for simplicity) the reflectivity field to be either uniform or to have a Gaussian-shaped profile with a peak that could be displaced from the vortex center (i.e., a toroidal reflectivity profile that allows us to account for radial inhomogeneities). We also account for reflectivity inhomogeneities by assuming a different level of reflectivity for each resolution volume location. Other parameters, such as tornado radius and position within the resolution volume, are also adjustable to best-fit data.

When centered on the beam axis, the Rankine vortex model predicts a bimodal spectrum, called a *tornado spectral signature* (TSS; the dashed line in Fig. 9.29b), which has been verified experimentally several times (Zrnić and Doviak, 1975; Zrnić *et al.*, 1977). The TSS is also predicted by the analytic formula (9.34) if the beam is centered on the vortex and the beamwidth is large compared to the vortex diameter. The spectral peaks are caused by scatterers outside the tornado radius r_t, where the velocity gradient is less and hence the scattering volume contribution per unit velocity interval is larger, as predicted by Eq. (5.52); farther from the center this increase in spectral power is attenuated by the antenna gain. The velocity at the peak is related to the parameters of the tornado and the beam and can be found by locating the maxima of Eq. (9.34).

In Fig. 9.29 are Rankine model vortex spectra (dashed lines), matched to the Del City tornado data by a least squares fit. A von Hann weight (raised cosine) was applied to the data prior to a discrete Fourier transform. The von Hann weight offers a good compromise between the width of the main spectral window and the size of the window sidelobes (Table 5.2). Specifically, the rapid sidelobe decay reduces contamination of high-velocity spectral coefficients by

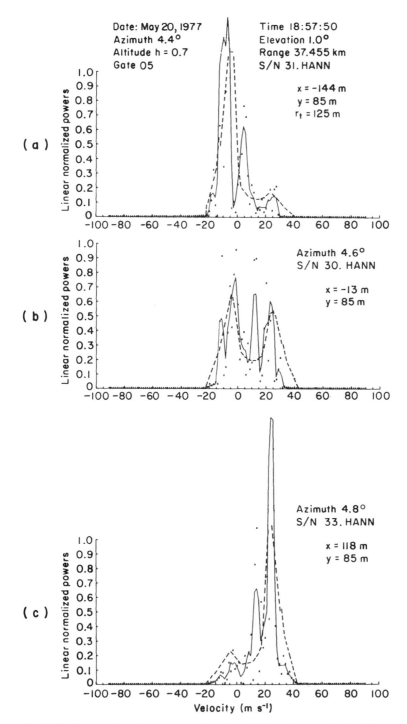

Fig. 9.29 Spectra from three consecutive azimuthal locations for the Del City tornado. The dots are squared magnitudes of Fourier coefficients for the recorded time series data weighted with a von Hann window. The solid lines are three-point running averages. Dashed lines are simulated spectra. The SNR (S/N) is in decibels; *x* (azimuthal distance) and *y* (range distance) are the coordinates of the tornado center with respect to the resolution volume center; the altitude *h* (km) is to the beam center from the ground.

strongly reflecting low-velocity ones (e.g., Fig. 5.10). The mean square difference between the data and the simulated spectra is simultaneously minimized for the four resolution volumes closest to the tornado. The resolution volume depth r_6 is 230 m, the range gate spacing 600 m, and the antenna beamwidth 0.8°. Although a 0.3-μs pulse was transmitted, the receiver filter bandwidth was 0.6 MHz, and thus a range resolution of 230 m is obtained from Eq. (4.26). The tornado is almost centered at azimuth 4.6°. The beam height above ground level is 640 m. Note that changes in resolution volume location by as little as 0.2° (i.e., one-quarter of the beamwidth) make large differences in the spectral shape.

The data from the Del City tornado were collected with an unambiguous velocity of ±91 m s^{-1}. Uniform and Gaussian-type reflectivity profiles were used in the spectra models. More often the toroidal profile resulted in a somewhat better fit because the two degrees of freedom (diameter and thickness) allow easier adaptation of the reflectivity to nonuniformities caused by debris. The model fitted in Fig. 9.29 had a toroidal reflectivity.

The tornado location determined from the spectral fit is superimposed in Fig. 9.30 on the damage path obtained by survey teams. Also in Fig. 9.30 are maximum measured Doppler velocities V_M, peak rotational (tangential) velocities V_t, and heights of the scans above ground. The tornado parameters deduced from the fit of the Del City tornado (20 May 1977) are (1) a diameter between 130 and 250 m, and (2) peak tangential speeds between 22 and 35 m s^{-1}. Estimated radii of wind maxima are drawn to scale in Fig. 9.30. The relative position of the tornado (whose spectra are shown in Fig. 9.29) with respect to the isodops of the mesocyclone is drawn in Fig. 9.25.

We emphasize that the observed Del City tornado produced moderate damage although it passed through a populated area. Corresponding peak rotational winds deduced from Doppler spectra were 35 m s^{-1}. These peak winds are deduced from plots of spectra such as those of Fig. 9.29 (Zrnić and Istok, 1980). Although higher wind speeds cannot be ruled out on the basis of spectral measurements, the damage survey indicates that radar-estimated values are quite realistic. Nonuniformities in reflectivity, tilting of the vortex with height, and scatterers in sidelobes are just a few effects that do occur but are not accounted for in the model used here.

Acquiring *in situ* data in and around the tornado is very difficult, but Doppler radar provides a remote and relatively safe method of surveying storms for tornadoes. The first scanning of a *violent tornado* (Fig. 9.31a) with a pulsed-Doppler radar having an unambiguous interval sufficiently large (180 m s^{-1}) to measure directly the high tornadic wind speeds was in the spring of 1981 at the National Severe Storms Laboratory in Norman, Oklahoma. The contour in Fig. 9.31a shows the cross section for each of 16 consecutive (in range) resolution volumes (see Fig. 9.28) from which weather signals were Fourier analyzed to produce the 16 Doppler spectra in Fig. 9.31b. These spectra were recorded within a minute of the time of the tornado photographed in Fig. 9.31a. Because the antenna was scanning in azimuth while data were

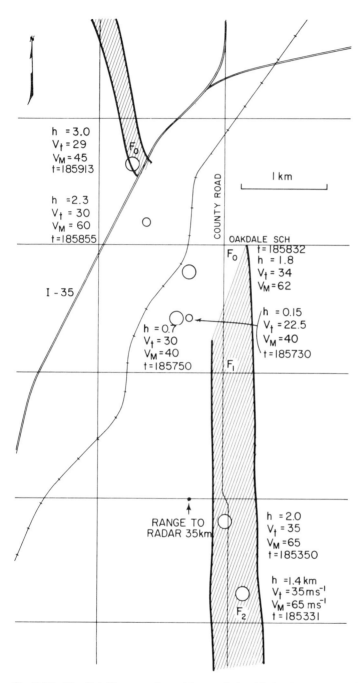

Fig. 9.30 The Del City tornado positions and size (circles drawn to scale), as deduced from the Doppler spectra, are superimposed onto the damage path (hatched areas). The height of the beam center with respect to the ground is h (km), V_t (m s^{-1}) is the speed of rotation, and V_M is the absolute maximum speed. The damage scale ($F_0 \rightarrow F_2$) is according to Fujita (1981). County roads (the square grid) are 1 mile apart.

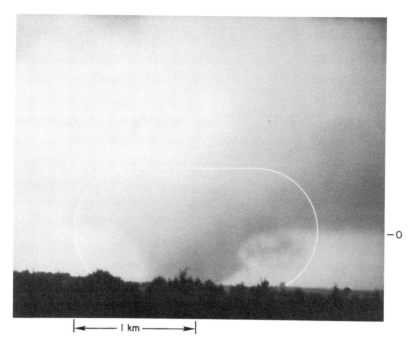

Fig. 9.31a The tornado near Binger, Oklahoma at ~1607 C.S.T. on 22 May 1981. View is to the northwest and the range from the photographer is 5.2 km. The white contour is the approximate size of the resolution volume for the data shown in (b). The radar beam is blocked by nearby terrain (not seen in this photo) up to an elevation angle of 0.1°. (Courtesy of R. Davies-Jones and D. Burgess.)

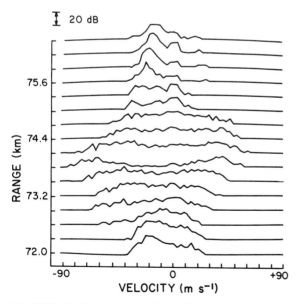

Fig. 9.31b Reflectivity spectral densities (in dB) versus the Doppler velocity at 16 range locations in the tornado's circulation. The elevation angle is 0.3°. The nearly horizontal lines at each end of the spectra are thresholds at 10 dB above the receiver noise level and correspond to a reflectivity spectral density of ~3.4 mm^6 m^{-1} s. The vertical scale is 20 dB per division.

collected to form these spectra, the effective azimuthal beamwidth is 1.5°, but the elevation beamwidth remains at 0.8° (Section 7.6). The range resolution $r_6 = 230$ m. The 16 volumes are aligned along the beam, but only two or three of these cross the tornado's circle of maximum wind. The asymmetry in the spectra (e.g., the peak positive velocity occurs in a resolution volume farther in range than the peak negative velocity) is caused by scatterers centrifuging out of the vortex.

In Fig. 9.31b the widest spectra denote the location of the tornado. Maximum tornadic wind speeds of ~90 m s^{-1} have been estimated from these data. Bluestein and Unruh (1991) have taken an FM cw transportable radar into the field to obtain close-in measurements of tornadic winds with high resolution; they report a peak wind of about 128 m s^{-1}.

The display of the spectrum of all detected radial velocities gives a definite indication of the tornado's presence because the peak circulating winds of the tornado are directly observed. Nevertheless, all large tornadoes are preceded by a larger scale circulation that starts at middle levels (6–8 km) and works its way toward the ground; this incipient tornado cyclone is larger, so its rotation can usually be resolved from fields of mean Doppler velocities. Spectrum width has not been a reliable indicator of tornadoes, because turbulent areas in storms often exhibit large widths.

A project involving the National Severe Storms Laboratory, National Weather Service (NWS), Air Weather Service (AWS), Federal Aviation Administration (FAA), and Air Force Geophysical Laboratory was established to conduct experiments to determine the improvement in tornado warnings over that obtained by a non-Doppler *system* (Burgess *et al.*, 1979). The non-Doppler system *included in situ observations by trained spotters* as well as the U.S. National Weather Services's non-Doppler WSR-57 radar, whereas warnings with the Doppler radar were based solely on radar observations. Operations were conducted during the spring of 1977 and 1978 (Table 9.1). Although the probability of tornado detection (≈ 0.64) did not improve significantly by using the criteria for mesocyclone recognition, the false-alarm rate decreased from 0.63 to 0.25. Moreover, the average lead time (the time between the issuance of a warning and the time of tornado occurrence) was about 20 minutes when using Doppler data alone, whereas the lead time for the non-Doppler system was a negative 0.8 minute.

Using criteria for tornado vortex signatures (TVS) discussed in Brown *et al.*, 1978, the project scientists were able to detect all tornadoes that occurred within a range of less than 115 km (Burgess *et al.*, 1979). At ranges beyond 115 km, signatures of small tornadoes are lost because of poor resolution; however, large, destructive tornadoes were detected up to 240 km with the 0.8° beam of the Doppler radar.

On the basis of those experiments the NWS, AWS, and FAA joined together to develop a network of 10-cm Doppler radars for the next generation of weather radars (i.e., NEXRAD, now WSR-88D). The operational tornado detection system involves forecasters interpreting products from computer algorithms that recognize velocity patterns of tornado cyclones (Zrnić et al., 1985). Although recognition of these patterns by trained observers is not complicated, there are good reasons to use automated techniques. It is well known that humans excel machines in recognizing signatures, yet when several signatures need to be examined, quantified, and remembered for correlations in space or time the machines become superior. Furthermore, a machine's performance is consistent regardless of the environmental conditions and is not subject to the boredom and fatigue that affect humans.

During 1991 operations in Oklahoma with the WSR-88D, the following scores for tornado warnings were obtained: probability of detection, 0.83; false alarm rate, 0.54; and critical success index, 0.42. The average lead time of 18 minutes for verified tornado warnings was almost the same as in the early experiments. These 1991 results are based on 42 tornadoes that occurred in the area of warning responsibility of the Norman, Oklahoma NWS forecast office. The false-alarm rate is higher because of the occurrence of several days of locally severe weather in quick succession without sufficient time between events to conduct an extensive verification. Warnings issued for such events are extremely difficult to verify over sparsely populated regions. The addition of the automated procedures, and its judicious use by the forecasters, is credited for the increase in the probability of detection over earlier results (Table 9.1).

9.6 Downdrafts and Outflows

Because most storms are observed with the radar's beam axis almost perpendicular to the vertical, direct observation and measurement of vertical motion are quite unlikely. Nevertheless, since the stratosphere impedes ascending air and the ground stops descending air, diverging outflows are created so that up- and downdrafts can often be seen generating telltale divergence signatures on plan position indicators (e.g., Fig. 9.23 with $\alpha = 90°$, an idealized signature; Fig. 9.32, an actual observation). Whereas strong updrafts can be located reasonably well by association with divergence signatures at the storm top (Color Plate 2a), downdrafts are not so easily detected because diverging outflows, confined to a shallow layer near the earth's surface, may not be sampled (ground clutter and earth's curvature obscure observation), and also because downdraft air is sometimes found in regions of weak reflectivity (note the RFD location in Fig. 9.2a and the precipitation-free volume in Fig. 9.2c).

An example of nearly vertical descent of air is seen in Fig. 9.33a showing rain embedded in the downdraft. The outrush of air as it deflects from the ground carries precipitation along with the air, producing relative symmetry on both sides of the rain shaft. Intense downdraft air impinging on the ground does not necessarily have precipitation embedded in it. Cloud particles and

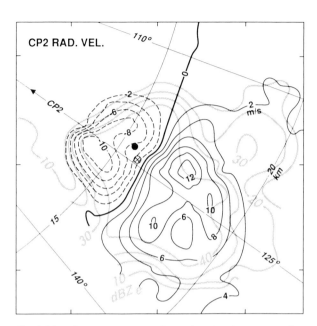

Fig. 9.32 The radial velocity field of a thunderstorm outflow near the ground. Dashed (solid) contours are isolines of constant negative (positive) radial velocities which are overlaid on the equivalent reflectivity factor (faint lines). The circled x is location of maximum divergence three minutes earlier and the solid dot is the estimated location of the divergent center based on damage surveys. (Kingsmill and Wakimoto, 1991.)

smaller drops may have completely evaporated before downdraft air reaches the ground; in Fig. 9.33b the presence of downdraft air is delineated by the streaks of dust made airborne by the outward-moving air.

9.6.1 Density Currents

Outward-moving air near the ground is more dense than the ambient air and it will continue to flow away from the storm as a density current. If the outflow is strong, intense horizontal shear (i.e., $\geq 3 \times 10^{-3}\ s^{-1}$; will be generated along the front of the current where the outflow meets the ambient air (e.g., Color Plate 3a). This intense horizontal convergence over narrow zones (i.e., less than a few kilometers) causes the ambient air to lift over the top of the thunderstorm density current.

If the ambient air is sufficiently moist, as it usually is in a thunderstorm environment, it condenses water vapor as it is lifted and forms a conspicuous arcus cloud that appears near the front of the current (Fig. 9.2c). If ambient conditions permit, the arcus cloud can propagate away from the density current front and become a solitary wave cloud (Section 9.7.2 and Fig. 9.40b).

Fig. 9.33a Downdraft in an Oklahoma thunderstorm with embedded precipitation. (Courtesy of H. B. Bluestein, University of Oklahoma.)

Fig. 9.33b Downdraft without precipitation near Denver, Colorado. (Courtesy of T. Fujita, University of Chicago.)

Outflow air is most easily detected when precipitation is entrained within it, but if the current moves far from the storm and precipitation falls out of the current, detection of the outflow is more difficult. To give an idea of the reflectivities found in intense outflows (i.e., microbursts, Section 9.6.3) we present the cumulative probability of reflectivity factors in the region of peak Doppler velocity for three climate regions in the United States (Fig. 9.34). Outflows in the high plains of the United States (i.e., Denver), have median reflectivity about 25 dB less than those found in mid-America (i.e., Kansas City) or in the subtropical environment of Orlando.

Lifting of aerosols from near the surface, or the generation of refractive index irregularities by mixing the different air masses at the front (Chapter 11), will create a narrow band of enhanced reflectivity (a reflectivity "thin line" in Figs. 9.35 and 9.36). Because winds there are relatively strong and brief (Color Plate 3a), this region is popularly called a "gust front," even though density current fronts are not the only phenomena that produce thin lines of reflectivity and wind gusts. Another example of a reflectivity thin line along a density current front is in Fig. 9.38a, but one not associated with a current front is in Fig. 9.40a.

All density current fronts produce abrupt changes (especially when viewed head-on) in Doppler velocity (i.e., a Doppler wind-shift line), but not all density currents produce a thin line of reflectivity. Associated thin lines of enhanced

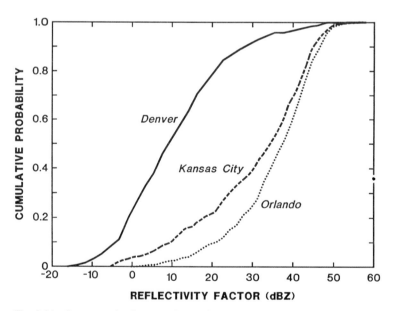

Fig. 9.34 Summer microburst outflow reflectivity factor at the time of maximum velocity change across the diverging outflow for three different regions in the United States. (Courtesy of P. J. Brion and T. Sen Lee, Lincoln Laboratories, Bedford, Massachusetts.)

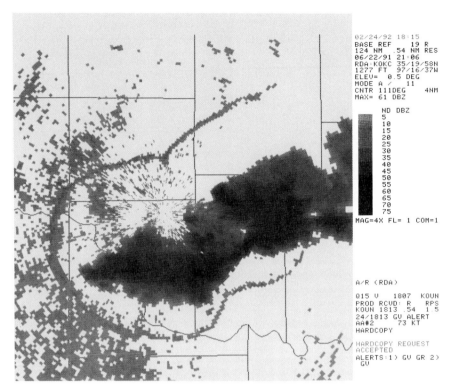

Fig. 9.35 Thin line of enhanced reflectivity surrounding thunderstorms, 22 June 1991, WSR-88D in Twin Lakes, Oklahoma. (Courtesy R. Murnan, Operational Support Facility (OSF)/NWS.)

spectrum width are often evident, however, on the PPI (e.g., Klingle *et al.*, 1987, Fig. 4).

Often there is a prominent region of circulating air (density current head; Goff 1976) just behind the leading edge of the density current, and this region is usually deeper than the current behind it. The circulation is toward the front near the ground and away from it above. The circulation could be closed, in which case the head is a horizontal vortex; or the circulation could be open, in which case the flow does not recirculate at the front. Some characteristics of density current fronts are presented in Table 9.2.

The velocity data in Color Plate 3a and the corresponding reflectivity field in Fig. 9.36 are from a thunderstorm and its associated density current at the lowest elevation angle (1°). From comparison of these two figures it becomes apparent that the current front was as much as 13 km away from the storm's precipitating regions at the data collection time (2108 L.S.T.). Later, the density current front was much farther away. The reflectivity thin line is seen to coincide with the density current front, and the large area of air advancing

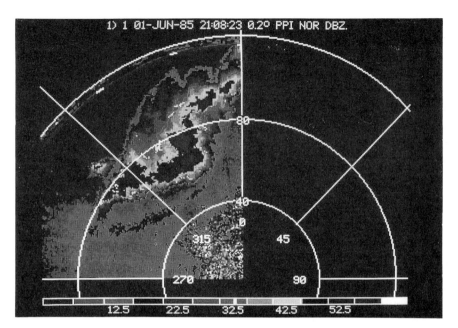

Fig. 9.36 The reflectivity field of a density current and the storms that generated it on 1 June 1985 (the Doppler velocity field is shown in Color Plate 3a). The reflectivity thin line that marks the density current front is about 13 km away, at its farthest point, from the high reflectivity regions of the storms. The brightness bar categorizes the reflectivity factors (dBZ); the reflectivity thin line has Z values >15 dBZ, and the storm has Z values >45 dBZ (i.e., the darkest areas).

toward the radar (the area of green in Color Plate 3a) is invading the ambient air (the region of red that tags positive radial velocities) flowing away from the radar.

Intense vertically oriented vortices, perhaps not intense enough to be called tornadoes, are often found along gust fronts (Carbone, 1982), and that is one more reason for aircraft to avoid or proceed with caution across fronts. The presence of a vortex can produce large decreases in headwind if the flight path is tangent to the circle of maximum wind (Doviak and Lee, 1985); documentation suggests that a vortex destroyed a commercial airliner in the Netherlands in 1981.

Because of the density current's typical long lifetime (>1 h) and large horizontal extent (>10 km), tracking of density current fronts is practical. A detection algorithm has been developed (Uyeda and Zrnić, 1986) that locates areas of convergence along a radial and pieces them together to locate lines of convergence, which are then identified as density current fronts. The alogrithm is able to identify fronts and then extrapolate their location up to 20 minutes in advance (Fig. 9.37). A derivative of this algorithm has been tested for use by Doppler radars for airport weather surveillance (Section 9.6.3).

Table 9.2

Characteristics of Density Current Fronts[a]

Height (km)	
Maximum	4.2
Average maximum[b]	2.2
Length (km)	
Maximum	197
Average maximum	89
Peak reflectivity factor (dBZ)[c]	
Maximum	11
Average maximum	7
Doppler speed difference (m s^{-1})[d]	
Maximum	40
Average maximum	32
Width (km)[e]	
Maximum	10.0
Average maximum	5.8

[a] Based on data taken from Klingle *et al.* (1987) and Doviak and Zrnić (1984, Table 9.2) for 14 density currents from severe storms in Oklahoma.

[b] Average maximum is the average of the maximum height, length, etc., of 14 fronts.

[c] The largest Z_e over a patch of several resolution volumes.

[d] The difference in the peak radial velocity along the front and the radial velocity of the environmental flow immediately in advance of the front.

[e] Measured from the peak gust velocity location to the leading edge of the front.

9.6.2 Convergence Bands

Convergence bands form at the density current fronts and are characterized by reflectivity factors between 0 and 20 dBZ and by zones of strong radial or azimuthal gradients of Doppler velocity. The bands can be several hundred kilometers long and have intense convergence concentrated in narrow (e.g., 0.5 to 5 km) zones.

The most common location of convergence bands is at the front of density currents associated with storm outflows (Fig. 9.38a; color plate 3a), but synoptic fronts (e.g., Fig. 11.22a), buoyancy waves (e.g., Fig. 9.40; Color Plate 3d), mountain outflows, and horizontal gradients in the heating of boundary-layer air (Segal *et al.*, 1984) can also cause convergence zones and thin lines of reflectivity (Wilson and Schreiber, 1986).

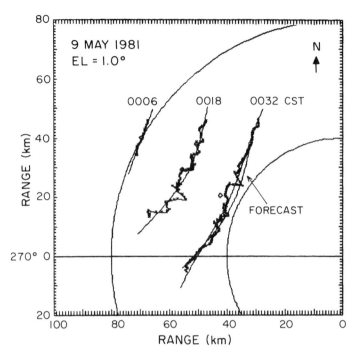

Fig. 9.37   Three consecutive positions of a thunderstorm outflow front on May 9, 1981. The smooth curves are least square fits to data, and the forecast position is indicated.

Radar reflectivity is an exceptionally good indicator of air mass boundaries, and all radar-observed thin lines (boundaries) are believed to be associated with convergence bands. Wilson and Schreiber (1986) used displays of reflectivity and Doppler velocity fields to find that a large percentage (79%) of storms around Denver, Colorado were initiated in close proximity to convergence bands, in agreement with conclusions of Purdom and Markus (1982) who used satellite observations of clouds over southeast United States to locate convergence bands.

Changes in temperature, humidity, and wind at the surface during the passage of three radar-detected boundaries are plotted in Figs. 9.38b,c. On examining this figure it is obvious that boundaries are not necessarily accompanied by changes in temperature or humidity. For example, boundary 2 gives no evidence of changes in temperature, but humidity (dew point temperature) and wind change significantly, whereas boundary 3 (the thin line of reflectivity in Fig. 9.38a) shows a small temporary change in temperature and no significant change in humidity; the sustained wind change is the only evidence of the passage of a density current front. Reflectivities in Fig. 9.38a were observed at about the time when the density current front passed station 15 (Fig. 9.38c). Boundary 6 produces a marked and sustained decrease in temperature and

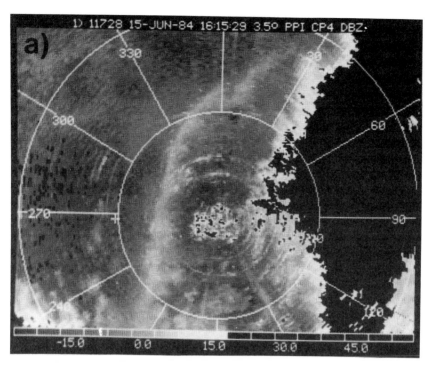

Fig. 9.38a   A NNE to SSE oriented reflectivity thin line about 5 km west (at its nearest point) of the radar. This thin line marks the location of the front of a density current that flowed out of a thunderstorm complex east of the radar (the darkened area, which denotes reflectivity factors larger than about 20 dBZ). Range circles are 10 km apart, elevation angle is 3.5°, and reflectivity brightness scale (in dBZ) is at the figure bottom. 15 June 1984 at 2215 G.M.T. in Denver, Colorado.

increase in humidity, also suggestive of a density current front. The exceptionally large velocity of short duration preceding the sustained wind change is suggestive of a cutoff vortex (e.g., Clarke, 1965; Droegemeier and Babcock, 1989) that has propagated out in front of the density current (see Section 9.7.2.2 for an example of a cutoff vortex that evolved into a solitary buoyancy wave; also Doviak *et al.*, 1991).

The results given by Wilson and Schreiber (1986) confirm that mesoscale convergence bands in the boundary layer play a major role in determining the regions where and when storms will likely form; moreover, they suggest that thunderstorm initiation, which often appears random, might be predictable, as found earlier from satellite observations by Purdom (1982).

9.6.3   Downbursts and Microbursts

A downburst is defined as a "strong downdraft that induces an outburst of damaging wind on or near the ground" (Fujita, 1981; Fujita, 1985). Downbursts

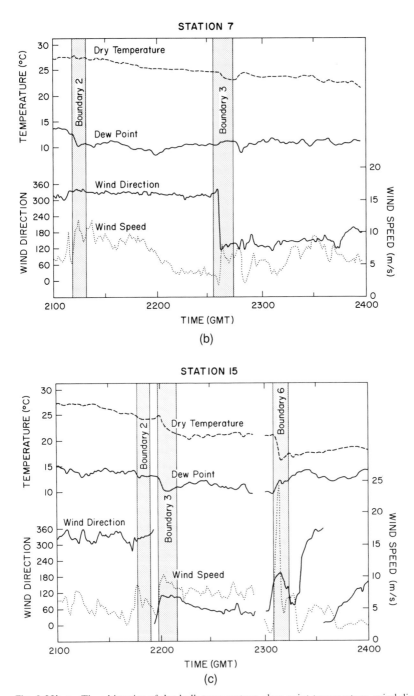

Fig. 9.38b,c Time histories of dry-bulb temperature, dew-point temperature, wind direction, and maximum wind speed at Portable Automated Mesonet (PAM) stations 7 and 15 as boundaries 2, 3, and 6 pass over them. The maximum wind speed is the highest reported 1-s average in a 60-s period. The time of boundary passages (shaded zones) is deduced from radar data. Station 7 is at about 20 km and 335° with respect to the radar, and station 15 is at the radar. (From Wilson and Schreiber, 1986).

are mostly associated with intense, severe storms but they also have been observed, especially in the high plains of the United States, below innocuous-looking (low reflectivity) virga (Eilts and Doviak, 1987; McCarthy and Wilson, 1984). Observations in regions where subcloud moisture contents are higher reveal that many microbursts have reflectivity factors larger than 40 dBZ (Fig. 9.34).

When the horizontal dimensions of a thunderstorm's density current are small but velocities are intense, the downdraft and its associated outflow is called a microburst (Fujita, 1981). More specifically, a microburst has been defined as a diverging outflow that exhibits anywhere a wind speed difference of at least 10 m s^{-1} across a distance of no more than 4 km (Campbell *et al*, 1989). Experiments indicate that intense small-scale divergent flow is more common than previously thought (Wilson *et al.*, 1984) and, as with vortices, it is the smaller-scale downdrafts that produce the most hazardous shear. The Doppler velocity field of a microburst is shown in Fig. 9.32 and Color Plate 3b. Strong divergence of horizontal flow can cause low-flying aircraft to markedly depart from their intended flight altitude, whch can be disastrous on approach to and departure from airports, even if winds are not intense enough to cause damage on the ground.

The level of threat to aircraft is conditioned by the response of the aircraft to a wind shear event. For example, a Boeing 727 aircraft has enhanced response to wind perturbations at frequencies near its phugoid frequency (i.e., 0.3 Hz; McCarthy and Blick, 1979). Using typical values of takeoff and landing speeds, we deduce that wind swirls of about 3-km wavelength can have a more delete-rious effect on the performance of the aircraft than other wavelengths of similar amplitude. To resolve the most hazardous shears a few hundred meters above the ground (where aircraft cannot timely recover the significant loss of altitude), a resolution of a few hundred meters or less is required. Thus, radars need high angular resolution and should be close to the microburst.

Compounding the difficulty in making accurate measurements of hazardous wind shear is the short lifetime of the phenomena. For example, lifetimes of microbursts are from 5 to 15 minutes, but the period of severe wind shear lasts only from 2 to 4 minutes with an average difference of 25 m s^{-1} across the divergent flow (McCarthy and Serafin, 1984). Although these small-scale hazards might be difficult to detect, they are usually embedded in larger-scale phenomena that have longer lifetimes and hence are more easily detected. Thus, detection of phenomena that might cause hazardous wind shears could be equally or more important than detection of the hazard itself (Doviak and Lee, 1985). Results of recent research reveal that there are a number of possible precursors before the formation of divergence at the surface. These include a rapidly descending reflectivity core, organized convergence near and above cloud base, and mid-altitude (2–5 km) rotation (Roberts and Wilson, 1986).

The U.S. Federal Aviation Administration (FAA) is deploying 5 cm wave-length Terminal Doppler Weather Radars (TDWR) at some airports to detect microbursts and density currents as well as heavy precipitation along the arrival and departure paths. A prototype TDWR has been tested at Denver, Kansas City, and Orlando. Furthermore, automated algorithms to estimate the expected loss (or gain) of airspeed that an aircraft might encounter are being developed and tested. Trial alerts were issued to air traffic controllers and/or pilots based on the following criteria: If the estimated airspeed loss (gain) is greater than 10 m s^{-1}, but less than 15 m s^{-1}, a negative (positive) *shear alert* is issued, but if the expected loss is more than 15 m s^{-1}, a *microburst alert* is issued (Bernella, 1991). The TDWR radars were also being tested for their capability to forecast the arrival of density currents at airports (Fig. 9.37). This could be of considerable value for aircraft controllers in planning changes to approach and departure corridors.

Radar measurements are always representative of wind at heights significantly above the ground, so care must be exercised when comparing these measurements with those from surface anemometers. Results from a recent study indicate that radar-estimated horizontal shear at heights between 50 and 600 m AGL in Oklahoma density currents are stronger on the average by 1.6 times than the shear measured 2 m above the surface (Eilts and Doviak, 1987). Wind profiles obtained from anemometers on a 500 m tall tower confirm that wind shear near the surface is weaker than the wind shear a few hundred meters aloft.

Changes in airspeed can be experienced when the aircraft descends (or ascends) through a vertically sheared horizontal flow. One of the earliest documented wind shear accidents in the United States was due to vertical shear of horizontal wind, which caused the crash of a commercial airliner (a DC-10-30) on its approach to Logan Airfield in Boston in 1973; and more recently it downed a small commuter-type aircraft on its approach into Little Rock, Arkansas in 1990 (Kessler, 1991). A vertical shear of $2 \times 10^{-1} s^{-1}$ in horizontal wind will cause the equivalent horizontal shear of head or tailwind equal to $10^{-2} s^{-1}$ along a 3° glide slope. A vertical shear of horizontal wind in excess of $7 \times 10^{-2} s^{-1}$ may perturb aircraft control, as do downdrafts in excess of 2 m s^{-1} (Table 9.3). Bowles and Targ (1988) define a factor F to describe the severity of the combined threat of a horizontal shear and vertical wind w. The F-factor is

$$F = \frac{1}{g} \times \frac{du}{dt} - \frac{w}{v_a},$$

where du/dt is the rate of change of the horizontal component of wind along the aircraft flight path, g is the acceleration due to gravity, and v_a is the airspeed of the aircraft. F values greater than about 0.1 (Targ and Bowles, 1988) are considered hazardous, which is consistent with the vertical shear hazard level in Table 9.3 for an airspeed of 75 m s^{-1} along the glide slope. Elmore *et al.* (1989) used dual Doppler-derived estimates of wind shear and vertical velocity to calculate the F parameter and found that vertical and horizontal components contribute almost equally at about 200 m AGL in microburst outflows.

Table 9.3
Proposed Classifications for Wind Shear along a 3° Glide Slope

Intensity of wind shear	Effect on aircraft control	Vertical shear of horizontal wind[a] (s^{-1})	Updraft or downdraft velocity $(m\ s^{-1})$
light	little	$(0–7) \times 10^{-2}$	0–2
moderate	significant[b]	$(7–13 \times 10^{-2}$	2–4
strong	considerable difficulty	$(1.3–2) \times 10^{-1}$	4–6
severe	hazardous	$>2 \times 10^{-1}$	>6

[a] Along a glide slope. This is based on criteria given by Brown (1982).
[b] Greene *et al.* (1977) choose a wind shear of 8.4×10^{-2} as significant to aircraft operation.

Although evidence supports the idea that microbursts are a very prominent factor in most fatal wind shear accidents, it has been proposed (Linden and Simpson, 1985) that wind shear and downdraft, at the rear of a vortex ring formed around the microburst and located at its leading edge (i.e., the density current front), are responsible for the danger of flying through thunderstorm outflows. Furthermore, an analysis of wind shear accident reports for 11 of 33 weather related accidents investigated by the National Transportation Safety Board suggests that waves or wave-induced turbulence might have been the cause or contributing factor (Doviak and Christie, 1989).

There is no apparent correlation between reflectivity factor in the divergent flow of microbursts and the maximum velocity difference across it (McCarthy *et al.*, 1984; Rinehart and Isaminger, 1986). Nevertheless, because all fatal thunderstorm-related accidents involving U.S. air carriers from 1970 to date have occurred in or near heavy precipitation, Kessler (1985) suggests that the most prudent action is to avoid high-reflectivity thunderstorms by a sufficient margin.

9.7 Buoyancy Waves

Buoyancy waves are meteorological phenomena with characteristics (i.e., speed, amplitude, etc.) that are often horizontally uniform over large areas and thus can be retrieved from single-Doppler data. Once initiated, the propagation of a buoyancy wave depends strongly on conditions of the ambient atmosphere that forms the waveguide (Gossard and Hooke, 1975; Lindzen and Tung, 1976). In particular, sustained propagation under windless conditions requires a statically stable layer of air bounded by a neutrally stable atmosphere.

9.7.1   Observation in a VAD

Testud *et al.* (1980) have analyzed the residuals (Fig. 9.39) of the least squares
fitted data obtained by subtracting the first three Fourier components, that is,
subtracting the linear wind from the measured Doppler velocities. They assume
a wave perturbation $\delta v(h)$ of the horizontal wind

$$\delta \mathbf{v}(h) = A(h)(\mathbf{K}/K)\exp[j\,\mathbf{K}\cdot\mathbf{d} + \psi(h) - \omega t], \qquad (9.35)$$

where $A(h)$ is the amplitude, which is a function of height h; $\mathbf{K}$ the horizontal
wave vector; $\psi(h) - \omega t$ a phase term; and $\mathbf{d}$ the horizontal distance vector whose
origin is the center of the measurement circle. The factor $\mathbf{K}/K$ expresses the fact
that $\delta \mathbf{v}$ is collinear with $\mathbf{K}$.

Under the assumption that the period of the wave is large compared to the
acquisition time, the corresponding perturbation in the radial velocity becomes

$$\delta v_r(h, \phi) = A(h)\cos[Kr\cos\theta_e\cos(\phi - \phi_0) + \psi(h) - \omega t]\cos(\phi - \phi_0)\cos\theta_e,$$
$$(9.36)$$

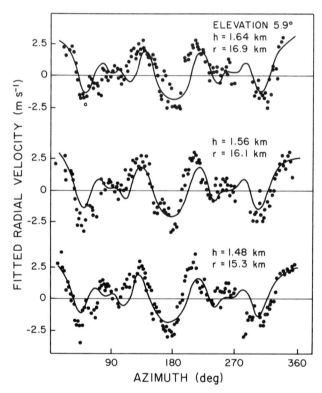

Fig. 9.39 Examples of the radial velocity fluctuations about a linear wind fitted to Doppler velocity
data. Dots are the observed fluctuations; the continuous lines are theoretical curves realizing the best
fit to the residuals. (From Testud *et al.*, 1980.)

where ϕ_0 is the azimuth in the $\mathbf{K}$ direction. This model can be extended to include vertical motions, which are ignored for small elevation angles θ_e.

The wind perturbation δv_r depends on the four parameters A, K, ϕ_0, and ψ. By using a least squares fit analysis, Testud et al. (1980) obtained the four parameters as a function of height. An example of their result for three ranges (Fig. 9.39) shows a remarkably good resemblance between the fitted curve and data. More important, a cursory examination of the residuals does not reveal the fact that these nonperiodic residuals are, for the most part, caused by a coherent meteorological phenomena; one might be led to believe the residuals are errors in the Doppler velocity field. For the 83 scans they analyzed, the standard deviation of the data from the theoretical curves never exceeded 1 m s^{-1}; errors in the radial velocity were estimated to be about 0.5 m s^{-1}. The authors conclude that the three parameters A, K, ϕ_0 can be precisely determined but that the error in the phase term ψ is very large ($\pm 100°$) so that the direction of propagation along the azimuth ϕ_0 cannot be determined (has an 180° uncertainty).

9.7.2 Large-Amplitude Buoyancy Waves

Density currents, interacting with stable boundary-layer air, have been observed to initiate large-amplitude buoyancy waves that can propagate as undular bores (Color Plate 3d and Mahapatra et al., 1991) and/or solitary waves (Fig. 9.40 and Doviak and Ge, 1984). A stable layer of air ahead of the density current front in Color Plate 3a was deposited by outflows from earlier storms (Doviak et al., 1989), and the interaction between this density current and the stable layer launched a pair of evolving solitary waves, one of which is displayed in Color Plate 3c. Because these waves have large amplitudes, nonlinear effects become important, and as a consequence the wave's characteristics (e.g., wavelength) are no longer spatially uniform.

9.7.2.1 Bores

A bore is a *propagating* disturbance that causes a steplike increase in the characteristics of the waveguide layer of stable air (e.g., thickness, wind, etc.; Fig. 9.41). If the bore is modulated by a sequence of waves that may evolve from it, it is called an undular bore. Most notable, the undular bore produces a propagating step increase in surface pressure (i.e., a pressure jump), atop which there are quasiregular undulations in pressure. In addition to a wind shift, an almost negligibly small temperature increase (e.g., $\approx 1°C$; Fig. 9.41) is seen at the ground, although larger changes (usually cooling) are observed aloft. The small temperature rise is caused by the compression of air, and cooling is due to the adiabatic lifting of stable air (i.e., air having a positive gradient of potential temperature). Crook and Miller (1985), Crook (1986), and Haase and Smith (1989) demonstrated through numerical computations that undular bores can be generated by density currents that invade stable layers. Fulton et al. (1990) reported observations (Doppler radar and a network of surface measurements)

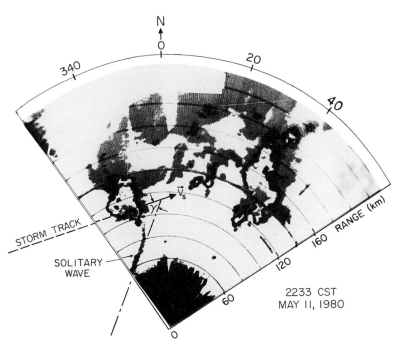

Fig. 9.40a Plan position indicator display of echo power contours (boundaries of shaded areas) in about 10-dB steps. The maximum reflectivity factor in the storm (at 110 km and 340°) is 55 dBZ. Elevation angle is 0.4° and the storm velocity is $\mathbf{V}_s \approx 12 \ \mathrm{m\,s^{-1}}$.

Fig. 9.40b A morning glory cloud formation at Burketown, Australia; a visual image of a solitary wave cloud. (Courtesy of D. R. Christie, Australian National Uni., Canberra, Australia.)

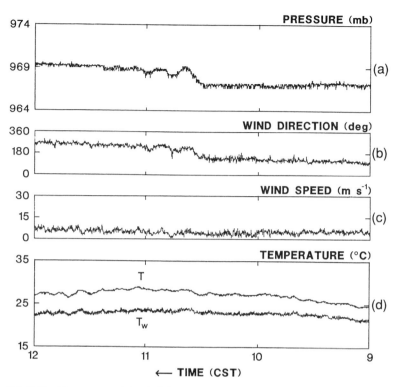

Fig. 9.41 The pressure (a), wind direction and speed (b, c), and temperature changes (both dry T and wet T_w bulb) at the ground during the passage of an undular bore over NSSL's instrumented 500-m tower on 22 June 1987. Time scale increases to the left.

that clearly reveal the evolution of an undular bore from a density current. Rottman and Simpson (1989) simulated, in the laboratory, thunderstorm out-flows that interacted with stable layers to generate undular bores.

9.7.2.2 Solitary Waves

Fornberg and Whitham (1978), and Christie (1989), showed how the evolution of an initially smooth atmospheric bore leads to the formation of an amplitude-ordered sequence of solitary waves. This evolution is illustrated in Fig. 9.42 for an internal bore propagating in a ground-based stable layer. Because the propagation speed of buoyancy waves is amplitude dependent, the waves progressively lag behind the front as they develop into independent entities. The evolution of each wave into a solitary one of permanent form is due to the competing effects of nonlinearity, whereby a wave grows and steepens, and frequency dispersion, which flattens a steep-sided wave. A wave steepens because the higher amplitude portions of the wave travel faster than those of lesser amplitude (e.g., Lamb, 1932, p. 281). Flattening occurs because the

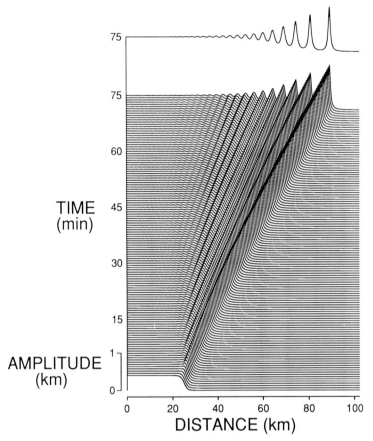

Fig. 9.42 A numerical solution illustrating the formation and evolution of an atmospheric undular bore. Each line in the plot corresponds to the wave-induced vertical displacement of air as a function of position along the direction of propagation. The amplitude of the unperturbed bore is 400 m and the speed c_0, for waves of infinitesimal amplitude and extremely long length, is 9.2 m s^{-1}. (Courtesy of D. R. Christie, Australian National Uni., Canberra, Australia.)

longer-wavelength Fourier components of the wave, distorted by nonlinearity, travel faster than the shorter-wavelength ones (i.e., the atmospheric waveguide is dispersive). A balance between the effects of nonlinearity and frequency dispersion leads to waves with exceptional stability. Thus, without losses, the natural end state for a disturbance of this type is a long sequence of amplitude-ordered solitary waves (Color Plate 3d; Christie, 1989; Mahapatra *et al.*, 1991).

The radar image of a solitary wave is a thin line of Doppler shift and reflectivity propagating away from a storm (e.g., Color Plate 3c and Fig. 9.40a). The thin line of Doppler velocities in Color Plate 3c is the first of two solitary waves that emerged from the density current depicted in Color Plate 3a (the second wave is not easily seen in this image because its amplitude is weaker and the radar beam is above it). It is notable that the Doppler velocities on both sides of the wave are representative of the ambient air, suggesting that the wave has emerged from the density current and is propagating freely in an atmospheric waveguide. The frequent observation of propagating reflectivity lines in the neighborhood of thunderstorms and the ease with which solitary waves can be generated in the laboratory (Maxworthy, 1980) suggest that solitary waves, and the closely related undular bores, might be commonly occurring features in the thunderstorm environment whenever suitable atmospheric stratification exists and sources are active.

9.7.2.3 Buoyancy Wave Effects

Apart from scientific curiosity, practical reasons have fostered interest in the study of buoyancy wave phenomena. In a conditionally unstable atmosphere, the lifting of air masses by buoyancy waves can trigger the formation of convective clouds and thunderstorm cells. Shapiro et al. (1985) suggest that a frontal head (i.e., the leading and elevated portion of a density current) may separate and propagate as a solitary wave that could initiate precipitation systems in the prefrontal environment. Waves of this type have large amplitudes and play a significant role in the initiation and organization of deep convection (Zhang and Fritsch, 1988; Carbone et al., 1990). They are also a source of intense localized wind shear near the ground, which can be hazardous to aircraft arriving at and departing from airports (Christie and Muirhead, 1983; Doviak and Christie, 1989), even after propagating several kilometers from their generating thunderstorms (Doviak and Christie, 1991).

Although periodic waves that are accompanied by organized cloud bands appear frequently in satellite photographs (e.g., Fig. 9.43), Doppler radar observations (Color Plate 3d) of buoyancy-wave phenomena are scarce. Large-amplitude waves can trap potentially cooler and/or drier air from the density current and carry it, at speeds in excess of any speed within the current, beyond what would have been the border of the current in absence of the wave (Haase and Smith, 1989). Thus, cooler air might be present at places where it would not have advected if closed waves were not generated; this could give the appearance that the density current front is propagating at speeds faster than predicted by theory (based on the current's depth and temperature deficit relative to the environment; Benjamin, 1968). It then becomes difficult to define the density current front.

The difficulty in resolving the front is compounded by the fact that the trapped fluid is continually leaking and being deposited behind as the wave propagates ahead of the current (this has been demonstrated in laboratory experiments; Maxworthy, 1980; Rottman and Simpson, 1989). With the understanding that the "true" location of the density current front can be obscured by the advection of outflow air trapped within the wave, Fulton et al. (1990) plotted

Fig. 9.43   Visible band GOES satellite images of the undular bore at 1531 G.M.T. (its Doppler velocity image is shown in Color Plate 3d). The star locates NSSL's Norman Doppler radar in central Oklahoma.

the speeds of the bore front (i.e., where the wind shift and pressure jump occur) and the density current front, defined as the location where *equivalent* potential temperature (Hess, 1959) begins its abrupt decrease. The plotted data imply that the bore propagated at nearly constant velocity, apparently determined by the characteristics of the atmospheric waveguide, whereas the density current's speed decreased with distance, apparently due to the decrease in current depth as the dense air spread away from the storms.

    These observations suggest a scenario for the evolution of undular bores and the formation of a sequence of solitary waves that results from the inter- action of a storm's downdraft with a stable layer (Doviak *et al.*, 1991 and Fig. 9.44). A rapidly moving thunderstorm (or one of short lifetime) generates locally a limited outflow current with a horizontal rotor at its leading edge. If the outflow current and rotor invade a stable layer, they excite one or more solitary wave disturbances that propagate at speeds determined by wave ampli- tude and the vertical profiles of wind and virtual potential temperature (Lalas and Einaudi, 1974) of the ambient environment. Support for this mechanism is provided by the recent numerical model results; Haase and Smith (1989) found that density currents, which invade stable layers under supercritical conditions (i.e., frontal speeds that are, in absence of a stable layer, faster than the speed

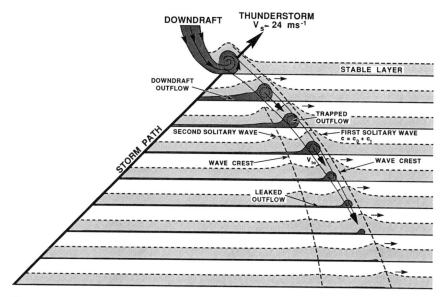

Fig. 9.44   A scenario for the evolution of thunderstorm-generated solitary waves.

of waves of infinitesimal amplitude), are transformed directly into nonlinear waves of large amplitude. A weak, dissipating density current may, however, be left behind in the wake of these waves. As the leading solitary wave propagates away from the storm path, it can drag a vortex of storm outflow with it (Fig. 9.44) which, under the influence of the Coriolis force, is guided along the wave's axis (Doviak *et al.*, 1991).

9.8   Large Weather Systems

The resolution (several hectometers), coverage (≈ 100 km) and data update rates (a few minutes) of a single-Doppler weather radar is well matched to thunderstorm scales (i.e, lifetimes of a few tens of minutes or more; dimensions of several kilometers). Although a single radar does not have the coverage to map the entire structure and evolution of larger weather systems (e.g., hurricanes, mesoscale convective systems, etc.), it can provide detailed measurements of the wind and water fields over a good portion of these systems. This section contains examples of mesoscale convective systems, hurricanes, cold fronts, and dry lines that are observed with Doppler radars.

9.8.1   Mesoscale Convective Systems

Mesoscale convective systems (MCSs) consist of interacting thunderstorm cells (Fig. 9.45) and a large area of stratiform precipitation (Houze, 1989). The hori-

Fig. 9.45   A series of thunderstorms along a squall line.

zontal extent of these systems is hundreds of kilometers and lifetimes are of the order of 10 hours. Considerable amounts of rainfall and severe weather in central USA are produced by the MCSs. Although MCSs were initially defined from satellite images (Maddox, 1980) some types such as squall lines had been investigated much earlier (Newton, 1950; Fujita, 1955). A squall line can take the form of a line of convective cells as in Fig. 7.3, or as a quasi-two-dimensional line of convection (Color Plate 4a and Fig. 9.46). The Doppler velocity and reflectivity fields of the squall line that passed through central Oklahoma in the early morning hours of 4 May 1989 exhibit a very thin line of high reflectivity factor (>50 dBZ) and enhanced velocities in the direction of squall-line propagation. The thin line of Doppler velocities (Color Plate 4a) lies at the leading edge of the intense line of reflectivity and suggests the development of a horizontal vortex and, if conditions allow (Section 9.7.2), the generation of a wave that could leave the zone of precipitation.

Environmental factors favorable for the formation of MCS have been summarized by Maddox (1983). The genesis of a MCS is characterized by mesoscale convergence and lifting usually associated with low-altitude temperature advection. Many MCSs develop during the evening in areas of preexisting, unorganized convective activity and transform, over a period of several hours, to a highly organized state, often with a squall line. Important substructures such as

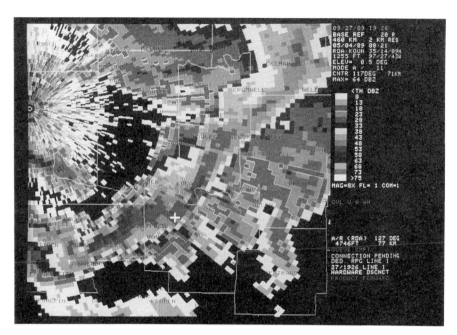

Fig. 9.46 The reflectivity of a nearly two-dimensional squall line observed with a WSR-88D radar in Norman, Oklahoma on 4 May 1989 (the Doppler field of this storm is in Color Plate 4a). (Courtesy of D. Purcell OSF/NWS.)

convective cells and lines, density currents, mesolows and mesohighs, tornadoes, etc., might occur within the MCS.

A conceptual model of the wind and water fields in a vertical cross section perpendicular to the line of convergence through a squall line with a trailing stratiform region is drawn in Fig. 9.47a, and two horizontal cross sections (i.e., without and with a mesoscale vortex) are shown in Fig. 9.47b and c. Front-to-rear flow begins in the boundary layer near the density current front (i.e, gust front) and extends through the convective region, where it is superposed over intense localized updrafts; the flow then slopes gradually into the trailing stratiform cloud at mid to upper levels. New cells form ahead of heavy convective showers, and decaying cells at the back of the squall line are advected over a layer of subsiding storm-relative rear inflow. The front-to-rear flow also advects ice particles (asterisk in Fig. 9.47a) which grow by vapor deposition and begin to fall. Eventually the particles reach warmer air and form aggregates, which produce an enhanced reflectivity layer (i.e., the bright band indicated by the horizontal stippled area in Fig. 9.47a; see also Section 8.5.3.2) just below the 0° isotherm. The rainfall from the stratiform region is about 40% of the total (McAnelly and Cotton, 1989). The storm-relative rear-to-front flow through and below the bright band could be continuation of a jet or one branch of a mid-level

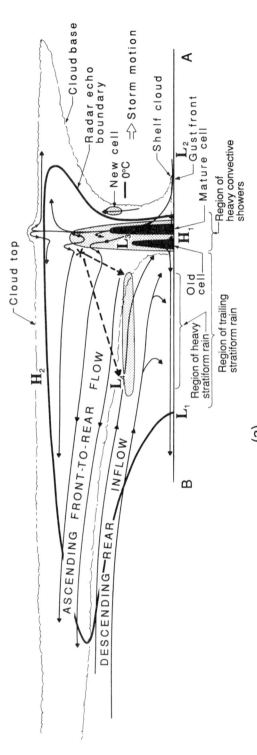

Fig. 9.47a A conceptual model of a squall line with a trailing stratiform rain area viewed in a vertical cross section oriented perpendicular to the convective line (i.e., parallel to its motion). (From Houze *et al.*, 1989.)

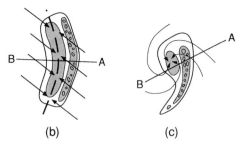

Fig. 9.47b,c   Conceptual model of a mid-level horizontal cross-section through (a) an approximately two-dimensional squall line and (b) a squall line with a well-defined mesoscale vortex in the stratiform region. In each case, the mid-level storm-relative flow is superposed on the low-level radar reflectivity. The AB lines represent the location of the cross section displayed in 9.47a. The stippling indicates regions of higher reflectivity. (From Houze *et al.*, 1989.)

mesoscale vortex (Fig. 9.47c; Brandes, 1990). As rear inflow subsides it reaches the convective line and enhances convergence.

A vertical cross section of reflectivity and storm-relative horizontal wind component normal to a squall line is presented in Fig. 9.48. Distinct characteristics such as the bright band, descending rear inflow, and stratiform and convective regions with a new cell in front are evident.

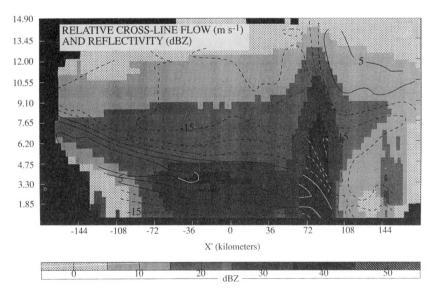

Fig. 9.48   The reflectivity factor field and storm-relative flow in a cross section perpendicular to a quasi-two-dimensional squall line observed with a 5-cm radar. The solid (dashed) contours are front to rear (rear to front) speeds in $m\,s^{-1}$. Line motion is to the right and the trailing stratiform region is centered at 0 km. (Courtesy of R. Houze, University of Washington, Seattle.)

9.8.2 Hurricanes and Typhoons

Hurricanes are large rotating storms (diameters of 100–300 km) that form over warm tropical seas (Gray, 1978). (In the western Pacific Ocean, these tropical cyclones are called typhoons). They have a warm core of low pressure associated with the rotation. In the mature stage, surface friction acts on the low-level air causing it to flow toward the center. As this low-level air flows inward toward lower pressure, latent heat energy is picked up from the sea, and is subsequently released by condensation in the updrafts of inwardly spiraling convective bands (Fig. 9.49).

Fig. 9.49a The reflectivity field of Typhoon Alex in a 0.5-km constant altitude plan position indicator at 1004 L.S.T. when it was over the East China Sea and about 70 km north of Taiwan. Peak reflectivity factors of about 40 dBZ are present and the first level of brightness corresponds to a reflectivity factor of 10 dBZ. The reflectivity factor scale is given on the brightness bar. Range to the outer circle is 120 km, and the northern part of Taiwan is outlined. (Courtesy of H.-Y. Tseng, Air Navigation and Weather Services, Taipei, Taiwan.)

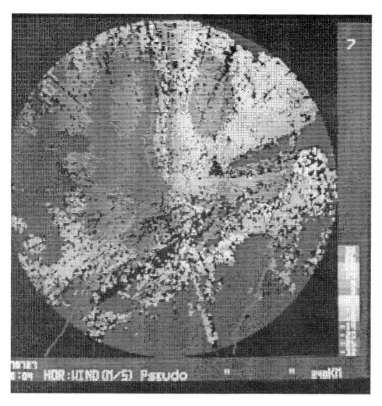

Fig. 9.49b As in (a) except the Doppler velocity field is displayed. The velocity in m s^{-1} is indicated. (Courtesy of H.-Y. Tseng, Air Navigation and Weather Services, Taipei, Taiwan.)

This increase in energy is sufficient to maintain a near-isothermal condition against the adiabatic cooling (3–5°C) induced by the lowering pressure as the air approaches the center (Byers, 1944). The low-level inflow does not penetrate to the center of the mature hurricane, but only to the outer edge of the eye. The eye is a region of low wind velocity generally free of clouds and precipitation. The convergence of low-level air supports vigorous convection in a band (i.e., the *eyewall*) surrounding the eye. Maximum winds are found within this band. The eye is maintained by the subsidence of upper tropospheric air in response to the vertical motion caused by convective heating in the eyewall (Shapiro and Willoughby, 1982). The eye may be from 5 to 50 km wide, and its translational speed (i.e., that of the hurricane) is usually slow (i.e., 5–10 m s^{-1}).

As the hurricane passes over land, it diminishes in intensity, since its source of moisture has been shut off and surface friction increases. Severe weather such as tornadoes, gust fronts, or hail is often a by-product of a hurricane landfall (Novlan and Gray, 1974; McCaul, 1991). The immediate coastal areas are most affected because of the strong winds (from 50 to 70 m s^{-1}), storm surge, and flooding. Because hurricanes have long life (several days or more) and have short-term (several hours) predictable trajectories, there is usually enough time to issue warnings. The large scale of hurricanes causes intense rain to persist for long periods, resulting in large accumulations of rainfall (typically 250 mm) and flooding. Typical rain rates within the largest convective cells are about 75 mm h^{-1} (Jorgensen and Willis, 1982).

Typhoon Alex, which formed over the Philippine Sea, was the first to be observed thoroughly by a ground-based Doppler weather radar over the open sea. A 5.4 cm wavelength Doppler radar, on Chiang Kai-Shek (CKS) airport in Taiwan, was operated in a dual PRF mode ($T_{s2} = 1.11$ ms and $T_{s1} = 0.833$ ms; Section 7.4.3) to give this radar an unambiguous velocity $v_m = \pm 48$ m s^{-1}, and an unambiguous range $r_a \approx 120$ km; this radar also has a non-Doppler mode for which the reflectivity fields can be mapped to ranges of 480 km with its 0.85° beamwidth. The typhoon was detected as it entered into the coastal regions of northeastern Taiwan during the early morning hours on 27 July 1987 (Wang and Tseng, 1991). The reflectivity field of this typhoon (after it passed over the East China Sea) is presented in Fig. 9.49a and the Doppler velocity in Fig. 9.49b. These fields are displayed on constant altitude plan position indicators (CAPPI) for an altitude of 0.5 km and to a range of 120 km. The reflectivity fields show the center of a nearly symmetric circulation about 70 km north of the radar, and the Doppler velocity field confirms this location, but in addition it shows peak tangential velocities of about 26 m s^{-1} in its western sector and about 32 m s^{-1} in its eastern sector. This asymmetry in peak velocities is caused, for the most part, by Alex's northerly speed of propagation (≈ 7 m s^{-1}). Reflectivity is about 25 to 30 dBZ with narrow bands of 35 dBZ principally within 50 km of the typhoon's eye.

A Doppler velocity field of hurricane Bob obtained by the TDWR radar (Section 9.6.3), located in Massachusetts, is displayed in Color Plate 4b. As the hurricane tracked northeastward, the southern branch of the zero isodop sector containing the eye rotated counterclockwise.

9.8.3 Cold Fronts and Dry Lines

A front is a relatively narrow zone of strong temperature gradients (i.e., $\gg 0.01°$/km) separating air masses; the motion of fronts designates them as cold or warm, cold fronts being those that advance cold air. (For more detailed description of fronts see, for example, Bluestein, 1986.) A dry line is simply the boundary between dry and moist air masses; in the central United States, northward-streaming moist air from the Gulf of Mexico is bounded on its western flank by extremely dry air of the high plains (Schaefer, 1986).

There is an obvious convergence (e.g., Color Plate 4c) of air masses as the cold air penetrates the ambient air flowing from the SSW. This convergence along the boundary creates a line of persistent vertical motion (a few meters/second) that can lift scattering aerosols, normally confined to the mixed layer (i.e., the region adjacent to the ground where low-level shear and convection induce turbulence) making them visible as a reflectivity thin line (Fig. 9.50). The convergence is clearly defined in Color Plate 4c because weather signals are returned from scatterers on both sides of the front.

There is no obvious convergence of air masses across the dry line (Color Plate 4c); this is attributed in part to the coarse quantization of the velocity field as well as weak convergence. Vertical motions forming the thin line could be caused by an ageostrophic response to an upper-level disturbance forcing moist air to flow out from beneath a capping inversion that was formed by dry and hot air originating over the Mexican Plateau (e.g., Carlson *et al.*, 1983); however there are other explanations (Parsons *et al.*, 1990). Because fronts can harbor strong gradients of moisture, mixing of the contrasting air masses creates irregularities in refractive index that can also contribute (in addition to particulates) to the thin-line reflectivity along the boundaries (Chapter 11 and Fig. 11.22a). The equivalent reflectivity factors along the thin line in Fig. 9.50 are relatively strong (15 dBZ) which, using Eq. (11.104), corresponds to the unrealistically large

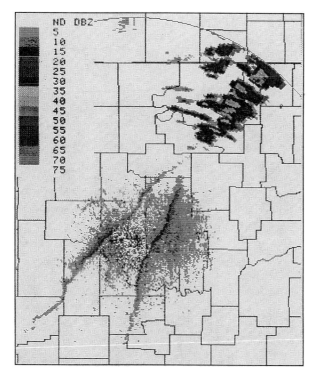

Fig. 9.50 Intersecting reflectivity thin lines in central Oklahoma on 30 April 1991 at 2249 U.T. The thin line, just to the west of the radar, is along a NE–SW oriented cold front, and the thin line immediately to the east is along a NNE–SSW oriented dry line (the Doppler velocity field is presented in Color Plate 4c). The reflectivity factor (dBZ) categories are indicated by the brightness bar. (Courtesy of Steve Smith OSF/NWS.)

refractive-index structure parameter $C_n^2 \approx 10^{-10}$ m$^{-2/3}$; data in Fig. 11.17 suggests that particulate scattering dominates in this case.

The intersection of cold fronts and dry lines is a favored location for the initiation of storms. As the cold front propagated to the southeast, the intersection of it and the dry line progressed southwestward, and storms were initiated behind this intersection.

9.9 Lightning

A lightning event is a propagating discharge characterized by relatively continuous and weak currents (i.e., approximate durations of hundreds of milliseconds and intensities of tens of amperes) and strong current pulses (i.e., durations from a fraction of to tens of microseconds and intensities of tens of kiloamperes). The highly branched lightning channels can be distributed inside huge volumes (e.g., hundreds of cubic kilometers) of a thunderstorm (see Uman, 1987 for a general description of lightning and its measurement). Most discharges propagate entirely within the thunderstorm cloud or between clouds and are called intracloud flashes (IC). Those flashes that have discharges to ground are called cloud-to-ground flashes (CG), which can be composed of several return strokes (R), as shown in Fig. 9.51. Lightning echoes are the result of backscattering from free electrons (see Table 3.1 for the cross section) in the ionized channels. Scientists first discovered echoes from lightning during radar observations of precipitation (e.g., Ligda, 1950).

9.9.1 Physical Characteristics Determined by Radar

The echo properties of a lightning flash include the variability of reflectivity in range and time (Fig. 9.51). It is not always possible to identify flash type by its echo signature, but sometimes variations of reflectivity correspond to characteristic changes in the electrostatic field that are produced by the current pulses (return strokes, R) in the channel to ground.

Well-known radar images of lightning (Ligda, 1956) suggest a tree-like structure. Segments of a lightning channel have been observed to extend as much as 160 km along the beam.

Median values of lightning reflectivity range from about 3×10^{-10} ($\lambda = 70$ cm) to 3×10^{-8} m^{-1} ($\lambda = 5$ cm) (Williams et al., 1989); but there is wide distribution of values at any wavelength. For example, at the weather radar wavelength of 11 cm, 97% of the reflectivities range from 2×10^{-10} to 2×10^{-7} m^{-1} (Williams et al., 1989).

Echoes from lightning last from a few tens to a few hundreds of milliseconds (Zrnić et al., 1982) and have backscattering cross sections that decay at a rate of about 0.2 dB/ms (Mazur et al., 1985). This decay rate is much slower than that of optical emission from lightning channels and can be explained by the presence

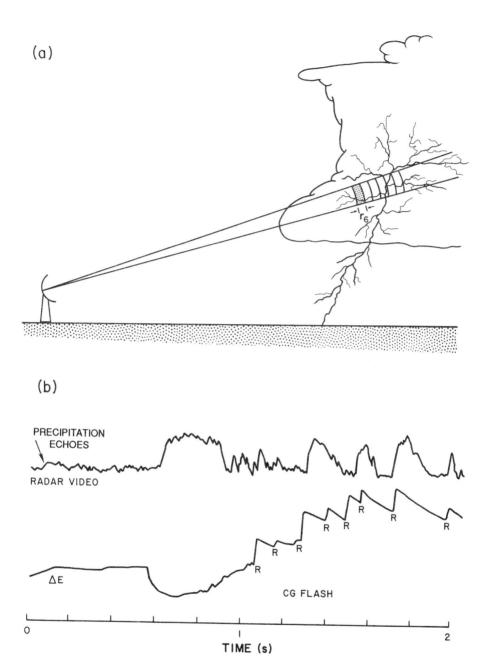

Fig. 9.51 Depiction of a radar technique for observing lightning, and an example of lightning echoes and corresponding electric field changes for a long CG flash. The stippled region in (a) represents one of the radar's resolution volumes. The top trace of (b) is the time history of the intensity of lightning echoes from a resolution volume at a range of 86 km and an altitude of 4 km. Although depicted as a continuous line, the trace is actually a series of points separated in time by the period T_s ($\simeq 1$ ms) between transmitted pulses. Precipitation echoes (0–0.6 s) denote pre-flash intensities associated with hydrometeor backscatter. The bottom trace in (b) is the electrostatic field change ΔE at the radar site produced by the same flash. (From Rust *et al.*, 1981.)

of small amplitude continuing currents that maintain the channel's ionization but do not produce significant luminosity (Holmes *et al.*, 1980; Mazur *et al.*, 1985).

Lightning signals (echoes), like weather signals, exhibit variability from pulse to pulse. These fluctuations can be attributed to interference effects caused by scatter from various segments of differentially moving ionized channels.

A fundamental and long-standing problem in determining the backscattering cross sections of lightning concerns plasma density. Laboratory arcs in air have temperatures of about 60,000 K (Cobine, 1941) and associated electron densities of 10^{19}–10^{20} m^{-3} (Yos, 1963). A steady current as small as 1 A is sufficient to produce this electron density. There is considerable evidence of quasisteady currents as large as 10–100 A during the interstroke period of both natural (Krehbiel *et al.*, 1979) and triggered (Hubert *et al.*, 1984) lightning. During the duration of such currents the plasma is overdense at all wavelengths (i.e., >3 cm) of meteorological radars (Williams *et al.*, 1989).

9.9.2 Lightning and Storm Structure

By fixing the radar beam for a few seconds, and then elevating it, one can obtain the height distribution of lightning flashes (Mazur *et al.*, 1984). Lightning activity, regardless of flash type, is measured by flash density (i.e., the number of flashes per minute per kilometer along the radar beam). Applying this technique, Mazur *et al.* (1986) found the flash density to be bimodal with altitude in storm cells that exhibit vigorous vertical growth, and unimodal otherwise. The lower and upper maxima of flash density are found at about 6 to 8 km, and 13 to 15 km (i.e., near the top of the storm cells).

Using 70 and 10 cm wavelength radars to observe lightning evolution in a squall line, Mazur *et al.* (1986) found that lightning concentrates near the 50-dBZ contour between growing and decaying cells.

An investigation of lightning channels with Doppler radars began only recently (Zrnić *et al.*, 1982; Lhermitte, 1982), and first results have shown that spectral peaks associated with precipitation and lightning are often distinct (Fig. 9.52). Therefore, spectral analysis can increase the dynamic range of lightning echo detection, even at radar wavelengths where the power of precipitation echoes is larger than that of lightning.

Mazur *et al.* (1985) analyzed Doppler spectra of lightning echoes and found that channels can move with accelerations several times that of gravity; the acceleration of a channel can be seen as a shift in the velocity of the spectral peak (dashed line in Fig. 9.52). Such accelerations are attributed to buoyancy and Lorentz forces (i.e., the earth's magnetic field acting on channels having quasicontinuous current).

Zrnić *et al.* (1982) found, as did Szymanski *et al.* (1980), that lightning in intense precipitation usually causes no discernible change in precipitation reflectivity. But Zrnić *et al.* (1982) did find that lightning altered the reflectivity of precipitation in regions of low Z and suggested that this was a result of changes

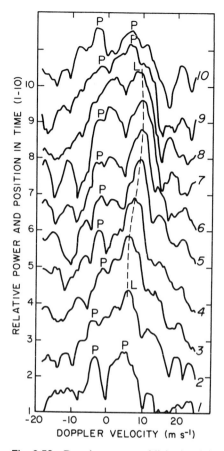

Fig. 9.52 Doppler spectra of lightning (L) and precipitation (P). The solid curves are five-point running averages of spectral powers. Time series, from which the consecutive spectra (1–10) were calculated, overlap by 75 ms and are 25 ms apart. The tic marks on the vertical axis are at the same power level for each spectrum. Intervals between marks are 10 dB. The shift of lightning peak velocity is seen in the curvature of the dashed line.

in hydrometeor orientation brought about by changes in the electrostatic field. These changes occurred within 0.1–0.5 s after the flash began. McCormick and Hendry (1979) give convincing evidence that both aerodynamic forces and the storm's electric field control the orientation of ice crystals.

Problems

9.1 Wind speed is constant with height but wind direction veers. Sketch the pattern of the zero Doppler velocity line that would be seen on a plan position indicator. Use any reasonable assumptions you like in constructing this diagram.

9.2   Given the VAD shown in Fig. P.9.1, taken at 30 km and $\theta_e = 2.0°$. Sketch the true (dealiased) VAD. What is the wind direction and speed?

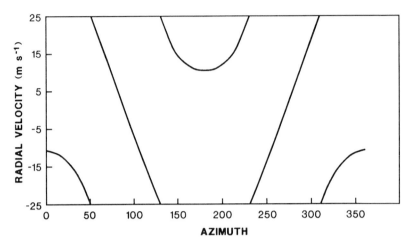

Fig. P.9.1

9.3   A thunderstorm is located at 160 km, azimuth 270°. Based on your observations over the past hour, it is moving eastward at 10 m s^{-1}. A profile of the (radar relative) radial velocity near the storm top for $\phi = 270°$ is in Fig. P.9.2. Assume the elevation angle is small, so radar beam height is constant. (a) Compute the true radial velocities at 150 km and 170 km and sketch the true radial velocity profile. (b) Compute the storm-relative radial velocities at 150 km and 170 km. (c) Make an estimate of the divergence at the top of the storm, assuming the outflow at the storm top to be circular.

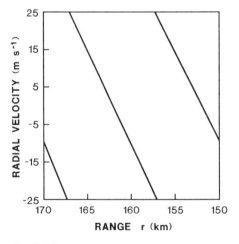

Fig. P.9.2

9.4 A cold front with a 10-km transition zone is 30 km northwest from a WSR-88 radar. The zone is oriented southwest to northeast and within it the winds are linear. On both sides of the zone the wind is uniform; 10 m s^{-1} from $180°$ on the south side and 20 m s^{-1} from $330°$ on the north side. (a) Calculate the horizontal divergence, deformation, and vertical vorticity in the transition zone. (b) Assume that Doppler velocities are measured at 60 km range and zero elevation; plot the Doppler velocity as a function of azimuth (VAD). Calculate the divergence, velocity, and deformation from the VAD. (c) Using information from (a), calculate the average divergence over a VAD circle.

9.5 Suppose that Doppler velocity measurements are made at three heights at a range of 60 km. The fitted harmonics are as follows:

$0 < h < 500$

$$v_r(\phi) = 0.1 + 10 \sin \phi + 5 \sin(2\phi + 10°) \quad \text{m s}^{-1}$$

For $500 \text{ m} < h < 1 \text{ km}$

$$v_r(\phi) = 0.5 + 20 \sin(\phi - 10°) + 7 \sin(2\phi)$$

For $1 \text{ km} < h < 1.5 \text{ km}$

$$v_r(\phi) = 0.02 + 30 \sin(\phi - 20°) + 10 \sin(2\phi - 10°)$$

(a) Calculate the average vertical velocity at the 1.5-km height. (b) What are the magnitudes and directions of the horizontal wind at each height?

9.6 Prove that a VAD analysis cannot distinguish between a uniform wind field and pure vorticity. That is, consider measurement of radial velocity at a range r on a circle (Fig. P.9.3). Assume that a pure vortex (solid body rotation) is centered at a distance $r + d$ from the radar and that its tangential velocity is $v = aR$ where a is a constant of proportionality and R is distance from vortex center. Show that this vortex and a uniform wind will give exactly the same radial velocities $v_r(\phi)$ as a function of azimuth ϕ. Find the magnitude and direction of such a uniform wind.

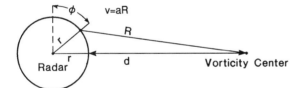

Fig. P.9.3

9.7 A simplified model of the storm updraft consists of a cylindrical column throughout which the vertical velocity w is a constant. The cylinder has a radius r_c. The tropopause deflects the air symmetrically in a layer of thickness h. Assume that the divergent velocity is a function only of distance r from the column center. (a) What is the profile of the velocity in the layer as a function of distance from the cylinder center. (b) Plot the isodop pattern of this divergent velocity field.

9.8 Assume a WSR-88D radar on a flat terrain and that the rms error in the mean Doppler velocities is 1 m s^{-1}. You are required to measure the average convergence on the circle of radius r where r varies from 10 to 100 km. How are you going to make the measurement? Find the error in the measured divergence as a function of distance from the radar. How many scans need to be averaged to reduce the error below 10^{-6} s^{-1}.

9.9 Show that if the wind is linear above a horizontal ground and satisfies the anelastic continuity equation then $w_x = w_y = 0$.

Measurements of Turbulence

It was pointed out in Chapter 5 that mean velocity and spectrum width measured by a Doppler radar are weighted averages of point velocities. Mean velocity measurements are quite adequate to depict motion on scales larger than the resolution volume V_6, and the spectrum width gives an estimate of the variance of motion on scales smaller than V_6; but we cannot infer the details of the flow inside the volume. In some special instances, such as when a vortex is within the resolution volume, certain attributes of motion can be deduced from the Doppler spectra (Section 9.5.3). Nevertheless, Doppler radar offers intriguing possibilities for the measurement and study of turbulence on scales larger and smaller than the resolution volume. The purpose of this chapter is to introduce the basic concepts of turbulence and to establish a firm connection between the physical (statistical) properties of the atmosphere and Doppler-derived measurements. We present relationships between turbulence and the mean Doppler velocity and spectrum width.

10.1 Statistical Theory of Turbulence

Thus far in our analysis of weather signals it has been assumed that statistical fluctuations are a result of the random motion of scatterers in the atmosphere, however, no characteristics of such motion were given. To relate measurements of the Doppler velocity and spectrum width to turbulent flow, it is necessary to introduce elements of the statistical theory of turbulence. Not only does this theory help establish a quantitative relationship between the Doppler spectrum width and turbulence, but it is also very useful for determining the intensity of echoes from fluctuations in the refractive index. Such echoes are treated in Chapter 11.

10.1.1 Turbulence Spectra and the Correlation Function

Properties of turbulence can be described in statistical terms. A random variable that is a function of position $\mathbf{r}$ and time defines a *random field*. If the random variable is a scalar quantity such as the refractive index $n(\mathbf{r}, t)$, the field is called a *scalar field*. When the random variable is a vector, it forms a *vector field*. Thus, the velocities in a turbulent flow form a random vector field such that at any point **r** (and time t) the velocity vector is a random variable **v**(**r**, t).

A random field is statistically homogeneous if its mean value is spatially uniform and if its correlation function does not change when the pair of points **r**, **r**' are both displaced by the same amount in the same direction. Thus, for a homogeneous random field in which the mean value has been removed,

$$\langle v_i(\mathbf{r}) \rangle = 0. \tag{10.1}$$

[In this and the next chapter angle brackets denote ensemble averages, whereas the overbar will signify a weighted spatial average, i.e., Eq. (5.48)]. Similarly the correlation function of two components v_i, v_j depends only on the distance $\boldsymbol{\rho} = \mathbf{r}' - \mathbf{r}$ between the two field points

$$R_{ij}(\mathbf{r}, \mathbf{r}') \equiv \langle v_i(\mathbf{r}) v_j(\mathbf{r}') \rangle = R_{ij}(\boldsymbol{\rho}). \tag{10.2}$$

The index $i = 1$, 2, or 3 identifies the time-dependent velocity component along one of three orthogonal directions.

If the correlation does not depend on the direction of the vector $\boldsymbol{\rho}$, the field is *isotropic*. Even though isotropic fields on scales larger than few hundred meters are an exception in the atmosphere, the mathematical handling of such fields is tractable, and general results are relatively simple to obtain. The concept of isotropy is significantly different for vector fields than for scalar fields, as we now illustrate. Consider the correlation of the velocity at **r** with that at **r**', and suppose the separation vector $\boldsymbol{\rho}$ is allowed to point in any direction from **r**. Construct a coordinate system that is rigidly attached to $\boldsymbol{\rho}$ and that, without loss of generality, has axes parallel and perpendicular to $\boldsymbol{\rho}$ (Fig. 10.1). A vector field is isotropic if the correlations of the various components in separation vector

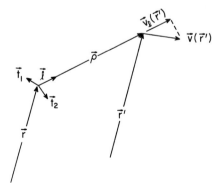

Fig. 10.1   Coordinate system of orthogonal unit vectors l, $\mathbf{t}_1$, $\mathbf{t}_2$ rigidly tied to the separation vector $\boldsymbol{\rho}$. The projection $v_l(\mathbf{r}')$ of the velocity vector **v**(**r**') is onto the coordinate axis l that is parallel to $\boldsymbol{\rho}$. Vectors are indicated by overarrows in figures and by bold letters in the text.

coordinates (e.g., $\langle v_l(\mathbf{r}')v_t(\mathbf{r})\rangle$) are independent of the orientation of $\boldsymbol{\rho}$. Although the component correlations in the coordinate system tied to $\boldsymbol{\rho}$ are independent of the direction of $\boldsymbol{\rho}$, it can be deduced that *the correlation of components in the nonrotated frame does, in general, depend on the direction of* $\boldsymbol{\rho}$. Thus for an isotropic field $\langle u(\mathbf{r})u(\mathbf{r}')\rangle$ does depend on the direction of $\boldsymbol{\rho}$, so u cannot be treated as a scalar, for which correlation is a function only of the magnitude $|\boldsymbol{\rho}|$.

A homogeneous field need not be isotropic. For example, the field having the correlation function

$$R_{ij}(\mathbf{r} - \mathbf{r}') = R_{ij}[\alpha(x - x') + \beta(y - y') + \gamma(z - z')] \tag{10.3}$$

is homogeneous but not isotropic. Although a random vector field can be homogeneous but not isotropic, it can be shown that an isotropic field must be homogeneous

For isotropic random vector fields, the correlation tensor is in general

$$R_{ij}(\mathbf{r}, \mathbf{r} + \boldsymbol{\rho}) = R_{ij}(\boldsymbol{\rho}) = P(\rho)\delta_{ij} + Q(\rho)\rho_i\rho_j/\rho^2 \tag{10.4}$$

(Tatarskii, 1971), which does not depend on the location of $\mathbf{r}$ but does depend on the direction and magnitude of $\boldsymbol{\rho}$, where δ_{ij} is the Kronecker delta function and ρ_i is the projection of $\boldsymbol{\rho}$ on the ith axis. $P(\rho)$ and $Q(\rho)$ can be any functions of the magnitude ρ. If $\boldsymbol{\rho}$ lies along the kth coordinate of the fixed frame, it can be deduced from Eq. (10.4) that $R_{ii} = R_{jj} = P(\rho)$, which is then called the *transverse correlation function* $R_{tt}(\rho)$, and $R_{kk} = P + Q$ is R_{ll}, the *longitudinal correlation*. Thus Eq. (10.4) can be expressed

$$R_{ij}(\boldsymbol{\rho}) = (1/\rho^2)(R_{ll} - R_{tt})\rho_i\rho_j + R_{tt}\delta_{ij}. \tag{10.5}$$

The Fourier transform in vector wave space of the correlation function of a statistically homogeneous field is known as the *spectral density tensor*.

$$\Phi_{ij}(\mathbf{K}) \equiv \frac{1}{(2\pi)^3} \int R_{ij}(\boldsymbol{\rho}) \exp(-j\mathbf{K}\cdot\boldsymbol{\rho}) \, dV_\rho. \tag{10.6}$$

Its inverse retrieves the correlation function

$$R_{ij}(\boldsymbol{\rho}) = \int \Phi_{ij}(\mathbf{K}) \exp(j\mathbf{K}\cdot\boldsymbol{\rho}) \, dV_K, \tag{10.7}$$

where V_ρ and V_K are volumes in separation distance $\boldsymbol{\rho}$ space and wave number[1] $\mathbf{K}$ space, and the integrals are over these volumes. Introducing spherical coordinates and executing the integrations along the angular coordinates for an assumed isotropic scalar field, the transform pair, Eqs (10.6) and (10.7), reduces

---

1. There are two definitions of wave number; in one $K = 1/\Lambda$ (m^{-1}) and the other $K = 2\pi/\Lambda$ $(rad\ m^{-1})$; throughout this book we use the latter.

to the following:
   For scalar fields,

$$\Phi(K) = \frac{1}{2\pi^2} \int_0^\infty \frac{\sin K\rho}{K\rho} R(\rho)\rho^2 \, d\rho, \tag{10.8}$$

$$R(\rho) = 4\pi \int_0^\infty \frac{\sin K\rho}{K\rho} \Phi(K)K^2 \, dK. \tag{10.9}$$

For an isotropic vector field, the spectral density tensor must, like the correlation tensor, be independent of rotation, so it can also be expressed in a form analogous to Eq. (10.4)

$$\Phi_{ij}(\mathbf{K}) = F(K)\delta_{ij} + G(K)K_iK_j/K^2, \tag{10.10a}$$

or to Eq. (10.5)

$$\Phi_{ij}(\mathbf{K}) = (1/K^2)(\Phi_{ll} - \Phi_{tt})K_iK_j + \Phi_{tt}\delta_{ij}, \tag{10.10b}$$

where Φ_{ll} and Φ_{tt} are functions of the magnitude of the wave number and are called the longitudinal and transverse spectral densities. They are related to R_{ll} and R_{tt} through Eqs. (10.5)–(10.7) and (10.10) and are not simply Fourier transform pairs of their corresponding correlation functions [e.g., $\Phi_{ll} \neq \mathcal{F}(R_{ll})$]. Panchev (1971, p. 102) gives the relations between Φ_{ll}, Φ_{tt} and R_{ll}, R_{tt} for isotropic vector fields. These relations show that, for example, Φ_{tt} is a function of both R_{ll} and R_{tt}.

   If the flow is incompressible, the continuity equation puts an additional constraint on the random vector field, resulting in

$$R_{tt} = \frac{1}{2\rho} \frac{d}{d\rho}(\rho^2 R_{ll}). \tag{10.11}$$

The condition of incompressibility results in $\Phi_{ll}(K) \equiv 0$ (Panchev, 1971, p. 108), so that Eq. (10.10) reduces to

$$\Phi_{ij}(\mathbf{K}) = \left(\delta_{ij} - \frac{K_iK_j}{K^2}\right)\frac{E(K)}{4\pi K^2}, \tag{10.12}$$

where

$$E(K) = 4\pi K^2 \Phi_{tt}(K) \tag{10.13}$$

is the single spectral density that characterizes isotropic incompressible turbulent flow. $E(K) \, dK$ is the contribution to the total kinetic energy per unit mass (or one half of the total velocity variance) from wave numbers in the interval K to $K + dK$.

   Measurements of three-dimensional spectra in the past were quite impractical because the sensing instruments, such as towers or even aircraft, could at best estimate the one-dimensional counterpart $S_{ij}(K_l)$, where K_l is the wave number along the direction of the $i = l$ coordinate. In contrast, radars can in a very short time (i.e., a few minutes) obtain a volume of velocity data. Even though the three-dimensional spectrum can be obtained from radar data, it is considerably simpler to compute spectra along lines; such spectra are the variances of velocities per unit wave number (i.e., measure of turbulence intensity). Moreover, one-dimensional spectra provide a convenient visual picture of the wave components along the line. By applying the formula $2\pi\delta(K) = \int \exp(jK\rho) \, d\rho$ to Eq. (10.6), we can derive the following relation:

$$S_{ij}(K_1) \equiv \frac{1}{2\pi} \int_{-\infty}^{\infty} R_{ij}(\rho_1, 0, 0)e^{-jK_1\rho_1} \, d\rho_1 = \int\int_{-\infty}^{\infty} \Phi_{ij}(\mathbf{K}) \, dK_2 \, dK_3, \quad (10.14)$$

where now the separation vector ρ is directed along the coordinate axis for which $i = 1$.

For isotropic scalar fields, the spectral density $\Phi(\mathbf{K})$ is a function of the magnitude of the wave number, so with a change to cylindrical coordinates and after integration the double integral in Eq. (10.14) reduces to

$$S(K_1) = 2\pi \int_0^{\infty} \Phi[(K_1^2 + K'^2)^{1/2}]K' \, dK' = 2\pi \int_{K_1}^{\infty} \Phi(K)K \, dK, \quad (10.15)$$

where $K' = (K_2^2 + K_3^2)^{1/2}$, or

$$\Phi(K) = -\frac{1}{2\pi K} \frac{dS(K)}{dK}. \quad (10.16)$$

For incompressible flow, the one-dimensional longitudinal and transverse spectra of isotropic vector fields are related to $E(K)$ through Eqs. (10.12)–(10.14), as

$$S_l(K_1) = \frac{1}{2} \int_{K_1}^{\infty} \left(1 - \frac{K_1^2}{K^2}\right) \frac{E(K)}{K} \, dK, \quad (10.17)$$

$$S_t(K_1) = \frac{1}{4} \int_{K_1}^{\infty} \left(1 + \frac{K_1^2}{K^2}\right) \frac{E(K)}{K} \, dK, \quad (10.18)$$

where $S_l(K_1) = S_{11}(K_1)$, whereas $S_t(K_1) = S_{22}(K_1) = S_{33}(K_1)$. Note that any wave number K_i can replace K_1 in Eqs. (10.17) and (10.18), in conformity with the fact that isotropic turbulence is independent of direction.

Of considerable value in turbulence studies is the correlation function, given by

$$R(\rho) = \frac{R(0)}{2^{\nu-1}\Gamma(\nu)} \left(\frac{\rho}{\rho_0}\right)^{\nu} K_{\nu}\left(\frac{\rho}{\rho_0}\right) \quad \text{for} \quad \nu > -\tfrac{1}{2}, \quad (10.19)$$

which has the one-dimensional spectrum

$$S(K) = R(0)\Gamma(\nu + \tfrac{1}{2})\rho_0 / \sqrt{\pi}\Gamma(\nu)[1 + (K\rho_0)^2]^{\nu + 1/2}, \quad (10.20a)$$

where $K_\nu(\rho/\rho_0)$ is the Bessel function of the second kind of order ν, and ρ_0 (the correlation length) is the distance beyond which variables can be considered uncorrelated. The Bessel correlation function is particularly attractive because it can, depending on the values selected for ν, ρ_0, fit quite well many experimental data for the correlations of the longitudinal and transverse components. In fact, the value $\nu = 1/3$ gives a spectrum that coincides with one predicted by turbulence theory (Batchelor, 1953, p. 114) for the range of K corresponding to $K\rho_0 \gg 1$ (i.e., the inertial subrange). The three-dimensional spectrum corresponding to the correlation (10.19) of a scalar field is given by

$$\Phi(K) = \frac{\Gamma(\nu + \frac{3}{2})}{\pi\sqrt{\pi}\Gamma(\nu)} \frac{R(0)\rho_0^3}{[1 + (K\rho_0)^2]^{\nu+3/2}}. \tag{10.20b}$$

10.1.2 Structure Functions, Locally Homogeneous Fields

The previous sections have dealt with homogeneous random fields. But quite often the velocity, temperature, pressure, and humidity change gradually in space, and therefore it becomes necessary to decide whether these changes should be regarded as slow changes in the statistical moments (e.g., the mean and correlation) or as large-scale irregularities. It is evident that the mean square value or intensity of the fluctuations and the form of the autocorrelation function and its corresponding power spectrum depend on this decision.

For these situations it has been found convenient (Kolmogorov, 1941) to define increments that are the differences in the random variable $v_i(\mathbf{r})$ taken at two points separated ρ units apart.

$$v_i(\mathbf{r} + \boldsymbol{\rho}) - v_i(\mathbf{r}), \tag{10.21}$$

which for ρ not too large (i.e., small compared to the length of gradual changes) is a random variable that can be considered homogeneous. For example, if a random variable $v_i(\mathbf{r})$ has the spatial distribution like that sketched in Fig. 10.2 and has a mean value that changes with r as indicated by the dashed line, it is not a homogeneous process. However, if the mean value is linear over the range of r

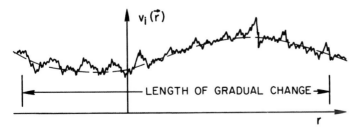

Fig. 10.2 Example of a random variable with a gradual change or trend.

under consideration, it can be shown that the difference (10.21) is a homogeneous random function. Under these conditions $v_i(\mathbf{r})$ is referred to as a *random variable with stationary first increments*. For example, with

$$\langle v_i(\mathbf{r}) \rangle = \mathbf{a} \cdot \mathbf{r} + b \tag{10.22}$$

it can be shown that

$$\langle v_i(\mathbf{r} + \boldsymbol{\rho}) - v_i(\mathbf{r}) \rangle = \mathbf{a} \cdot \boldsymbol{\rho}, \tag{10.23}$$

which is independent of $\mathbf{r}$. If the autocorrelation of Eq. (10.21) is only a function of the difference $\boldsymbol{\rho} = \mathbf{r}' - \mathbf{r}$, then Eq. (10.21) is a statistically (in the wide sense; see Papoulis, 1965, p. 305) homogeneous process.

We can represent the correlation function of the increment as

$$\begin{aligned}
\langle [v_i(\mathbf{r} + \boldsymbol{\rho}) &- v_i(\mathbf{r})][v_i(\mathbf{r}' + \boldsymbol{\rho}) - v_i(\mathbf{r}')] \rangle \\
&= \tfrac{1}{2}\langle [v_i(\mathbf{r} + \boldsymbol{\rho}) - v_i(\mathbf{r}')]^2 \rangle + \tfrac{1}{2}\langle [v_i(\mathbf{r}) - v_i(\mathbf{r}' + \boldsymbol{\rho})]^2 \rangle \\
&\quad - \tfrac{1}{2}\langle [v_i(\mathbf{r} + \boldsymbol{\rho}) - v_i(\mathbf{r}' + \boldsymbol{\rho})]^2 \rangle - \tfrac{1}{2}\langle [v_i(\mathbf{r}) - v_i(\mathbf{r}')]^2 \rangle.
\end{aligned} \tag{10.24}$$

The function

$$D_{ii}(\mathbf{r}_m, \mathbf{r}_n) \equiv \langle [v_i(\mathbf{r}_m) - v_i(\mathbf{r}_n)]^2 \rangle \tag{10.25}$$

that forms each term of Eq. (10.24) is called the *structure function* of the random process. If Eq. (10.24) depends only on the difference $\mathbf{r}_m - \mathbf{r}_n$, the random variable $v_i(\mathbf{r})$ is then said to form a *locally homogeneous process*. The structure function

$$D_{ii}(\boldsymbol{\rho}) = \langle [v_i(\mathbf{r} + \boldsymbol{\rho}) - v_i(\mathbf{r})]^2 \tag{10.26}$$

is the *basic characteristic* of a random process having stationary first increments. From Eq. (10.26) we see that $D_{ii}(\boldsymbol{\rho})$ is proportional to the intensity of the fluctuations with spatial length $\boldsymbol{\rho}$, small compared to the length of a gradual change (Fig. 10.2). If the random process under consideration has structure function (10.26) independent of $\mathbf{r}$ for the range $\boldsymbol{\rho}$ under consideration, then $v_i(\mathbf{r})$ is said to be locally homogeneous.

For a completely homogeneous random process (i.e., one without large-scale changes), there is a simple direct relation between the structure function $D_{ii}(\boldsymbol{\rho})$ and the correlation function $R_{ii}(\boldsymbol{\rho})$. This is made evident by expanding Eq. (10.26) to obtain

$$D_{ii}(\boldsymbol{\rho}) = 2[R_{ii}(0) - R_{ii}(\boldsymbol{\rho})]. \tag{10.27}$$

Because v_i is assumed to have zero mean so that $R_{ii}(\infty) = 0$ [see Eq. (5.15)], it follows from Eq. (10.27) that we can express the auto correlation function in terms of the structure function.

$$R_{ii}(\boldsymbol{\rho}) = \tfrac{1}{2}[D_{ii}(\infty) - D_{ii}(\boldsymbol{\rho})]. \tag{10.28}$$

The structure function is widely used in turbulence studies because large-scale irregularities are neither homogeneous nor isotropic. Thus, in beginning a study of a random process that we are not sure is completely homogeneous, it is prudent to construct the structure function and to test its statistics rather than to use directly the correlation function.

10.1.3   Structure and Spectral Functions, Locally Isotropic Fields

In this section the general relationships between the structure function and the spectra of the correlation function for locally isotropic scalar fields and those between the longitudinal and transverse components of vector fields are discussed. Formal relationships among the structure function, the correlation function, and their spectra developed in previous sections are used to derive the spectral densities. Often in applications of the statistical theory of turbulence that are relevant to the atmosphere, the structure function corresponding to the correlation function given by Eq. (10.19) is useful. Applying Eq. (10.27), we obtain

$$D_{ii}(\rho) = 2R_{ii}(0)\left[1 - \frac{1}{2^{\nu-1}\Gamma(\nu)}\left(\frac{\rho}{\rho_0}\right)^{\nu}K_{\nu}\left(\frac{\rho}{\rho_0}\right)\right], \qquad (10.29)$$

where now the indices ii must denote either the transverse or the longitudinal component with respect to the separation vector $\boldsymbol{\rho}$. For small values of $\rho \ll \rho_0$ and $|\nu| < 1$, Eq. (10.29) reduces to

$$D_{ii}(\rho) \approx 2R_{ii}(0)[\Gamma(1-\nu)/\Gamma(1+\nu)](\rho/2\rho_0)^{2\nu}. \qquad (10.30)$$

With $\nu = 1/3$ this becomes

$$D_{ii}(\rho) = C_{ii}^2\rho^{2/3}, \qquad (10.31)$$

where

$$C_{ii}^2 = [3 \times 2^{1/3}\Gamma(\tfrac{2}{3})/\Gamma(\tfrac{1}{3})]R_{ii}(0)\rho_0^{-2/3}. \qquad (10.32)$$

A relationship exactly analogous to Eq. (10.11) ties D_{tt} to D_{ll}. From this relationship it follows that $C_{tt}^2 = \frac{4}{3}C_{ll}^2$. That is, for isotropic turbulence the structure parameter C_{tt}^2 of the transverse velocity component is larger than the parameter C_{ll}^2 for the longitudinal component. Furthermore, because $R_{ll}(0) = R_{tt}(0)$, the ρ_0 for the longitudinal structure function differs from the one for the transverse.

Equation (10.31) was derived by Kolmogorov, who used dimensional analysis. Briefly, he hypothesized that there is a range of eddy sizes in which there is no creation or dissipation of energy. Only a cascade (flow) of energy from larger eddies to smaller ones occurs as they fragment. In equilibrium this continuous flux of energy numerically equals the energy dissipation rate ε due to viscosity at very small sizes. Dimensional analysis then reveals that for the longitudinal structure function $D_{ll}(\rho)$ to depend only on ε it must have the form (Tatarskii, 1971, p. 54).

$$D_{ll}(\rho) = C^2\varepsilon^{2/3}\rho^{2/3}, \qquad (10.33)$$

where C^2 is a dimensional constant with a value of about 4. The sizes ρ for which Eq. (10.33) holds satisfy.

$$\rho_i < \rho < \rho_0, \tag{10.34}$$

where ρ_0 is the outer scale and ρ_i is the inner scale of turbulence. It can be shown (Tatarskii, 1971) from turbulence theory that for $\rho < \rho_i$

$$D_{ll}(\rho) = \tfrac{1}{15}(\varepsilon/\nu)\rho^2. \tag{10.35}$$

The range given by Eq. (10.34) for which Eq. (10.33) is applicable is known as the *inertial subrange*. The region for which $\rho < \rho_i$ is known as the *dissipation range*. Note that because of Eq. (10.27) the correlation function follows a power of two-thirds law if turbulence is completely homogeneous.

The power spectrum for the inertial subrange, where $K\rho_0 \gg 1$, is obtained by substituting Eq. (10.32) in Eq. (10.20a).

$$S_i(K) = C_{ii}^2 \Gamma(\tfrac{5}{6}) K^{-5/3}/3 \times 2^{1/3} \Gamma(\tfrac{2}{3})\sqrt{\pi}. \tag{10.36}$$

Thus, the one-dimensional two-sided spectra of the velocity components parallel to and perpendicular to the separation vector ρ have a $K^{-5/3}$ dependence. C_{ii}^2 is a measure of the intensity of the fluctuations having scale sizes within the inertial subrange.

The general form of the correlation in the inertial subrange can be written

$$R_{ii}(\rho, \tau_1 = 0) = R_{ii}(0)[1 - (\rho/\rho_0)^{2/3}], \tag{10.37}$$

where $\rho \ll \rho_0$ and the time lag τ_1 is shown explicitly in Eq. (10.37) to enunciate a possible time dependence, which will be treated shortly. This well-known relationship assumes steady-state conditions. Equation (10.37) is valid for isotropic turbulence in the inertial subrange, where viscous forces are negligible and energy cascades from large-scale eddies toward dissipation by viscosity at small scales (Tennekes and Lumley, 1972). In the lower atmosphere the inertial subrange consists of eddies ranging in size from centimeters to several tens of meters. Nevertheless, the correlation function of the longitudinal and transverse components in planes parallel to the earth's surface sometimes exhibits the two-thirds behavior to much larger scales (Vinnichenko and Dutton, 1969; Doviak and Berger, 1980).

For isotropic scalar fields having a correlation of the form (10.19), the spectrum of variance can be found by substituting Eq. (10.32) into Eq. (10.20b).

$$\Phi_n(K) = \frac{\Gamma(\tfrac{11}{6}) C_n^2 K^{-11/3}}{\pi^{3/2} 3 \times 2^{1/3} \Gamma(\tfrac{2}{3})} = 0.033 C_n^2 K^{-11/3}, \tag{10.38}$$

where the subscript ii has been replaced by n to represent the scalar field of, for example, the refractive index.

10.1.4 Chandrasekhar's Theory

Section 10.1.3 gave the spatial behavior of the correlation function and the spectra for isotropic turbulence. The verified theory predicts a power of two-thirds law for spatial separations corresponding to scales in the inertial subrange but makes no prediction about the temporal dependence. A completely satisfactory and accepted theory of the joint temporal and spatial laws that the correlation function should follow does not exist.

Chandrasekhar (1955, 1956) has derived the partial differential equation

$$\frac{\partial^3 R_{ll}}{\partial^2 \tau_n \partial \rho_n} = R_{ll}^\nu \frac{\partial}{\partial \rho_n} (D_5 R_{ll}), \tag{10.39}$$

which the longitudinal correlation function must satisfy (for zero viscosity). It describes the space–time velocity correlation function $R_{ll}(\rho_n, \tau_n)$ for isotropic turbulence. D_5 is the Lagrangian operator $\partial^2/\partial^2 \rho_n + (4/\rho_n)(\partial/\partial \rho_n)$, and the parameters $\rho_n = \rho/\rho_0$ and $\tau_n = \tau_1 R_{ll}^{1/2}(0)/\rho_0$ are normalized space and time coordinates, respectively. The parameter ρ_0 represents a characteristic length associated with the largest eddies present in the medium. Similarly, it is possible to define a parameter τ_0 to represent the characteristic lifetime of the eddies. In addition to assuming an isotropic, nonviscous, and incompressible medium in deriving Eq. (10.39), Chandrasekhar (1955, 1956) assumed that various fourth-order moments can be expressed as sums of products of second-order moments; he then produced the following solution to Eq. (10.39):

$$R_{ll}(\rho_n, \tau_n) = [1 - \rho_n^\nu \psi(\tau_n/\rho_n)]R_{ll}(0,0) \tag{10.40}$$

applicable when changes in $R_{ll}(\rho_n, \tau_n)/R_{ll}(0,0)$ are small compared to unity and $\tau_n \ll 1$. In Eq. (10.40),

$$\psi(x) = \frac{1}{2(\nu + 2)} [|x - 1|^{\nu+2}(\nu + 2 + x) + |x + 1|^{\nu+2}(\nu + 2 - x)], \tag{10.41}$$

where ν is a constant to be assigned a particular value and $x = \tau_n/\rho_n$.

There are two limiting cases of Eq. (10.41)—one in which $\tau_n \to 0$ when $\rho_n \neq 0$; the other such that $\tau_n \neq 0$ when $\rho_n \to 0$. These cases are

$$R_{ll}(\rho_n, 0) = (1 - \rho_n^\nu)R_{ll}(0,0), \qquad \tau_n \to 0, \tag{10.42}$$

and

$$R_{ll}(0, \tau_n) = [1 - \tfrac{1}{3}(\nu + 1)(\nu + 3)\tau_n^\nu]R_{ll}(0,0), \qquad \rho_n \to 0, \tag{10.43}$$

respectively.

With the presence of uncorrelated measurement errors, the correlation functions will contain both a signal and a noise portion at $\rho_n = 0$, $\tau_n = 0$. Then $R_{ll}(0,0)$ must be computed from

$$\sigma^2 = R_{ll}(0,0) + \sigma_e^2, \tag{10.44}$$

where σ^2 is the measured variance and σ_c^2 is the variance due to measurement error.

Smythe and Zrnić (1983) report on the spatial and temporal behavior of the autocorrelation function. They give (Fig. 10.3) the spatial dependence of the correlation coefficients for velocity and reflectivity fields. The data used to generate Fig. 10.3 were obtained in clear air, and each measured point is an average obtained from more than 20 estimates. A single autocorrelation estimate is calculated by lagging an array (9 km × 12°) of radial velocities (or reflectivities) on itself. The azimuthal spacing was 0.5° and range gate separation was 150 m, so the array contained 60 × 25 velocity or reflectivity estimates. The lagging was along radials in regions where radials were oriented parallel to the mean wind. The orientation of observed horizontal rolls was parallel to the mean wind on that day, so contamination by azimuthally directed wavelike components associated with the rolls was significantly reduced. The least squares fitted curve [Eq. (10.42) for $\nu = 2/3$] to the velocity autocorrelation matches the data well.

Most striking in comparing the two correlation functions is the considerably higher value of the autocorrelation for the velocity field. This implies that characteristic scale sizes are larger in the velocity field than in the reflectivity

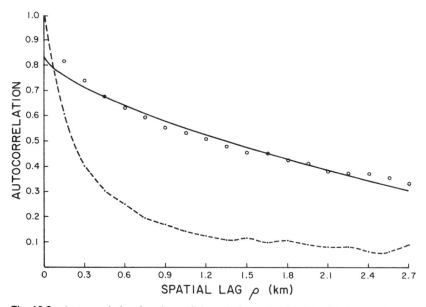

Fig. 10.3 Autocorrelation functions of the velocity and reflectivity (dashed line) obtained from single-radar data in the planetary boundary layer. Velocity data (filled circles) are least squares fitted with a two-thirds power law (solid line). (27 April 1977 from 1427:07 to 1427:31.) (From Smythe and Zrnić, 1983.)

field. A possible explanation for the difference in scale sizes may be found in the mechanism generating these eddies. The reflectivity fluctuations may be primarily due to plumes and convective bubbles of small scale emanating from the solar-heated earth's surface, whereas velocity perturbations might be caused by additional larger scale effects.

The temporal behavior of the correlation functions (Fig. 10.4) is obtained from the maxima of cross correlation between arrays (6 km $\times$ 8°) of velocities at two different times. Data points in Fig. 10.4 are averages of such correlation maxima. Note again that the velocity field has higher correlation values. A least squares fit of velocity correlation [Eq. (10.43)] for $\nu = 2/3$ is made from which the correlation time $\tau_0 = (27/55)^{3/2} \rho_0 / R_{ll}^{1/2}(0)$ is found. We have no firm theoretical basis to choose a two-thirds power law time dependence, but the differential equation (10.39), which describes the time dependene of the velocity correlation, does admit a solution of the selected form (10.43) at small time lags.

Reflectivity was observed to lose correlation in space much more rapidly than radial velocity. As a consequence, the average correlation lengths have a ratio of 0.176. If scale sizes are defined by the length at which the correlation

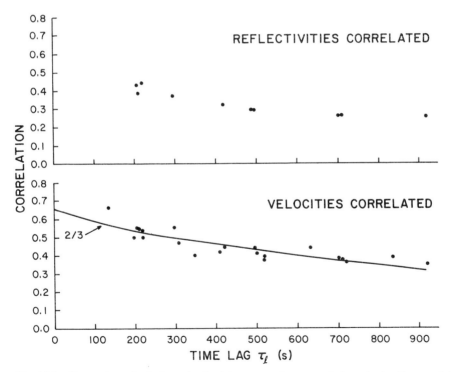

Fig. 10.4 Comparison of velocity and reflectivity temporal autocorrelations obtained by correlating 6 km $\times$ 8° arrays.

coefficient first drops to one-half of its extrapolated value at zero, then the velocity scales are about 2 km versus 300 m for reflectivity. Corresponding lifetimes are about 900 s for velocities and less than 200 s for reflectivities.

10.2 Spatial Spectra of Point and Average Velocities

10.2.1 Filtering by the Weighting Function

In radar measurements the spatial spectra of turbulence are filtered by the weighting function (5.40), and this produces attenuation in the observed turbulence intensity. To find quantitative relationships, we follow Srivastava and Atlas (1974), who use Eq. (5.49) as a starting point and assume that the normalized weighting function $I_n(\mathbf{r}, \mathbf{r}_1)$ [Eq. (5.46b)], depends only on $\mathbf{r} - \mathbf{r}_1$. With this, Eq. (5.49) becomes

$$\bar{v}(\mathbf{r}) = \int v(\mathbf{r}_1) I_n(\mathbf{r} - \mathbf{r}_1) \, dV_1 . \tag{10.45}$$

It is assumed that the reflectivity $\eta(\mathbf{r}_1)$ is constant. Because Eq. (10.45) is a convolution product, the spatial spectrum $\Phi_{\bar{v}}(\mathbf{K})$ of averaged radial velocities equals

$$\Phi_{\bar{v}}(\mathbf{K}) = (2\pi)^6 \Phi_v(\mathbf{K}) |F(\mathbf{K})|^2, \tag{10.46}$$

where $F(\mathbf{K})$ is the Fourier transform of the weighting function I_n. A three-dimensional Gaussian shape for $|F(\mathbf{K})|^2$ is a good approximation to many weighting functions. Thus we shall use

$$|F(\mathbf{K})|^2 = (2\pi)^{-6} \exp[-(K_2^2 + K_3^2) r^2 \sigma_\theta^2 - K_1^2 \sigma_r^2]. \tag{10.47}$$

Equation (10.47) assumes that the wavenumber $\mathbf{K}_1$ is along the beam axis, σ_θ^2 is the second central moment of the two-way antenna pattern Eq. (5.75), and σ_r^2 is the second central moment of the range weighting function $|W(r)|^2$. For a rectangular transmitted pulse and Gaussian receiver frequency response under matched conditions, the relationship between pulse duration and σ_r is given by Eq. (5.76).

The spectrum $\Phi_v(\mathbf{K})$ of point radial velocities can be related to the spectral density tensor Φ_{ii}. Assume that the size of the resolution volume is small compared to the range, so that radial velocities can be considered parallel to $\mathbf{K}_1$ everywhere within the resolution volume. Then

$$\Phi_v(\mathbf{K}) \equiv \Phi_{ll}(\mathbf{K}). \tag{10.48}$$

The combination of Eqs. (10.12), (10.17), or (10.18) and (10.46) produces the measured (filtered by the weighting function) spectra S_l^f and S_t^f of the longitudinal and transverse velocities.

$$S_l^f(K_1) = \frac{(2\pi)^6}{2} \int_{K_1}^{\infty} \left(1 - \frac{K_1^2}{K^2}\right) \frac{E(K)}{K} |F(K_1, K')|^2 \, dK, \qquad (10.49)$$

$$S_t^f(K_1) = \frac{(2\pi)^6}{4} \int_{K_1}^{\infty} \left(1 + \frac{K_1^2}{K^2}\right) \frac{E(K)}{K} |F(K_1, K')|^2 \, dK, \qquad (10.50)$$

where we have explicitly separated K_1 from $K' = (K^2 - K_1^2)^{1/2}$ in $F(K_1, K')$. This dependence on only two wave numbers is caused by the circularly symmetrical antenna beam.

The spatial spectra (10.49) and (10.50) may be obtained from single or dual Doppler measurements. To help visualize the analysis for a single radar, Fig. 10.5 is presented. Mean radial velocities are obtained along and perpendicular to the mean wind, so that subsequent spectral analysis of data along these two directions yields $S_l^f(K_1)$ and $S_l^f(K_2)$. A third spectral component $S_t^f(K_1)$ can be computed from any one of the range locations along radials perpendicular to the mean wind, assuming turbulent eddies are advected by the mean wind (i.e., Taylor's hypothesis); it suffices to record the temporal change of the velocity, transform time to space, and perform spectral analysis. Using Eq. (10.17) or (10.18), the energy spectrum $E(K)$ of isotropic turbulence can be obtained from either of the two one-dimensional measured spectra for scale sizes whose observed variance is not strongly attenuated by the weighting function.

Filtering effects on one-dimensional spectra of isotropic scalar fields smoothed by a spherically symmetric three-dimensional weighting function can be assessed by expressing the filtered $S^f(K_1)$ and unfiltered $S(K_1)$ one-dimensional spectra in terms of their three-dimensional counterparts, as in

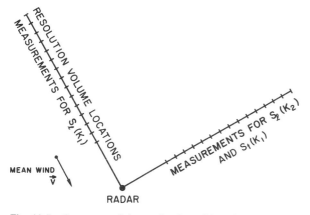

Fig. 10.5 Geometry of data collection with a single Doppler radar in order to compute $S_l(K_1)$, $S_l(K_2)$, and $S_t(K_1)$.

Eq. (10.16). Substitution of Eq. (10.46) and inversion then produces

$$S(K_1) = -\frac{1}{(2\pi)^6} \int_{K_1}^{\infty} \frac{1}{|F(K)|^2} \frac{dS^f(K)}{dK} dK. \qquad (10.51)$$

Whereas the relationship between filtered and point three-dimensional spectra involves a simple multiplication with the transfer function [Eq. (10.46)], the corresponding one-dimensional spectra have a more complicated functional dependence, Eqs. (10.49)–(10.51), which must be evaluated for each case if one wants to determine the effects of the filter.

A comprehensive illustration of the effects of filtering on the longitudinal and transverse one-dimensional spectra of turbulence is given by Srivastava and Atlas (1974). They consider a Kolmogorov–Obukhov spectrum with cutoff at wave number K_0.

$$E(K) = \begin{cases} \pi K^{-5/3} & \text{for} \quad K > K_0, \\ 0 & \text{for} \quad K < K_0. \end{cases} \qquad (10.52)$$

Then, from Eq. (10.17), they obtain the one-dimensional spectra of point velocities.

$$S_l(K_1) = \begin{cases} \frac{9}{55}\pi K_1^{-5/3} & \text{for} \quad K_1 > K_0, \\ 0.3\pi K_0^{-5/3}[1 - \frac{5}{11}(K_1/K_0)^2] & \text{for} \quad K_1 < K_0. \end{cases} \qquad (10.53)$$

Even though the three-dimensional spectrum $E(K)$ has no energy for $K < K_0$, the one-dimensional spectrum does have values for $K_1 < K_0$. The energy at wave numbers smaller than the cutoff ($K_1 < K_0$) is contributed by scales (K_1, K_2, K_3) such that $K^2 > K_0^2$; that is, by wave numbers larger than K_0 whose projection along the K_1 axis is smaller than K_0. (This is also why dashed curves in Fig. 10.6 with the larger cutoff wave number have less power than the solid curves at long wavelengths.)

Filtered spectra are obtained from Eq. (10.49) after Eq. (10.47) is substituted.

$$S_l^f(K_1) = \tfrac{1}{2}\exp[-K_1^2(\sigma_r^2 - r^2\sigma_\theta^2)] \int_a^{\infty} \left(1 - \frac{K_1^2}{K^2}\right) \frac{E(K)}{K} \exp(-K^2 r^2 \sigma_\theta^2) \, dK, \quad (10.54)$$

where $a = K_1$ for $K_1 > K_0$ and $a = K_0$ for $K_1 < K_0$.

Observe that the filtered one-dimensional spectrum is attenuated in a direction along which there may have been no filtering; in other words, even when $\sigma_r = 0$ in Eq. (10.54) there is degradation in the K_1 direction. This is due to filtering in the orthogonal (K_2, K_3) directions and the superposition (integration) of the resulting attenuated three-dimensional spectral density.

A plot of $S_l^f(K_1)$, from Srivastava and Atlas (Fig. 10.6), illustrates the filtering effect. Equation (10.54) is graphed with $\sigma_r = 0$ for various $r\sigma_\theta$ and for $K_0 = \pi$ and 2π. Results for other σ_r are obtained by multiplying the curves by the

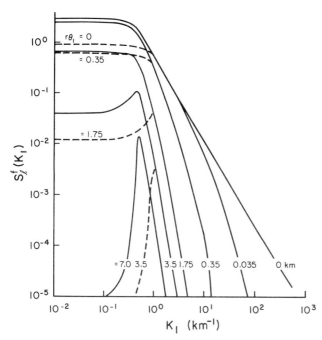

Fig. 10.6   Filtered longitudinal one-dimensional spectrum function. Equation (5.67) relates θ_1 to σ_θ. ———, $K_0 = \pi$ rad km^{-1}, ---, $K_0 = 2\pi$ rad km^{-1}; $\sigma_r = 0$. (From Srivastava and Atlas, 1974.)

Gaussian function $\exp(-K_1^2\sigma_r^2)$. Unfiltered spectra, from Eq. (10.53), are curves with $r\theta_1 = 0$ obeying the power of five-thirds law for $K_1 > K_0$.

An illustration of the one-dimensional spatial spectrum calculated from dual Doppler synthesized wind fields in the planetary boundary layer (Doviak and Berger, 1980) is shown in Fig. 10.7, together with the spectrum calculated from tower measurements. The only filter—other, of course, than the weighting function—acting on the velocity field is the Cressman interpolation filter (Section 9.2.1).

The S_l for the radar data in Fig. 10.7 is an average of 32 individual spectra obtained by a discrete Fourier transform of velocities parallel to the mean wind direction. Smoothed tower spectra were computed by weighting the auto-covariences with an 85-lag (850 s) Tukey window (Jenkins and Watts, 1968), and space-to-time conversion (Taylor hypothesis) was used to compare the scales observed by radar and tower. We note in these data that the $-5/3$ slope extends to long wavelengths well beyond those expected for the inertial subrange.

Plots of the tower spectra have shapes similar to the spectra of wind synthesized from radar data. Also, spectral densities and the rms of the horizontal wind from both tower and radar data are comparable: $\sqrt{R(0)}$ (tower) $= 2.1$ m s^{-1}, and $\sqrt{R(0)}$ (radar) $= 1.7$ m s^{-1}. The lower variance observed

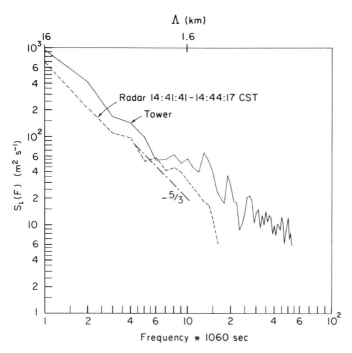

Fig. 10.7 Longitudinal spectra from KTVY TV tower and radar data. An advection speed of 15 m s^{-1} gives the wavelength scale on the upper part of the plot. Level 7 (444 m); 1333:00–1555:40 CST.

with the radar is partly due to interpolation filtering. The unfiltered spectra for the radar observations have more variance density at all wavelengths; however, the increase will be largest at shorter wavelengths, where the attenuation is strongest, and hence there would be even better agreement between radar and tower spectra. An example of how to account for the filtering effects of the radar's weighting function and the interpolation filters is presented by Doviak and Berger (1980).

First measurements of turbulent velocity spectra were reported by Lhermitte (1968), who estimated the up- and downwind spectra from mean Doppler velocities along a radial aligned with the wind. His results in a snowstorm showed good agreement with the five-thirds power law. The up- and downwind spectra were quite similar, suggesting that turbulence was homogeneous.

Chernikov *et al.* (1969) have compared spatial spectra obtained by a Doppler radar to spectra measured *in situ* by an aircraft. The experiment consisted of a Doppler radar, instrumented aircraft, and tethered balloon for chaff release. The radar antenna beam was directed along the mean wind, and time variations of the radial velocity in a single resolution volume were recorded. Longitudinal spectra in the frequency domain were converted to the spatial domain to com-

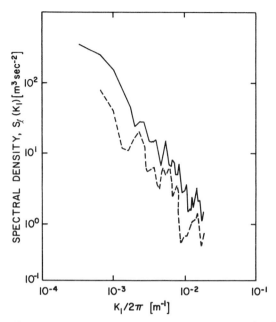

Fig. 10.8   An example of the comparison of radar (---) and aircraft-obtained (———) spectra. (The radar spectrum is plotted 4 dB below its real position.) (Adapted from Chernikov *et al.*, 1969, © American Geophysical Union.)

pare the results with those obtained by an aircraft that flew up- and downwind collecting spatial spectra. The authors report good agreement in general, with resemblance even between small details (Fig. 10.8).

10.2.2   Variance of Point and Average Velocities

In this section we define the relationships between the variance of velocities at a point and the spectrum width (Rogers and Tripp, 1964). Let the variance of the velocity v at a point be σ_p^2. It is obtained from the ensemble average

$$\sigma_p^2 = \langle v^2 \rangle - \langle v \rangle^2. \tag{10.55}$$

The Doppler spectrum width σ_v is given by Eq. (5.51),

$$\sigma_v^2 = \overline{v^2} - (\overline{v})^2, \tag{10.56}$$

where the overbar denotes a spatial average of velocities. The variance of the mean Doppler velocity $\overline{v}$ is, by definition,

$$\sigma_{\overline{v}}^2 \equiv \langle (\overline{v})^2 \rangle - \langle \overline{v} \rangle^2. \tag{10.57}$$

Note that $\sigma_{\bar{v}}^2$ does not include the variance associated with the statistical uncertainty of the estimate $\bar{v}$. Assuming turbulence to be locally homogeneous, the weighting functions symmetrical, and recognizing that ensemble and spatial averages commute, we can rewrite Eq. (10.57) as

$$\sigma_{\bar{v}}^2 = \langle (\bar{v})^2 \rangle - \langle v \rangle^2. \tag{10.58}$$

Finally, the ensemble average of Eq. (10.56) added to Eq. (10.58), after commuting ensemble and spatial averages, produces

$$\sigma_{\mathrm{p}}^2 = \langle \sigma_v^2 \rangle + \sigma_{\bar{v}}^2, \tag{10.59}$$

which indicates that the variance of the velocity at a point is equal to the sum of the ensemble average of the square of the Doppler spectrum width and the variance of the velocities spatially weighted by $I_n(\mathbf{r}_0, \mathbf{r})\eta(\mathbf{r})$.

This very general result is independent of the weighting function but requires turbulence to be locally homogeneous, although not isotropic. In principle, statistical properties of an ensemble can be obtained by repeating the measurement under the same circumstances, but this is practical only if the phenomena are stationary for the duration of the experiments. Another approach to estimate the ensemble average of a statistically homogeneous media is to take measurements at points spaced far enough apart that samples are uncorrelated.

In addition to being proportional to the turbulent kinetic energy, the two variances $\langle \sigma_v^2 \rangle$ and $\sigma_{\bar{v}}^2$ have relative magnitudes that describe how the kinetic energy is partitioned between subresolution volume scales and scales larger than the resolution volume.

10.2.3 Turbulence Parameters from a Single Radar

A variation of a VAD technique (Section 9.3.3) can be applied to turbulent velocities to deduce some characteristics if turbulence is horizontally homogeneous. We shall examine two regimes of turbulent eddy scales: (1) scales large compared to the resolution volume V_6 and (2) scales small compared to V_6.

10.2.3.1 Large-Scale Eddies

Equation (9.20) also applies to the weighted radial velocities Eq. (5.48). The variance of the weighted velocities is obtained directly from Eq. (9.20):

$$\begin{aligned}
\mathrm{var}(\bar{v}_r) = {}& \sigma_{\bar{u}}^2 \cos^2 \theta_e \sin^2 \phi + \sigma_{\bar{v}}^2 \cos^2 \theta_e \cos^2 \phi + \sigma_{\bar{w}}^2 \sin^2 \theta_e \\
& + \mathrm{cov}(\bar{u}\bar{v}) \cos^2 \theta_e \sin 2\phi + \mathrm{cov}(\bar{v}\bar{w}) \sin 2\theta_e \cos \phi \\
& + \mathrm{cov}(\bar{u}\bar{w}) \sin 2\theta_e \sin \phi, \tag{10.60}
\end{aligned}$$

where

$$\sigma_{\bar{u}}^2 \equiv \langle (\bar{u})^2 \rangle - \langle \bar{u} \rangle^2,$$
$$\mathrm{cov}(\bar{u}\bar{v}) \equiv \langle \bar{u}\bar{v} \rangle - \langle \bar{u} \rangle \langle \bar{v} \rangle$$

are typical forms of the variances and covariances. When turbulence is horizontally homogeneous such that the variances and covariances are independent of ϕ, the various terms of Eq. (10.60) can be estimated using the techniques outlined in Section 9.3. If turbulence is superimposed on a linear wind field, the radial velocity of the linear wind needs to be subtracted from the estimate $\hat{v}_r$ at each point. The residuals then would have a variance with angular dependence given by Eq. (10.60). Before the residuals can be fitted by an assumed homogeneous turbulent field, the variance in the estimates due to statistical uncertainty (Section 6.5) must be removed, as must velocity perturbations due to waves (Section 9.7.1).

Lhermitte (1969) first suggested the difference in $\text{var}(\hat{v}_r)$ at $\phi = 0$ and π could be used to estimate directly $\text{cov}(\overline{v}\overline{w})$, which is related to shearing stress.

10.2.3.2 Eddies Smaller Than the Resolution Volume

Taking the spectrum width expression (5.50) and inserting Eq. (9.20) one obtains an expression for the spectrum width σ_v exactly analogous to Eq. (10.60).

$$\sigma_v^2 = \sigma_{u'}^2 \cos^2 \theta_e \sin^2 \phi + \sigma_{v'}^2 \cos^2 \theta_e \cos^2 \phi + \sigma_{w'}^2 \sin^2 \theta_e$$
$$+ \text{cov}(u'v') \cos^2 \theta_e \sin 2\phi + \text{cov}(v'w') \sin 2\theta_e \cos \phi$$
$$+ \text{cov}(u'w') \sin 2\theta_e \sin \phi, \tag{10.61}$$

which is in terms of variances and covariances on scales small compared to V_6 and where, for example, $\sigma_{u'}^2 = \overline{(u - \bar{u})^2}$ and $\text{cov}(u'v') = \overline{(u - \bar{u})(v - \bar{v})}$. If turbulence is homogeneous, the various terms in Eq. (10.61), as in Eq. (10.60), can be determined by applying the techniques outlined in Section 9.3. A more instructive form of Eq. (10.61) is

$$\sigma_v^2 = \tfrac{1}{2}(\sigma_{u'}^2 + \sigma_{v'}^2) \cos^2 \theta_e + \sigma_{w'}^2 \sin^2 \theta_e + \text{cov}(u'v') \cos^2 \theta_e \sin 2\phi$$
$$+ \text{cov}(v'w') \sin 2\theta_e \cos \phi + \text{cov}(u'w') \sin 2\theta_e \sin \phi$$
$$+ \tfrac{1}{2}(\sigma_{v'}^2 - \sigma_{u'}^2) \cos^2 \theta_e \cos 2\phi. \tag{10.62}$$

Therefore, harmonic analysis of $\sigma_v^2(\phi)$ (similar to that for v_r in Section 9.3) can be used to yield turbulent velocity parameters on subresolution volume scales. At low-elevation angles the first term in Eq. (10.62) is an estimate of the horizontal turbulent kinetic energy. The last term is a measure of the horizontal isotropy of the turbulent field. Harris (1975) has applied these concepts to stratiform precipitation and has obtained realistic values for the variances and covariances.

In Fig. 10.9a are profiles of the zonal covariances derived from a single Doppler radar (wavelength 1 cm) measurements using Eqs. (10.60) and (10.62). The data were collected on 22 June 1984 near Boulder, Colorado at about 1240 Mountain Standard Time (MST). The mean surface wind at this time was from the east, and above 2 km the wind was westerly; the easterly wind had a maximum at the top of the surface layer ($\sim$100 m) and decreased to zero below.

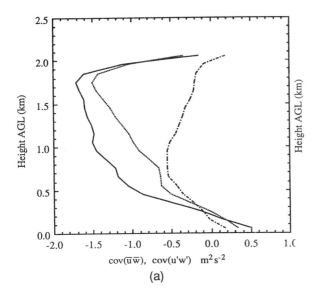

(a)

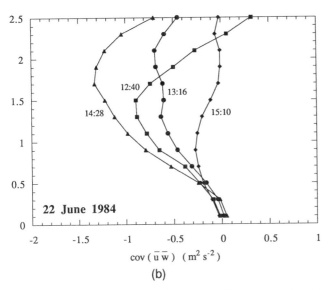

(b)

Fig. 10.9 (a) Profiles of the zonal covariances from single-radar data for velocity scales larger [i.e., cov($\overline{u}\overline{v}$), dotted line] and smaller [i.e., cov($u'w'$), dashed-dotted line] than the resolution volume V_6. The solid line is the sum of cov($\overline{u}\overline{v}$) and cov($u'w'$). (Courtesy of Mei Xu, School of Meteorology, University of Oklahoma.) (b) The evolution of zonal covariances obtained using data from two radars. The beginning of the 20-minute averaging period is labeled (Mountain Standard Time). (Courtesy Jean Schneider, Cooperative Institute for Mesoscale Meteorology, University of Oklahoma.)

By about 1500, winds had veered to be from the southeast at low altitudes and northwest at the higher altitudes. Shear and covariances of opposite sign, as in this case, implies a positive eddy diffusion coefficient (e.g., Sorbjan, 1989, p. 61). Thus turbulent diffusion would smooth vertical gradients of wind if large-scale forces did not maintain the shear. These data also show that the momentum transport by scales smaller than the resolution volume could be substantial at some heights, comprising about one-third of the total transport (Xu and Gal-Chen, 1992).

10.2.4   Turbulence Parameters from Two Doppler Radars

As outlined in Section 9.2, a pair of Doppler radars can be used to determine the wind field at all points on a common analysis grid. A number of researchers, including Rabin *et al.* (1982), Gal-Chen and Kropfli (1984), and Schneider (1991), have used two spaced Doppler radars to observe the turbulent planetary boundary layer (PBL) and have produced horizontally averaged mean wind and velocity covariances. The most recent and comprehensive measurements (Schneider, 1991) are of sufficient accuracy and resolution to test models of PBL evolution.

Dual Doppler analysis domains in PBL studies are typically 6 to 25 km on a side, by 1 to 2 km deep, with grid spacings of a few hundred meters. Each complete scan of the observational volume requires from 2 to 5 minutes. The resulting velocity fields are representative of motions between the smallest resolvable scale (no finer than twice the beamwidth of the more distant radar) and the largest observable scale (the horizontal extent of the analysis domain). Because the wind components are synthesized from interpolated Doppler velocities (Section 9.2.1), the smallest scales represented in $\bar{u}$, $\bar{v}$ are determined by the influence domain of the interpolation filter (Doviak and Berger, 1980). Time resolution is determined by the volume scanning time; total observing period of several volume scans is needed to obtain representative covariances of the velocity field in the PBL (Schneider, 1991, p. 56). Contributions to covariances from eddies smaller than the interpolation domain, or with lifetimes shorter than twice the scan period, are not included. Other observing platforms (towers or aircraft) have finer space and time resolution but have limited spatial coverage, producing statistics that might not be representative of the turbulent flow. Dual Doppler radar observations are space filling, sampling more of the flow, and thus lead to more representative statistics over their resolved range of scales.

Figure 10.9b shows an example of the zonal covariances derived from analysis of data from two 3-cm radars that operated in about the same region and time (i.e., for the 1240 profile) as the 1-cm radar used for data presented in Fig. 10.9a. Each profile represents a 20-minute mean of the grid-point covariances averaged over a domain 9 km by 9 km, with a 0.2-km grid spacing and a two-minute scan period. Thus, at the altitudes where data fields nearly filled the analysis domain, about 15,000 covariance estimates were averaged for each data point plotted in Fig. 10.9b. On this day turbulent kinetic energy [i.e., $(\overline{u'^2} + \overline{v'^2} + \overline{w'^2})/2$], buoyancy flux, covariances, and the depth of the boundary layer all increased during the first half of the afternoon, reached maximum values by about 1430, and then decreased as surface heating weakened.

Note that the covariances from one and two radars (for the common 1240 time) have similar profiles. The smaller covariances calculated from the two-radar data are probably caused by the interpolation used to synthesize the wind field, but some of the differences in covariances might have been caused by incomplete overlap of the averaging domains.

10.3 Doppler Spectrum Width and Eddy Dissipation Rate

Equation (10.46) states that the spatial spectrum of the point Doppler velocity is filtered by the weighting function. This, together with Eq. (10.59), will be used to obtain a relationship between the Doppler spectrum width and the spatial spectrum of turbulence. First, we express the variances of point and averaged velocities in terms of their spectra.

$$\sigma_p^2 = \int \Phi_v(\mathbf{K}) \, dV_K, \tag{10.63}$$

$$\sigma_{\bar{v}}^2 = \int \Phi_{\bar{v}}(\mathbf{K}) \, dV_K. \tag{10.64}$$

Now, Eq. (10.59) shows that the Doppler spectrum width $\langle \sigma_v^2 \rangle$ is the difference between Eqs. (10.63) and (10.64). After this difference is taken and $\Phi_{\bar{v}}(\mathbf{K})$ is substituted from Eq. (10.46), we get the following formula, which connects the Doppler spectrum width to $\Phi_v(\mathbf{K})$ (Frisch and Clifford, 1974):

$$\langle \sigma_v^2 \rangle = \int [1 - (2\pi)^6 |F(\mathbf{K})|^2] \Phi_v(\mathbf{K}) \, dV_K. \tag{10.65}$$

If some expression is found to relate $\Phi_v(\mathbf{K})$ to the turbulence parameters of the medium and Eq. (10.65) is inverted, those parameters can be estimated. This implies knowledge of $\mathbf{I}_n(\mathbf{r} - \mathbf{r}_1)$. Frisch and Clifford (1974) assume that the reflectivity within the resolution volume is uniform and that $\mathbf{I}_n(\mathbf{r} - \mathbf{r}_1)$ is a three-dimensional Gaussian function with one width parameter transverse to and the other along the beam, so that the squared magnitude of its Fourier transform is given by Eq. (10.47).

To treat Eq. (10.65) further, we consider the turbulence of an incompressible fluid so that Eqs. (10.12) and (10.13) are valid. Furthermore, we invoke Kolmogorov's hypothesis, which implies that kinetic energy is provided to the medium by the space Fourier components corresponding to large scales (for thunderstorm studies these are about a few hundred meters and larger) and is dissipated by viscosity at very small scales (of the order of a few centimeters to millimeters). Between these ranges there exists an *inertial subrange* in which energy cascades from the large to the smaller scales, with eventual dissipation by viscous forces. Under these conditions the spectrum function $E(K)$ becomes

$$E(K) = A\varepsilon^{2/3}K^{-5/3}. \tag{10.66}$$

A is a universal dimensionless constant between 1.53 and 1.68 (Gossard and Strauch, 1983, p. 262), and ε is the turbulent energy dissipation rate, normalized to unit mass.

To tie Eq. (10.66) to the Doppler spectrum width, it must be assumed that all σ_v contributions come from velocity scales within the inertial subrange. This is approximately true provided contributions from scales larger than V_6, which are not part of the inertial subrange, are removed (Istok and Doviak, 1986). If the outer scale of inertial subrange turbulence is known, it is possible to obtain a relationship between the dissipation rate and spectrum width. Such a calculation has been made by Bohne (1982), who has also considered the influence of falling precipitation through isotropic turbulence. He concluded that the imperfect response of precipitation to turbulent motion must be accounted for at short range (<20 km) and when the turbulent outer scale is ≤ 0.5 km. At longer range an estimate of the outer scale is necessary if one is concerned with classification of the severity of turbulence.

If the outer scale of the inertial subrange is much larger than the dimensions of the resolution volume, Eqs. (10.12) and (10.66) can be substituted into Eq. (10.65) and the integration performed to obtain an expression giving ε as a function of σ_v and the parameters describing $\mathbf{I}_n(\mathbf{r} - \mathbf{r}_1)$ (see Labitt, 1981; Gossard and Strauch, 1983). For a range resolution smaller than the beamwidth ($\sigma_r \leq r\sigma_\theta$), this expression is

$$\sigma_v^2 = A\Gamma\left(\tfrac{2}{3}\right)(\varepsilon r\sigma_\theta)^{2/3}F\left(-\tfrac{1}{3}, \tfrac{1}{2}; \tfrac{5}{2}; 1 - \sigma_r^2/r^2\sigma_\theta^2\right), \tag{10.67}$$

where F is the hypergeometric function bounded between 0.918 and 1. For $\sigma_r \ll r\sigma_\theta$, $F = 0.918$, and the dissipation rate is very well approximated by

$$\varepsilon \approx 0.72\sigma_v^3/r\sigma_\theta A^{3/2}. \tag{10.68}$$

If the beamwidth is smaller than the range resolution ($r\sigma_\theta \leq \sigma_r$), σ_v^2 has the same form as (10.67), but $r\sigma_\theta$ must be interchanged with σ_r. The hypergeometric function is then bounded by

$$\tfrac{27}{55} < F\left(-\tfrac{1}{3}, 2; \tfrac{5}{2}; 1 - r^2\sigma_\theta^2/\sigma_r^2\right) < 1 \tag{10.69}$$

(Labitt, 1981). Using the series expansion for F, the dissipation rate can be approximated to first order in $r^2\sigma_\theta^2/\sigma_r^2$ by

$$\varepsilon \approx [\sigma_v^3/\sigma_r(1.35A)^{3/2}](\tfrac{11}{15} + \tfrac{4}{15}r^2\sigma_\theta^2/\sigma_r^2)^{-3/2}. \tag{10.70}$$

10.4 Doppler Spectrum Width in Severe Thunderstorms

The two major broadening mechanisms of the Doppler spectrum width due to meteorological factors are shear and turbulence. It was shown in Chapter 5 that, because these spectral broadening mechanisms are independent of one another, the total spectrum width squared can be considered a sum of the variances contributed by each.

The radial-velocity shear across the radar resolution volume can be determined directly from the spatial dependence of the mean radial velocity $\bar{v}$. Then, to arrive at the component due to turbulence, one must extract the shear part from the total spectrum width.

Thunderstorms contain continuum of turbulence scales. Turbulence with scales larger than the radar resolution volume is most likely to consist of anisotropic eddies, which would appear in the data as radial-velocity shears. An

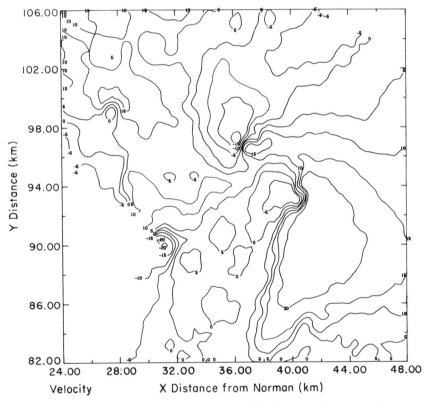

Fig. 10.10a Doppler velocity field at a height of 1.5 km for the Stillwater tornadic storm at 1742 on 13 June 1975. The grid spacing is 400 m. Velocities are in meters per second and contours (isodops) are in 5 m s^{-1} steps.

example of radial-velocity field in a thunderstorm exhibits areas of large shear (Fig. 10.10a). This is the shear that contributes to the measured spectrum width. A large shear region is about the mesocyclone centered at 97 km north and 36 km east of the Norman radar. Another shear region, starting at 94 km north and 40 km east and extending to the bottom of the field, was identified from dual Doppler data to be the low-level boundary (density current front) between ambient air to the east of the shear line and storm outflow to the west (also see Fig. 9.2a). Increased spectrum widths coincident with these larger shear regions are evident in Fig. 10.10b. Large widths farther north are where the tornado mesocyclone formed; the other region of large width is embedded in the density current front.

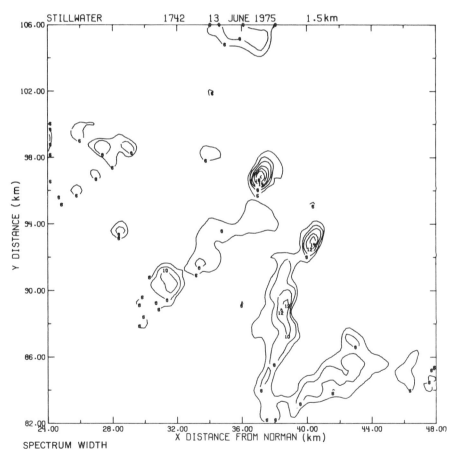

Fig. 10.10b Contours of the spectral width at 1.5 km above ground (13 June 1975 at 1742). Values greater than or equal to 6 m s^{-1}, in steps of 2 m s^{-1}, are displayed for visual clarity.

Measurements of spectrum widths σ_v observed in three severe tornadic storms (Fig. 10.11), one of whose field is shown in Fig. 10.10b (13 June 1975), exhibit a median value of about 4 m s^{-1}, and 20% of widths are larger than 6 m s^{-1}. These probabilities are derived from about 15,000 sample points for each storm; the sample points extend from near the ground to over 15 km. Because these data do not have the range dependence predicted for isotropic turbulence [Eq. (10.68)], the widths are probably due to nonisotropic turbulence.

An estimate of the angular shear across the resolution volume can be obtained by least squares fitting a plane surface that relates the mean radial velocity to the azimuth and elevation position surrounding the resolution volume. The limitation of this assumption is that not all shear is removed. That is, if data were fitted to a quadratic or higher-order polynomial surface, a broader range of turbulence scales larger than the resolution volume would be represented by the surface.

The equation of the linear surface with superposed deviations e_i reads

$$v_i = v_0 + k_\phi(\phi_i - \phi_0)r \cos \theta_i + k_\theta(\theta_i - \theta_0)r + e_i, \qquad (10.71)$$

where (θ_0, ϕ_0) is the origin of the surface.

The parameters of the surface (i.e., v_0, k_ϕ, k_θ) are determined through a matrix operation described by Neter and Wasserman (1974).

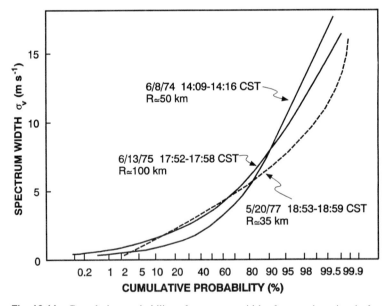

Fig. 10.11 Cumulative probability of spectrum widths for weather signals from three tornadic storms. Spectrum widths are derived from spectra computed using discrete Fourier transforms of signal samples having SNR ≥ 15 dB. Antenna beam width $\theta_1 = 0.8°$ and pulse width $\tau = 1$ μs.

$$\begin{bmatrix} v_0 \\ k_\phi \\ k_\theta \end{bmatrix} = \begin{bmatrix} M & \Sigma l_{\phi i} & \Sigma l_{\theta i} \\ \Sigma l_{\phi i} & \Sigma l_{\phi i}^2 & \Sigma l_{\phi i} l_{\theta i} \\ \Sigma l_{\theta i} & \Sigma l_{\phi i} l_{\theta i} & \Sigma l_{\theta i}^2 \end{bmatrix}^{-1} \begin{bmatrix} \Sigma v_i \\ \Sigma v_i l_{\phi i} \\ \Sigma v_i l_{\theta i} \end{bmatrix}, \quad (10.72)$$

where

$$l_\phi = r(\phi_i - \phi_0)\cos\theta_i, \qquad l_\theta = r(\theta_i - \theta_0)$$

are the arc lengths from the origin of the fitted surface to a data point location (θ_i, ϕ_i). M is the number of data points used to determine the fitted surface. Once shears k_ϕ, k_θ are obtained, the spectrum width due to these linear shears can be computed using Eq. (5.74).

Istok and Doviak (1986) have applied the procedure described here to data from a severe storm. One example from the analysis is shown in Fig. 10.12. The circle in Fig. 10.12 indicates a mesocyclone in which the contribution of the shear is significant; there is one more region, 20 km east and 52 km north of Norman, where width due to shear exceeds 3 m s^{-1}. Examination of dual Doppler wind fields revealed that this region is at the transition between updraft and downdraft. Otherwise, the contribution of shear is minimal, 0–1 m s^{-1}.

Figure 10.13 is presented to show the correlation between turbulence and

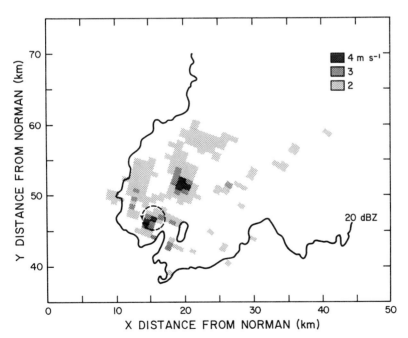

Fig. 10.12 Spectrum width due to radial velocity shear. (8 June 1974 at 1420 CST, Elevation angle: 5.1°.)

the up-(down-) drafts in another tornadic storm (Lee, 1977). The maximum reflectivity (not shown) is north of the updrafts. It is apparent in this case that areas of large spectrum width are on the edges of the updraft, with a preference for higher values when in close proximity of downdrafts. Thus, in this example, turbulence is probably produced by horizontal shear of the vertical wind.

Figure 10.14 shows the width due to turbulence for the same data as in Fig. 10.12. Spectrum widths were calculated using stored time-series I, Q samples, on which the pulse pair algorithm Eq. (6.27) was applied; the contribution of the shear was then subtracted. At the edge of the 20-dBZ contour, the signal-to-noise ratios are small and spectrum widths are large because receiver noise biases have not been removed (Section 6.7). The mesocyclone region has only moderate widths due to turbulence whose contribution is comparable to that from shear. The reflectivity field had a strong core (from 55 to over 60 dBZ) about 2 km to the east of the mesocyclone. We note large widths near this region as well as near the 20-dBZ contour at the same range. These are from sidelobes that have illuminated the 60-dBZ core (Section 7.9). Doppler spectra for the two locations contained several peaks of comparable powers that origi-

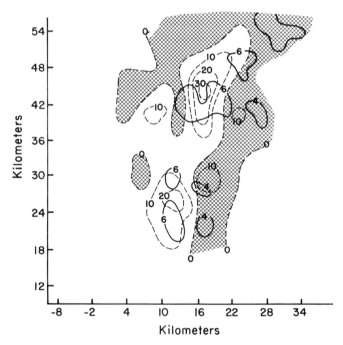

Fig. 10.13 Storm of 8 June 1974; the spectrum width (m s^{-1}) contours are solid lines (———); updraft (m s^{-1}) contours are dashed lines (– – –). The cross-hatched areas are the downdrafts. The origin of the coordinate system is at the radar site. The height is 5 km. (From Lee, 1977.)

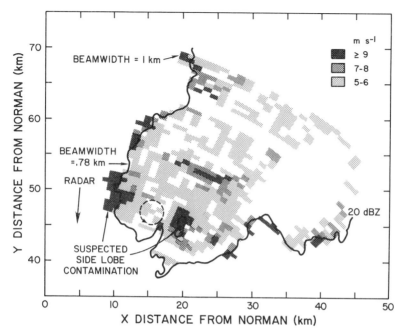

Fig. 10.14   Spectrum width due to turbulence. (8 June 1974 at 1420 CST, elevation angle: 5.1°.)

nated from the region where reflectivities and sidelobes are large. We believe that most of the other broad spectra from Fig. 10.14 are due to eddies that were shed off the strong updraft by environmental winds (from the southwest).

According to Lee (1977), there is a strong connection between the spectrum width and aircraft penetration measurements of turbulence. His data show that when aircraft-derived gust velocities exceeded 6 m s^{-1}, corresponding to moderate or severe turbulence, the spectrum width exceeded 5 m s^{-1} in every case for aircraft within 1 km of the radar resolution volume. The cumulative probability of the total spectrum width and the width due to shear and turbulence, from Istok (1981), is presented in Fig. 10.15. Spectrum widths in excess of 5 m s^{-1} exist in about 50% of the storm volume, suggesting that one-half of the volume contains moderate or severe turbulence. Most areas of large widths are found within the upper regions of the storm.

Large shear by itself is not necessarily very dangerous to aircraft except during takeoff or landing, because at higher altitudes there is ample space and time for the aircraft to recover. Nevertheless, large shear may produce extreme turbulence, which is dangerous, and hence pilots should avoid it.

The cumulative probability of shears in azimuth k_ϕ and elevation k_θ from the least squares fit is plotted in Fig. 10.16. In general, a greater proportion of the storm contains larger shear in the elevation direction than in the azimuth direction. Only 19% of all azimuthal shears in the storm are in excess of 3×10^{-3}

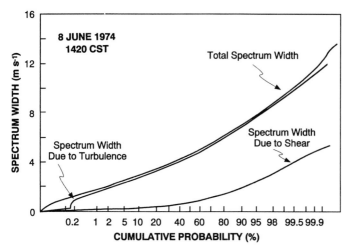

Fig. 10.15 Cumulative probability of the total spectrum width and the width due to linear radial velocity shear and turbulence. (8 June 1974 at 1420 CST.)

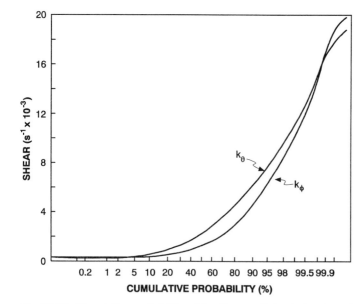

Fig. 10.16 Cumulative probability of the linear radial velocity shear in the elevation (k_θ) and azimuthal (k_ϕ) directions. (8 June at 1420 CST.)

s^{-1}, while 35% of all elevation shears are larger than $3 \times 10^{-3} \, s^{-1}$. But the largest values of shear are in the azimuthal direction. These shears ($16-20 \times 10^{-3} \, s^{-1}$) are associated with the mesocyclone.

There is an inherent problem with using the Doppler spectrum width as a measure of turbulence: The spectrum width is a function of the radar resolution-volume size in addition to the intensity of turbulence. The turbulent kinetic-energy dissipation rate [Eq. (10.68 or 10.70)] that represents the energy flow across any given wave number is a more useful quantity for characterizing the intensity of turbulence within the inertial subrange. Within this subrange, turbulence is locally isotropic. Sinclair (1974) has observed the upper wavelength of the inertial subrange in a severe storm to vary from 150 to about 2000 m. It appeared to Sinclair that this variability is related to the storm intensity and the measurement altitude. That is, the outer scale of the inertial subrange is largest in the upper half of the storm, where vertical velocities are usually largest and turbulence is most intense. This variability of the outer scale is not uncommon. Other investigators (Rhyne and Steiner, 1964; MacCready, 1962, 1964; Reiter and Burns, 1960; Reiter, 1970) have shown or suggested that the outer scale of the inertial subrange may vary from 300 to 800 m, depending on the phenomena and the location of measurement.

The dissipation rate of the turbulent kinetic energy can be estimated from the Doppler spectrum width due to turbulence if the largest scale of the inertial subrange is larger than the largest dimension of the radar resolution volume. Such estimates were made by Brewster and Zrnic (1986) with a vertically pointing Doppler radar; the energy dissipation rates computed from spatial spectra of velocity fields agreed very well with ε computed from Doppler spectral widths (Fig. 10.17). Furthermore they determined that longitudinal (S_l) and transverse (S_t) spatial spectra of velocities followed a theoretical relationship expected for the inertial subrange up to 2.4 km wavelengths (Fig. 10.18). The slightly larger spectral values for S_t are also consistent with the fact that the structure parameter $C_{tt}^2 = \frac{4}{3} C_{ll}^2$ (Section 10.1.3).

If the largest dimension of the radar resolution volume is larger than the outer scale of the inertial subrange, then the spectrum width (after removal of the shear contribution) would still be due to turbulence within the energy input and inertial subranges. That is, although the total spectrum width correctly characterizes the spread of reflectivity-weighted radial velocities within the radar resolution volume, one cannot always relate these to eddy dissipation rate. This is because, for the input energy containing range, eddies might not be isotropic; they may have no known spectral form, and hence relating σ_v to ε can result in large errors. Furthermore, total spectrum width may become aspect sensitive, so that radars viewing the volume from different directions will detect different magnitudes of σ_v^2. From simultaneous measurements of σ_v with two spaced Doppler radars surveying few storms (Fig. 10.19), Lee and

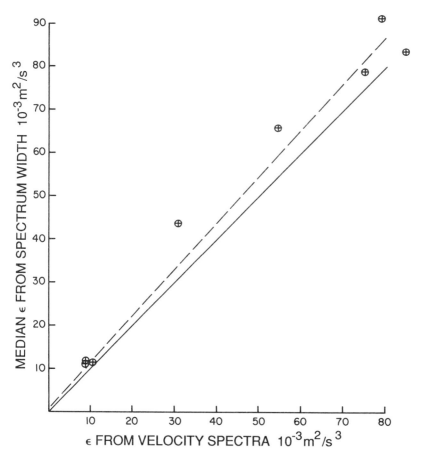

Fig. 10.17 Median ε from spectrum width versus ε from spatial spectra of velocities for several vertically incident measurements. A least-squares line (---) is fitted to the data; (——) is a 45° line.

Thomas (1989) conclude that Doppler spectral width is practically independent of viewing angle. For one of these storms in which the beams intersected at about 45° the difference between the two independently measured spectral widths is centered at zero and has a spread of only a few meters per second (Fig. 10.20).

Dissipation rates for the 8 June 1974 storm were computed from Eq. (10.68), which is valid at all ranges in this storm. Presented in Fig. 10.21 is a plot of the cumulative probability of ε. Dissipation rates less than 1 m² s⁻³ exist in 99% of the storm volume, and in 50% of the storm they are less than 0.1 m² s⁻³.

Frisch and Strauch (1976) observed maximum dissipation rates of 0.06 m² s⁻³ in a Colorado convective storm. (This value is 5.2 times smaller than the one

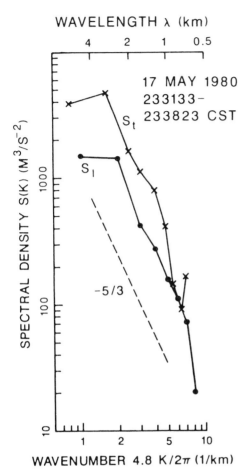

Fig. 10.18 Spatial spectral densities. S_t is the transverse and S_l is the longitudinal spectrum of vertical velocity variance. A dashed line represents the $-5/3$ law expected from isotropic turbulence in the inertial subrange.

they actually reported, because we use a more recent estimate of the universal constant $A = 1.6$.) Such large rates were often noted between updrafts and downdrafts. It seems reasonable to expect larger dissipation rates in the Oklahoma storm, which had an updraft of 40 m s^{-1}, than in the Colorado storm with an updraft of 20 m s^{-1}. It is also likely, however, that some of the largest dissipation rates are due to anisotropic eddies that are not part of the inertial subrange. In such cases the derived dissipation rate could be larger than the true dissipation rate of turbulent kinetic energy.

Figure 10.22 compares the squares of the Doppler spectrum widths measured by radar and those deduced from aircraft penetrations. Radar-measured

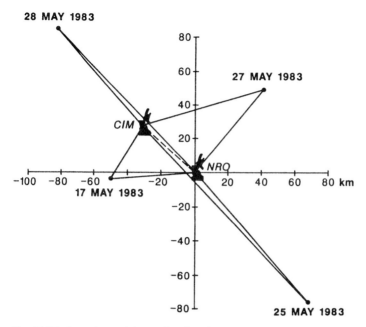

Fig. 10.19 Location and dates of analyzed storms relative to the radar site at Norman (NRO) and Page Field, Oklahoma City, Oklahoma (CIM).

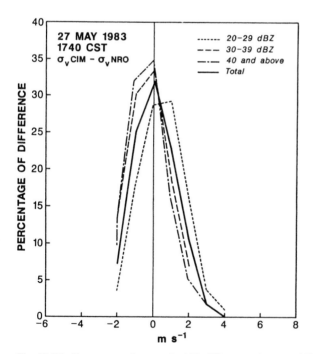

Fig. 10.20 Percentage of spectral width differences between NRO and CIM for all altitudes combined. The distribution of the 8500 data points is quantized by reflectivity at 20–29 dBZ, 30–39 dBZ, and over 40 dBZ. In addition the combined distribution is shown with a heavy black line.

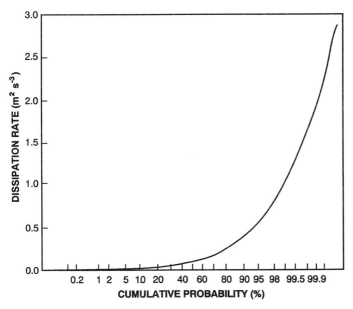

Fig. 10.21 Cumulative probability of the eddy dissipation rate ε (8 June 1974 at 1420 CST). (From Istok, 1981.)

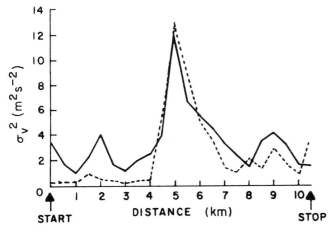

Fig. 10.22 Aircraft (dashed) and radar (solid) estimates of the square of the Doppler spectrum width (variance) at grid point locations, along the best-correlated aircraft tracks for one penetration. (From Bohne, 1981.)

widths were computed from the Fourier transform of time-series data, and an adaptive threshold proposed by Hildebrand and Sekhon (1974) was used to reduce noise effects. The contribution of the linear shear to the width has been taken into account. To obtain Fig. 10.22 the aircraft path was shifted until a maximum correlation between the two spectrum widths was achieved. The data deduced from aircraft were obtained by weighting, with the radar resolution function, the point velocities the aircraft had recorded. To avoid contamination by scales larger than the outer scale of turbulence, Bohne (1981) subjectively detrended segments of aircraft data. He then obtained a correlation of 0.89 between the two curves. In this and other instances, agreement is best at large widths, whereas at smaller widths the radar indicates larger values. If this proves true in general, some false alarms could be expected.

Problems

10.1 Why do components of velocity transverse to the lag vector ρ have more variance per unit wave number interval (i.e., per Δk) than do the components of velocity parallel to ρ when the velocity components are associated with scales of turbulence within the inertial subrange?

10.2 Using the hypothesis of Kolmogorov (i.e., in the inertial subrange the structure function D_{ll} is only a function of ρ and ε, and in the dissipative range it is a function of ρ, v, and ε) sketch R_{tt} and R_{ll}. Show that the velocity components transverse to the lag vector ρ are less correlated than those parallel to it. Discuss, in the best way you can, why this is so. Determine the lag at which the square dependence of the dissipative range intersects the 2/3 law curve of the inertial subrange.

10.3 Given the equation of motion

$$\frac{\partial v_i}{\partial t} + v_j \frac{\partial v_i}{\partial x_j} = \frac{1}{\rho} \frac{\partial \sigma_{ij}}{\partial x_j}, \tag{1}$$

where σ_{ij} is the stress tensor

$$\sigma_{ij} = P\delta_{ij} + 2\mu S_{ij} \tag{2}$$

δ_{ij} is the Kronecker delta, P the pressure, ρ the density, μ the viscosity, and

$$S_{ij} = \frac{1}{2} \left(\frac{\partial v_i}{\partial x_j} + \frac{\partial v_j}{\partial x_i} \right) \tag{3}$$

is the rate-of-strain tensor. The convention of summation over repeated indices is used in Eqs. (1)–(3). The equation of continuity is

$$\frac{\partial v_i}{\partial x_i} = 0.$$

(a) Express the x component of the equations of motion in terms of u, v, w and their derivatives along x, y, z. (b) Show that for incompressible fluids, Eq. (1) is equivalent to the Navier–Stokes equation

$$\frac{dv_i}{dt} = -\frac{1}{\rho} \frac{\partial P}{\partial x_i} + v \frac{\partial^2 v_i}{\partial x_j \partial x_j}, \tag{4}$$

where v is the kinematic velocity. (c) Assume that the velocity components can be expressed in terms of a mean value $\langle v_i \rangle$ plus a turbulent component v_i, the same holds for P, and that

$\langle \rho \rangle = \rho$. Substituting into Eq. (1) show that the equations of motion for the mean flow are

$$\langle v_j \rangle \frac{\partial \langle v_i \rangle}{\partial x_j} + \left\langle v_j' \frac{\partial v_i'}{\partial x_j} \right\rangle = \frac{1}{\rho} \frac{\partial \Sigma_{ij}}{\partial x_j},\tag{5}$$

where

$$\Sigma_{ij} \equiv -\langle P \rangle \delta_{ij} + 2\mu \langle S_{ij} \rangle$$

is the stress tensor of the mean flow. (d) Show that

$$\left\langle v_j' \frac{\partial v_i'}{\partial x_j} \right\rangle = \left\langle \frac{\partial v_i' v_j'}{\partial x_j} \right\rangle = \frac{\partial \langle v_i' v_j' \rangle}{\partial x_j}.\tag{6}$$

(Hint: Use the continuity equation and consider differentiation as the limit of the differencing process) so that the equations of motion for the mean flow can now be written as

$$\langle v_j \rangle \frac{\partial \langle v_i \rangle}{\partial x_j} = \frac{1}{\rho} \frac{\partial \langle T_{ij} \rangle}{\partial x_j},\tag{7}$$

where the Reynolds stress tensor is $T_{ij} = -\rho \langle v_i' v_j' \rangle$ and is the contribution that turbulence makes to the forces acting on the mean flow. (e) Now show that the rate of change of energy due to advection is

$$\rho \langle v_j \rangle \frac{\partial (\langle v_i \rangle \langle v_i \rangle / 2)}{\partial x_j} = \frac{\partial (\langle T_{ij} \rangle \langle v_i \rangle)}{\partial x_j} - \frac{\langle T_{ij} \rangle \partial \langle v_i \rangle}{\partial x_j},\tag{8}$$

which is the mean flow budget energy equation. Why do we write $\langle v_i \rangle \langle v_i \rangle$ and not $\langle v_i \rangle^2$? (f) Demonstrate that $\langle S_{ij} \rangle \langle S_{ij} \rangle = \langle S_{ij} \rangle \, \partial \langle v_i \rangle / \partial x_j$ so that Eq. (8) can be written as

$$\rho \langle v_j \rangle \frac{\partial}{\partial x_j} (\langle v_i \rangle \langle v_i \rangle / 2) = \frac{\partial}{\partial x_j} (\langle T_{ij} \rangle \langle v_i \rangle) - \langle T_{ij} \rangle \langle S_{ij} \rangle\tag{9}$$

(g) Substitute for $\langle T_{ij} \rangle$ into (9) and give a physical interpretation for each term. (h) Using the principles applied in the above exercises, derive the equation for the turbulent energy budget:

$$\langle v_j \rangle \frac{\partial}{\partial x_j} (v_i' v_i' / 2) = -\frac{\partial}{\partial x_j} \left(\frac{1}{\rho} \langle v_i' P' \rangle + \frac{1}{2} \langle v_i'^2 v_j' \rangle - 2\nu \langle v_i' S_{ij}' \rangle \right)$$

$$- \langle v_i' v_j' \rangle \langle S_{ij} \rangle - 2\nu \langle S_{ij}' S_{ij}' \rangle,\tag{10}$$

where S_{ij}' is the rate of strain of the turbulent components. Give physical interpretations to each of the terms in Eq. (10).

11

Observations of Fair Weather

Although weather radar's principal function is to identify and track precipitating storms, these radars also detect echoes from scatterers in fair weather. Sometimes the spatial distribution of reflectivity in clear air can be associated with meteorological phenomena such as waves, turbulent layers, fronts, etc. Echoes from clear air have been seen almost from the inception of radar observations (e.g., Colwell and Friend, 1936). These "angel echoes" were at first mystifying and were often associated with birds and insects. Clear-air echoes not related to any visible object were conclusively proven to emanate from refractive index irregularities in experiments with the National Aeronautics and Space Administration's (NASA) multiwavelength radars at Wallops Island, Virginia (Hardy *et al.*, 1971). It is the purpose of this chapter to develop, from basic theory, a relation between the characteristics of the refractive index irregularities and the Doppler-shifted signals sensed by radar. Wind profilers and the Radio Acoustic Sounding Systems (RASS), which use Doppler radars, are also discussed.

11.1 Reflection, Refraction, and Scatter: Coherence

The permittivity $\varepsilon(\mathbf{r}, t)$ of the troposphere has, in general, both spatial and temporal variation. Even though the mean permittivity $\langle \varepsilon(\mathbf{r}, t) \rangle$ may not contain small-scale spatial irregularities, it always exhibits large-scale spatial inhomogeneities. The spectrum of scales that remain is a function of integration time, so the longer the latter, the smoother is the average permittivity. But in the troposphere, because of the omnipresence of gravity and solar heat, there is a limit beyond which, no matter how long we integrate, or at least for any practical time interval, $\varepsilon(\mathbf{r})$ does not become smoother and spatial variations, particularly along the vertical, remain. As discussed in Chapter 2, these spatial inhomogeneities of the average ε cause ray paths to be refracted from their normal straight-line course.

Studies of the reflection and refraction of waves in inhomogeneous nonturbulent media have been carried out by many authors (Budden, 1961; Wait, 1962; Brekhovskikh, 1960). Others (Pekeris, 1947; Tatarskii, 1961; Wheelon,

1959; Booker and Gordon, 1950) have considered the scattering properties of turbulent media, but most ignore the spatial inhomogeneity of the average permittivity. DuCastel *et al.* (1962) have studied the reflection process when both turbulent fluctuations and inhomogeneity in average values are present. These give rise to waves comprising (1) a coherent component caused by reflection or refraction and (2) an incoherent component caused by scatter.

Waves that are reflected from sharp quasipermanent changes in ε, as well as those refracted from a transmitter to a remote receiver by gradual spatial changes in ε, form the coherent component of the signal, whereas waves scattered from turbulent media give incoherent signals. Signals are time coherent if the scattering medium does not modulate the amplitude or phase of the wave. Nevertheless, it should be understood that spatial variations of the amplitude and phase can exist, and furthermore, these spatial variations can also be described in terms of spatial coherence. In reality, even the nonturbulent atmosphere is nonstationary, and signals change, although slowly, in amplitude and phase.

We can distinguish between reflection or refraction and scatter if the signal spectrum exhibits two distinct distributions—one peaked and narrow, the other broad, as sketched in Fig. 11.1a. The narrow distribution is associated

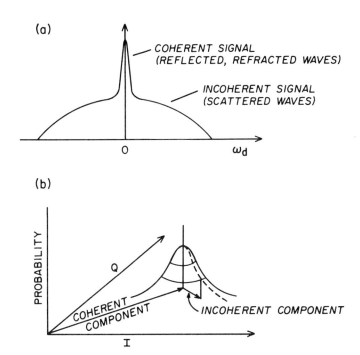

Fig. 11.1   (a) The spectrum of signals from waves simultaneously reflected or refracted and scattered. (b) The distribution of samples of I and Q. Samples are taken over a time interval small compared to the coherence time of the coherent signal.

with coherent signals and usually with refracted or reflected waves, and the broad distribution is associated with incoherent turbulent scatter. If irregularities in refractive index do not move relative to one another (i.e., no turbulence) the velocity spectra of scattered signals will also be narrow. Over-the-horizon communication links contain both coherent and incoherent signals. Vertical gradients of the mean refractive index can refract the transmitted signal to the receiver, as shown in Chapter 2. This refracted signal will exhibit slow changes in amplitude and phase. Simultaneously, irregularities in the refractive index, which change much more quickly, will scatter signals to the receiver, and these signals have rapid variations in amplitude and phase superimposed on the slow variations caused by the gradual changes in refractive index. If many independent scattering elements are effective (Section 4.3), there is a Gaussian distribution centered about a mean value (Fig. 11.1b).

The following development, which describes the propagation of waves through inhomogeneous turbulent media, is framed so as to embrace refraction, reflection, and scatter. The amplitude and phase of refracted waves can be found through exact solution of Maxwell's equations, if certain vertical profiles (e.g., linear, parabolic, or exponential) of ε are considered, or by geometric-optics ray-tracing calculations if the spatial variations cannot be described by simple functions but are sufficiently gradual, as demonstrated in Chapter 2. The random or scatter contribution to the fields are usually solved by invoking the Born approximation. In this approximation it is assumed that contribution to the electromagnetic field by scattering is small compared to the unperturbed refracted field (i.e., the field that would exist if turbulence were nonexistent).

11.2 Formulation of the Wave Equation for Inhomogeneous and Turbulent Media

Starting with Maxwell's equations,

$$\nabla \times \mathbf{E} = -\mu_0\, \partial \mathbf{H}/\partial t, \tag{11.1}$$

$$\nabla \times \mathbf{H} = \partial \varepsilon \mathbf{E}/\partial t, \tag{11.2}$$

for media in which the permeability μ is a constant μ_0, and taking the curl of Eq. (11.1) and substituting into Eq. (11.2), we obtain

$$\nabla(\nabla \cdot \mathbf{E}) - \nabla^2\mathbf{E} = -\mu_0\varepsilon \frac{\partial^2\mathbf{E}}{\partial t^2} \tag{11.3}$$

when the time scales from changes in the permittivity $\varepsilon(\mathbf{r}, t)$ are long compared to changes in $\mathbf{E}$ at the transmitted frequency. Taking the divergence of Eq. (11.2),

$$\nabla \cdot \nabla \times \mathbf{H} \equiv 0 = \frac{\partial}{\partial t}\nabla \cdot (\varepsilon\mathbf{E}) = \frac{\partial}{\partial t}(\varepsilon\nabla \cdot \mathbf{E} + \mathbf{E} \cdot \nabla\varepsilon), \tag{11.4}$$

we find that

$$\varepsilon \nabla \cdot \mathbf{E} + \mathbf{E} \cdot \nabla \varepsilon = \text{const},$$

independent of time. Because we seek solutions that are wavelike and therefore time varying, we can set the constant equal to zero and solve the resulting equation to obtain

$$\nabla \cdot \mathbf{E} = -(\mathbf{E} \cdot \nabla \ln \varepsilon/\varepsilon_0). \tag{11.5}$$

Substitution of Eq. (11.5) into Eq. (11.3) produces the wave equation.

$$\nabla^2 \mathbf{E} - \mu_0 \varepsilon \frac{\partial^2 \mathbf{E}}{\partial t^2} = -\nabla(\mathbf{E} \cdot \nabla \ln \varepsilon/\varepsilon_0), \tag{11.6}$$

in which the right side is considered a source distribution determined by the resultant electric field and permittivity. Equation (11.6), a formulation of the wave equation for an inhomogeneous random medium, is a set of three coupled linear differential equations with nonconstant coefficients and, as such, has no known general analytic solution. To find an acceptable approximation, let us express the relative permittivity as

$$\varepsilon/\varepsilon_0 = \varepsilon_a/\varepsilon_0 + \Delta\varepsilon/\varepsilon_0, \tag{11.7}$$

where ε_a is an ensemble average of the permittivity taken at any point. In general ε_a may be a function of the coordinates of the point. But if its spatial changes are sufficiently gradual, then geometric-optics ray-tracing solutions outlined in Chapter 2 can be used to solve for the refracted fields. $\Delta\varepsilon$ represents a zero-mean random process and is the fluctuation produced by atmospheric turbulence. The refractive index will, for convenience, be defined as follows:

$$n^2 = n_a^2 + 2n_a \, \Delta n = \varepsilon_a/\varepsilon_0 + \Delta\varepsilon/\varepsilon_0, \tag{11.8}$$

where

$$n_a \equiv (\varepsilon_a/\varepsilon_0)^{1/2}, \qquad \Delta\varepsilon/\varepsilon_0 = 2n_a \, \Delta n, \tag{11.9}$$

and Δn is, to a good approximation if $\Delta n \ll n_a$, the deviation of the refractive index from its average value n_a. Substituting Eqs. (11.7) and (11.8) into Eq. (11.6), we obtain

$$\nabla^2 \mathbf{E} - \frac{n_a^2}{c^2} \frac{\partial^2 \mathbf{E}}{\partial t^2} = \frac{2n_a \, \Delta n}{c^2} \frac{\partial^2 \mathbf{E}}{\partial t^2}$$

$$-2\nabla[\mathbf{E} \cdot \nabla \ln(n_a)] - \nabla\left[\mathbf{E} \cdot \nabla \ln\left(1 + \frac{2\,\Delta n}{n_a}\right)\right]. \tag{11.10}$$

We seek a solution of the form

$$\mathbf{E} = \mathbf{E}_0 + \mathbf{E}_1, \tag{11.11a}$$

where $\mathbf{E}_0$ is the solution in the absence of turbulence (i.e., for $\Delta n = 0$).

Substituting Eq. (11.11a) into Eq. (11.10) after expanding

$$\ln\left(1 + \frac{2\Delta n}{n_a}\right) = \frac{2\Delta n}{n_a} - \frac{1}{2}\left(\frac{2\Delta n}{n_a}\right)^2 + \frac{1}{3}\left(\frac{2\Delta n}{n_a}\right)^3 - \cdots$$

in a power series, we obtain

$$\nabla^2 \mathbf{E}_0 + \nabla^2 \mathbf{E}_1 - \frac{n_a^2}{c^2}\frac{\partial^2}{\partial t^2}(\mathbf{E}_0 + \mathbf{E}_1)$$

$$= \frac{2n_a \Delta n}{c^2}\frac{\partial^2}{\partial t^2}(\mathbf{E}_0 + \mathbf{E}_1) - 2\nabla\{(\mathbf{E}_0 + \mathbf{E}_1)\cdot[\nabla \ln(n_a)]\}$$

$$- \nabla\left\{(\mathbf{E}_0 + \mathbf{E}_1)\cdot\nabla\left[\frac{2\Delta n}{n_a} - \frac{1}{2}\left(\frac{2\Delta n}{n_a}\right)^2 + \cdots\right]\right\}. \tag{11.11b}$$

Because we assume $\Delta n \ll n_a$, we shall ignore higher-order terms in $\Delta n/n_a$. Furthermore, we shall also ignore terms that contain products of $\mathbf{E}_1$ and Δn or its derivatives. In this way we are in effect neglecting multiple scattering, that is, the interaction of the scattered field $\mathbf{E}_1$ with refractive index perturbations producing secondary scattered fields. Usually these are small compared to the first-order scattered fields for $\Delta n \ll n_a$. Under these assumptions, the previous equation reduces to

$$\nabla^2 \mathbf{E}_0 + \nabla^2 \mathbf{E}_1 - \frac{n_a^2}{c^2}\frac{\partial^2}{\partial t^2}(\mathbf{E}_0 + \mathbf{E}_1)$$

$$= \frac{2n_a \Delta n}{c^2}\frac{\partial^2 \mathbf{E}_0}{\partial t^2} - 2\nabla[\mathbf{E}_0 \cdot \nabla \ln(n_a) + \mathbf{E}_1 \cdot \nabla \ln(n_a)]$$

$$- \nabla\left[\mathbf{E}_0 \cdot \nabla\left(\frac{2\Delta n}{n_a}\right)\right]. \tag{11.12}$$

Note that $\mathbf{E}_1$ is not necessarily small compared to $\mathbf{E}_0$. There can be regions of space where the scattered field $\mathbf{E}_1$ is larger than $\mathbf{E}_0$, but this occurs only outside the main lobe of the transmitter beam. Zero-order terms are obtained for $\Delta n \to 0$ (i.e., $\mathbf{E}_1 \to 0$), and from Eq. (11.12) we have $\mathbf{E}_0$ satisfying

$$\nabla^2 \mathbf{E}_0 - \frac{n_a^2}{c^2}\frac{\partial^2}{\partial t^2}\mathbf{E}_0 = -2\nabla[\mathbf{E}_0 \cdot \nabla \ln(n_a)]. \tag{11.13a}$$

If the transmitter produces harmonic waves with radian frequency ω_0, the refracted wave $\mathbf{E}_0(\mathbf{r}, t)$ will also be harmonic, and Eq. (11.13a) can be written

$$\nabla^2 \mathbf{E}_0(\mathbf{r}) + k_0^2 n_a^2 \mathbf{E}_0(\mathbf{r}) = -2\nabla[\mathbf{E}_0(\mathbf{r}) \cdot \nabla \ln(n_a)], \tag{11.13b}$$

where $\mathbf{E}_0(\mathbf{r}, t) = \mathbf{E}_0(\mathbf{r})e^{j\omega_0 t}$ and $k_0^2 \equiv \omega_0^2/c^2$.

Because $\Delta n(\mathbf{r}, t)$ is time varying, it produces a scattered field $\mathbf{E}_1(\mathbf{r}, t)$ that is not a pure harmonic signal. Thus, we cannot in general express the second-order time derivative of $\mathbf{E}_1$ as a product of it and ω_0^2.

Outwardly Eq. (11.13) is similar to Eq. (11.6), but, by definition, $\mathbf{E}_0$ of Eq. (11.13) is the coherent part of the solution. Subtracting Eq. (11.13a) from Eq. (11.12), we obtain the wave equation whose solution gives the incoherent field that results from scatter of the incident refracted field $\mathbf{E}_0$.

$$\nabla^2 \mathbf{E}_1 - \frac{n_a^2}{c^2} \frac{\partial^2 \mathbf{E}_1}{\partial t^2} = -2k_0^2 n_a \, \Delta n \, \mathbf{E}_0 - 2\nabla\left[\mathbf{E}_1 \cdot \nabla \ln(n_a) + \mathbf{E}_0 \cdot \nabla\left(\frac{\Delta n}{n_a}\right)\right]. \quad (11.14)$$

Thus, from Eqs. (11.13) and (11.14) we see that $\mathbf{E}_0$ is the field due to the refractive effects of n_a, and $\mathbf{E}_1$ is the contribution from the turbulent irregularities in the structure of n.

Solutions of Eq. (11.13) for plane-stratified media are discussed extensively in the literature by many authors (Budden, 1961; Brekhovskikh, 1960; Wait, 1962). In their texts exact analytic expressions are obtained for specialized cases of the wave polarization and spatial variations of n_a. However, if the change of n_a within a wavelength is small (Appendix D), that is,

$$|\nabla n_a|/2\pi \ll 1/\lambda \quad \text{everywhere,} \quad (11.15)$$

then Eq. (11.13b) can be approximated by

$$\nabla^2 \mathbf{E}_0 + k_0^2 n_a^2 \mathbf{E}_0 = 0. \quad (11.16)$$

It should also be noted from Eq. (11.13) that for media in which the stratification is in one direction, as may occur whenever a field of mechanical force along a dominant direction is exerted on the medium, Eq. (11.16) becomes exact for polarization of $\mathbf{E}_0$ perpendicular to the direction of stratification. Thus, we would expect Eq. (11.16) to be exact for horizontally polarized waves propagating in the atmosphere. At short radio wavelengths, such as those in the decimetric-to-centimetric bands, it is unlikely that large gradients in n_a would be sustained to violate Eq. (11.15). Thus, Eq. (11.16) should be valid for UHF and microwave weather radars.

The solution of Eq. (11.14) is known as the Born approximation and is valid whenever multiple scattering [i.e., higher-order terms in Eq. (11.11b)] can be neglected. In a medium in which the average refractive index profile varies smoothly [i.e., where Eq. (11.15) holds], the term

$$2\nabla[\mathbf{E}_1 \cdot \nabla \ln(n_a)] \quad (11.17)$$

is negligible with respect to $(n_a^2/c^2) \, \partial^2 \mathbf{E}_1/\partial t^2$, as can be deduced from the arguments put forth in Appendix D. We need only make the added assumption that the time rate of change of the amplitude and phase of $\mathbf{E}_1$ is much slower than the harmonically related changes (i.e., changes at the rate ω_0). Thus, the differential equation that gives the first-order solution to the fields scattered by irregularities in an otherwise smoothly changing medium is

$$\nabla^2 \mathbf{E}_1 - \frac{n_a^2}{c^2} \frac{\partial^2 \mathbf{E}_1}{\partial t^2} = -2k_0^2 n_a^2 \left(\frac{\Delta n}{n_a}\right) \mathbf{E}_0 - 2\nabla\left[\mathbf{E}_0 \cdot \nabla\left(\frac{\Delta n}{n_a}\right)\right], \quad (11.18)$$

where $\mathbf{E}_0$ is the solution of Eq. (11.16). It will be shown in Section 11.3 that the second term on the right-hand side, $2\nabla[\mathbf{E}_0 \cdot \nabla(\Delta n/n_a)]$, does not contribute to the first-order transverse fields but does cancel a first-order longitudinal field arising from the first term. Even though we are usually interested only in the transverse field components, because they are associated with power transfer, we cannot ignore the second term on the right side of Eq. (11.18) unless we ignore the first-order longitudinal fields.

11.3 Solution for Fields Scattered by Irregularities

Although the focus of this book is on weather radars, in this section we consider the transmitter to be separate from the receiver. This will allow us (1) to demonstrate the relation between the scales of refractive index irregularities effective in the scattering process and transmitter–receiver separation, (2) to determine the transmission loss of a forward-scatter link given the spectrum of turbulence in the illuminated volume, and (3) to examine the application of bistatic Doppler radar to map the reflectivity and wind fields in fair weather (Doviak *et al.*, 1972; Doviak, 1972) and to measure temperature profiles (Frankel and Peterson, 1976).

Because narrow-beam antennas are typically used in remote sensing and also in forward-scatter communication links, the dimensions of the volume contributing to the scattered field are sufficiently small that variations in r_t and r_r over this volume of integration are much smaller than the respective distances r_{t0} and r_{r0} (Fig. 11.2). This condition allows simplifying assumptions to be

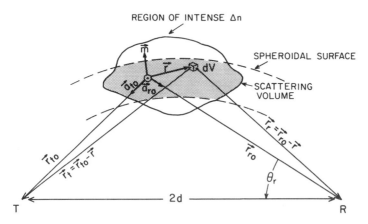

Fig. 11.2 Illustration of the scattering geometry and the spheroidal shell within which the scattering irregularities contribute to the received signal sampled at some time delay after the emission of a transmitter pulse. Line TR is the axis of rotation for the spheroid. $\mathbf{a}_{t0}$ and $\mathbf{a}_{r0}$ are unit vectors at the origin. Vectors are indicated by overarrows in figures and by bold letters in text.

applied to the development of a solution to Eq. (11.18). We also assume the origin of our coordinate system to be at the intersection of the beam axes, which lie along $\mathbf{r}_{t0}$, $\mathbf{r}_{r0}$.

In the troposphere $n_a \simeq 1$, so Eq. (11.18) can be approximated by

$$\nabla^2 \mathbf{E}_1(\mathbf{r}, t) - \frac{1}{c^2} \frac{\partial^2 \mathbf{E}_1(\mathbf{r}, t)}{\partial t^2} = -\mathbf{f}(\mathbf{r}, t), \qquad (11.19)$$

where

$$\mathbf{f}(\mathbf{r}, t) \equiv (2k_0^2 \Delta n(\mathbf{r}, t)\mathbf{E}_0(\mathbf{r}) \\ + 2\nabla\{\mathbf{E}_0(\mathbf{r}) \cdot \nabla[\Delta n(\mathbf{r}, t)]\})e^{j\omega_0\tau} \qquad (11.20)$$

and $\mathbf{r}$ is the coordinate of any observation point, which in this case would be $\mathbf{r}_{r0}$ (Fig. 11.2).

Pulse-modulated transmitters generate

$$\mathbf{E}_0(\mathbf{r}, t) = \mathbf{E}_0(\mathbf{r})e^{j\omega_0(t - r/c)} U(t - r/c), \\ U(t^*) = 1, \qquad 0 \le t^* \le \tau \\ = 0 \qquad \text{otherwise}, \qquad (11.21)$$

where τ is the pulse width, $U(t^*)$ is a unit pulse function, and $t^* \equiv t - r/c$ is a retardation time. To account for pulse modulation we shall now make an assumption that represets a compromise between the ease of following a reasonable physical argument and the pursuit of a less-tractable mathematical method. In Section 4.4.2 it was shown that sampled backscattered signals came, for all practical purposes, from a thin spherical shell of scatters whose location and contribution to the signal are determined by the sample time delay, the transmitted pulse shape, and the receiver frequency transfer function. It can be shown analogously that for a spaced transmitter-receiver the scatter volume that contributes to the sampled received signal is confined to a prolate spheroidal shell whose foci are at the transmitter and receiver locations. The range r_r from the receiver to an element of the scatter volume V is

$$r_r = \frac{1}{2} \frac{(ct_s)^2 - (2d)^2}{ct_s - 2d \cos \theta_r}, \qquad t_s \ge 2d/c, \qquad (11.22)$$

where θ_r is the receiver elevation angle and $2d$ is the distance between transmitter and receiver. The sample time t_s is the time at which $\mathbf{E}_1$ is sampled after the transmitter pulse is radiated. Obviously we cannot receive echoes for $t_s < 2d/c$. Analogous to the backscatter case, the weighting is uniform across the spheroidal shell if the pulse is rectangular and $B_6 \gg \tau^{-1}$. The separation Δs between the inner and outer boundaries of the spheroidal shell is determined by the pulse width τ and t_s.

$$\Delta s \simeq (c\tau/2)[1 + 4d^2 \sin^2 \theta_r/(ct_s - 2d \cos \theta_r)^2]^{1/2}. \qquad (11.23)$$

If $d \to 0$ (i.e., for backscatter), $\Delta s \to c\tau/2$, as it should. The shell has maximum thickness (largest scatter volume) when the beams intersect at a point equidistant from the transmitter and receiver. Furthermore, the shell thickness for forward scatter is always larger than for backscatter if the pulse widths are the same.

We can therefore dispense with the unit pulse functions in Eq. (11.21) if, when computing $\mathbf{E}_1$ sampled at t_s, we add only the signals from scatterers within the spheroidal shell of thickness Δs. To determine the signal at the output of the receiver, we need also to consider the receiver transfer function. For example, to compute the output signal for a matched filter receiver, we must weight the source function $\mathbf{f}(\mathbf{r}, t)$ with a normalized triangular function whose base equals $2\Delta s$.

It can be shown that if the period of the harmonic vibrations is much shorter than the time scales for changes in Δn, the scattered field $\mathbf{E}_1(\mathbf{r}, t)$ is a narrowband signal spread about ω_0, and Eq. (11.19) can be then approximated by

$$\nabla^2 \mathbf{E}_1(\mathbf{r}, t) + k_0^2 \mathbf{E}_1(\mathbf{r}, t) = -\mathbf{f}(\mathbf{r}, t). \tag{11.24}$$

Using Green's function, developed in Appendix E, for the general time-varying function, we obtain the solution of the scattered field at the receiver location.

$$\mathbf{E}_1(\mathbf{r}_{r0}, t) = \frac{1}{4\pi} \int_{-\infty}^{+\infty} \int_V \frac{\mathbf{f}(\mathbf{r}, t')\,\delta(t - t' - r_r/c)}{r_r} dV\, dt', \tag{11.25}$$

where $r_r \equiv |\mathbf{r}_{r0} - \mathbf{r}|$.

The integration with respect to t' can be carried out to obtain

$$\int_{-\infty}^{+\infty} \mathbf{f}(\mathbf{r}, t')\delta(t - t' - r_r/c)\, dt'$$

$$= (2k_0^2 \mathbf{E}_0(\mathbf{r})\, \Delta n(\mathbf{r}, t^*) + 2\nabla\{\mathbf{E}_0(\mathbf{r}) \cdot \nabla[\Delta n(\mathbf{r}, t^*)]\}) e^{j\omega_0 t - jk_0 r_r}$$

$$= \mathbf{f}(\mathbf{r}, t^*), \tag{11.26}$$

where $t^* \equiv t - r_r/c$ is a retardation time. For typical scattering volume dimensions, the time shifts r_r/c relative to r_{r0}/c are assumed to be short compared to the time scales for changes in Δn. Then

$$\Delta n(\mathbf{r}, t^*) \simeq \Delta n(\mathbf{r}, t - r_{r0}/c) \equiv \Delta n(\mathbf{r}, t_0^*),$$

so the solution to the scattered field is

$$\mathbf{E}_1(\mathbf{r}_{r0}, t) = \frac{1}{4\pi} \int_V \frac{\mathbf{f}(\mathbf{r}, t_0^*)e^{-jk_0 r_r}}{r_r} dV. \tag{11.27}$$

With the proviso that receivers are in the far field of the scatter volume (the conditions for this are given in Section 11.4), the signals at two receivers spaced along the direction $\mathbf{a}_{r0}$ will have the same time dependence, except that one will be delayed with respect to the other by a time difference equal to the difference in retardation times. However, receivers located at the same range r_{r0} but along a different direction $\mathbf{a}_{r0}$ will not necessarily have the same time dependence because the exponential term in Eq. (11.27) will have different phases for the same $\mathbf{r}$; hence the integral (11.27) can have different values.

All Doppler shifts are contained in the complex function $\mathbf{E}_1(\mathbf{r}_{r0}, t)$. We show in Section 11.4 how irregularities of Δn that drift with uniform velocity produce a mean Doppler shift. If there is no mean motion, then the time variations of $\mathbf{E}_1$ produce a spectrum whose broadness about ω_0 depends on the rapidity of changes in $\mathbf{E}_1$ due to changing Δn, which is the only parameter in $\mathbf{f}(\mathbf{r}, t)$ that has other than harmonic time variation.

For narrow-beam antennas and the typical ranges used in remote sensing, we can use the approximation

$$r_r \approx r_{r0} \tag{11.28}$$

in the denominator of Eq. (11.27). For $\mathbf{E}_0$ in Eq. (11.26) we shall use

$$\mathbf{E}_0 = \mathbf{A}_0(\mathbf{r})e^{-jk_0 r_t}/r_t, \tag{11.29}$$

where $\mathbf{A}_0$ contains the polarization and angular dependence (i.e., the radiation pattern) of the transmitted field. For reasons already cited, we can also approximate the denominator of Eq. (11.29) by

$$r_t = |\mathbf{r}_{t0} - \mathbf{r}| \approx r_{t0}. \tag{11.30}$$

Substitution of Eqs. (11.29) and (11.30) into Eq. (11.20) and the result into Eq. (11.27) gives the scatter field intensity.

$$\mathbf{E}_1(\mathbf{r}_{r0}, t) = \frac{1}{2\pi r_{t0} r_{r0}} (\mathbf{I}_1 + \mathbf{I}_2), \tag{11.31a}$$

where

$$\mathbf{I}_1 = k_0^2 \int_V \Delta n \, \mathbf{A}_0 e^{-jk_0(r_t + r_r)} \, dV, \tag{11.31b}$$

$$\mathbf{I}_2 = \int_V \nabla[e^{-jk_0 r_t} \mathbf{A}_0 \cdot \nabla(\Delta n)]e^{-jk_0 r_r} \, dV. \tag{11.31c}$$

We now apply the identity (Johnson, 1965, p. 444)

$$\int_V u \, \nabla \phi \, dV \equiv \int_S u\phi \, d\mathbf{S} - \int_V \phi \, \nabla u \, dV \tag{11.32}$$

to Eq. (11.31c), where we define

$$u \equiv e^{-jk_0 r_t}, \tag{11.33a}$$

$$\phi \equiv \mathbf{A}_0 \cdot \nabla(\Delta n)e^{-jk_0 r_t}. \tag{11.33b}$$

The surface integral can be set equal to zero by letting the surface of integration move beyond the limits of the scattering volume (i.e., the region of space that contributes significantly to the scattered field). Equation (11.31c) is therefore reduced to

$$I_2 = -\int_V \{\mathbf{A}_0 \cdot \nabla_r(\Delta n)\} e^{-jk_0 r_t} \nabla_r e^{-jk_0 r_r} \, dV, \tag{11.34}$$

where, to emphasize that ∇ operates only on the $\mathbf{r}$ and not the $\mathbf{r}_{r0}$ dependence of the functions, we have added the subscript r. Now,

$$\nabla_r e^{-jk_0 r_r} = -jk_0 e^{-jk_0 r_r} \nabla_r(r_r), \tag{11.35}$$

and it can be shown that, without approximation,

$$\nabla_r(r_r) = \frac{r_{r0}}{2} \left(\frac{2r\mathbf{a}_r}{r_{r0}^2} - \frac{2\mathbf{a}_{r0}}{r_{r0}} \right) \left(1 - \frac{2\mathbf{a}_{r0} \cdot \mathbf{r}}{r_{r0}} + \frac{r^2}{r_{r0}^2} \right)^{-1/2}, \tag{11.36}$$

where $\mathbf{a}_r$ is the unit vector in the direction of $\mathbf{r}$. Because we have assumed a scattering volume dimensions small compared to r_{r0} (e.g., $r/r_{r0} \ll 1$),

$$\nabla_r(r_r) \approx -\mathbf{a}_{r0}. \tag{11.37}$$

Thus I_2 can be approximated by

$$I_2 = -jk_0\mathbf{a}_{r0} \int_V [\mathbf{A}_0 \cdot \nabla_r(\Delta n)] e^{-jk_0(r_t + r_r)} \, dV. \tag{11.38}$$

Using the identity (11.32) and the argument that follows it, the equation

$$\int_V \mathbf{u} \cdot \nabla \phi \, dV = -\int_V \phi \nabla \cdot \mathbf{u} \, dV \tag{11.39}$$

can be derived. Assigning

$$\mathbf{u} = \mathbf{A}_0 e^{-jk_0(r_t + r_r)}, \tag{11.40a}$$

$$\phi = \Delta n, \tag{11.40b}$$

and applying Eq. (11.39) to Eq. (11.38), we obtain

$$I_2 = j\mathbf{a}_{r0}k_0 \int_V \Delta n \, \nabla_r \cdot [\mathbf{A}_0 e^{-jk_0(r_t + r_r)}] \, dV. \tag{11.41}$$

Because the pattern weighting function varies slowly over distances comparable to a wavelength, Eq. (11.41) can be reduced to

$$I_2 = -\mathbf{a}_{r0}k_0^2(\mathbf{a}_{r0} + \mathbf{a}_{r0}) \cdot \int_V \mathbf{A}_0 \Delta n \, e^{-jk_0(r_t + r_r)} \, dV \tag{11.42}$$

by applying Eqs. (11.35) and (11.37).

For the condition of a narrow beam of transmitted radiation, $\mathbf{A}_0$ is nearly perpendicular to $\mathbf{a}_{t0}$ over the region where $\mathbf{A}_0$ is significantly large. Therefore, the second term of Eq. (11.42) is negligibly small. Furthermore, observe that the first portion of Eq. (11.42) exactly cancels the longitudinal component (i.e., the component in the direction of $\mathbf{a}_{r0}$) of Eq. (11.31b). Therefore only the transverse component (i.e., transverse to $\mathbf{a}_{r0}$) of Eq. (11.31b) contributes to the scattered field, whereas all the contribution of $\mathbf{I}_2$ is canceled. This transverse component is obtained from Eq. (11.31b) by forming the vector product $\mathbf{a}_{r0} \times \mathbf{A}_0$ with $\mathbf{a}_{r0}$. Therefore the solution (11.27) for the scattered field reduces to

$$\mathbf{E}_1(\mathbf{r}_{r0}, t) = -\frac{k_0^2 \mathbf{a}_{r0}}{2\pi r_{r0} r_{t0}} \times \int (\mathbf{a}_{r0} \times \mathbf{A}_0) \, \Delta n(\mathbf{r}, t_0^*) e^{-jk_0(r_t+r_r)} \, dV. \quad (11.43)$$

Although Eq. (11.43) is our basic solution to the scattered field the integration is not easily performed, because Δn is usually a random function whose statistical properties only are known. Because we shall be considering scatter from random media, the detailed relation between time change in Δn and the corresponding time change in $\mathbf{E}_1$ is of no concern; rather, it is the statistical properties of the echoes that count. Therefore, the constant retardation time delay r_{r0}/c in Eq. (11.43) can be ignored without affecting the statistics of the received signal.

In Section 11.4 we make simplifying assumptions to show clearly how the electromagnetic wave samples certain scales from the spectrum of scales contained in the field of irregularities Δn. Furthermore, these assumptions simplify the development of the solution for the statistical properties of the random variable $\mathbf{E}_1$.

11.4 Small Volume Scatter

We now introduce conditions that will specify that the phase fronts of the refracted incident wave are essentially planar over the scattering volume and that the receiver antenna is in the far field of this volume. These conditions are derived by expanding the exponential terms of Eq. (11.43) in a Taylor series.

$$r_r = |\mathbf{r}_{r0} - \mathbf{r}| = r_{r0} - \mathbf{a}_{r0} \cdot \mathbf{r} + (1/2r_{r0})[r^2 - (\mathbf{a}_{r0} \cdot \mathbf{r})^2] + \cdots \quad (11.44)$$

and retaining only the first-order terms. If we are to neglect the second-order terms, it is necessary that the third term of Eq. (11.44) multiplied by k_0 satisfy

$$(k_0/2r_{r0})[r^2 - (\mathbf{a}_{r0} \cdot \mathbf{r})^2] \ll 1. \quad (11.45)$$

Condition (11.45) stipulates that the maximum dimension of the scattering volume transverse to $\mathbf{a}_{r0}$ must satisfy

$$r_{\max,r} \ll \sqrt{\lambda r_{r0}/\pi}. \quad (11.46)$$

Equation (11.46) specifies that the receiver is in the far field of the scattering volume. In a like manner we obtain, on expansion of r_t, the condition

$$r_{\max,t} \ll \sqrt{\lambda r_{t0}/\pi}. \tag{11.47}$$

which limits the incident phase fronts to be essentially planar over the scattering volume.

Often $r_{\max,r}$ is determined by the field of view $r\theta_1$ of the receiving antenna (Fig. 11.3), and not by the spatial extent of Δn, then

$$r_{\max,r} \approx \theta_1 r_{r0}/2 \approx 0.6\lambda r_{r0}/D, \tag{11.48}$$

which is always larger than $\sqrt{\lambda r_{r0}/\pi}$ for the r_0 in the far field of the receiving antenna having diameter D. If V_c were in the near field of the antenna, the dimension $r_{\max}$ would be equal to the diameter D of the antenna, and even then condition (11.46) would be violated.

It seems that we have derived conditions for which the ensuing solutions have limited applicability. Nevertheless, it can be shown (Section 11.5.1) that if V_c has dimensions much larger than the longest correlation length of Δn, then V_c may be divided into smaller subvolumes (i.e., scattering "blobs" having dimensions roughly the size of the correlation lengths). The scattering signals from blobs add incoherently and hence the average received power is an integral of the power contributed by each.

Assuming that condition (11.46) is satisfied,

$$e^{-jk_0 r_r} \approx e^{-jk_0 r_{r0} + jk_0 \mathbf{a}_{r0} \cdot \mathbf{r}}. \tag{11.49}$$

Similar arguments can be developed for r_t, and so

$$e^{-jk_0 r_t} \approx e^{-jk_0 r_{t0} + jk_0 \mathbf{a}_{t0} \cdot \mathbf{r}}, \tag{11.50}$$

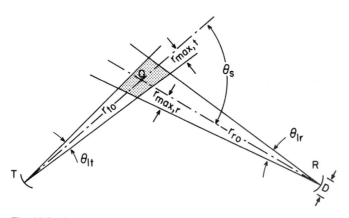

Fig. 11.3 Scattering volume size determined by the beamwidths θ_{1t}, θ_{1r}. The common beam volume V_c is shaded, and θ_1 is the one-way 3-dB beamwidth of an assumed circularly symmetric antenna aperture.

which implies that the phase fronts of the incident field are planar (to within a fraction of a wavelength) across the scatter volume V_s. In these approximations, V_s is small compared to V_c, so $\mathbf{A}_0(\mathbf{r})$ is nearly constant over the scatter volume. This, and substitution Eqs. (11.49) and (11.50) into Eq. (11.43), result in the field intensity at the receiver location.

$$\mathbf{E}_1(\mathbf{r}_{r0}, t) = -\frac{k_0^2 \mathbf{a}_{r0} \times (\mathbf{a}_{r0} \times \mathbf{A}_0)}{2\pi r_{r0} r_{t0}} e^{-jk_0(r_{t0}+r_{r0})} C_1, \tag{11.51}$$

where

$$C_1 = \int_V \Delta n(\mathbf{r}, t) e^{-jk_0 \mathbf{m} \cdot \mathbf{r}} \, dV, \tag{11.52}$$

$$\mathbf{m} = -\mathbf{a}_{t0} - \mathbf{a}_{r0}, \qquad |\mathbf{m}| = 2\sin(\theta_s/2), \tag{11.53}$$

θ_s, the scatter angle, is the angle between $-\mathbf{a}_{t0}$ and $\mathbf{a}_{r0}$, and $k_0\mathbf{m}$ is termed a mirror wave number. In developing this solution, we noted that the term $2\nabla[\mathbf{E}_0 \cdot \nabla \Delta n(\mathbf{r}, t)]$ on the right-hand side of Eq. (11.20) contributes only a component of the field intensity parallel to $\mathbf{a}_{r0}$, canceling a like component resulting from the first term. Because the term $2\nabla[\mathbf{E}_0 \cdot \nabla \Delta n(\mathbf{r}, t)]$ does not contribute to the first-order transverse field components at $\mathbf{r}_{r0}$, we may write, for a first-order scatter theory, the approximate differential equation

$$\nabla^2 \mathbf{E}_1 + k_0^2 \mathbf{E}_1 = -2k_0^2 \Delta n \, \mathbf{E}_0, \tag{11.54}$$

provided one ignores the longitudinal component of the resulting solution.

11.4.1 Bragg Scatter

Equation (11.52) is the basic integral for Fraunhofer scatter. For the purpose of better understanding this integral, we shall discuss some of its properties by considering simplified examples. The vector $\mathbf{m}$ is dimensionless, and its magnitude is given by Eq. (11.53). The direction of $\mathbf{m}$ defines a mirror direction. Planes of discontinuity or of rapid change of Δn perpendicular to this direction are able to reflect the incident ray of direction $-\mathbf{a}_{t0}$ into its mirror direction $\mathbf{a}_{r0}$. The integral (11.52) gives the Fourier composition of Δn along this direction; hence the intensity of the scattered signal depends directly on the Fourier content of the irregularities in the mirror direction.

In a rectangular coordinate system chosen such that the direction of the z axis lies along the mirror direction, Eq. (11.52) reduces to

$$C_1 = \int_V \Delta n(\mathbf{r}, t) e^{-j2k_0 z \sin(\theta_s/2)} \, dx \, dy \, dz. \tag{11.55}$$

Thus in order that C_1 be relatively large, the irregularities should have significant Fourier components with a spatial wavelength (structure wavelength)

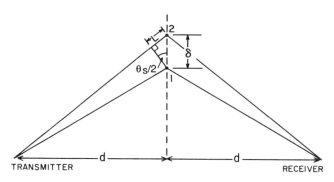

Fig. 11.4 Geometry showing the distance between scattering centers that produces constructive interference.

in the mirror direction. From Eq. (11.55) it can be deduced that the structure wavelength Λ that produces constructive interference at the receiver is

$$\Lambda = \lambda/2\sin(\theta_s/2) \tag{11.56}$$

measured along the vector direction **m**. To give this result a physical interpretation, consider a scatterer located at a point 1 and another located at a point 2, a distance δ directly above 1. For the fields scattered from 2 to interfere constructively *at the receiver* with the fields scattered from 1, the path length difference $2L$ must be equal to λ or to some integer multiple of λ. From the geometry depicted in Fig. 11.4 we find that

$$\delta = L/\sin(\theta_s/2) = \lambda/2\sin(\theta_s/2), \tag{11.57}$$

which corresponds to Eq. (11.56). This equation is Bragg's law, which gives the condition for waves reflected from adjacent scattering planes to be in phase.

Example 1: Frozen Irregularities

Let us now consider a simple example in which $\Delta n(\mathbf{r}, t)$ is given by

$$\Delta n(\mathbf{r}) = \Delta n \cos(2\pi x/\Lambda_x)\cos(2\pi z/\Lambda_z), \tag{11.58}$$

where Λ_x, Λ_z are the structure wavelengths of $\Delta n(\mathbf{r})$ along the coordinate axes x, z, respectively, and Δn is the amplitude of variation. Assume a symmetrical forward-scatter link for which $\mathbf{m} = 2\mathbf{a}_z \sin\theta_s/2$, where $\mathbf{a}_z$ is a unit vector in the z direction. Consider that $\Delta n(\mathbf{r})$ is contained in a box of dimensions as shown in Fig. 11.5 and that outside the box $\Delta n(\mathbf{r}) = 0$. For these stipulated conditions, Eq. (11.55) becomes

$$C_1 = (4\,\Delta n\, L_y\Lambda_x/\pi)\sin(2\pi L_x/\Lambda_x)\int_0^{L_z}\cos(K_z z)\cos(k_0' z)\,dz, \tag{11.59}$$

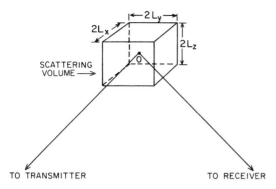

Fig. 11.5 Scattering from irregularities in a rectangular box.

where $K_z \equiv 2\pi/\Lambda_z$ and $k_0' \equiv 2k_0 \sin(\theta_s/2)$. We note immediately from Eq. (11.59) that C_1 has a periodic variation of intensity as L_x changes. Let us assume a fixed value for L_x that gives a nonzero value to C_1 and define

$$A = (4\,\Delta n\, L_y \Lambda_x/\pi)\sin(2\pi L_x/\Lambda_x). \tag{11.60}$$

Performing the integration in Eq. (11.59) yields

$$C_1 = \frac{AL_z}{2}\left[\frac{\sin(K_z + k_0')L_z}{(K_z + k_0')L_z} + \frac{\sin(K_z - k_0')L_z}{(K_z - k_0')L_z}\right]. \tag{11.61}$$

The dependence of the two terms in the brackets is sketched in Fig. 11.6 using k_0' as the independent variable. It is immediately obvious that the maximum contribution to the scattered signal occurs if

$$k_0' = K_z \quad \text{or} \quad \lambda = 2\Lambda_z \sin(\theta_s/2), \tag{11.62}$$

in agreement with Eq. (11.56). We note that if Eq. (11.62) is satisfied, then the scatter intensity is proportional to L_z, and not periodic in L_z as it is for the case in which we vary L_x.

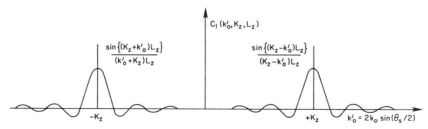

Fig. 11.6 Scatter intensity $C_1(k_0', K_z, L_z)$ dependence on the electromagnetic wave number k_0 for sinusoidal refractive index fluctuations.

11.4.2   Radio Acoustic Sounding System (RASS)

In this section use of Bragg scatter for remote measurement of temperature is described. By combining radar and acoustic techniques it is possible to obtain temperature profiles of the atmosphere. Briefly the RASS consists of an acoustic source that beams sound waves toward the zenith; these waves cause modulations of the refractive index that scatter or reflect radio waves (Doviak and Zrnić (1984b) show that reflection formulation is equivalent to the scattering formulation). Although RASS was proposed in the early 1960s (Fetter, 1961; Smith, 1961) researchers begun to use it in the late 1970s (Marshall et al., 1972; Frankel and Peterson, 1976) at about the time that Doppler weather radar techniques matured. Early attempts were limited in altitude to about 1 km by strong acoustic attenuation, by horizontal advection of the sound wave and by relatively primitive signal processors. Advances in signal processing techniques and sophisticated beam steering have allowed temperature measurements to as high as 20 km (Masuda, 1988). It is significant that RASS can be easily added to existing wind profiling radars (see Section 11.8.1.1); because such addition is economical, it could supplement temperature soundings by radio sondes.

In the RASS technique the local speed of sound generated by an acoustic source is estimated with a coherent radar. The virtual temperature T_v is related to the speed of sound c_a (m s^{-1}) by

$$T_v = (c_a/20.047)^2 \quad \text{(K)}. \tag{11.63}$$

There are two methods that can be used to determine c_a; one is applicable to acoustic bursts that are much shorter than the range extent of the radar's resolution volume (Section 4.4.2); the other is used if the acoustic source is continuous.

In the first technique acoustic pulses of relatively short length (i.e., $c_a \tau_a \ll c\tau$) are transmitted at a slow rate (<1 Hz). A monostatic or bistatic (pulsed or continuous wave) radar can be used to observe the acoustic disturbance as it passes through the radar's resolution volume. The disturbance backscatters the radio waves, creating echoes somewhat like those from a point scatterer traveling at the local velocity of sound. Then spectral analysis of the signal (at a fixed range delay) reveals the Doppler shift at the local speed of sound.

In the second technique a superposition of acoustic sinusoids (e.g., white noise) is continuously transmitted; then spectral analysis retrieves the acoustic frequency f_a, which has a local (i.e., in V_6) wavelength Λ matched to one-half the radar wavelength. Thus the retrieved f_a and known Λ yield c_a and virtual temperature. The essence of both methods will be illustrated with realistic but simplified examples.

Example 2:   RASS with a Short Acoustic Pulse

A tutorial explanation of this and other examples can be found in May et al. (1990). Consider intense pressure fluctuations of the acoustic wave that modulate the radio refractive index, producing a traveling pulsed sinusoidal perturbations in n.

$$\Delta n(\mathbf{r}, t) = \Delta n \cos[K_z(z - c_a t)]U(z - c_a t), \qquad (11.64)$$

where

$$U(z - c_a t) = \begin{cases} 1 & \text{for } c_a t - L_z < z < c_a t + L_z \\ 0 & \text{otherwise,} \end{cases}$$

and $K_z = 2\pi/\Lambda$ is the acoustic wave number. In this formulation the acoustic pulse of length $2L_z$ is at the center of V_6 at time $t = 0$; we assume that the radar transmitter and receiver and the acoustic source are collocated.

Spectral analysis of the radar signal is done along the sample time axis ($t = nT_s$) for a fixed range time. We substitute Eq. (11.64) into Eq. (11.55) and execute the integration to obtain

$$C_1 = \frac{1}{2}V \Delta n \left[\frac{\sin(K_z - 2k_0)L_z}{(K_z - 2k_0)L_z} + \frac{\sin(K_z + 2k_0)L_z}{(K_z + 2k_0)L_z} \right] e^{-j2k_0 c_a t} \qquad (11.65a)$$

where V is the volume of the disturbance contained within V_6 and $\omega_d = 2k_0 c_a$ is the Doppler frequency, which is the same as for a point scatterer [Eq. (3.30)] moving at a velocity c_a. Thus spectral processing of signals backscattered as the acoustic pulse traverses V_6 produces a power spectrum with a peak at c_a. As shown in Fig. 11.6 (where $k_0' = 2k_0$) the scattered field at the receiver is maximized if $2k_0 = K_z$. In general the scattered field obtained from Eq. (11.51) is

$$E_1(\mathbf{r}_0, t) \approx \left[\frac{\sin(K_z - 2k_0)L_z}{(K_z - 2k_0)L_z} \right] e^{j(\omega_0 - 2k_0 c_a)t - j2k_0 r_0} \qquad (11.65b)$$

for $2k_0$ near K_z. Therefore a negative Doppler shift is expected for refractive index perturbations moving away from the radar. To have large SNR the acoustic wavelength should be set to produce maximum return (i.e., K_z near $2k_0$). But in the presence of large vertical gradients of temperature this match may not be possible at all heights of interest; hence a change in acoustic frequency is required to probe various heights. RASS that use continuous-wave bistatic radars operate in this mode of short acoustic pulse (Bonino et al., 1979). Range information is obtained by performing Fourier transforms on contiguous blocks of backscattered signals and referencing each block to the transmit time of the acoustic pulse.

The conclusions drawn from this example are not restricted to a short rectangular acoustic pulse but are valid for short acoustic pulses with an arbitrary shape of the envelope (see Problem 11.1).

Example 3: RASS with a Continuous Acoustic Source

In this example we consider again a collocated radar and an acoustic source that is continuously emitting white noise within a band of frequencies centered on $k_0 \bar{c}_a/\pi$ where $\bar{c}_a$ is the nominal speed of sound. Thus, on the average, the traveling wave of refractive index perturbations [Eq. (11.64)] would be a superposition of sinusoids that have equal mean square fluctuations (i.e., $E[(\Delta n_i)^2] =$ constant) but random phase ψ_i.

$$\Delta n(\mathbf{r}, t) = \sum_i \Delta n_i \cos[K_{zi}(z - c_a t) + \psi_i]. \tag{11.66a}$$

Assuming a uniform range weighting function of width $c\tau/2$ (i.e. $L_z = c\tau/4$), substitution of Eq. (11.66a) into Eq. (11.55) and integration produces C_1.

$$C_1 \approx \frac{1}{2} V_6 \sum_i \Delta n_i \frac{\sin(K_{zi} - 2k_0)L_z}{(K_{zi} - 2k_0)L_z} e^{-j(K_{zi} c_a t - \psi_i)}. \tag{11.66b}$$

The term $\sin(K_{zi} + 2k_0)L_z/(K_{zi} + 2k_0)L_z$ has been neglected because K_{zi} is near $2k_0$. It has been assumed that the beamwidth of the acoustic source is larger than that of the radar.

It is important to note that each sinusoid of the acoustic band generates, in the power spectrum of the backscattered signals, a spectral component at its transmitted frequency ($K_{zi} c_a/2\pi$); in other words, there is no apparent Doppler shift! That is, if a continuous acoustic sinusoid were transmitted there would be no way to estimate the temperature (unless one had the capability to adjust the radar wavelength to obtain a peak in backscattered power at $\lambda = 2\Lambda$). But, because a band of acoustic sinusoids is transmitted; it is possible to identify the one that produces the strongest echo from each V_6. This is the sinusoid of frequency f_a equal to the Doppler shift at the spectral peak, and it has a corresponding wavelength that is Bragg-matched to the radar wavelength. For example, spectral analysis of Eq. (11.66b) produces mean power spectral coefficients S_i proportional to

$$S_i \sim E[(\Delta n_i)^2] \left| \frac{\sin(K_{zi} - 2k_0)L_z}{(K_{zi} - 2k_0)L_z} \right|^2, \tag{11.67}$$

where $E[(\Delta n_i)^2]$ is the mean square amplitude of the refractive index fluctuation at the acoustic wavelength Λ_i. A sharp peak of S_i [Eq. (11.67)] occurs where Λ_i is matched to the Bragg electromagnetic wavelength (i.e., $\Lambda_i = \lambda/2$). Thus the peak of the power spectrum would indicate the acoustic frequency for which this condition occurs. From that estimated frequency and the K_{zi} (known from k_0) the speed of sound is found, leading to temperature via Eq. (11.63). Our derivation assumed a rectangular EM pulse in order to produce analytic expressions for the electric field and power spectrum coefficients. This assumption is not necessary (see Problem 11.2) and would not be precise in practice because the range weighting function $W(r)$ has a more complicated shape (Chapter 4).

Attenuation of the acoustic wave can be severe, and it increases dramatically with frequency (Harris, 1966). Advection of the spherical wave fronts by the horizontal wind causes misalignment of the acoustic and radio wave fronts,

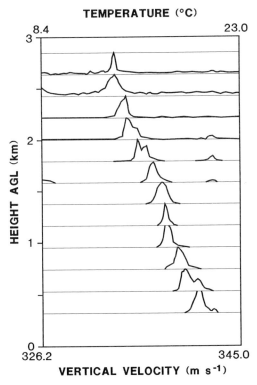

TEMPERATURE (°C)

VERTICAL VELOCITY (m s⁻¹)

Fig. 11.7 Doppler spectra and corresponding temperatures measured by a RASS. Power scale is linear and data were obtained on 10 September 1991 in Erie, Colorado. (Courtesey D. Strauch, NOAA.)

reducing considerably the coherent addition of echoes from each wavelet of the acoustic pulse; moreover if the wind is sufficiently strong it could move the acoustic waves outside of the radar beam (Peters *et al.*, 1983). Vertical winds produce additive frequency shifts, which must be removed from the estimated acoustic shifts to avoid bias in estimates of temperature (May *et al.*, 1989). With profiling radars this can be done because the vertical velocity of air is directly measured (Section 11.8.1.1).

An example of the Doppler shifts and corresponding temperatures at 12 heights obtained from a RASS in which the radar frequency was 404 MHz is shown in Fig. 11.7. A decrease with height is evident and isothermal layers appear between 1 and 1.4 km, and above 2.4 km.

11.4.3 Expected Scattered Power Density

Refractive index irregularities usually do not have such well-behaved deterministic variations as illustrated in the example of Section 11.4.1. It is more likely that the statistical properties of the irregularities will only be known. Therefore in this section we relate the expected (i.e., time or ensemble average) power density incident at the receiver location $\mathbf{r}_{r0}$ to these statistical properties. The power density (Section 2.1), averaged over a cycle of the radio wave, is

$$S = \tfrac{1}{2}\mathrm{Re}(\mathbf{E}_1 \times \mathbf{H}_1^*),\qquad(11.68\mathrm{a})$$

where, from Eq. (11.1),

$$\mathbf{H}_1 = -\nabla \times \mathbf{E}_1/j\omega\mu_0.\qquad(11.68\mathrm{b})$$

In computing $\mathbf{H}_1$ we assume that, at the receiver location, the primary field $\mathbf{E}_0$ is negligible compared to $\mathbf{E}_1$. Since the $\mathbf{E}_1$ and $\mathbf{H}_1$ fields are transverse to $\mathbf{a}_{r0}$ and to each other, as can be seen from (11.51) and (11.68b), it can be shown that

$$|\mathbf{H}_1| \approx |\mathbf{E}_1|\sqrt{\frac{\varepsilon_0}{\mu_0}}.$$

Furthermore,

$$|-\mathbf{a}_{r0} \times (\mathbf{a}_{r0} \times \mathbf{A}_0)| = A_0 \sin\chi,\qquad(11.69)$$

where χ is the angle between the vectors $\mathbf{A}_0$ and $\mathbf{a}_{r0}$. Thus the Poynting vector $\mathbf{S}$ is directed along $\mathbf{a}_{r0}$ and has intensity

$$S = \tfrac{1}{2}|\mathbf{E}_1|^2/\eta_0,\qquad \eta_0 \equiv \sqrt{\mu_0/\varepsilon_0}.$$

Substituting Eq. (11.51) for $\mathbf{E}_1$ results in

$$S = (A_0^2 k_0^4 \sin^2\chi/8\eta_0\pi^2 r_{r0}^2 r_{t0}^2)C_1 C_1^*,\qquad(11.70)$$

which, upon substitution for C_1, becomes

$$S = \frac{A_0^2 k_0^4 \sin^2\chi}{8\eta_0\pi^2 r_{t0}^2 r_{r0}^2}\left|\int_V \Delta n(\mathbf{r},t)e^{-jk_0\mathbf{m}\cdot\mathbf{r}}\,dV\right|^2.\qquad(11.71)$$

Since $\Delta n(\mathbf{r},t)$ is a random variable, so is S. We shall determine its expected or mean value $\langle S\rangle$, which is related to the statistical characteristics of Δn. Before we proceed, however, let us write

$$C_1(t)C_1^*(t') = \int_V \Delta n(\mathbf{r},t)e^{-jk_0\mathbf{m}\cdot\mathbf{r}}\,dV \int_V \Delta n(\mathbf{r}',t')e^{jk_0\mathbf{m}\cdot\mathbf{r}'}\,dV',$$

where $\mathbf{r}$ and $\mathbf{r}'$ are variables of integration and $C_1(t)$, $C_1(t')$ are samples of the complex (i.e., I and Q) signal at times t, t', respectively. Taking an ensemble average, we obtain

$$C_1(t)C_1^*(t') = \int_V\int_V \langle\Delta n(\mathbf{r},t)\Delta n(\mathbf{r}',t')\rangle e^{-jk_0\mathbf{m}\cdot(\mathbf{r}-\mathbf{r}')}\,dV\,dV'.\qquad(11.72)$$

The term

$$\langle\Delta n(\mathbf{r},t)\Delta n(\mathbf{r}',t')\rangle \equiv R(\mathbf{r},\mathbf{r}',t,t')\qquad(11.73\mathrm{a})$$

is the autocorrelation of the random variable $\Delta n(\mathbf{r}, t)$. In the case of temporally and spatially homogeneous turbulence,

$$R(\mathbf{r}, \mathbf{r}', t, t') = R(\mathbf{r} - \mathbf{r}', t - t') = R(\boldsymbol{\rho}, \tau_1). \qquad (11.73b)$$

We briefly digress to discuss the relation of $C_1(t)$ to the Doppler spectrum and correlation functions developed in earlier chapters. The received complex signals $V(t)$ and $V(t')$ discussed in Chapter 5 are directly proportional to $C_1(t)$ and $C_1(t')$. Hence $\langle C_1(t)C_1^*(t') \rangle$ is proportional to the received signal correlation $\langle V(t)V^*(t') \rangle$, and its Fourier transform is the Doppler spectrum. Thus, given the time dependence of the correlation function for refractive index irregularities, we can directly derive the Doppler spectrum. However, relating this spectrum to the turbulent velocity field is not so straightforward. In Chapter 10 we discussed elements of turbulence theory to derive the statistical characteristics of velocity irregularities under the assumption of isotropic turbulence. We illustrated that the space–time correlation function for the velocity field requires solving a rather difficult differential equation, Eq. (10.39). Even though we may find a solution to this equation, we are still confronted with the problem of relating the statistical properties of the velocity field to those for the refractive index irregularities generated by the turbulence.

Although we offer no rigorous solution that relates the statistical properties of the velocity field to the space–time correlation of Δn, we can, through heuristic arguments and assumptions, relate our earlier results for point scatterers in turbulent flow to this case. Later in this section we demonstrate that, given a broad spectrum of irregularities, only those having scales near $\Lambda_0 = \lambda/2$ $\sin(\theta_s/2)$ contribute to the scattered field. We assume that these irregularities act as tracers of larger-scale motion. Hence, we can apply our earlier results (e.g., Section 5.2) that relate Doppler spectra to the distribution of velocities within the resolution volume.

Even though the refractive index irregularities have finite lifetimes and thus do not possess the long coherence time typical of point scatterers such as hydrometeors, the finiteness of the lifetimes can be ignored if they are longer than the period required to decorrelate signals by the reshuffling of Λ_0 scales. To estimate the lifetimes of irregularities on the scale of Λ_0, we shall assume that velocity scales larger than Λ_0 transport (reshuffle) smaller scales and that scales $\Lambda \le \Lambda_0$ deform the shape and destroy irregularities of size Λ_0. The lifetime τ_0 of irregularities on the scale of Λ_0 is roughly equal to Λ_0/v, where v is the rms velocity of scales $\le \Lambda_0$. To calculate v we integrate the variance spectral density $2E(K)$ from $K_0 = 2\pi/\Lambda_0$ to infinity, which gives the velocity variance of scales smaller than Λ_0. If $K_0 \ll K_i$ where K_i is the wave number of the inner scale (i.e., where dissipation of kinetic energy due to viscosity begins), we can estimate v approximately by taking the square root of twice the integral of $E(K)$ given by Eq. (10.66) between K_0 and infinity assuming $A = 1.6$. Thus $\tau_0 \approx 0.84\Lambda_0^{2/3}\varepsilon^{-1/3}$, in agreement with the estimate made by Kristensen (1979). Thus the stronger the turbulence (i.e., the larger ε), the shorter is the lifetime of irregularities. For example, if $\Lambda_0 = 0.05$ m and turbulence is moderate ($\varepsilon \approx 10^{-2}$ m^2 s^{-3}), then $\tau_0 \simeq 0.8$ s, which is much longer than the time usually required to reshuffle irregularities a distance $\Lambda_0/2 \simeq 0.02$ m. Thus, the formulas previously derived that relate the velocity field to the Doppler spectrum (e.g., spectrum widths and eddy dissipation rate) also apply to the case in which refractive index irregularities are the scatterers. Henceforth we assume $\tau_1 = 0$ and proceed to work with $R(\rho, 0)$ alone to determine the average power scattered by Δn.

Consider for the sake of illustration a parallelopiped of dimensions $2L_x$, $2L_y$, $2L_z$ along the x, y, z axes, respectively. The integral (11.72) can be then written

$$\int\limits_{-L_x}^{L_x} \int\limits_{-L_y}^{L_y} \int\limits_{-L_z}^{L_z} R(\rho)\exp[-jk_0(\mathbf{m}\cdot\rho)]\, dx\, dx'\, dy\, dy'\, dz\, dz', \qquad (11.74)$$

where

$$\rho = \mathbf{a}_x(x - x') + \mathbf{a}_y(y - y') + \mathbf{a}_z(z - z') = \mathbf{a}_x\alpha + \mathbf{a}_y\beta + \mathbf{a}_z\gamma, \qquad (11.75)$$

and $\mathbf{a}_x$, $\mathbf{a}_y$, $\mathbf{a}_z$ are unit vectors along x, y, z, respectively. On substituting Eq. (11.75) into Eq. (11.74), we obtain

$$\langle|C_1|^2\rangle = \int\limits_{-L_x}^{L_x} \int\limits_{-L_y}^{L_y} \int\limits_{-L_z}^{L_z} R(\alpha, \beta, \gamma)e^{-jk_0(m_x\alpha + m_y\beta + m_z\gamma)}\, dV\, dV', \qquad (11.76)$$

where $k_0 m_x$, etc., are the component directions of the mirror wave number. Consider first the integration over x and x'. Using the identity (Papoulis, 1965)

$$\int\limits_{-L_x}^{L_x} R(x - x')\{\exp[-js(x - x')]\}\, dx\, dx'$$

$$= \int_{-2L_x}^{2L_x} R(\alpha)e^{-js\alpha}(2L_x - |\alpha|)\, d\alpha, \qquad (11.77)$$

we can reduce the double integral over $2L_x$ to a single integral over α, which results in Eq. (11.76) taking the form,

$$\langle|C_1|^2\rangle = \int\limits_{-L_y}^{L_y} \int\limits_{-L_z}^{L_z} \left[\int_{-2L_x}^{2L_x} R(\alpha, \beta, \gamma)e^{-jk_0m_x\alpha}(2L_x - |\alpha|)\, d\alpha \right]$$

$$\times e^{-jk_0m_y\beta - jk_0m_z\gamma}\, dy\, dy'\, dz\, dz'. \qquad (11.78)$$

Interchanging the order of integration and applying Eq. (11.77) again, to the y and z integrals, Eq. (11.78) reduces to

$$\langle |C_1|^2 \rangle = \iiint_{V_2} \exp(-jk_0\mathbf{m}\cdot\boldsymbol{\rho})R(\boldsymbol{\rho})(2L_z - |\alpha|)$$
$$\times (2L_y - |\beta|)(2L_z - |\gamma|)\,dV_\rho, \tag{11.79}$$

where $dV_\rho \equiv d\alpha\,d\beta\,d\gamma$. If the dimensions in any direction of the volume are much larger than the correlation lengths α_c, β_c, and γ_c, the product terms $(2L_x - \alpha)(2L_y - \beta)(2L_z - \gamma)$ can be approximated by the volume $8L_xL_yL_z = V$ and the limits reduced from V_2 to V; the expected power density is then

$$\langle S \rangle = \frac{A_0^2 k_0^4 \sin^2\chi}{8\eta_0\pi^2 r_{t0}^2 r_{r0}^2} V \int_V \exp(-jk_0\mathbf{m}\cdot\boldsymbol{\rho})R(\boldsymbol{\rho})\,dV_\rho. \tag{11.80}$$

The correlation length is taken to be that value of α (or β or γ) for which $R(\alpha, \beta, \gamma)$ is insignificant.

It is instructive to compare Eq. (11.80) to Eq. (11.71), the latter giving the power density of any sample of the ensemble or, in other words, the power density averaged over a cycle of ω_0 at any instant of time. Thus from Eq. (11.71) we note that S depends on the square of the total volume V of integration, but $\langle S \rangle$ is linearly dependent on the product of V and some fraction of V over which the autocorrelation $R(\boldsymbol{\rho})$ is significant.

Let us consider the case in which dimensions of V satisfy the conditions (11.46) and (11.47), although we shall assume that the volume size is large enough that Eq. (11.80) is a good approximation to Eq. (11.79). We now use the spectral representation of the correlation function, as defined in Chapter 10.

$$R(\boldsymbol{\rho}) = \iiint_{-\infty}^{+\infty} \exp(j\mathbf{K}\cdot\boldsymbol{\rho})\Phi_n(\mathbf{K})\,dV_K.$$

Substituting this into the following integral, obtained from Eq. (11.80),

$$\langle |\mathcal{I}_1|^2 \rangle \equiv \frac{1}{8\pi^3} \int_V R(\boldsymbol{\rho})\exp(-jk_0\mathbf{m}\cdot\boldsymbol{\rho})\,dV_\rho,$$

we have

$$\langle |\mathcal{I}_1|^2 \rangle = \iiint_{-\infty}^{+\infty} \Phi_n(\mathbf{K})\left[\frac{1}{8\pi^3}\int_V \exp[j(\mathbf{K}-k_0\mathbf{m})\cdot\boldsymbol{\rho}]\,dV_\rho\right]dV_K. \tag{11.81}$$

Let us now examine the inner integral and define a sampling function

$$F(\mathbf{q}) \equiv \frac{1}{8\pi^3} \int_V \exp(j\mathbf{q} \cdot \boldsymbol{\rho}) \, dV_\rho, \qquad \mathbf{q} \equiv \mathbf{K} - k_0\mathbf{m}. \qquad (11.82)$$

When the limits of integration can be extended to infinity, the integral in Eq. (11.82) becomes a delta function,

$$F(\mathbf{q}) = \delta(\mathbf{q}), \qquad (11.83)$$

and so Eq. (11.81) reduces to

$$\langle |\mathcal{I}_1|^2 \rangle = \Phi_n(k_0\mathbf{m}). \qquad (11.84)$$

Thus the expected scatter intensity is proportional to the power spectrum of refractive turbulence at the structure wave number $\mathbf{K}$ equal to the mirror wave number $k_0\mathbf{m}$.

In the usual case of finite scatter volume V, the function $F(\mathbf{q})$ has its maximum values in the region near the point $\mathbf{q} = 0$, and outside this region its amplitude oscillates about zero and diminishes in intensity (Fig. 11.8). Again as an example, consider the finite volume to be a parallelopiped with sides $2L_x$, $2L_y$, $2L_z$. In this case

$$F(\mathbf{q}) = L_x L_y L_z \frac{\sin q_x L_x}{\pi q_x L_x} \frac{\sin q_y L_y}{\pi q_y L_y} \frac{\sin q_z L_z}{\pi q_z L_z}, \qquad (11.85a)$$

whose first zeros are at

$$q_{x0} = \pi/L_x, \qquad q_{y0} = \pi/L_y, \qquad q_{z0} = \pi/L_z. \qquad (11.85b)$$

The region of wave vector space $\mathbf{q}$ over which $F(\mathbf{q})$ is appreciable is of the order of

$$q_{x0} q_{y0} q_{z0} \approx 8\pi^3/V \equiv \Upsilon. \qquad (11.86)$$

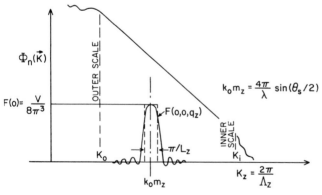

Fig. 11.8 Spectrum of the refractive index variations with imposed sampling function $F(\mathbf{q})$.

Of course, the behavior of $F(\mathbf{q})$ in each direction of wave vector space depends on the shape of the real space volume V. From the foregoing discussion we recognize that the integral $\langle|\mathscr{I}_1|^2\rangle$ can then be approximated by

$$\langle|\mathscr{I}_1|^2\rangle \approx \iiint_Y \Phi_n(\mathbf{K}) F(0)\, dV_K = \int_Y \Phi_n(\mathbf{K})(V/8\pi^3)\, dV_K$$

$$= \frac{1}{Y}\int_Y \Phi_n(\mathbf{K})\, dV_K \equiv \overline{\Phi_n(k_0\mathbf{m})}, \tag{11.87}$$

where $\overline{\Phi_n(k_0\mathbf{m})}$ is the mean value of the function $\Phi_n(\mathbf{K})$ obtained by averaging over a region of wave-number space with a volume $8\pi^3/V$ surrounding the point $k_0\mathbf{m}$. In most cases $\Phi_n(\mathbf{K})$ can be considered constant over the small volume $8\pi^3/V$ of wave-number space. In arriving at Eq. (11.87) we have essentially approximated $F(\mathbf{q})$ by a pulse-type sampling function of width $q_{i0}(i = x, y, \text{or } z)$ for each direction of wave vector space and integrated $\Phi_n(\mathbf{K})$ over this range of $\mathbf{K}$ centered about the point $k_0\mathbf{m}$. We now substitute Eq. (11.87) into Eq. (11.80) to obtain the expected average power density (i.e., the average over a cycle of ω_0)

$$\langle S\rangle = (\pi A_0^2 k_0^4 \sin^2 \chi V/\eta_0 r_{t0}^2 r_{r0}^2)\overline{\Phi_n(k_0\mathbf{m})}. \tag{11.88}$$

It follows from Eq. (11.88) that the power scattered in a particular direction $\mathbf{a}_{r0}$, where $\mathbf{a}_{r0}$ makes an angle θ_s with respect to the incident direction $-\mathbf{a}_{t0}$, is determined only by a narrow portion of the spectrum of irregularities near the wave number $\mathbf{K} = k_0\mathbf{m}$. Thus only a small group of irregularity scales scatter energy in a given direction and these scales are usually smaller than the correlation lengths (see Section 11.5.3 for more discussion).

11.4.3.1 Scattering Cross Section per Unit Volume (Reflectivity)

The reflectivity of precipitation defined in Chapter 4 is simply the sum of the scattering cross section of each particle within an elemental volume. Although there are no point scatterers in clean air, we can nevertheless still define reflectivity. The power density arriving at the receiver site a distance r_r from the scattering volume dV is dS, which is defined as the power dP incident per unit area subtended by the solid angle $d\Omega$ (see Fig. 11.9). That is,

$$dS = dP/dA. \tag{11.89}$$

The scattering cross section $d\sigma_s$ of dV is an equivalent area that intercepts a power $S_i d\sigma_s$ (where S_i is the magnitude of the incident Poynting vector), which, if radiated isotropically, produces in the direction of the receiver a power density

$$dS = S_i\, d\sigma_s/4\pi r_r^2 \tag{11.90}$$

equal to that actually observed. Thus

$$d\sigma_s = (4\pi)^2 r_r^2 r_t^2\, dS/P_t g_t f_t^2(\theta_t, \phi_t), \tag{11.91}$$

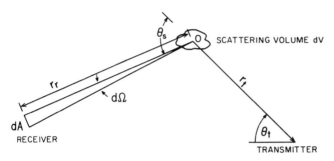

Fig. 11.9 Geometry used to define reflectivity.

where $f_t^2(\theta, \phi)$ is the normalized angular pattern of the transmitted power density. After substituting the peak signal amplitude squared,

$$A_0^2 = P_t g_t f_t^2 \eta_0 / 2\pi, \tag{11.92}$$

into Eq. (11.88) and using Eq. (11.88) to compute dS (where V is replaced by dV), which in turn is substituted into Eq. (11.91), the expected scattering cross section per unit volume (reflectivity) is obtained.

$$\langle \eta \rangle \equiv d\sigma_s / dV = 8\pi^2 k_0^4 \, \overline{\Phi_n(k_0\mathbf{m})} \, \sin^2 \chi. \tag{11.93}$$

11.4.3.2 Discussion and an Example

To clarify how the sampling function (11.82) determines $\Phi(\mathbf{K})$, consider as an example the case of statistically homogeneous and isotropic refractive index irregularities. We choose rectangular coordinates such that z lies parallel to $\mathbf{m}$. Thus,

$$m_z = |\mathbf{m}| = 2\sin(\theta_s/2), \qquad \Phi_n(\mathbf{K}) = \Phi_n(K_x, K_y, K_z). \tag{11.94}$$

Because $m_x = m_y = 0$, $K_x = K_y = 0$, and $K_z = k_0|\mathbf{m}|$ locate the position where $F(\mathbf{q})$ is maximum, as can be deduced from Eqs. (11.82) and (11.83). Assuming that the width of $\Phi_n(\mathbf{K})$ is large compared to the width of $F(\mathbf{q})$ transverse to the K_z axes, we can replace K with

$$K = |\mathbf{K}| = K_z, \tag{11.95}$$

and $\overline{\Phi_n(k_0 m_z)}$ in the expression for reflectivity (11.93) is then

$$\overline{\Phi_n(k_0 m_z)} = \frac{1}{Y} \int_Y \Phi_n(K_z) \, dK_x \, dK_y \, dK_z. \tag{11.96}$$

Integrating over K_x, K_y (see Fig. 11.8), we obtain

$$\overline{\Phi_n(k_0 m_z)} \approx \frac{\pi^2}{Y L_x L_y} \int_{a^-}^{a^+} \Phi_n(K_z) \, dK_z, \tag{11.97}$$

where $a^{\pm} \equiv k_0 m_z \pm \pi/2L_z$. Substituting for Y from Eq. (11.86), this reduces to

$$\overline{\Phi_n(k_0 m_z)} = \frac{L_z}{\pi} \int_{a^-}^{a^+} \Phi_n(k_z) \, dK_z. \tag{11.98}$$

Equation (11.98) shows that, in addition to the prime contribution from the spectral component $k_0 m_z$, spectral components with structure wavelengths in the interval

$$\frac{1}{k_0 m_z + \pi/2L_z} \le \frac{\Lambda}{2\pi} \le \frac{1}{k_0 m_z - \pi/2L_z} \tag{11.99}$$

also contribute significantly to the power scattered in the direction θ_s. Thus, the function $F(\mathbf{q})$ samples a band of spectral components of Δn centered about $k_0 m_z$. The width of the band is π/L_z; in the case of scatter at microwave frequencies, $\lambda/4 \sin(\theta_s/2) \ll L_z \sim \Delta s$ [Eq. (11.23)], and the width is much smaller than the central value $k_0 m_z$. This is particularly true if $k_0 m$ is at structure wave numbers larger than the reciprocal of the correlation length along z and if the scatter volume is large compared to this length. This is consistent with our basic assumption that the dimensions of V are much larger than the correlation length. Only for these cases can we approximate $\overline{\Phi_n(k_0 \mathbf{m})}$ by $\Phi_n(k_0 \mathbf{m})$.

For the purpose of estimating the order of magnitude of some of the parameters we have defined, consider a forward-scatter link of $2d = 250$ km having at each terminal an 18 m diameter antenna pointed at an angle of $4°$ above the straight line connecting the terminals (Fig. 11.2). Assuming a transmitted wavelength of 10 cm, the beamwidth would be about $0.5°$ and $g_r = g_t \simeq 1.6 \times 10^5$.

Assume the beam axis are in a vertical plane $\mathbf{a}_{t0}$, $\mathbf{a}_{r0}$, and x, y, z, are coordinates with origin along the bisector of TR (Fig. 11.2). For the sake of simplicity and to satisfy the conditions of our formula, we shall assume the size of the scatter volume V to be equal to limits imposed by the conditions (11.46) and (11.47). Thus, the lengths are about 200 m along y, 3000 m along x, and 200 m along z. Furthermore, a pulse width τ is assumed to be sufficiently large that the distance Δs [Eq. (11.23)] within the spheroidal shells of constant time delay is larger than the vertical thickness of V. Assume that this volume contains refractive index irregularities having a variance spectrum [e.g., Eq. (10.38)] as predicted by turbulence theories. To obtain some idea of the magnitude of C_n^2, consider the following categories, defined by Davis (1966) to be associated with different levels of turbulence:

$$C_n^2 \approx 6 \times 10^{-17} \quad m^{-2/3} \quad \text{(weak)}, \tag{11.100a}$$

$$C_n^2 \approx 2 \times 10^{-15} \quad m^{-2/3} \quad \text{(intermediate)}, \tag{11.100b}$$

$$C_n^2 \approx 3 \times 10^{-13} \quad m^{-2/3} \quad \text{(strong)}. \tag{11.100c}$$

As we shall see in Section 11.6, however, the value of C_n^2 depends not only on the level of turbulence but also on the temperature and humidity gradients where turbulence occurs. Thus we consider Eqs. (11.100) to define categories of refractive index structure parameter that have been measured in the atmosphere. Substituting Eq. (10.38) into Eq. (11.98) gives

$$\overline{\Phi_n(k_0 m_z)} = C_n^2 \frac{0.033 L_z}{\pi} \int_{a^-}^{a^+} K_z^{-11/3} \, dK_z, \tag{11.101}$$

where, for our example, $k_0 m_z = 8.8 \text{ m}^{-1}$. As mentioned previously, we note that the range of k_z over which the integration in Eq. (11.101) is performed is much smaller than the central value $k_0 m_z$. Thus Eq. (11.101) is, to a good approximation, equal to

$$\overline{\Phi_n(k_0 m_z)} \simeq 3.3 \times 10^{-2} C_n^2 (k_0 m_z)^{-11/3}. \tag{11.102}$$

Substituting Eq. (11.102) into Eq. (11.93) and assuming $\chi = \pi/2$, we obtain the reflectivity of this scattering volume:

$$\eta = 2.6 C_n^2 \, k_0^4 [2k_0 \sin(\theta_s/2)]^{-11/3}. \tag{11.103a}$$

For strong C_n^2

$$\eta = 4.2 \times 10^{-9} \quad \text{m}^{-1}. \tag{11.103b}$$

From Eq. (11.56) we see that refractive index irregularities having a structure wavelength $\Lambda = 71$ cm contribute most to η in this forward-scatter link, whereas for backscatter (at $\lambda = 10$ cm) the irregularities having $\Lambda = 5$ cm contribute most to the scattered signal.

Applying Eq. (11.103a) to the backscatter case, it can be shown that

$$\eta = 0.38 \lambda^{-1/3} C_n^2. \tag{11.104}$$

This wavelength dependence was first verified by experiments with high-resolution multiwavelength radars located at Wallops Island, Virginia (Atlas *et al.*, 1966). In these experiments simultaneous backscatter measurements were made from a volume at an altitude of 12 km common to three radars operating on three different wavelengths.

Equation (11.104) was verified by Kropfli *et al.* (1968) in a carefully controlled experiment in which an airborne microwave refractometer measured nearly simultaneously the C_n^2 along the flight path through the volume in which η was measured by a high-resolution radar.

11.5 Common Volume Scatter

So far our derivation of reflectivity has followed that of Tatarskii (1961, Section 4.2) where it is assumed that the scatter volume V_s has dimensions that satisfy the conditions (11.46) and (11.47). These conditions could imply that the dimensions of V_s cannot be determined by the volume V_c common to the intersecting beams of the transmitting and receiving antennas and the spheroidal shell whose location and thickness are given by Eqs. (11.22) and (11.23). These are highly restrictive conditions, and therefore we shall extend the formulations.

11.5.1 Correlation Length Shorter Than the Fresnel Length

In a later publication Tatarskii (1971) derived a fairly general solution for the scattered field with less restrictive assumptions concerning the geometry of the incident wave, the size of the scattering volume, and its distance from the observation point. Equation (11.88) can also be used to derive an expression for the total scattering cross section for the case in which the scattering volume is determined by the common volume V_c. The assumptions needed are that the dimensions of V_c are much larger than the correlation lengths (i.e., the dimensions of the scattering blob, Fig. 11.10) and that the correlation lengths transverse to **m** must satisfy

$$\rho_{c,\parallel} \ll \frac{1}{\sin(\theta_s/2)} \left[\frac{\lambda r_t r_r}{\pi(r_t + r_r)} \right]^{1/2} \equiv \frac{f_\parallel}{\sqrt{\pi}}, \tag{11.105}$$

$$\rho_{c,\perp} \ll \left[\frac{\lambda r_t r_r}{\pi(r_t + r_r)} \right]^{1/2} \equiv \frac{f_\perp}{\sqrt{\pi}}, \tag{11.106}$$

where r_t, r_r are ranges to the origin of each blob, $\rho_{c,\parallel}$ is the correlation length perpendicular to the mirror direction **m** but in the plane of $\mathbf{a}_t$, $\mathbf{a}_r$; $\rho_{c,\perp}$ is the correlation length perpendicular to **m** and the plane of $\mathbf{a}_t$, $\mathbf{a}_r$. The lengths $f_\parallel$, $f_\perp$ are the Fresnel lengths perpendicular to **m** and along the respective orthogonal directions. Inequalities (11.105) and (11.106) are derived by imposing the condition

$$\frac{k_0}{2r_r} [r^2 - (\mathbf{a}_r \cdot \mathbf{r})^2] + \frac{k_0}{2r_t} [r^2 - (\mathbf{a}_t \cdot \mathbf{r})^2] \ll 1, \tag{11.107}$$

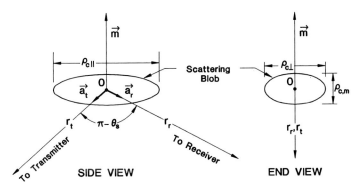

Fig. 11.10 A scattering blob with a size determined by the correlation lengths. The blob is assumed to be located in the vertical plane containing the transmitter and receiver.

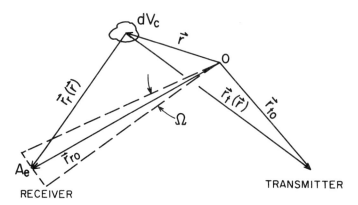

Fig. 11.11 Scattering geometry for a transmitter, a receiver, and a scattering blob dV_c.

which causes the phase term in Eq. (11.43) to be linear in $\mathbf{r}$ within the blob. Note that the phase is quadratic in $\mathbf{r}$ for $\mathbf{r}$ in the plane perpendicular to $\mathbf{m}$ and linear in $\mathbf{r}$ along $\mathbf{m}$. Because the field scattered from each blob is incoherent, the total received power is given by the sum of the contributions from each blob. If there are many blobs within V_c, then the amplitude distribution of I and Q will be Gaussian. The case of either $f_\parallel$ or $f_\perp \leq \rho_c$ is solved in Section 11.5.2.

With reference to Fig. 11.11, $\mathbf{r}_{t0}$ and $\mathbf{r}_{r0}$ are the respective distances from the transmitter and receiver to the intersection O of the axes of the main beams of each respective antenna, and $\mathbf{r}$ is the vector distance from O to the scattering blob of volume dV_c. The elemental power dP received at $\mathbf{r}_{r0}$ because of scatter from dV_c located at $\mathbf{r}$ is

$$dP = A_e(\mathbf{r}) \, dS(\mathbf{r}), \tag{11.108}$$

where $A_e(\mathbf{r})$ is the effective reception area of the receiver antenna for energy incident from the direction of dV_c, and $dS(\mathbf{r})$ is the Poynting vector magnitude due to scatter from volume element dV_c. An equivalent scattering cross section $d\sigma_s$ that produces the same dP is assumed to be located at the origin O. Thus

$$[S_i(O) \, d\sigma_s/4\pi r_{r0}^2] A_e(\mathbf{r}_{r0}) = dP = A_e(\mathbf{r}) \, dS(\mathbf{r}),$$

where $S_i(O)$ is the magnitude of the incident Poynting vector at the origin. Substituting Eqs. (3.4) and (3.21) into Eq. (11.109) and solving for the mean value $\langle d\sigma_s \rangle$, we obtain

$$\langle d\sigma_s \rangle = \frac{(4\pi)^2 r_{t0}^2 r_{r0}^2 f_r^2(\theta, \phi)}{P_t g_t} \langle dS(\mathbf{r}) \rangle. \tag{11.110}$$

The power density scattered from dV_c and incident on the receiver is given by Eq. (11.88) with V replaced by dV_c, r_{t0} and r_{r0} replaced by $r_t(\mathbf{r})$ and $r_r(\mathbf{r})$, respectively, and Eq. (11.92) substituted for A_0^2. When all this is substituted into

Eq. (11.110) and the result is integrated over the common volume V_c, we obtain the total scattering cross section of the common volume V_c.

$$\langle \sigma_s \rangle = 8\pi^2 r_{t0}^2 r_{r0}^2 k_0^4 \int_{V_c} \frac{f_r^2(\mathbf{r}) f_t^2(\mathbf{r}) \sin^2 \chi(\mathbf{r})}{r_t^2(\mathbf{r}) r_r^2(\mathbf{r})} \, \overline{\Phi_n(k_0 \mathbf{m}(\mathbf{r}), \mathbf{r})} \, dV, \quad (11.111)$$

where V_c is the spheroidal shell whose location and thickness is given by Eqs. (11.22) and (11.23). Most contribution to σ_s come from the region about the main beam axis of each antenna. Sidelobes can contribute significantly only if the intensity of Φ_n at the sidelobe location is strong relative to the intensity of Φ_n at the intersection of the main beams, as observed by Doviak et al. (1971).

11.5.2 Correlation Length Comparable to or Larger Than the Fresnel Length

Although Eq. (11.111) is useful, it does contain the implicit assumption that the correlation lengths of the refractive index irregularities must be small compared to the Fresnel lengths [Eqs. (11.105), (11.106)]. There is evidence that scattering irregularities can be highly anisotropic with correlation lengths along the horizontal that can be comparable to or larger than is stipulated by the conditions (11.46) and (11.47) (Gage et al., 1981). This is especially true for wind profiling radars (Section 11.8.1.1) because the range is often short (several kilometers), so that the Fresnel lengths are comparable to horizontal correlation lengths. On the basis of angle-of-arrival observations of satellite emissions at 7.3 GHz, Crane (1976) deduced that the spectral density of refractive index variance along K_z was several orders of magnitude larger than along K_x or K_y at the same scale size ($\Lambda \approx 70$ m) in the stably stratified inversion layer capping the planetary boundary layer. This deduction and propagation-path geometry suggested the presence of highly anisotropic irregularities. We shall focus on backscatter because (1) it simplifies the mathematics and (2) it covers the case of most importance in remote sensing.

The magnitude of the backscattered field intensity, obtained from Eq. (11.43), is

$$E_1(\mathbf{r}_0, t) = \frac{k_0^2}{r_0^2} \left[\frac{P_t \eta_0 g}{(2\pi)^3} \right]^{1/2} \int_{V_s} f_\theta(\mathbf{r}) \, \Delta n(\mathbf{r}, t) e^{-j2k_0 r_t} \, dV, \quad (11.112)$$

where Eq. (11.92) has been substituted for A_0, and $f_\theta(\mathbf{r})$ is the one-way electric field angular pattern, which will be assumed to be circularly symmetric about the beam axis. Using Eq. (3.21), which expresses the dependence of the effective area of the receiving antenna on the position of the elemental scattering volume dV, the increment of current $|dI|$ produced by dE in a matched-filter receiver having an internal resistance R is

$$|dI| = |dE_1| \lambda |W(\mathbf{r})| [g f_\theta^2(\mathbf{r})/4\pi\eta_0 R]^{1/2}, \quad (11.113)$$

where dE_1 is the increment of field intensity scattered from dV. $|W(\mathbf{r})|$ is the range-weighting function (Section 4.4.2), which we now introduce explicitly into the equation for the scattered field. The integration extends over all $\mathbf{r}$ for which $|W|f_\theta \Delta n$ is significant.

The time-averaged received power (i.e., averaged over a cycle of ω_0) is

$$P_r = \tfrac{1}{2} II^* R, \tag{11.114}$$

where from the integral of Eq. (11.113)

$$I = \frac{\lambda k_0^2 g}{(2\pi)^2 r_0^2} \left(\frac{P_r}{2R}\right)^{1/2} \int |W(\mathbf{r})| f_\theta^2(\mathbf{r}) \, \Delta n(\mathbf{r}, t) e^{-j2k_0 r_t} \, dV. \tag{11.115}$$

Substituting Eq. (11.115) into Eq. (11.114), taking the ensemble average, and using Eq. (11.73a), we obtain for the expected received power

$$\langle P_r \rangle = \frac{P_t g^2}{4\lambda^2 r_0^4} \int\!\!\int R(\mathbf{r}, \mathbf{r}') \, W(\mathbf{r}) \, W^*(\mathbf{r}') f_\theta^2(\mathbf{r}) f_\theta^2(\mathbf{r}') e^{-j2k_0(r_t - r_t')} \, dV \, dV'. \tag{11.116}$$

For a circularly symmetric antenna, the two-way electric field pattern $f_\theta^2(\mathbf{r})$ (equal to the one-way power density pattern) is

$$f_\theta^2(\mathbf{r}) = e^{-\alpha^2/4\sigma_\theta^2}, \tag{11.117a}$$

where σ_θ^2 is the second central moment [Eq. (5.75)] of the two-way power density pattern and α is the angular displacement, measured at the radar site, of $\mathbf{r}$ from the origin centered in V_6 (Fig. 11.12). On the assumption of narrow

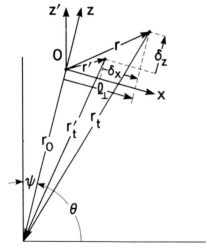

Fig. 11.12 Geometry and parameters used in deriving formulas for Fresnel backscattering.

beams, Eq. (11.117a) can be approximated by

$$f_\theta^2 \approx e^{-(x^2+y^2)/4r_0^2\sigma_\theta^2},$$ (11.117b)

where $(x^2 + y^2)^{1/2}$ is the projection of $\mathbf{r}$ onto the plane perpendicular to the beam axis, which is directed along z at a zenith angle ψ.

A Gaussian filter matched to a transmitted Gaussian pulse provides the best resolution of all receivers having the same bandwidth (Zrnić and Doviak, 1978). Because of this, and because practical matched filters used in some Doppler weather radars have a Gaussian-shaped frequency response, we assume that $|W(\mathbf{r})|$ is well approximated by

$$|W(\mathbf{r})| \approx e^{-(\mathbf{a}_0\cdot\mathbf{r})^2/4\sigma_r^2},$$ (11.118)

where σ_r^2 is the second moment (5.76) of the weighting function $|W^2(\mathbf{r})|$.

Next we use the Taylor series expansion [Eq. (11.44), with $\mathbf{r}_t \equiv \mathbf{r}_r$]:

$$r_t \equiv |\mathbf{r}_0 - \mathbf{r}| \approx r_0 - \mathbf{a}_0 \cdot \mathbf{r} + (1/2r_0)[r^2 - (\mathbf{a}_0\cdot\mathbf{r})^2]$$ (11.119a)

for terms up to second order in r^2. We note that the second and third terms of this expansion are the projection of $\mathbf{r}$ in the $\mathbf{a}_0$ direction and onto the x, y plane, respectively (Fig. 11.12). Thus, in terms of the Cartesian coordinates centered in the resolution volume

$$r_t \approx r_0 + z + (1/2r_0)(x^2 + y^2).$$ (11.119b)

This quadratic expansion about the origin is valid provided the phase contributed by higher-order terms is negligible. In order that the phase contributed by the next term [i.e., the term in fourth order of $(x^2 + y^2)^{1/2} \equiv l_\perp$] in the series is less than one radian, the scattering location $l_\perp$ in the plane $z = 0$ must be confined to

$$l_\perp < (8r_0^3/k_0)^{1/4}.$$ (11.119c)

Along the z axis the expansion is exact, and it can be shown (by expanding in a plane $z \neq 0$) that there is a weak dependence of $l_\perp$ on z (Problem 11.4).

Scattering volumes are usually determined by the resolution volume V_6 or, in the case of a thin layer of scattering irregularities, both by V_6 and the vertical thickness of the layer. For a 6 m wavelength radar pointed vertically at a 6 km high scattering layer, the beam width would have to be less than 22° for relation (11.119c) to be satisfied. Most remote-sensing radars have beamwidths much smaller than this value. Therefore the second-order expansion yields a solution even if irregularities are correlated across the beam (Gurvich and Kon, 1992); for such a condition the first-order expansion (Section 11.4) does not apply. At ranges where $r_0\theta_1$ is larger than the right side of (11.119c) the second-order expansion applies to blobs having transverse correlation lengths satisfying condition (11.119c).

Comparing (11.119c) with (11.107) we deduce that the solution based on the second-order expansion applies to scattering media with transverse correlation lengths larger by a factor of $2(\pi r_0/\lambda)^{1/4}$, than those lengths for which the first-order expansion is valid; for the example given in the previous paragraph this factor is larger than 10!

Substituting (11.119b), the integral in Eq. (11.116) becomes

$$\Im = \int\int R(\mathbf{r},\mathbf{r}')W(z)W(z')\exp\left[-(x^2+x'^2)/2\sigma_\perp^2 - (y^2+y'^2)/2\sigma_\perp^2\right.$$
$$\left. -j2k_0\left(z-z'+\frac{x^2-x'^2+y^2-y'^2}{2r_0}\right)\right]dV\,dV',\qquad(11.120)$$

where $\sigma_\perp \equiv \sigma_\theta r_0\sqrt{2}$. We shall now find it convenient to define new coordinate axes.

$$x-x'\equiv\delta_x,\qquad y-y'\equiv\delta_y,\qquad z-z'\equiv\delta_z,$$
$$\tfrac{1}{2}(x+x')\equiv\sigma_x,\qquad \tfrac{1}{2}(y+y')\equiv\sigma_y,\qquad \tfrac{1}{2}(z+z')\equiv\sigma_z.$$

Consistent with this coordinate transformation, we could separate $R(\mathbf{r},\mathbf{r}')$ into a product of two functions, $R(\mathbf{r},\mathbf{r}')=\langle\Delta n^2(\boldsymbol\sigma)\rangle R_n(\boldsymbol\rho)$ where $\boldsymbol\rho\equiv\mathbf{a}_{x0}\delta_x + \mathbf{a}_{y0}\delta_y + \mathbf{a}_{z0}\delta_z$. This allows the variance of Δn to be spatially dependent, a feature that would be especially useful if scattering irregularities did not fill the resolution volume. For sake of simplicity, however, assume $R(\mathbf{r},\mathbf{r}')=R(\boldsymbol\rho)$ to be statistically homogeneous. The x,x' component of Eq. (11.120) can then be written as

$$\Im_x = \int\int R(\boldsymbol\rho)\exp[-(\sigma_x^2+\delta_x^2/4)/\sigma_\perp^2 - j2k_0\sigma_x\delta_x/r_0]\,d\sigma_x\,d\delta_x.\quad(11.121a)$$

The transformation from Eq. (11.120) to Eq. (11.121a) is valid provided the limits of integration cover the entire volume where the integrand has significant value. Thus executing the integration over σ_x,

$$\Im_x = \sigma_\perp\sqrt{\pi}\int R(\boldsymbol\rho)\exp[-(k_0\delta_x\sigma_\perp/r_0)^2 - \delta_x^2/4\sigma_\perp^2]\,d\delta_x.\quad(11.121b)$$

Applying similar procedures to the integrations along y and z coordinates, yields

$$\Im = \sqrt{2}\sigma_r\sigma_\perp^2\pi^{3/2}\int\int\int_{-\infty}^{+\infty} R(\boldsymbol\rho)$$
$$\times\exp\left[-\underbrace{\left(\frac{\delta_x^2+\delta_y^2}{4\sigma_\perp^2}+\frac{\delta_z^2}{8\sigma_r^2}\right)}_{\text{resolution volume weight}}-\underbrace{\frac{\pi^2\sigma_\perp^2}{f^4}(\delta_x^2+\delta_y^2)}_{\text{Fresnel term}}-j2k_0\delta_z\right]dV_\rho,\quad(11.122)$$

where $f = \sqrt{\lambda r_0 / 2}$ is the radius of the first Fresnel zone for backscatter [Eq. (11.105) or (11.106)].

Now compare the integral in Eq. (11.122) to (11.80). We have the same form but the correlation function is multiplied by two additional exponential functions:

1. The "resolution volume" term weights $R(\rho)$ in production to the beamwidth and range resolution (the larger are $\sigma_\perp$ or σ_r compared to the correlation lengths transverse or parallel, respectively, to the beam axes, the less important is the beamwidth or range resolution to the integral).

2. The "Fresnel" term gives a weight along the x, y direction that depends on the ratio $f/\sigma_\perp$ of the Fresnel radius to the beamwidth.

Because we have assumed V_6 to be in the far field (i.e., $r_0 \geq 2D^2/\lambda$), it can be shown that the Fresnel term in Eq. (11.122) has more weight than the beam width part. Thus situations that allow us to neglect the Fresnel term will also permit us to ignore beamwidth influence.

Consider that $R(\rho)$, transverse to the beam axis, is described by a Gaussian function $R(0)\,\exp[-(\delta_x^2 + \delta_y^2)^2/2\rho_{c\perp}^2]$ where $\rho_{c\perp}$ is the correlation length. Then by comparing exponents of $R(\rho)$ with other exponents in Eq. (11.122) we find that only when the radius of the Fresnel zone is large compared to $2^{1/4}\sqrt{\pi\sigma_\perp\rho_{c\perp}}$ can the Fresnel term be ignored. Alternatively, since $\sigma_\perp$ is [from Eqs. (3.2b) and (5.75)],

$$\sigma_\perp = \frac{0.45}{\sqrt{\ln 2}}\frac{\lambda r_0}{D} = \frac{0.9}{\sqrt{\ln 2}}\frac{f^2}{D}, \tag{11.123}$$

the condition on $\rho_{c\perp}$ under which the Fresnel term can be ignored is

$$\rho_{c\perp} \ll \frac{D}{0.9\pi}\sqrt{\frac{\ln 2}{2}}. \tag{11.124}$$

If the correlation length $\rho_{c\perp}$ is larger than the beamwidth, the Fresnel term in Eq. (11.122) needs to be retained. Gurvich and Kon (1992) have also recognized that beamwidth often determines the transverse size of the scatter volume and that (11.119c) can easily be satisfied with narrow-beam radars. Furthermore, they have generalized the solution, for the case of a vertically pointed beam, to include both the near (Fresnel) region of the antenna and the far (Fraunhofer) region considered in this section.

Hodara (1966) showed that within the lower troposphere, the horizontal correlation length has the following height dependence:

$$\rho_{ch} \approx 0.4h/(1 + 0.01h) \quad \text{m,} \tag{11.125}$$

where h is in meters. Furthermore VHF backscatter data fitted to the theory in Section 11.5.4 suggest that $\rho_{ch} \approx 20$ m for irregularities in the lower stratosphere. Thus, unless the antenna diameter is of the order of 100 m or more, the

Fresnel term will be important in determining the field scattered by refractive irregularities.

If relation (11.124) is satisfied, Eq. (11.122), substituted into Eq. (11.116), yields a P_r identical to that derivable from Eq. (11.80) by extension to scatter from V_6. We call this Fraunhofer scatter because the Fresnel term in Eq. (11.122) can be ignored. If condition (11.124) is not satisfied, Eq. (11.122) accounts for the quadratic phase shift across the scattering blob's transverse dimension and thus describes Fresnel scatter. Note that conditions (11.105) and (11.106) (for backscatter 11.105 is the same as 11.106) differ from (11.124). Condition (11.106) specifies when the phase is linear across the blob, whereas condition (11.124) specifies when the Fresnel term is important. If $\rho_{c\perp}$ is much larger than the transverse dimension of a vertically pointed beam, the refractive index can be considered to be horizontally stratified. In this case a Gaussian distribution of I and Q will be produced only if there are many strata within the resolution volume and their spacings change randomly; otherwise the return signal would be coherent and the echo can be said to be from a reflecting layer. In this case reflection formulas (e.g., Brekhovskikh, 1960) can be used to determine echo power; it can be shown (Doviak and Zrnić, 1984b) that this approach is equivalent to the scattering approach employed here.

11.5.3 The Spectral Representation

Next we shall use spatial spectra to examine the effects that the weighting functions in Eq. (11.122) have on received power. Application of Eq. (10.7) to scalar fields allows Eq. (11.122) to be written as

$$\Im = \sqrt{2\sigma_r \sigma_\perp^2}\, \pi^{3/2} \int\limits_{-\infty}^{+\infty}\!\!\int \Phi_n(\mathbf{K}) \left[\int\limits_{-\infty}^{+\infty}\!\!\int\!\!\int \exp[j(\mathbf{K} - 2k_0 \mathbf{a}_{z0}) \cdot \boldsymbol{\rho}] H(\boldsymbol{\rho})\, dV_\rho \right] dV_K,$$

$$(11.126)$$

where

$$H(\boldsymbol{\rho}) = \exp\left[-\frac{\delta_z^2}{8\sigma_r^2} - \frac{\pi^2 \sigma_\perp^2}{f^4}(\delta_x^2 + \delta_y^2) \right] \qquad (11.127)$$

is the weighting function in lag space. We have omitted the beamwidth term because, in the antenna's far field, it is significantly smaller than the Fresnel term. The bracketed term in Eq. (11.126) is a sampling function, similar to that of Eq. (11.81) but modified by the weighting function $H(\boldsymbol{\rho})$. We therefore define a normalized spectral sampling function

$$F(\mathbf{K}) \equiv \frac{1}{8\pi^3} \iiint \exp[j(\mathbf{K} - 2k_0 \mathbf{a}_{z0}) \cdot \boldsymbol{\rho}] H(\boldsymbol{\rho}) \, dV_\rho, \qquad (11.128)$$

which has a peak value at $\mathbf{K} = 2k_0 \mathbf{a}_{z0}$. As the width of $H(\boldsymbol{\rho})$ becomes broader, the spectral sampling function becomes narrower. The integrations along ρ can be performed to give

$$F(\mathbf{K}) = \frac{0.44 D^2 \sigma_r \ln 2}{\pi^{7/2}} \exp\left[-2\sigma_r^2 (K_z - 2k_0)^2 - \frac{D^2(K_x^2 + K_y^2) \ln 2}{3.24\pi^2} \right], \qquad (11.129)$$

in which we have substituted Eq. (11.123) for $\sigma_\perp$. Thus, the larger the antenna diameter D, the narrower is the sampling function. A surprising result is that the sampling function shape is independent of r_0. For a given antenna diameter, the spectrum $\Phi_n(\mathbf{K})$ of irregularities is weighted equally for all resolution volumes in space. Condition (11.119c) however, must be satisfied.

To provide physical insight, we now consider a spectral description for the angular dependence of scatter from anisotropic irregularities. From Eqs. (11.126), (11.128), and (11.116), the backscattered power becomes

$$\langle P_t \rangle = \frac{2\sqrt{2}(0.45)^2 P_t g^2 \sigma_r \pi^{9/2}}{r_0^2 D^2 \ln 2} \iiint \Phi_n(\mathbf{K}) F(\mathbf{K}) \, dV_K. \qquad (11.130)$$

Equation (11.130) shows that the echo power is proportional to the integral of the product of the spectral intensity $\Phi_n(\mathbf{K})$ of the refractive index irregularities and the normalized sampling function $F(\mathbf{K})$. Now, if the horizontal correlation lengths are large compared to the vertical ones, $\Phi_n(\mathbf{K})$ will be sharply peaked in the K_x, K_y directions and less peaked along the K_z axis. If the irregularities can be roughly described as oblate spheroids, then the correlation $R(\rho)$ has a similar form, but $\Phi_n(\mathbf{K})$ is prolate spheroidal in shape (Fig. 11.13a).

In deriving $F(\mathbf{K})$ we assumed the z axis to be along the beam axis but made no assumption as to the direction of z in space relative to the earth. Furthermore, the shape of the sampling function depends only on the range resolution and antenna diameter and is thus independent of the location of the resolution volume V_6 in real space. Therefore, in K space $F(\mathbf{K})$ is invariant under rotation of the K coordinate axes, and the sampling function (11.129) can be formulated more generally as

$$F(\mathbf{K}) = \frac{0.44 D^2 \sigma_r \ln 2}{\pi^{7/2}} \exp\left[-2\sigma_r^2 (K_\parallel - 2k_0)^2 - \frac{D^2 K_\perp^2 \ln 2}{3.24\pi^2} \right], \qquad (11.131)$$

where $K_\parallel$ is the wave-number magnitude along $\mathbf{k}_0$, and $K_\perp$ is the perpendicular distance of $\mathbf{K}$ from $\mathbf{k}_0$. Thus, if z is along the vertical and the beam axis direction

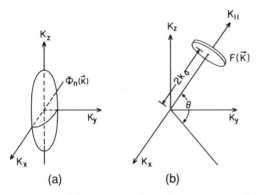

Fig. 11.13 (a) Contour surface of constant spectral intensity $\Phi_n(\mathbf{K})$ for irregularities having symmetric correlation lengths along x and y that are longer than the correlation length along z. (b) Contour surface of spectral sampling function $F(\mathbf{K})$ for a beam axis at elevation angle θ.

(i.e., $\mathbf{k}_0/k_0$) is rotated by $\pi/2 - \theta$ from it, then $F(\mathbf{K})$ must be rotated by $\pi/2 - \theta$ from the K_z axis. Equation (11.131) reveals that whenever the range resolution $r_6 = 3.33\sigma_r > 0.34D$, a common situation, $F(\mathbf{K})$ will be narrower along $K_\parallel$ than along $K_\perp$. Figure 11.13b shows the contours of $F(\mathbf{K})$.

Obviously, we obtain maximum $\langle P_r \rangle$ for $F(\mathbf{K})$ centered where $\Phi_n(\mathbf{K})$ is a maximum. But, we do not have the freedom to choose arbitrarily the location of $F(\mathbf{K})$, because the radar wavelength centers the $F(\mathbf{K})$ peak at a wave-number magnitude $2k_0$. For the conditions shown in Fig. 11.13, maximum $\langle P_r \rangle$ occurs when $\theta \rightarrow \pi/2$. The angular dependence of $\langle P_r \rangle$ on θ is strongly influenced by the sharpness of $\Phi_n(\mathbf{K})$ and $F(\mathbf{K})$. If $\Phi_n(\mathbf{K})$ is highly anisotropic (as in Fig. 11.13a) then $\langle P_r \rangle$ will be linearly dependent on the range resolution but proportional to the square of the antenna diameter D, because g^2 is proportional to D^4 and the integral in Eq. (11.130) is independent of D.

To measure $\Phi_n(\mathbf{K})$ accurately with radar requires $F(\mathbf{K})$ to be sharp compared to the shape of $\Phi_n(\mathbf{K})$. Thus, for anisotropic irregularities we need $D > \rho_h$ and $\sigma_r > \rho_v$, where ρ_v is the vertical correlation length. For spectral samples along the horizontal direction, we must have $D > \rho_v$ and $\sigma_r > \rho_h$. That the antenna diameter instead of the linear beamwidth $(r_0 \theta_1)$ enters into Eq. (11.131) illustrates the importance of the second-order expansion (11.119a) and the Fresnel term. Only when this term is negligible does the sampling function narrow as the linear beamwidth widens. Furthermore, the Fresnel term results because the irregularities have their axis of symmetry along Cartesian coordinates, whereas the phase fronts of the radar are spherical. If the irregularities had shapes that were concave downward with radius of curvature equal to r_0 (this occurs in an ideal RASS, Section 11.4.2), then there would be no Fresnel term, and the width of $F(\mathbf{K})$ would be inversely proportional to the linear beamwidth as well as to the range resolution.

11.5.4 Scattering from Anisotropic Irregularities, an Example

Let us consider the angular dependence of signal power for the scattering medium that has a strong anisotropic component. We shall assume that the correlation function can be decomposed into a part R_i due to isotropic turbulence and an anisotropic component R_a (i.e., $R = R_i + R_a$). We shall further assume that R_a is two dimensional and isotropic in a plane perpendicular to z' (Fig. 11.12). To obtain order-of-magnitude effects, we shall take R_a to be of the form

$$R_a = \langle \Delta n_a^2 \rangle \exp(-\delta_x'^2/2\rho_h^2 - \delta_y'^2/2\rho_h^2)\delta(\delta_z'), \qquad (11.132)$$

where ρ_h is the horizontal correlation length and $\delta(\delta_z')$ is a Dirac delta function.

The resolution volume coordinates δ_x, δ_y, δ_z are related to the primed coordinates aligned with the axis of the correlation function by

$$\delta_x' = \delta_x, \qquad \delta_y' = \delta_y \cos \psi - \delta_z \sin \psi, \qquad \delta_z' = \delta_y \sin \psi + \delta_z \cos \psi. \quad (11.133)$$

After introducing Eqs. (11.133) into Eq. (11.132) and the correlation (11.132) into Eq. (11.122), integration along δ_z is performed so that the three-dimensional integral reduces to a two-dimensional one. For anisotropic turbulence alone (i.e., $R_i = 0$) and for far-field conditions, this leads to the ratio of powers at an angle ψ and at zenith ($\psi = 0$).

$$\frac{\langle P_a(\psi) \rangle}{\langle P_a(0) \rangle} = \cos \psi \sqrt{\frac{Q(0)}{Q(\psi)}} \exp\left[-\frac{k_0^2 \tan^2 \psi}{Q(\psi)} \right], \qquad (11.134)$$

where

$$Q(\psi) = \frac{\sec^2 \psi}{2\rho_h^2} + \frac{\tan^2 \psi}{8\sigma_r^2} + \frac{\pi^2 \sigma_\perp^2}{f^4}.$$

The term $\cos \psi$ accounts for the decrease in power at constant height h due to the range and for the increase in power with tilt away from the vertical due to $\sigma_\perp(\psi) = \sqrt{2} h \sigma_\theta / \cos \psi$. To $\langle P_a(\psi) \rangle$ we must add the power due to isotropic turbulence. Assume $A = P_i(0)/P_a(0)$ is the ratio of the powers due to isotropic and anisotropic scatter at $\psi = 0$. The theoretical formula for normalized power is then

$$(\langle P_a(\psi) \rangle / \langle P_a(0) \rangle + A \cos^2 \psi)/(1 + A), \qquad (11.135)$$

which was subjectively fitted to data from Röttger et al. (1981) (Fig. 11.14). Pertinent parameters for these data are a wavelength $\lambda = 6.4$ m, $h = 17.5$ km, beamwidth $\theta_1 = 1.7°$, and range resolution $= 300$ m. We find the value $\rho_h = 20$ m for the width of the correlation function, and $A = 0.04$ fits these data well. Although the Fresnel term does not contribute significantly in this case, we see that scattering from turbulent anisotropic irregularities can account for the observed angular dependence of the returned power.

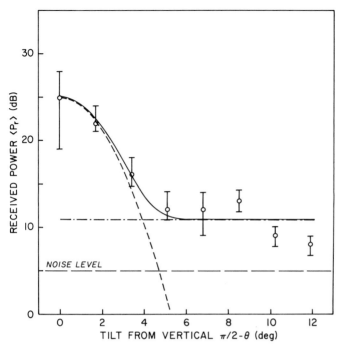

Fig. 11.14 Angular dependence of the observed mean backscatter power (open circles) from anisotropic irregularities as the radar beam axis is tilted from the vertical. (From Röttger, 1981.) Fitted to the data is a model that consists of anisotropic irregularities of n with a two-dimensional (horizontal) correlation function in an isotropic background. $-\cdot-$, isotropic; $---$, anisotropic; ———, total power. Height = 16.9–18.1 km; $\theta_1 = 1.7°$; $\rho_h = 20$ m; $A = 0.04$ (-14 dB); and $\lambda = 6.4$ m. (Reprinted with permission from *J. Atmos. Terr. Phys.* **43**, Röttger *et al.* Copyright 1981, Pergamon Press, Ltd.)

11.6 Characteristics of Refractive Index Irregularities

In Section 11.5 a relation between the spatial spectrum of refractive index irregularities and the scattered field has been derived. In the inertial subrange of turbulence the spectral density of Δn variance has the well-known eleven-thirds power law dependence on K and is proportional to the refractive index structure parameter C_n^2 (Section 10.1.3). (The span of wavelengths for which this assumption of an inertial range is valid is discussed in Section 11.6.2.) Because turbulent mixing creates irregularities in refractive index it is instructive to know how C_n^2 depends on the intensity of turbulence and the state variables of the atmosphere. In the following discussion we assume that condensation does not occur, so that water remains in the gaseous state during displacements of air parcels.

Tatarskii (1971, p. 73) showed that the structure parameter C_p^2 of any conserved passive additive is

$$C_p^2 = a^2 \varepsilon^{-1/3} K_p \left(\frac{d}{dz} \langle p \rangle \right)^2, \qquad (11.136)$$

where $\langle p \rangle$ is the ensemble average of the additive, which is assumed to have only a height dependence; K_p is its coefficient of turbulent diffusion; ε is the rate at which turbulent energy per unit mass is being dissipated, and a^2 is a dimensionless constant that needs to be found from experiments. It appears at present that a^2 is in the range from 3.2 to 4.0 (Gossard et al., 1982).

Equation (11.136) is valid for any conserved passive additive p mixed by turbulence. An additive is considered conserved if its characteristics (e.g., number of chaff needles) within an air parcel do not change when the parcel is advected. It is passive if it does not affect the flow. For example, in the absence of condensation, potential temperature θ, specific humidity q (Hess, 1959), and scattering particulates such as chaff are conserved passive additives. Equation (11.136) has been derived under the assumption that the turbulence is in a state of equilibrium. That is, the energy required to produce large-scale turbulence is equal to the rate at which viscous forces remove the energy associated with small-scale turbulence. Although turbulent mixing tends to reduce the gradient $\nabla \langle p \rangle$, it is assumed that external sources maintain this gradient. The eddy coefficient of diffusion K_p for the additive is analogous to the molecular diffusion coefficient. The latter is a measure of the rate at which gradients of additives are smoothed out by the thermal motion of the molecules, whereas K_p is a measure of the smoothing of gradients by turbulence. K_p usually exceeds the molecular diffusion coefficient by several orders of magnitude.

The refractive index in N units [Eq. (2.19)] is

$$N = (77.6/T)(P + 4810 P_w/T), \qquad (11.137)$$

where P is the total atmospheric pressure in millibars, P_w is the water vapor pressure in millibars, and T is the absolute temperature. But it is well known that as parcels of air are displaced vertically, their pressure is continuously brought into equalization with that of the surrounding air, resulting in a change of parcel temperature and water vapor pressure (a rising parcel of air cools 1 K for every 100 m of altitude change). Therefore, T and P_w, and thus N, are not conserved properties. Although the refractive index of an air parcel changes while the parcel moves vertically, irregularities Δn are produced only when there is change relative to the surrounding environment, not necessarily when there is change in the parcel's N. For example, if the environment had an adiabatic lapse rate, and P_w decreased with height at the same rate as P, then N of a displaced parcel would always be the same as the surrounding N, and hence no irregularities in refractive index would be created even though there is a gradient of $\langle N \rangle$. Thus radio scientists were led to define the potential refractive index modulus ϕ (Bean and Dutton, 1966, p. 17), which we designate as the potential refractive index.

$$\phi \equiv (77.6/\theta)(P_0 + 4810P_{w0}/\theta), \tag{11.138}$$

where

$$\theta = T(P_0/P)^\alpha, \qquad \alpha = (C_p - C_v)/C_p = 0.286, \tag{11.139a}$$

$$P_{w0} = P_w(P_0/P) \tag{11.139b}$$

are the potential temperature and potential water-vapor pressure at the reference pressure $P_0 = 1000$ mbar; and C_p, C_v are the specific heat capacities at constant pressure and volume. The mixture of water vapor with dry air does not significantly alter the heat capacity of the mixture from the value for dry air (Hess, 1959, p. 42), so α can be considered a constant. Because θ and P_{w0} are conserved properties of the air parcel, ϕ is also a conserved property. Thus we can use Eq. (11.136) to determine the structure parameter C_ϕ^2 from gradients of ϕ.

$$C_\phi^2 = a^2 \varepsilon^{-1/3} K_\phi \left(\frac{d}{dz} \langle \phi \rangle \right)^2. \tag{11.140}$$

Because the scattered field intensity is a function of C_n^2, it is necessary to relate C_ϕ^2 to C_N^2 where $C_N^2 = 10^{12} C_n^2$. This relation can be obtained by dividing Eq. (11.137) by Eq. (11.138), and substituting Eqs. (11.139) for T and P_w so that

$$N = \phi \left[\frac{1 + \dfrac{4810P_{w0}}{\theta P_0} \left(\dfrac{P}{P_0} \right)^{-\alpha}}{1 + \dfrac{4810P_{w0}}{\theta P_0}} \right] \left(\frac{P}{P_0} \right)^{1-\alpha}. \tag{11.141}$$

If one is to relate C_ϕ^2 to C_N^2, the variance of N must be expressed in terms of the variance of ϕ and covariances of ϕ and P_{w0}, of ϕ and θ, and of P_{w0} and θ using Eq. (11.141). This relation can be simplified by assuming a reference height at the level of the scattering layer; then $P/P_0 \approx 1$, and hence

$$C_N^2 \approx C_\phi^2. \tag{11.142}$$

In the remainder of this chapter we assume the reference level to be at the height of observation, so the variables θ, P_{w0}, and ϕ [the generalized potential temperature, potential water-vapor pressure, and potential refractive index (Ottersten, 1969b)] will be taken to be at this height as well. With this understanding,

$$C_n^2 = a^2 \varepsilon^{-1/3} K_\phi \times 10^{-12} \left(\frac{d}{dz} \langle \phi \rangle \right)^2. \tag{11.143}$$

Under the assumption of steady-state turbulence, horizontally homogeneous perturbation statistics, and a mean velocity horizontally directed and dependent only on height, the turbulent energy budget equation (Tennekes and Lumely, 1972, p. 97) can be reduced to

$$\varepsilon = \langle -u'w' \rangle \frac{\partial \langle u \rangle}{\partial z} + \langle -v'w' \rangle \frac{\partial \langle v \rangle}{\partial z} + \frac{g}{\theta_0} \langle w'\theta' \rangle \qquad (11.144)$$

by neglecting pressure perturbations (parcels are assumed to be continuously brought into equilibrium with the environment so that pressure fluctuations are not generated, or are negligibly small) and terms of order higher than the second in the perturbation quantities. The equations

$$\langle -u'w' \rangle \equiv K_m \partial \langle u \rangle / \partial z, \qquad \langle -v'w' \rangle \equiv K_m \partial \langle v \rangle / \partial z, \qquad (11.145)$$

$$\langle -\theta'w' \rangle \equiv K_H \partial \langle \theta \rangle / \partial z \qquad (11.146)$$

define the diffusion coefficients of momentum K_m and heat K_H given the mean gradients and the transfer rates of turbulent momentum $\langle -\rho u'w' \rangle$ and heat $\langle \rho C_p \theta'w' \rangle$, where C_p is the specific heat at constant pressure. We assume that the diffusion coefficients for u' and v' are equal, although this may not always be the case (Rabin et $al.$, 1982). Because the reference level is at the height of observation, the potential temperature θ is simply the mean temperature T_1 of the turbulent layer. Combining Eqs. (11.144)–(11.146), we obtain

$$\varepsilon = K_m |d\mathbf{v}_h/dz|^2 (1 - R_f), \qquad (11.147)$$

where R_f is the flux Richardson number

$$R_f = \left(g \frac{\partial \langle \theta \rangle}{\partial z} \middle/ T_1 \left| \frac{d \langle \mathbf{v}_h \rangle}{dz} \right|^2 \right) \left(\frac{K_H}{K_m} \right) = R_g \left(\frac{K_H}{K_m} \right), \qquad (11.148)$$

and R_g is the gradient Richardson number. The eddy diffusion coefficient K_ϕ of the refractive index can be defined analogously to Eqs. (11.146) for heat. But, because K_ϕ is expected to be nearly equal to K_H, we write Eq. (11.143) in an alternative form using Eqs. (11.147) and (11.148).

$$C_n^2 = \frac{a^2 \varepsilon^{2/3} T_1 R_f}{(1 - R_f) g \, d \langle \theta \rangle / dz} \left(\frac{K_\phi}{K_H} \right) \times 10^{-12} \left(\frac{d \langle \phi \rangle}{dz} \right)^2. \qquad (11.149a)$$

Although the ratio K_ϕ / K_H might be well approximated by unity, the ratio K_H / K_m is much larger than one if buoyancy-generated turbulence dominates shear production, as can be expected in the boundary layer during sunny afternoons (Tennekes and Lumley, 1972, p. 102).

Sometimes C_θ^2, C_w^2 of θ, P_{w0} and the structure parameter associated with the product θP_{w0} can be measured separately or estimated in the convective boundary layer using parameterizations based on observation (Gossard, 1960; Lenschow and Wyngaard, 1980). In this case (Gossard, 1977; Gossard and Strauch, 1983, p. 167)

$$C_n^2 = a^2 C_\theta^2 + b^2 C_w^2 - 2ab C_{\theta w}^2, \qquad (11.149b)$$

where the constants depend on the properties of the air mass. For tropical summer maritime air at a height of 500 m, for example (Gossard, 1977),

$$a^2 = 2.24 \times 10^{-12} \qquad b^2 = 17.8 \times 10^{-12}, \qquad 2ab = 12.6 \times 10^{-12},$$

whereas for warm continental air

$$a^2 = 1.25 \times 10^{-12}, \qquad b^2 = 16.2 \times 10^{-12}, \qquad 2ab = 9.01 \times 10^{-12}.$$

Carefully controlled laboratory experiments with statically stable shear flow in aqueous solutions (Thorpe, 1969) and wind tunnels (Scotti and Corcos, 1969) have established that disturbances become dynamically unstable when R_g is reduced to 1/4, at which value Kelvin–Helmholtz (K–H) waves appear at a wavelength having a maximum growth rate. Comparison of typical values of quantities in the laboratory, ocean, and atmosphere shows that the wavelength for maximum growth rate is about $2\pi d\sqrt{2}$, where d is the thickness of the turbulent layer (Stoeffler, 1972). Most interestingly, Thorpe's data suggest how fine-scale three-dimensional irregularities might be generated from large-scale two-dimensional waves. After all, it is the fine-scale irregularities that are responsible for scatter at microwave frequencies. As the wave amplitude increases, waves roll up, producing a spiraling layer of fluid, and then break. The transition to turbulence occurs after the rolling up has begun, probably resulting from gravitational instability as the denser air becomes superimposed over the lighter air in the spiral structure of the breaking wave. The spiraling structure generates patches of turbulence, which are then elongated in the shear flow and coalesce, eventually forming a quasiuniform layer. This layer contains turbulent motion and refractive index irregularities that are observed with radar and airborne refractometers (Doviak and Berger, 1980). The experiments of Thorpe (1973) suggest not only $R_g = 1/4$ is required for generation of unstable waves but also that $R_f \approx 1/4$ during the transition from initial to final flow after the turbulence has smoothed the initial gradients of $\langle \theta \rangle$ and $\langle v_h \rangle$.

Clearly Eq. (11.149a) stresses that C_n^2 is not only a function of the turbulence intensity K_m [through ε in Eq. (11.147)] but also depends on the gradient of $\langle \phi \rangle$. The turbulence can be strong and yet C_n^2 might be relatively small. In a dry atmosphere above the first kilometer or two, where the contribution of water vapor to the refractive index irregularities can often be ignored, C_n^2 can be expressed as

$$C_n^2 \approx \frac{a^2 R_f (77.6 P_1)^2 \varepsilon^{2/3} \times 10^{-12}}{(1 - R_f) g T_1^3} \left(\frac{K_\phi}{K_H} \right) \frac{d\langle \theta \rangle}{dz}. \tag{11.150}$$

Because the upper atmosphere is, on the average, stable, $d\langle \theta \rangle / dz$ is positive. Under the assumptions used to derive relation (11.150), both C_n^2 and ε can be estimated from Doppler radar measurements; the relation between ε and σ_v is given by Eqs. (10.68) or (10.70), and C_n^2 and η are related by Eq. (11.109). The mean layer temperature T_1 can be estimated with reasonable accuracy from climatological tables. The values $R_f \approx 1/4$, $3.2 \leq a^2 \leq 4.0$, and $K_\phi/K_H \approx 1$ specify all parameters except for $d\langle\theta\rangle/dz$ and the radar-measurable ones, C_n^2 and ε. Therefore measurements of C_n^2 and ε can be used in relation (11.150) to estimate $d\langle\theta\rangle/dz$, the stability of the atmosphere, wherever moisture contributions to Δn are insignificant. The locations of the temperature inversions (i.e., layers of large $d\langle\theta\rangle/dz$) are useful in retrieving temperature profiles from radiometric measurements (Westwater and Grody, 1980). Gossard *et al.* (1982) have used radar measurements of ε and C_n^2 to estimate the mean gradients of potential refractive index in a moist boundary layer. These results compare well with those calculated from rawinsonde observations. Turbulence usually occurs in thin layers (e.g., 10–1000 m), which might be small compared to the resolution volume size. Thus, unless a high-resolution (large-bandwidth) radar is used pointing nearly in a vertical direction, $d\langle\theta\rangle/dz$ could be underestimated because gradients of C_n^2 are smoothed by the weighting function.

On the other hand, Weinstock (1981) points out that ε (and hence K_m if $d\langle v_h\rangle/dz$ is also measured) can be estimated solely from C_n^2 with relatively high accuracy (i.e., to within a factor of 2) if C_n^2 is accurately measured. This is a result of a combination of factors:

1. The relation between shear and the $\langle\theta\rangle$ gradient required for the shear production of turbulence causes a weak dependence of C_n^2 on the gradients of θ.

2. The layer thickness, although smaller than the resolution volume, is approximately equal to the outer limit of the inertial subrange scales, which in turn is dependent on ε and $d\langle\theta\rangle/dz$.

3. The range of variation in $d\langle\theta\rangle/dz$ in the stably stratified layers above the lower troposphere is limited.

Thus the layer of turbulence need not fill the resolution volume in order to make reasonably accurate estimates of ε within the layer.

11.6.1 Dependence of the Structure Parameter on the Height

When there is an abundance of moisture near the earth's surface, the reflectivity of the lower atmosphere can be significantly larger than for the drier air of the upper troposphere because of the contribution to C_n^2 from gradients of P_{w0}. Where there is a marked decrease of P_{w0} accompanied by an increase in θ, (i.e., a stable layer) reflectivity can be relatively large even if turbulence is weak. This is because ϕ depends linearly on P_{w0} but inversely on $\langle\theta\rangle$, so opposite gradients of P_{w0} and $\langle\theta\rangle$ in this transition region result in large values of $\partial\langle\phi\rangle/\partial z$ and hence enhanced values of C_n^2. The layer of reflectivity at about 1500 m altitude (Fig. 11.15) is probably caused by weak turbulence (i.e., weak because the layer is thin and waves are not resolved) generated by K–H waves. Thus turbulent mixing in regions of opposing gradients of P_{w0} and θ results in the layer of

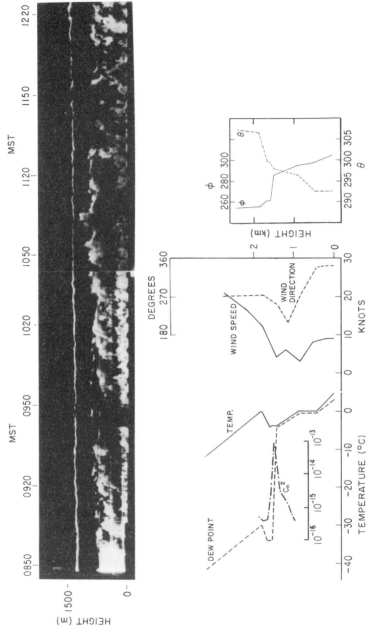

Fig. 11.15 Vertically pointing FM cw radar and balloon-sounding data near Denver, Colorado. (From Gossard, 1981.)

enhanced reflectivity seen in Fig. 11.15. The reflectivity field associated with convective plumes appears also in Fig. 11.15. Note that the reflectivity is enhanced at the top of the plumes (for explanation see Section 11.7).

Above the convectively mixed boundary layer (CBL), which is often moist, the refractive index irregularities are found in layered zones of relatively dry air where shear generates turbulence and potential temperature gradients are slightly positive. These zones usually have a small vertical extent (several tens of meters) separated by regions of little or no turbulent activity (Fig. 11.16a). The data in Fig. 11.16a were obtained with a 10.7-cm radar, whereas for Fig. 11.16b a 70-cm radar was used. The differences between these data are discussed in Section 11.6.2.

Whenever vertical profiles of C_n^2 are shown, they usually represent large horizontal and/or temporal averages, whereas C_n^2 is highly variable and changes by orders of magnitude over a few meters in altitude. Nevertheless, it is useful to estimate expected values of C_n^2 versus height in order to gauge the feasibility of using weather radars to detect backscatter with sufficient strength to enable measurements of wind and turbulence.

Gossard (1977) has determined the dependence of C_n^2 on height for different air masses using the mean properties of the air for each category (e.g., maritime air). He assumed that $\varepsilon^{-1/3} K_m$ is neither systematically related to height nor correlated with $\nabla \langle \phi \rangle$ and he estimated the mean value of the unknown factor $a^2 \varepsilon^{-1/3} K_m$ from the mean values of the measured optical structure constant, which is relatively insensitive to the amount of moisture in the air. Figure 11.17 shows the range of C_n^2 deduced by Gossard for summer and winter maritime air of both tropical and polar origin. Maritime air originates over oceans and is particularly significant because it contains substantial moisture, which is one of the essential ingredients for severe storm development. Continental air originating over land has slightly lower (by about a factor of 4) values of C_n^2 below 2 km, but there is a dramatic decrease in C_n^2 above. During midday, when surface heating and convection is vigorous in the CBL, C_n^2 might increase by an order of magnitude.

Burk (1978) used a second-moment turbulence closure model to deduce that the maritime boundary layer can have C_n^2 larger than $10^{-12} \, \mathrm{m}^{-2/3}$ for all heights up to 1 km. Aircraft and radar measurements in maritime air over Oklahoma compared well and gave a mean $C_n^2 = 5 \times 10^{-13}$ m and peak values as high as $3 \times 10^{-12} \, \mathrm{m}^{-2/3}$ for mid-afternoon clear skies (Doviak and Berger, 1980). Chadwick and Moran (1980) made long-term measurements of C_n^2 in the boundary layer with an FM cw 10-cm wavelength radar. For the thunderstorm months between March and October, they found a median value of C_n^2 of about $6 \times 10^{-16} \, \mathrm{m}^{-2/3}$ for data 800 m above the high plains of Colorado. This value is plotted as the star in Fig. 11.17. On the other hand C_n^2 median value of $2.4 \times 10^{-11} \, \mathrm{m}^{-2/3}$ was measured during daytime summer conditions with a 10-cm prototype Terminal Doppler Weather Radar (TDWR) in Denver (Biron and Sen Lee, 1990); this is more than four orders of magnitude larger than deduced

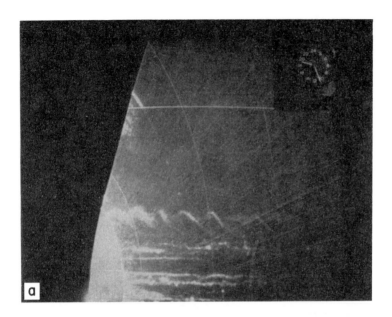

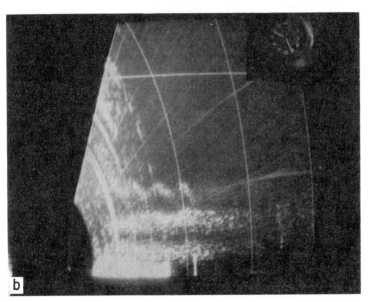

Fig. 11.16   (a) Range height indicator (RHI) photograph at 120° azimuth taken at 1426 E.S.T. on 17 March 1969 with the 10.7-cm radar at Wallops Island, Virginia. (b) The same situation except that data are obtained with a 70-cm radar. The height mark is at 12.2 km and the range marks are at 9.3-km intervals. More than seven separate horizontally stratified clear-air layers are visible below 6 km. Note the large-amplitude wave in the top layer. (Courtesy of Jack Howard, NASA Wallops Island, Virginia.)

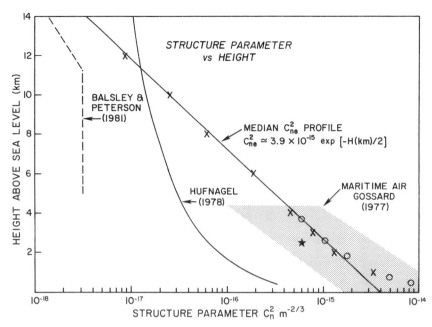

Fig. 11.17 Structure parameter C_n^2 estimates versus height. x are values computed by $L^{(101)} = L^t$ where $L^{(101)}$ is the median transmission loss measured from forward-scatter data and L^t is the loss computed assuming that Δn variance is uniform within the scatter volume and that Eq. (11.104) applies. The star denotes results from Chadwick and Moran (1980) and open circles are estimates computed using formulas of Rabin and Doviak (1989).

by Chadwick and Moran (1980), and about twice that measured with a 10-cm radar by Doviak and Berger (1980) in maritime air. These larger values suggest that point scatterers are a dominant contributor to the reflectivity; Chadwick and Moran (1980) used the superior resolution of the FM cw radar to edit reflectivity contributions from point scatterers. Measurements with the TDWR in Kansas City, Kansas gave a median reflectivity value even an order of magnitude higher than in Denver (Biron and Sen Lee, 1990), which again indicates the importance of point scatterer's contributions to reflectivity at centimetric wavelengths.

The reflectivity is considerably less in the upper troposphere, on the average, than in the CBL, and weather radars rarely receive echoes from refractive index irregularities at these high altitudes. In the late 1960s, high-power large-aperture incoherent 10-cm radars established that refractive index irregularities in the upper troposphere are usually located in widely spaced stratified layers and are associated with shear induced turbulence (Hardy and Katz, 1969). Doppler radars and especially wind profilers (Section 11.8.1) have increased capability over incoherent ones and should sense more of these layers, so that nearly continuous wind profiles can be obtained with reasonable vertical resolution (e.g., 1 km).

VHF (6-m wavelength, Fig. 2.1) radar measurements in the upper tropo-sphere suggest that C_n^2 may, on the average, be about 10^{-17} m$^{-2/3}$ or larger (Green *et al.*, 1979). Gage *et al.* (1980) measured C_n^2 larger than 10^{-18} m$^{-2/3}$ throughout the upper troposphere and lower stratosphere (<15 km). Neverthe-less, caution should be exercised before extending these observations to the UHF wavelength (i.e., $\lambda = 10$ cm, Fig. 2.1) of weather radars because the larger-scale refractive irregularities that scatter the VHF waves are likely to be anisotropic (Section 11.5.4), in which case Eq. (11.104) would not be applic-able. Also the UHF sampling wavelength may be outside the inertial subrange (Section 11.6.2).

The earliest reflectivity measurements in the UHF band were made during the 1940s and 1950s using troposcatter communication links. The National Bureau of Standards's Technical Note 101 (Rice *et al.*, 1966) synthesizes data from hundreds of links and gives the median condition of atmospheric reflectiv-ity for various link geometries. By equating the measured median transmission loss for $\lambda = 10$ cm, with the transmission loss computed assuming the resolution volume to be uniformly filled with refractive index fluctuations (having a spectral distribution corresponding to the inertial subrange), the height dependence of an effective structure parameter C_{ne}^2 can be derived (Fig. 11.17). But, because forward-scatter angles are usually less than $10°$, refractive irregularities effective in forward scattering are at scales significantly larger than $\lambda/2$ and hence can lie within the inertial subrange, whereas the 5-cm scales effective in 10-cm backscat-ter could fall into the dissipation range where the intensity of refractive index variations are much smaller (Section 11.6.2). If these C_{ne}^2 values are used in Eq. (11.104), the median radar reflectivity may be overestimated. The C_{ne}^2 val-ues (Fig. 11.17) correspond quite closely to those computed by Gossard (1977) and Rabin and Doviak (1989), who used structure parameter data from airborne measurements, and also to the mean values measured with a 10-cm radar by Chadwick and Morgan (1980). This suggests that, at least in the first kilometer or two of the troposphere, the inertial subrange extends to scales as small as a few centimeters.

Values of C_n^2 obtained from an empirical formula deduced by Hufnagel (1978) for the structure constant of the optical refractive index are also plotted, for comparison, in Fig. 11.17. It is evident that the values derived from 10-cm median transmission-loss data are larger in the lower troposphere, consistent with the fact that moisture irregularities, more intense in the first few kilometers, contribute significantly to the C_n^2 deduced from UHF observations while giving negligible contribution at optical frequencies.

Radar measurements of the structure parameter in the upper troposphere were made at a wavelength of 23 cm by Crane (1977), who observed $C_n^2 \geq 10^{-17}$ m$^{-2/3}$ for heights below 15 km on 13 days of clear air between January and July in New England. Balsley and Peterson (1981), using the 23-cm radar in Chatanika, Alaska, found that $C_n^2 \geq 10^{-18}$ m$^{-2/3}$ more than 50% of the time over several days at heights below 16 km. A subjective fit of their median data is plotted in Fig. 11.17. The large dispersion of median C_n^2 values in Fig. 11.17 may be a result of (1) the different locations of the radar sites and thus different meteorological conditions, (2) seasonal variation, and (3) insufficiently large data samples to produce a statistically stable median value.

11.6.2 Inertial Subrange

Equation (11.104) is valid only if the refractive index scales, of size Λ equal to half the radar wavelength, lie within the inertial subrange of turbulence scales. In this range the rate at which turbulent energy is transferred to smaller scales as large eddies fragment depends only on the dissipation rate ε of turbulent energy. As the scales become smaller, the air's viscosity directly affects the intensity of turbulence, and the variance density spectrum decreases more rapidly than for scales within the inertial subrange. The small-scale eddies affected by viscous forces are then within the dissipation range of the scales, and it is there, ultimately, that turbulent energy is dissipated as heat. Therefore, we need to examine the conditions under which radiation scatters from irregularity scales that fall into the dissipative spectral range where reflectivity would be significantly less than that deduced by application of Eq. (11.104).

Hill (1978) has developed a model that determines the behavior of the spectra for the transition region between the inertial and dissipation range of scales. For refractive index spectra dominated by temperature fluctuations, Hill's spectra show that if

$$\lambda/2 \geq \Lambda_t \equiv 5\pi(\nu^3/\varepsilon)^{0.25} \tag{11.151}$$

(where ν is the kinematic viscosity), the spectral density at $K = 4\pi/\lambda$ is at least as large as that predicted for the inertial subrange scales. The use of Eq. (11.104) and C_n^2 data to estimate the radar performance is then justified. The kinematic viscosity ν has a well-defined height dependence, but the eddy dissipation rate ε may vary by three or more orders of magnitude at any height. Based on model predictions of Hill that lead to the identity (11.151), the maximum height $H_{\max}$ to which Eq. (11.104) is applicable in a standard atmosphere (Shea, 1965) is plotted in Fig. 11.18 versus the transition wavelength λ_t (i.e., the *radar* wavelength below which reflectivity is less than that given by Eq. (11.104)). The radar receives echoes from refractive index irregularities at wavelengths $\Lambda = \lambda/2$ that lie within the dissipative range whenever $H_{\max}$ is exceeded. In this range the intensity of the irregularities, and hence the radar reflectivity η, decrease exponentially as $K = 2\pi/\Lambda$ increases (Tatarskii, 1971, p. 65). Thus, for radars operating at wavelengths to the right of the line of constant ε, η is proportional to $\lambda^{-1/3}$, whereas for those operating to the left η decreases exponentially.

Shear-generated turbulence occurs in layers having various levels of turbulence intensity ε, and if $\lambda > \lambda_t$, η is proportional to $\varepsilon^{2/3}$ [e.g., Eq. (11.150)]. But,

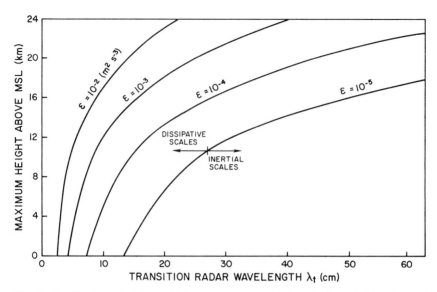

Fig. 11.18 Maximum height to which echoes are returned from Δn irregularities of size within the inertial subrange for a given level of the eddy dissipation rate ε. The values of ε corresponding to various levels of turbulence are gauged by its effects on aircraft: severe turbulence, $\varepsilon = 6.75 \times 10^{-2}$; moderate turbulence, $\varepsilon = 8.5 \times 10^{-3}$; light turbulence, $\varepsilon = 3.0 \times 10^{-3} \, \mathrm{m^2 \, s^{-3}}$ (Trout and Panofsky, 1969). (Courtesy of R. G. Strauch, WPL/NOAA.)

the strong exponential decrease of η for $\lambda \leq \lambda_t$ might give the appearance that C_n^2 is more strongly stratified than it really is if Eq. (11.104) is used to estimate C_n^2 from η. This may explain the layered appearance of reflectivity (Fig. 11.16) seen by radars operating at the upper end (i.e., 10 cm) of the UHF band.

The appearance of height continuum of echoes at VHF wavelengths may be a result of their capability to detect layers of weaker ε for which $\lambda_{\mathrm{VHF}} > \lambda_t$, so that irregularities of n are still within the inertial subrange. Nevertheless, it is unlikely that all layers are turbulent all the time, a deduction supported by the study of Lilly *et al.* (1974), although their results apply to turbulence measurements above 14 km. In a stable atmosphere, thermal stratification damps the vertical exchange of heat and momentum; however, velocity perturbation might be reduced more quickly than heat (and consequently refractive index irregularities) because the turbulent Prandtl number (i.e., the ratio of the eddy diffusivities of momentum and heat) is larger than one (Kundu, 1990). Similar arguments also apply to water vapor. Thus the apparent height continuum of echoes observed with long-wavelength radar may signify the presence of fossil "turbulence" in which irregularities in n remain although velocity perturbations have vanished (Woods *et al.*, 1969). On the other hand, if radar does not have adequate resolution, the intermittent turbulence occurring in various layers could appear continuous in height. High-resolution VHF observations by Rött-ger and Schmidt (1979) clearly show the layered structure of η.

Plots by Lilly *et al.* (1974) suggest that the cumulative probability of turbulence intensity for lower stratospheric data (14 km $\leq H \leq 21$ km) decreases exponentially with increasing turbulence. Extrapolation of their results indicates that no turbulence occurs 90% of the time. Although extrapolation is risky, the data show that $\varepsilon \geq 10^{-4}$ m^2 s^{-3} less than 2% of the time. Nevertheless, turbulence is expected to be much more frequent in the troposphere, where the air is less stable, and R_g is more likely to fall to values less than the critical 1/4.

Inferences of ε from the diffusion of smoke puffs, direct measurements of turbulent shearing stress and wind shear, and measurements of the spectra of turbulent velocities (e.g., Ball, 1961) suggest $\varepsilon < 10^{-5}$ m^2 s^{-3} for heights above 3 km. But, radar-inferred values appear to be larger than 10^{-5} or even 10^{-4} for most heights up to 12 km on a day of strong shear (Gage *et al.*, 1980). Radar-inferred values of greater than 10^{-3} m^2 s^{-3} were found to be common for all heights up to 12 km over a period of 12 h during which the atmosphere was disturbed by the passage of a polar jet stream. If prestorm atmospheres have such large values of ε, then Doppler weather radar (i.e., $\lambda \geq 10$ cm) has the potential to give profiles of winds preceding significant weather. Numerical models give evidence that the vertical shear of horizontal wind influences the intensity of storms that develop in a sheared environment (Weisman and Klemp, 1982; Schlesinger, 1982).

11.6.3 Criteria for Detection of Refractive Index Irregularities

Criteria for the detection of scatter from turbulent air under the assumption that refractive index irregularities are statistically homogeneous, stationary, and isotropic have been given by Hennington *et al.* (1976). The criteria are exactly the same as those specified by Gage and Balsley (1978). We make similar assumptions but derive criteria that incorporate the precision of the Doppler velocity estimates.

The SNR is a parameter of paramount importance to the measurement of the Doppler velocity. For a Gaussian receiver-filter matched to a rectangularly shaped transmitted pulse ($r_6 \approx 1.17\, c\tau/2$), the per-pulse SNR referred to the receiver input is

$$\text{SNR} = \frac{0.21 P_t \varepsilon_a A r_6^2}{4\lambda^{1/3} \pi c r_0^2 l_t kT_{sy}} C_n^2, \tag{11.152}$$

where P_t is the peak power at the transmitter; l_t is the total loss factor, equal to the product of the transmitter and receiver transmission-line losses (including) radome loss) as well as the atmospheric losses; $\varepsilon_a = A_e/A$ is the antenna efficiency relating the aperture's physical area A to A_e [Eq. (3.12)] and accounting for nonuniform illumination (Section 3.1.1) as well as ohmic losses in the reflector; k is the Boltzman constant; c is the speed of light; T_{sy} is the system noise temperature (Section 3.5.1); and r_6 is the range resolution (Section 4.4.4). In deriving Eq. (11.152), the weather radar equation (4.16) is used, but line and radome losses are deleted from the gain g. Then g is related to θ_1 as $g = (16 \ln 2)/\theta_1^2$ and combining Eq. (4.16) with Eqs. (3.21), (3.32), (3.39), and (11.104) leads to Eq. (1.152).

Note that SNR is independent of the system gain (g_s) because signal and noise are measured at the same point in the receiver and because we assume all electronic noise is effectively introduced in the receiver's front end. The SNR is weakly dependent on λ. Thus, the only radar parameters in Eq. (11.152) that can produce a proportional increase in the SNR without changing the range resolution are P_t, A, and T_{sy}. In the UHF band, T_{sy} is usually most dependent on the receiver noise figure, so the use of low-noise preamplifiers can significantly reduce T_{sy}. At the high elevation angles suitable for obtaining wind profiles, a representative value of T_{sy} is 195 K. Longer-wavelength radars have higher T_{sy} because of intense radiation from space at those wavelengths. P_t can be effectively increased by transmitting longer coded pulses so that r_6 remains the same (Section 7.6), but this compromises observation at the nearest ranges.

When signals are weaker than noise and the spectrum width is narrow compared to the unambiguous interval, the minimum SNR that can be tolerated to produce velocity estimates with a precision of SD($\hat{v}$) is obtained from Eq. (6.22a) by retaining only the term that is multiplied by N^2/S^2.

$$\min(\text{SNR}_{pp}) = \frac{\lambda \exp[8(\pi\sigma_v T_s/\lambda)^2]}{4\pi\sqrt{2T_d T_s} \text{SD}(\hat{v})}. \tag{11.153}$$

A minimum of minima is obtained if the system period T_s is selected to be

$$T_s = \lambda/4\pi\sigma_v\sqrt{2}. \tag{11.154}$$

In deriving Eq. (11.154) we assumed constant dwell time and P_{av} (at the highest permissible level). If r_6 is constrained to be a constant, however, and if the radar is peak-power limited, pulse-coding techniques (Section 7.6) may be necessary to maintian P_{av} at its level when T_s is set by Eq. (11.154).

Doppler spectra of clear-air echoes observed with the NSSL's radar usually have σ_v ranging between 1 and 2 m s^{-1}. Crane (1980) observed Doppler spectra of clear-air turbulence over the Marshall Islands and estimated spectrum widths on the order of 1 m s^{-1}. Thus, for 10 cm wavelength radars, T_s needs to be about 3 ms, with a corresponding unambiguous velocity interval of 16 ms^{-1}.

11.7   Observations of Clear-Air Reflectivity

Although clear air does not contain such awesome phenomena as tornadoes or baseball-sized hail found in severe storms, its structure is nevertheless rich in meteorological events that can lead to the development of these storms. The spectra of clear-weather phenomena span scales from centimeters to thousands of kilometers and embrace such occurrences as convective cells that spring from heated surfaces, K–H waves generated by shear instabilities, and buoyancy waves. Many of these can be identified from observations of the reflectivity and or Doppler fields.

In the 1960s ultra-sensitive, high-resolution radars at Wallops Island, Virginia routinely mapped reflectivity of the convective boundary layer and shear zones at higher altitudes. Moderately sensitive Doppler weather radars, which have the advantage of coherent processing, can also map the reflectivity of clear air. For example the structure of reflectivity fields clearly shows the presence of convective cells in conditions of light wind (Fig. 11.19) and reflectivity streets (i.e., alignment of enhanced reflectivity along parallel lines) associated with convective rolls in cloudless skies are clearly depicted in

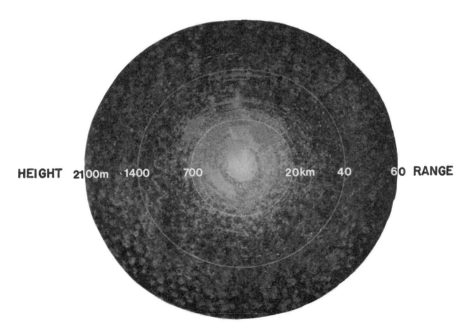

Fig. 11.19 Plan position display of a uniform pattern of convective cells observed with NSSL's 10-cm Norman Doppler weather radar on 17 August 1976 at 1312 C.S.T. (cumulus humilis clouds). Elevation: 2.0°.

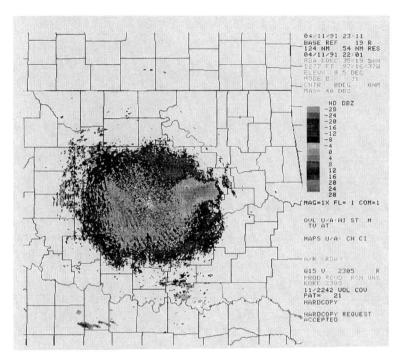

Fig. 11.20   The WSR-88D reflectivity factor display delineates reflectivity streets aligned parallel to the mean wind in central Oklahoma; streets are separated by about 7 km, and range to the edge of the reflectivity field is about 100 km; counties are outlined.

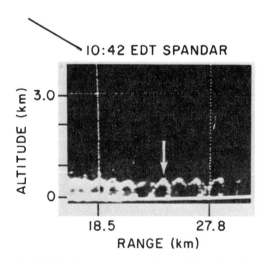

Fig. 11.21   Vertical cross section of the reflectivity structure produced by convective cells at Wallops Island, Virginia. (From Rowland and Arnold, 1975.)

Fig. 11.20. Curved flow (vertical gradients of vertical shear) limits convection to two dimensions in a plane perpendicular to the direction of the mean wind; the spacing of the rolls is determined by the depth of circulation (Doviak and Berger, 1980). Convective cells are delineated by 10-cm radar in the range height indicator (RHI) presentations (Fig. 11.21). Note the enhanced intensity of echoes at the cell top and sides, which can cause doughnut-shaped patterns when the radar beam scans azimuthally, as seen in Fig. 11.19. This increase in C_n^2 at the upper boundaries can be explained using Eq . (11.149b) and the following argument.

As air near the ground is warmed by the heated surface and moistened by evaporation, it becomes buoyant; thus it rises and cools adiabatically. But the warmed cell continues its upward acceleration until it passes the level of zero buoyancy (the altitude where the temperature of the cell air equals the ambient temperature). Because the ambient air has a positive gradient of θ, and P_{wo} usually decreases with height, air parcels carried above the level of zero buoyancy will be cool and moist relative to the air mixed in from the surrounding environment at the cell boundaries. The decrease in temperature and associated increased moisture causes the potential temperature and vapor pressure to be negatively correlated so that $C_{\theta w}^2$ in Eq. (11.149b) adds to the first two terms. Inside the cell, θ and P_{wo} are positively correlated (i.e., an increase in the temperature is accompanied by increased moisture), so $C_{\theta w}^2$ is subtracted from the first two terms. Therefore, C_n^2 inside the cell is expected to be less than along its boundaries, especially those above the level of zero buoyancy.

The location of strong gradients of $\langle \phi \rangle$ can be detected if turbulence is present. The thin continuous line of enhanced reflectivity seen on the time–height indicator display (Fig. 11.15) and the corresponding sounding data show the coincidence between the gradient of potential refractive index $\langle \phi \rangle$ and the peak of C_n^2. In this case, the peak gradient of $\langle \phi \rangle$ is principally controlled by $\langle P_{wo} \rangle$ and thus the thin line of reflectivity delineates the top of a moist layer.

The presence of frontal boundaries can also be resolved by observing the reflectivity structure of clear air. Figure 11.22a shows the reflectivity field with enhanced echoes along the boundary of an advancing cold front. The reflectivity enhancement was evident within the radar range (100 km) two hours before the clouds (Fig. 11.22b) developed. Notice that reflectivity to the north of the cold front is significantly higher than in the warm sector. Because air in the warm sector is well mixed, $\langle \theta \rangle$ and P_{wo} are constant with height (i.e., the air is neutrally stable, e.g., Fig. 11.15). Thus, from Eq. (11.138), we deduce that gradients of $\langle \phi \rangle$ are small. On the contrary, air in the cold sector is relatively stable and has larger gradients of $\langle \phi \rangle$. Furthermore, cold air over the warm ground enhances convective mixing (i.e., increases ε) which, together with the increased gradient of $\langle \phi \rangle$, generates larger C_n^2, as can be deduced from Eq. (11.149a). The markedly enhanced reflectivity along the frontal boundary might be associated with more vigorous mixing of the air between the two contrasting air masses, across which gradients of $\langle \phi \rangle$ are large; in addition, the gusts associated with fronts have been

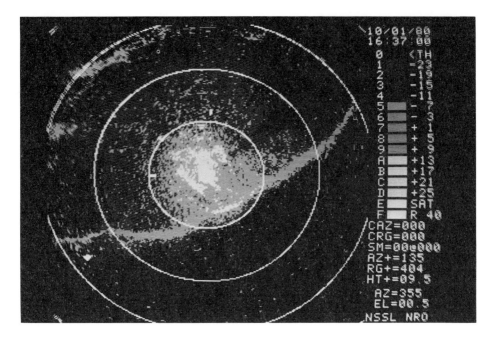

Fig. 11.22a The reflectivity field with enhanced backscatter on the north (cold) side and along a frontal boundary of a cold front.

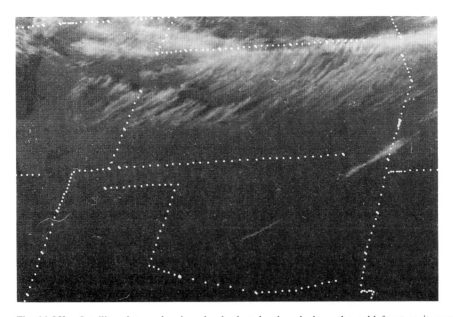

Fig. 11.22b Satellite picture showing clouds that developed along the cold front as it passed through central Oklahoma.

observed to sweep up substantial amounts of particles that also contribute to the reflectivity. There is no way that single-wavelength single-polarization radar can distinguish between returns from refractive index irregularities and particles if both are densely distributed throughout the resolution volume. Only when particles are so sparse that there is at most one or two scatterers per resolution volume can these be distinguished from refractive index irregularities; the amplitudes of echoes from these sparsely distributed scatterers will no longer have the Rayleigh distribution.

The ultra-sensitive, high-resolution radars at Wallops Island regularly detect horizontally stratified reflective layers in the troposphere (Fig. 11.16). These multiwavelength radars showed the wavelength dependence of the reflectivity within the layers to be associated with scatter from irregularities having scales within the inertial subrange (Hardy and Glover, 1966). The scattering layers are often quite thin; the vertical depth can be as small as a few meters and rarely exceeds a few hundred meters. Although water-vapor fluctuations are principal contributors to backscatter in the lower troposphere, there is mounting evidence that the layers of enhanced reflectivity in the upper troposphere are associated with layers of enhanced stability so that temperature fluctuations there may be the dominant contributors (Green and Gage, 1980).

The thin-layered reflectivity in Fig. 11.16a exhibits a wavelike shape. Similar layers have been reported by Ottersten (1969a) at the tropopause height. The tropopause marks the location of the base of the stratosphere at about 10 km, which is a region of persistent and intense static stability (Fig. 2.4). The breaking of K–H waves in the lower atmosphere is well illustrated by the reflectivity field of high-resolution FM cw radars (Fig. 11.23).

The evolution of convective plumes is also well observed with high-resolution 10-cm radars; an example from Wallops Island taken on 25 June 1970 is shown in Fig. 11.24. At 0736 Local Standard Time (L.S.T.; the clock in the upper right indicates G.M.T.) there was no evidence of convection, but two prominent reflectivity layers were observed (Fig. 11.24a). Examination of rawinsonde data implies that the lowest reflectivity layer is at the top of an inversion layer that extends from the surface to about 300 m AGL. The second layer is at the base of another inversion that extends from about 2.1 to 2.9 km. At 1023 L.S.T. the layer at 300 m shows evidence of breaking and diffusing, most likely caused by vigorous convection in the lowest 300 m of the atmosphere (Fig. 11.24b). Many more layers are also present in this figure, and these are probably caused by shear instabilities. Small-amplitude K–H waves (not resolved by this radar) created by these instabilities break up and mix $\langle \phi \rangle$ gradients. At 1253 L.S.T. the reflectivity layer at the top of the 300 m thick inversion layer has completely dissipated, and convective plumes have penetrated to altitudes of at least 2 km (Fig. 11.24c); the bright plume extending to about 4.3 km is probably associated with a cloud in which more vigorous vertical displacement are attained because of the release of latent heat of condensation, thus increasing the plume's buoyancy.

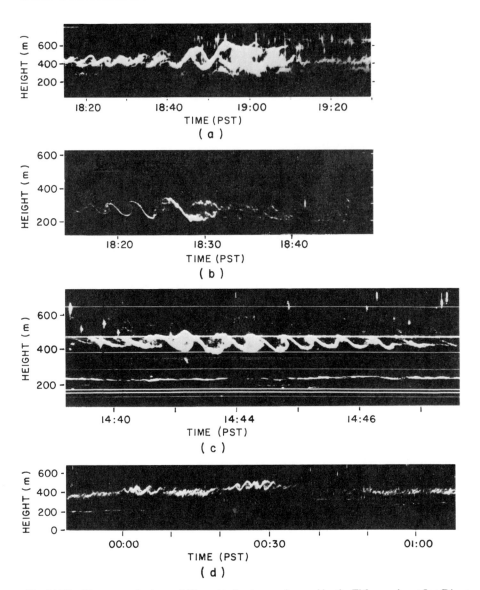

Fig. 11.23 Wave perturbations of thin scattering layers observed by the FM cw radar at San Diego. The remarkable resolution of this radar is evident from the vertical scale. (a) 28 September 1971. (b) 6 August 1969. (c) 24 August 1970. (d) 14 July 1979. (From Gossard and Hooke, 1975.)

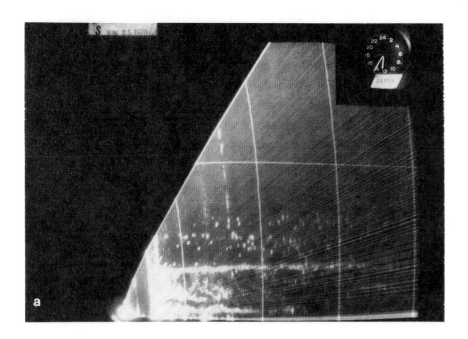

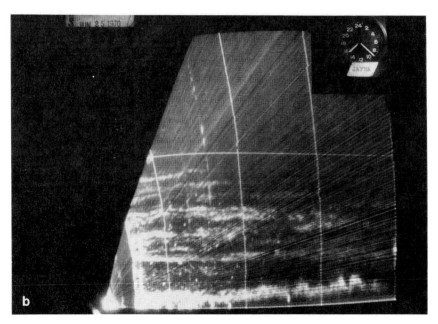

Fig. 11.24 A series of RHI reflectivity displays showing the evolution of convection. The range arcs are at 9.3 km and the height mark is at 6.1 km. The cross section is north of Wallops Island, Virginia. (a) 0736 L.S.T.; (b) 1023 L.S.T.; and (c) 1253 L.S.T. (*Figure continues.*)

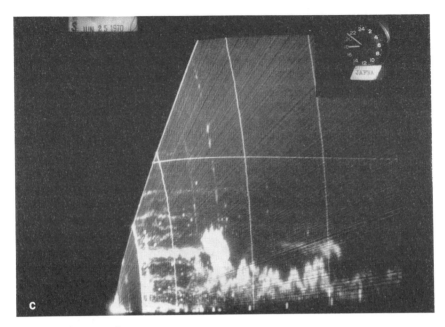

Fig. 11.24 (*Continued*)

11.8 Observations of Wind, Waves, and Turbulence in Clear Air

Although the reflectivity field obtained by high-resolution radars facilitates the identification of many normally invisible meteorological phenomena, the Doppler radar provides two additional fields of information, the mean Doppler velocity and spectrum width, which can be used to infer the wind and turbulence and hence to measure important kinematic properties of the atmosphere.

There are two observational modes that can generate data on winds.

1. With beam scanning (or with measurements at several fixed beam positions) at high-elevation angles (i.e., $>45°$), the upper regions of the troposphere, which have weak reflectivity, are close to the radar and thus can be visible. By assuming a linear wind within the small circles in which the beam intersects the layers of constant height, vertical profiles of horizontal wind, divergence, and deformation can be obtained (Section 9.3).

2. By beam scanning at low elevation angles (i.e., $<10°$) in regions of the more reflecting CBL, the spatial dependence of the radial (Doppler) wind on the radar's resolution volume position can be analyzed to produce maps of wind speed, direction, divergence, and deformation (e.g., Figs. 9.11 and 9.12).

11.8.1   Wind Profiling

The term wind profiling has become synonymous with remote measurements of winds aloft using radar techniques. Such measurements can be made with Doppler weather radars, lidars, or special-purpose profiling radars called profilers.

11.8.1.1   Profilers

The wind profiler is a ground-based Doppler radar that can measure vertical profiles of horizontal and vertical wind in nearly all meteorological conditions. Routine wind soundings in the troposphere by clear-air radar was first demonstrated in 1979 (Ecklund et al., 1979), and soon thereafter a wind-profiling network was proposed (Strauch et al., 1984). The operation frequencies of profiling radars are in the VHF and UHF bands (Fig. 2.1), notably near 50 MHz, 400 MHz, and 900 MHz, and the particular choice depends on the economy and desired altitude coverage. Eddies that create the larger scale irregularities of refractive index (which scatter the lower-frequency electromagnetic waves) are less likely to be in the dissipative range of turbulence; hence, everything else being equal coverage to greater heights is expected (Fig. 11.18). But antenna size for lower frequencies must be larger to preserve resolution; furthermore, minimum range of measurement is limited by receiver recovery time (longer at lower frequencies) and by bandwidth, which must be less at lower frequencies. Fifty-megahertz radars have been built for height coverage to 20 km, 400-MHz radars obtain echoes from regions as high as 16 km, and 900-MHz radars to 10 km depending on transmitter-power–aperture product. Seasonal variations in height coverage of some of these is discussed by Frisch et al. (1986).

Common to most of the profilers is the antenna type, usually a phased array. Some, like the MU (Middle Upper atmospheric radar) in Japan (Fukao et al., 1985), have electronic control of the beam position, but others have only a few fixed beams. A typical fixed-beam geometry consists of three beams (Fig. 11.25). The north and east beams are used to sample radial velocities that have contributions from both the vertical and horizontal wind components. But, because the vertical wind is directly estimated along the vertical beam, the horizontal components of the vector wind at a height of interest can be computed. Obviously the wind field needs to be locally uniform during the measurement to avoid bias errors (Koscielny et al., 1984). Other adverse meteorological conditions that bias the wind measurements are precipitation (Fig. 8.36, see also Wuertz et al., 1988) and gradients of reflectivity (e.g., vertical gradients of reflectivity due to very thin layers) within the resolution volume (Section 5.2).

The choice of elevation angles for the off-zenith beams is dictated by conflicting requirements. An elevation angle of $73.7°$ is used by the NOAA experimental network, and it represents an acceptable compromise between opposing effects (Strauch et al., 1984; Koscielny et al., 1984). High elevation angles are desirable to (a) keep the resolution volume for off-zenith and zenith-pointing antennas at nearly the same height, (b) minimize the range to a given

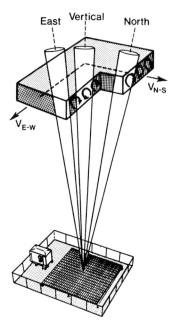

Fig. 11.25   Wind profiler with three beam positions.

height, (c) minimize the effects of horizontal gradients of wind, (d) make altitude resolution less dependent on antenna beamwidth, and (e) reduce contamination by ground clutter. Advantages of low elevation angles are (a) random errors of radial velocity estimates induce smaller errors in the estimates of horizontal winds, (b) wind estimate biases caused by antenna pointing errors are smaller, and (c) there are no biases due to anisotropic scattering or reflections (Section 11.5.3).

Because clear-air signal power is often less than receiver noise power, profilers use relatively long dwell times to estimate the wind. For example, the NOAA experimental profiling radar (Table 11.1) has a one-minute dwell time to estimate velocities along each beam for both high and low modes of data collection (Fig. 11.26). The high mode uses a long transmitted pulse to increase the SNR [e.g., see Eq. (4.36) or (11.152)] so that observations can be made, albeit with coarser resolution, at higher altitudes. Thus, a total of 6 minutes is required to obtain a profile of u, v, w for both height intervals.

The one-minute dwell time is partitioned for spectral-moment processing as follows. First, several consecutive echoes are coherently summed (i.e., I and Q samples are averaged) to reduce the number of time samples for spectrum computations. This reduces the unambiguous velocity interval, but does not compromise radar performance because the Doppler velocities associated with winds are usually smaller than the reduced unambiguous interval. After samples

Table 11.1
Wind Profiler Parameters [a]

Wavelength	74 cm
PRT	
High mode	150 μs
Low mode	100 μs
Pulsewidth	
Coded high	20 μs
Coded low	3.3 μs
Compressed high	6.67 μs
Compressed low	1.67 μs
Antenna type	Coaxial collinear array
Gain	$\simeq$ 32 dB
Beamwidth	$\simeq$ 4.5°
Elevation N/E beams	73.7°
Sidelobe level	< -20 dB
Peak power	
High mode	13.7 kW
Low mode	7.7 kW
Receiver	
System noise temperature	213 K
Bandwidth (3 dB)	
High mode	120 kHz
Low mode	350 kHz

[a] NOAA's Demonstration Network

are coherently summed, discrete Fourier transforms are computed and several spectra (collected over a period of one minute) are averaged for subsequent processing to estimate the three principal moments. (Note that the "spectral" SNR, the ratio of signal to noise spectral densities, is not altered by coherently summing time samples provided the dwell time is the same as with spectral processing of the original samples.) Ten mean Doppler velocities over the one-hour period are compared, and if a consensus among four or more of these is found, the ones that satisfy this test are averaged. This method of consensus averaging gives reliable estimates of mean Doppler velocities (standard error less than 1 m s^{-1}) at signal-to-noise ratios per sample as low as -35 dB (May and Strauch, 1989).

An example of a time–height cross section of horizontal wind measured with a NOAA profiler (which also had a RASS, Section 11.4) is shown in Fig. 11.27a. These data were collected during the time a strong synoptic-scale (~ 2500 km) trough containing frontal and jet-stream structures propagated southeastward across southwest USA (Neiman *et al.*, 1992). The 500-mb pressure height and temperature fields for 1200 UTC, 11 December (Fig. 11.27b) show the trough and cold (-35°C) polar air (thermal trough) over Colorado and an associated polar frontal zone circumscribing Colorado to the south and west.

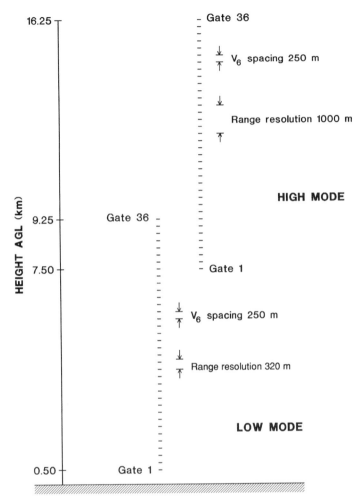

Fig. 11.26 Height coverage, resolution volume spacings, and range resolution of the wind profilers (NOAA's demonstration network).

The segment of the front to the south and southeast of Colorado is referred to as the leading branch of the polar front, and the segment to the west of the state is referred to as the trailing branch.

The profiler and RASS data clearly depict the leading and trailing branches of the polar frontal zone (Fig. 11.27a) surrounding the deep (at least 8 km) polar air mass, and also the shallow (below 650 mb) arctic air mass. A zone of enhanced static stability ($\partial\theta_v/\partial p < 0$) and local cooling near the surface at 0600 UTC, 10 December represent the passage of the leading branch, which ascends to 400 mb by 1600 UTC. The frontal passage below 600 mb is characterized by

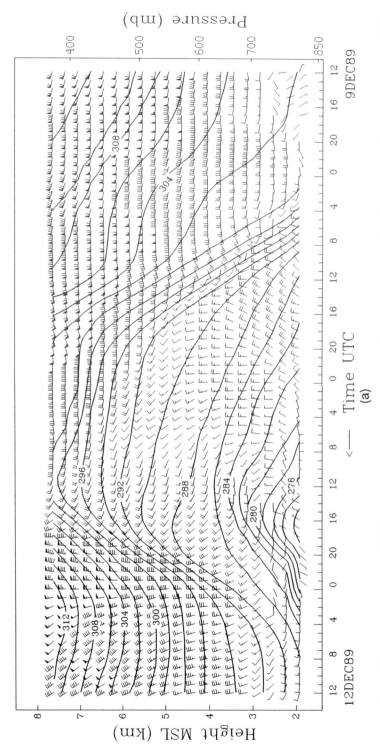

Fig. 11.27 (a) Time–height analysis of RASS virtual potential temperature (K) with accompanying wind profiler wind-velocity vectors, between 1200 UTC, 9 December and 1200 UTC, 12 December 1989. Wind vectors flags = 25 m s^{-1}, barbs = 5 m s^{-1}, and half barbs = 2.5 m s^{-1}. (b) 500-mb temperature (°C, dark contour) and its height (dam, lightly shaded contours), at 1200 UTC, 11 December 1989. Thin dashed lines bracket the frontal boundaries. The profiler's location is indicated by the star. (From Neiman *et al.*, 1992.) (*Figure continues.*)

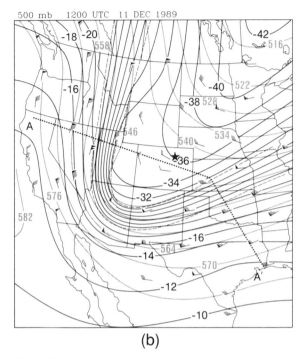

500 mb   1200 UTC   11 DEC 1989

Fig. 11.27 (*Continued*)

wind-direction shift from westerly to northerly. Assuming steady-state fields the wind-direction shift marks a concentrated region of cyclonic vorticity. The frontal passage above 500 mb is characterized by a decrease in the westerly wind speed followed by a smaller post-frontal change in θ_v within the polar air mass. Below 600 mb, post-frontal cooling occurs until 1600 UTC, 11 December in association with a southward influx of arctic air. An easterly (upslope) wind velocity component is confined to the shallow arctic air mass. The arctic air migrated to the east of Colorado between 1600 and 2300 UTC, 11 December together with the polar air in the middle troposphere.

For studies of the boundary layer an economical wind profiler has been developed (Ecklund *et al.*, 1988). Because boundary-layer reflectivities are usually orders of magnitude larger than in the upper troposphere, smaller transportable antennas can be used. Furthermore, the small antenna can easily be shrouded to reduce ground clutter, and with short-duration transmitted pulses (e.g., 0.5 μs) winds can be measured to altitudes as low as 100 m. Table 11.2 lists the characteristics of this radar.

11.8.1.2 Profiling with Doppler Weather Radars

Doppler weather radars have a capability to measure winds in clear air. Operation of the WSR-88D in a long-pulse mode (Table 3.1) sacrifices range

Table 11.2
Typical Boundary Layer Profiler Parameters

Wavelength	32.8 cm
Peak power	500 W
Average power	10 to 30 W
Range resolution	60, 100, 250, 500 m
Pulse repetition period	10 to 100 μs
Antenna type	Flat panel microstrip array (fixed beam and electronically steered to three or five beam positions)
Beamwidth	6 to 9 deg one-way
Number of range gates	50 to 100
Range sample spacing	minimum 0.3 μs (0.1-μs steps)
Maximum radial velocity	3 to 400 m s^{-1}
Number of spectral points	64 to 2048
RASS	Up to 27 gates with simultaneous measurement of sound speed and vertical velocity of air

resolution for improved detection capability. Furthermore, having a longer T_s, one near the optimum [Eq. (11.154)], allows autocovariance processing at lower SNR to produce quality estimates of velocity, whereas longer τ and less bandwidth B_6 increases the SNR [Eq. (4.36)]. In summary, the pulse width and T_s are increased, B_6 is decreased, but the averaged power P_{av} is kept constant.

Using the WSR-88D parameters for the long-pulse mode (Table 3.1) with a dwell time $T_d \approx 100$ s, and assuming $l_t = 1.25$, $C_n^2 = 10^{-18}$ m$^{-2/3}$, the maximum range to which wind can be measured is less than 4.5 km. This dwell time implies that velocity estimates are made with SNR = -13 dB and that $\sigma_v = 1.5$ m s^{-1} and SD(v) = 1 m s^{-1}. Although T_d can, in principal, be increased indefinitely to improve detection sensitivity, it becomes difficult to obtain a radar system free of artifacts that could allow for the processing of signals at extremely low SNRs.

Assume that consensus averaging and beam elevation angles are as with the NOAA profiler radar, and that velocities could be estimated at SNRs as low as -35 dB (May and Strauch, 1989). Then the WSR-88D could provide hourly profiles of winds to about 15 km above ground if C_n^2 values of at least 10^{-18} m$^{-2/3}$ filled the resolution volume, and if $\lambda \geq \lambda_t$. Recall that this last condition stipulates the scales of refractive index irregularities to be within the inertial subrange. As discussed in Section 11.6.2 this is expected to hold for light turbulence up to heights of about 14 km; thus Doppler weather radars should be able to measure wind wherever light turbulence is present in the troposphere. Under these conditions the profiling performance of the WSR-88D radar is

CLEAR AIR WINDMAPPING

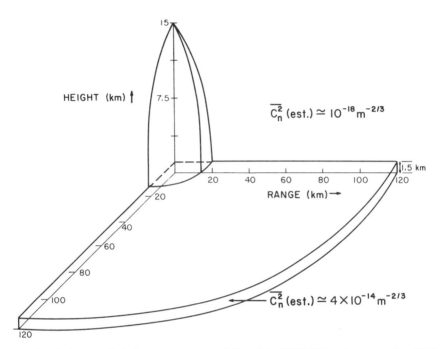

Fig. 11.28 Range of wind measurement capability of the WSR-88D radar system (see Table 3.1) for the two-step C_n^2 profile shown.

shown in Fig. 11.28 together with the range for clear-air boundary-layer wind mapping.

The C_n^2 of $4 \times 10^{-14} \, \mathrm{m}^{-2/3}$ in the boundary layer (Fig. 11.28) is obtained by adding 10 dB to the value in Fig. 11.17, as suggested by Gossard (1977), to account for midday convection; all other parameters are the same as for profiling except that the dwell time is 70 ms. With this dwell time and antenna rotation rate of 2 rpm the effective beamwidth (Fig. 7.25) would be 1.25°. Thus, Doppler weather radar might serve three functions: (1) storm observations, (2) obtaining winds to supplement those from wind profiler/rawinsondes, and (3) mapping the kinematic structure of the CBL.

An example (Fig. 11.29) of data collected by the WSR-88D (long-pulse mode, Table 3.1) on 18 April 1991 in clear skies demonstrates the potential for wind profiling. Detectable scatterers in precipitation-free air almost completely filled the volume from the surface up to 14,000 ft. These data were obtained shortly after passage of a cold front, which is evident from the northeasterly winds up to 5000 ft. Above this height the winds back quickly to westerlies suggestive of cold-air advection.

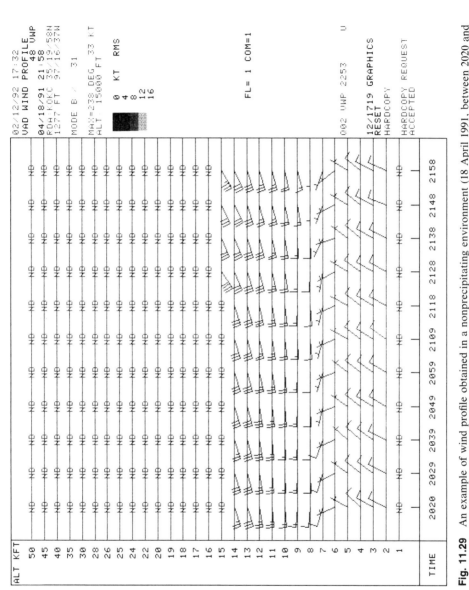

Fig. 11.29 An example of wind profile obtained in a nonprecipitating environment (18 April 1991, between 2020 and 2158 G.M.T., WSR-88D). The height in kilofeet is above sea level. (Courtesy S. Smith, OSF/NWS.)

11.8.2 Kinematic Structure of the Convective Boundary Layer

The second mode of clear-air radar surveillance pertains to observations of wind fields in the CBL. Measurements to ranges of about 120 km, nearly the practical limit set by earth's curvature, would certainly enhance the weather radar's capability to sense precursors to the development of thunderstorms. Assuming, as before $C_n^2 \approx 4 \times 10^{-14}$ m$^{-2/3}$ and NSSL radar parameters (Doviak and Zrnić, 1984a, Table 11.1), the maximum range would be about 40 km. But, during the 1980 storm period, a continuum of scatterers was observed to ranges in excess of 100 km on 10 of the 12 prestorm days (15 April–19 June; Doviak *et al.*, 1983). Therefore C_n^2 in the Oklahoma prestorm CBL appears to be somewhat larger than that depicted in Fig. 11.28, or appreciable contribution comes from point scatterers such as insects; etc. These observations usually began 2–3 h prior to the forecast time for storm initiation, and so most occurred under mid-afternoon sunny skies (if storms are forecast, an observation is prestorm, even if storms did not form).

Based on these preliminary observations it appears that weather radar could be capable of mapping the kinematic structure of the CBL during prestorm weather condition.

Figure 11.30 shows the locations of the radars at Cimarron Airport in Oklahoma City and at Norman, Oklahoma. Synthesized dual Doppler radar winds were analyzed in detail in a 625-km^2 region (the square region in Fig. 11.30), although winds were mapped over a 150-km square (Berger and Doviak, 1979). For comparison it would have been preferable to synthesize the wind in an area over the 444 m tall meteorologically instrumented tower, but ground clutter from Oklahoma City excessively contaminated our air-velocity estimates in that region. Ridges to the south of Cimarron and west of Norman shadowed the ground beneath the synthesis area and provided velocity data relatively free of ground-clutter contamination.

The winds were fairly uniform from the southwest on 27 April 1977, but there were small perturbations from the mean wind of about one order of magnitude less. As is evident in Fig. 11.30, the x direction and u component of the wind are along the mean wind, and the y direction and v component are normal to it. The mean convective boundary layer (CBL) wind in Fig. 11.30 is the vector average of the horizontal wind over the height interval 0–1.25 km in the synthesis area during the period 1426–1450 C.S.T. The wind versus height displayed in Fig. 11.31 represents the spatial average of u and v over the synthesis area (625 km^2). The two lower curves are determined from averaging winds at the KTVY tower for 29 minutes. The dotted curve corresponds to the same time at which the radar winds are presented, yet the winds appear discontinuous near 0.5 km. Because the air over the synthesis area at radar observation time only reaches the tower 75 minutes later, comparison must be made with that time lag. The solid curve corresponds to this later time, and winds

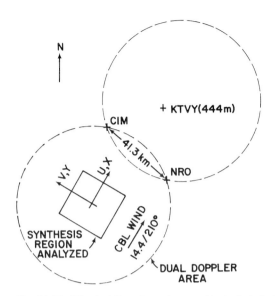

Fig. 11.30   The dual Doppler radar area (dashed circles) within which the angle subtended by the radials from the Cimarron (CIM) and Norman (NRO) radars lies between 30° and 150°. The outlined box is the region from which the Doppler velocities were synthesized for wind analyses shown in Figs. 11.31 and 11.32. The CBL wind speed and direction are a mean over 1.25-km depth of the planetary boundary layer.

appear nearly continuous with height. Although the change in wind direction is not large, considerable speed shear is noticeable in the lowest 400 m.

Caution must be exercised in making a comparison between a spatial average at an instant and a temporal average at a point because the existence of stationary helical circulations can make the wind comparison at a point unrepresentative for an area. Although the spatial variations due to turbulence average out in time, longitudinal waves have circulations with axes nearly parallel to the mean wind, and a temporal average might not erase their spatial variation. Brown (1970) used a numerical model to deduce significant spatial variability in the time-averaged wind following flow disturbances caused by convective rolls. The model shows that time-averaged speed at a point can be as much as 1.6 times the areal average, and the direction changes up to 10° when an observation point is moved 1 km perpendicular to the longitudinal roll circulations. Hence secondary circulations in the field can contribute to differences between spatially averaged radar winds and time-averaged winds measured at the tower.

The Norman and Cimarron Doppler velocities at common grid points are combined vectorially to synthesize the wind components u and v. The vertical velocity contributions are negligible because beam elevation angles are less than 4°. Variation in the mean flow are small but rich in structure, as the perturbation

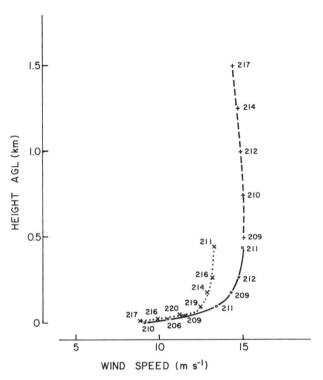

Fig. 11.31 Mean wind speed and direction above ground level (AGL) from tower and Doppler data. Wind obtained from two-Doppler radar data (———) is a spatial average u and v over the synthesis region analyzed at time 1430–1433 C.S.T.; ···, 1415–1445 C.S.T. (tower); ———, 1530–1600 C.S.T. (tower).

winds at two times separated by 3 minutes demonstrate (Fig. 11.32). Arrow lengths measure the wind speed relative to the scale at the upper left corner. Distances x and y are measured from the Norman radar site. The CBL mean wind has a magnitude twice that indicated by the arrow at the upper-right corner. The perturbation fields appear noiselike and random, but if one overlays the two fields after shifting the earlier one downstream by an amount that the eddies would have drifted owing to advection, one can immediately see that the fields are well correlated. These two fields were numerically correlated by lagging one data field with respect to the other in increments of 500 m. The correlation as a function of lag in the x and y directions is plotted in Fig. 11.33, where it is seen that the correlation peaks at a lag equal to the distance that perturbations would be advected by the mean wind in the time difference between the two data fields.

To evaluate the dominant scales of motion and to resolve waves in the turbulent wind, spectral power densities of the synthesized u and v fluctuations

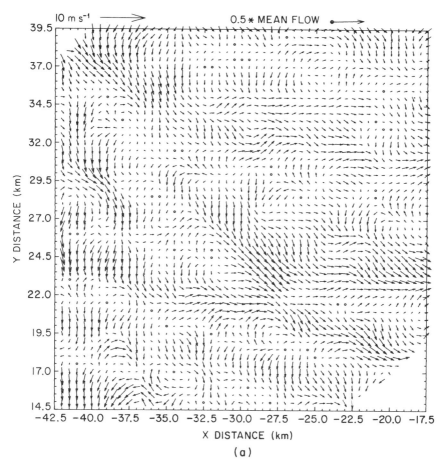

Fig. 11.32 Dual Doppler radar winds at (a) 1438 C.S.T. and (b) 1441 C.S.T. on 27 July 1977 with mean CBL wind removed. Circles indicate wind speeds of less than 0.1 m s^{-1}. Height 1.0 km. (*Figure continues.*)

along both the x and y directions were examined for scales of 1–16 km wavelength. One-dimensional spatial spectra S as a function of wave number K at the indicated heights are displayed in Fig. 11.34 for synthesized data from one volume scan. The only filter acting on the data set used for spectral analyses is the Cressman interpolation filter [Eq. (9.3)]. Because interpolation alters the spectral shape, it is necessary to estimate its influence before comparisons can be made with *in situ* point sensors. This was done by Doviak and Berger (1980), who showed good agreement between radar- and tower-derived spectra (Fig. 10.7).

The values of each $S(K)$ in Fig. 11.34 are averages from 32 individual spectra obtained by a discrete Fourier transform along the K_x and K_y directions.

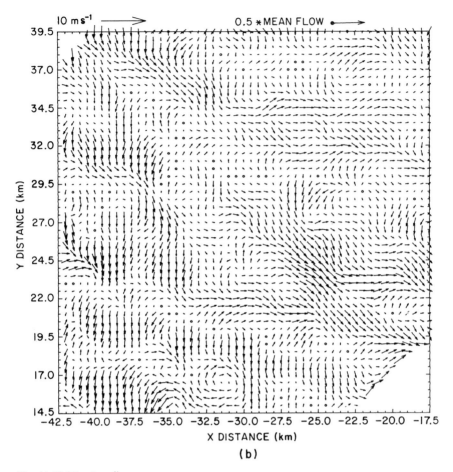

Fig. 11.32 (*Continued*)

Here $S(K)$ is defined as the velocity variance at a wave number $K = n \Delta K$, divided by the wave-number interval ΔK, where n is an integer and $\Delta K = (\pi/8) \times 10^{-3}$ m^{-1}. Although we have placed a power of $-5/3$ line in Fig. 11.34, we are not implying that these spectra should necessarily follow this law of the inertial subrange. On the other hand, Panofsky (1969) has observed that the law often applies to wavelengths as long as five times the measurement height. As might be expected, our data follow the law better at the highest wave numbers.

There is a hint of a spectral peak at a wavelength of 4 km for v in the K_y direction at a height of $H = 1$ km. Supportive evidence of this wave was seen in the reflectivity streets, which were aligned roughly parallel to the mean wind in the CBL and had a separation of about 4 km. To get a better indication of which scale sizes in the x and y directions are significant, $KS(K)$ was plotted versus log

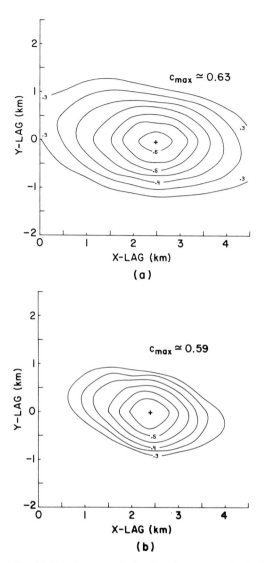

Fig. 11.33 Cross-correlation function c versus the horizontal spatial lag of (a) the u component and (b) the v component for two wind fields 3.5 minutes apart at 1.0 km AGL. The mean wind speed is 14.8 m s^{-1}, and the median times for the two fields are 1448:10 and 1451:40 C.S.T. (27 July 1977).

K for six tilt sequences covering a time interval of about one-half hour. These data are presented by Berger and Doviak (1979). The most consistent and prominent feature found was the large power of v in the y direction centered about a wavelength $\Lambda = 4$ km at a height of 1.0 km, a result that was confirmed

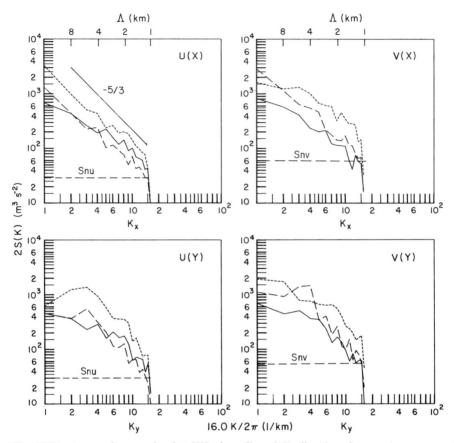

Fig. 11.34 Averaged power density $S(K)$ along K_x and K_y directions for u and v components synthesized from Doppler velocities. Spectra are for winds at heights of 0.5 (---), 1.0 (——), and 1.5 (———) km. The horizontal lines through the base of the spectra are estimated noise levels at 1.0 km. Λ is the wavelength of the eddies (1442–1444, April 1977).

by *in situ* aircraft observations (Reinking *et al.*, 1981). As can be seen from Fig. 11.34, this feature does not appear at $H = 1.5$ or 0.5 km.

The 4-km wave is probably an indication of horizontal roll vortices having axes parallel to the mean wind and spaced 4 km apart. In fact, the absence of the 4-km wave in the crosswind v spectra at heights of 0.5 and 1.5 km suggests that the roll is centered at about 0.5 km and has peak v wind at about 1 km, with another peak near the surface where radar data are unavailable. Other peaks appear in these spectral plots but with less consistency in time or height and are thus difficult to interpret.

11.9   Other Fair-Weather Observations

The radar reflectivity factor Z of clouds depends on the concentration, size, and phase of hydrometeors. Reflectivity factors of cumulus congestus clouds and storm cells are quite large (>10 dBZ) and therefore are easily seen with 10 cm wavelength radars. Although the measured values of radar reflectivity factor in nonprecipitating clouds range from -15 dBZ to -30 dBZ (Section 8.1.1), a good percentage of clouds should be detected by 10-cm weather radars to a range of about 20 km. Note that gamma distributed droplets in a cumulus cloud with total concentration of 750 cm^{-3} and median volume diameter of 14 μm

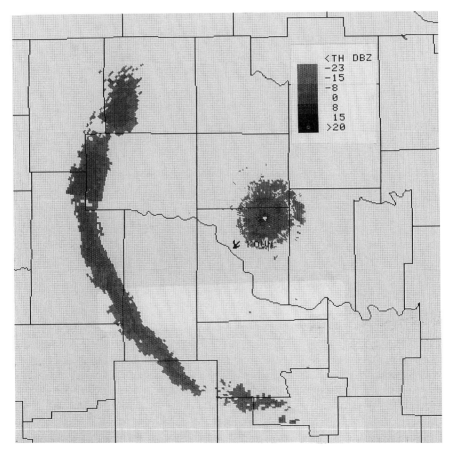

Fig. 11.35   A layer of cirrus clouds observed by the WSR-88D radar. The height above ground is 8 km and range is about 90 km. Two categories of measured reflectivity factor are displayed (-8 to -15 dBZ) and (-15 to -23 dBZ). (Courtesy S. Smith OSF/NWS.)

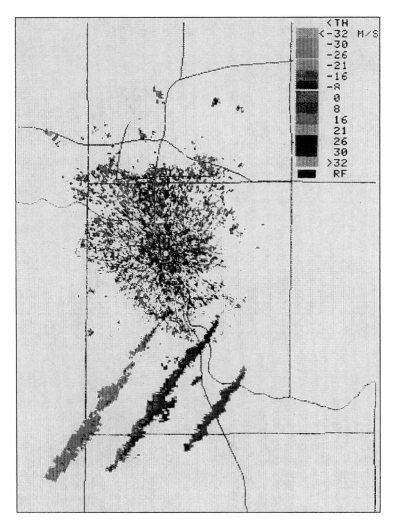

Fig. 11.36 Doppler velocity display (WSR-88D) of three smoke plumes in central Oklahoma. Elevation angle is 0.5° and the middle plume is 36 km long.

would produce a reflectivity factor of about -20 dBZ. Individual cumulus clouds are small and cannot be resolved with current satellite observations; yet they are numerous over large areas and significantly affect radiation transfer. Like deep convective clouds, they significantly reduce the absorption of short-wave energy at the surface; however, unlike deep convective clouds, they have little effect on outgoing long-wave radiation. (The reduction of short-wave radiation by deep cumulus tends to be offset by their shielding of long-wave radiation, normally emitted to space).

Besides observation of cumulus clouds weather radars can detect high-altitude cirrus clouds, which is important because cirrus blanket vast areas of land and sea and could also have pronounced effects on radiation budget. An example of a cirrus cloud deck approaching, from the southwest, a WSR-88D radar in central Oklahoma is shown in Fig. 11.35. The arcus shape is caused by the beam intersecting the leading portion of the cloud layer.

Not only does weather radar detect echoes from clouds, refractive index irregularities, insects, and birds (Fig. 7.35), but it can also trace particulates injected deliberately (e.g., chaff) or accidently such as smoke plumes from fires. Three grass fires in Oklahoma are the cause of streaks in Fig. 11.36. From the Doppler velocity we deduce that the northeasterly wind is about 18 m s^{-1} a few hundred meters above the ground. Scatterers are most likely carbonized blades of grass, but some contribution from refractive index perturbations could also be present.

Problems

11.1 Show that the Doppler shift measured with a cw radar corresponds to the speed of sound of an acoustic pulse of arbitrary envelope. Hint, assume that the traveling perturbation of refractive index is given by $\Delta n(z, t) = \Delta n \, f(z - c_a t) \cos K_z(z - c_a t)$. Evaluate your result for $f(z - c_a t) = \exp[-(z - c_a t)^2/2\sigma^2]$.

11.2 (a) Derive an expression analogous to relation (11.67) for an arbitrary range weighting function $|W(z)|^2$. Assume the receiver voltage is proportional to

$$\sum \Delta n_i \int W(z) \cos K_{zi}(z - c_a t) \exp(-j2k_{0z}) \, dz,$$

and that $W(z)$ is positive. (b) Evaluate your expression for a Gaussian-shaped range-weighting function with width σ_z.

11.3 A RASS operates as follows. It uses a continuous white-noise acoustic source and a pulsed Doppler radar. The radar wavelength is 74 cm and the range-weighting function is Gaussian with a width $\sigma_z = 300$ m. It measures a mean speed of sound of 330 m s^{-1} at a fixed height. What is the virtual temperature? Use the result from problem 11.2b to plot the power spectrum measured by this RASS.

11.4 Show, by expanding the range r_t [relation (11.119a)] in a Taylor series about the z axis, that for second-order expansion to be valid the condition on the transverse dimension $l_\perp$ of the scattering volume is

$$l_\perp < [(8r_0 + z)^3/k_0]^{1/4}$$

where z is the displacement, along the beam axis, of the scattering element from the resolution volume center.

Geometric Relations for Rays in the Troposphere

A.1 Integral Solution for Ray Path in a Spherically Stratified Medium

In this appendix we derive an integral equation for the path of a ray in a spherically stratified medium. The determination of electromagnetic energy paths can be obtained from ray theory under the condition (Stratton, 1941, p. 343) that

$$\lambda \, dn/dR \ll 1, \tag{A.1}$$

where λ is the wavelength of the electromagnetic radiation and dn/dR is the vertical gradient of refractive index n. One can easily show for most meteorological conditions existing in the troposphere inequality (A.1) is well satisfied at weather radar wavelengths. Consider the ray path geometry in Fig. A.1. Snell's law for spherically stratified media (Bean and Dutton, 1966, p. 87) is

$$Rn(R)\cos\theta = \text{const} \equiv C, \tag{A.2}$$

which applies to any point along the ray path $R(\psi)$. Now,

$$dl = [(R\,d\psi)^2 + (dR)^2]^{1/2}$$

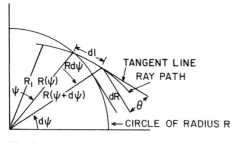

Fig. A.1 Ray path in a spherically stratified medium.

and

$$\cos^2 \theta = \frac{(R\, d\psi)^2}{(R\, d\psi)^2 + (dR)^2} = \frac{R^2}{R^2 + (dR/d\psi)^2}. \tag{A.3}$$

Solving for $d\psi/dR$, we obtain

$$d\psi/dR = (1/R)\cot\theta.$$

Therefore

$$\psi(R) = \int_{R_1}^{R} [(1/R)\cot\theta]\, dR + \psi(R_1), \tag{A.4}$$

where $\psi(R_1)$ is any known ray path location (e.g., the transmitter location). Using Eq. (A.2) we can express the integrand of Eq. (A.4) entirely in terms of functions dependent on R. For a transmitter on the surface of the earth, whose true radius we label a, the integral solution of Eq. (A.4) becomes

$$s(h) = \int_0^h \frac{aC\, dh}{(a+h)[(a+h)^2 n^2(h) - C^2]^{1/2}}, \tag{A.5a}$$

where $s(h)$ is the great circle distance from the transmitter and

$$C = an(0)\cos\theta_e \tag{A.5b}$$

is specified by the ray's initial elevation angle θ_e and refractive index $n(0)$ at the transmitter. It can be shown that Eq. (A.5a) satisfies

$$\frac{d^2h}{ds^2} - \left(\frac{2}{R} + \frac{1}{n}\frac{dn}{dh}\right)\left(\frac{dh}{ds}\right)^2 - \left(\frac{R}{a}\right)^2\left(\frac{1}{R} + \frac{1}{n}\frac{dn}{dh}\right) = 0, \tag{A.6}$$

given by Hartree *et al.* (1946) as the exact differential equation that specifies the path of a ray in a spherically stratified medium.

A.2 Relating a Scatterer's Apparent Range and Elevation Angle to Its True Height and Great Circle Distance

Although Eq. (A.5) is useful for plotting ray paths in the troposphere, it does not explicity relate scatterer height to radar measurements of apparent range r_a and elevation angle θ_e. The great circle distance s is not known, but we can seek a solution to h in terms of the measured parameters r_a and θ_e.

In a time interval t an electromagnetic packet (pulse) of energy travels a distance

$$l = vt = ct/n.$$

The radar measures time intervals, and we assign an apparent range r_a assuming that the electromagnetic packets travel with the velocity of light. Thus

$$r_a = ct$$

is the apparent range that energy travels in time t. Therefore

$$dr_a = n \, dl, \quad \text{or} \quad r_a = \int_0^l n \, dl \tag{A.7}$$

is the *relation between the true path length l of the ray and the radar-measured length or range r_a*.

Now, referring to Fig. A. 1, we deduce that $dh = dl \sin \theta$, so that

$$r_a = \int_0^h \frac{n(h) \, dh}{\sin \theta}, \tag{A.8}$$

and as before we apply Snell's law [(A.2)] to Eq. (A.8) to obtain

$$r_a = \int_0^h \frac{(a + h)n^2(h) \, dh}{[(a + h)^2 n^2(h) - a^2 n^2(0) \cos^2 \theta_e]^{1/2}}. \tag{A.9}$$

Solving Eq. (A.9) gives a formula that relates the scatterer height h to the two measurable parameters, the (apparent) range r_a and elevation angle θ_e. Of course, one needs to know $n(h)$. Once we find h for a given r_a and θ_e, we can substitute it into Eq. (A.5) to obtain the great circle distance to the scatterer.

Correlation between Signal Samples as a Function of Sample Time

The in-phase and quadrature-phase signal components obtained from Eq. (4.1) are

$$V(\tau_s, 0) = I_0 + jQ_0 = \frac{1}{\sqrt{2}} \sum_i A_i(0) W_i e^{-j\phi_i} \tag{B.1}$$

at a sample time 0, where $\phi_i = 4\pi r_i/\lambda$. These components for subsequent signal samples at the same range time τ_s but mT_s seconds later are

$$V(\tau_s, mT_s) = I_m + jQ_m = \frac{1}{\sqrt{2}} \sum_i A_i(mT_s) W_i e^{-j\zeta_i}, \tag{B.2}$$

where $\zeta_i = \phi_i + 4\pi v_i mT_s/\lambda$. It is important to bear in mind that the ϕ_i have a uniform distribution over the interval of width 2π, whereas the v_i are usually concentrated around the mean velocity. Strictly, v_i is an average radial velocity that takes the ith scatterer to its new position in time mT_s. Furthermore, we assume the complex amplitudes A_i are independent of the phases ϕ_i and velocities v_i (i.e., the scatterer's size and phase shift due to scattering should not be dependent on its position or velocity), and that the weight of the antenna and range-weighting functions applied to any hydrometeor does not change appreciably while it moves during the time the signal correlation function is estimated. The complex amplitudes are random variables because drops may oscillate or change their canting angle (i.e., the angle ψ in Fig. 8.15). Thus the autocorrelation function is

$$R(mT_s) = E[V^*(\tau_s, 0)V(\tau_s, mT_s)]$$

$$= \frac{1}{2} \sum_i \sum_k E[A_i^*(0) A_k(mT_s) W_i^* W_k] E\{\exp[j(\phi_i - \phi_k - 4\pi v_k mT_s/\lambda)]\}$$

$$= \frac{1}{2} \sum_k C_k \sigma_{bk}(mT_s) |W_k|^2 E(e^{-j4\pi v_k mT_s/\lambda}). \tag{B.3}$$

The double summation reduces to a single one because the expectations with respect to the exponential argument are zero except at $i = k$. In Eq. (B.3) we have defined the correlation of A_i's as

$$\sigma_{bk}(mT_s) = E[A_k^*(0)A_k(mT_s)]/C_k,$$

where C_k is given by

$$C_k = \frac{2P_t g^2 \lambda^2}{(4\pi)^3 r_k^4 l_k^2} |f(\theta_k, \phi_k)|^4.$$

For $m = 0$ we recognize that $\sigma_{bk}(0)$ is the expected backscatter cross section. As we can see, the autocorrelation is independent of the initial phases ϕ_i. Had we not used the conjugate of $V(\tau_s)$, we would have obtained

$$E[V(\tau_s, 0)V(\tau_s, mT_s)] = 0 \tag{B.4}$$

because $E(e^{-j2\phi_i}) = 0$.

Expressing Eq. (B.3) in terms of the in-phase and quadrature components, we obtain

$$E(I_0 I_m + Q_0 Q_m) = \text{Re}[R(mT_s)], \tag{B.5a}$$

$$E(I_0 Q_m - I_m Q_0) = \text{Im}[R(mT_s)]. \tag{B.5b}$$

We shall prove that

$$E(I_0 I_m) = E(Q_0 Q_m), \tag{B.6a}$$

$$E(I_0 Q_m) = -E(I_m Q_0). \tag{B.6b}$$

Start with

$$E(I_0 I_m) = \frac{1}{2} \sum_i \sum_k |W_i W_k| E(|A_i(0)A_k(mT_s)|)$$

$$\times E(\cos \gamma_{i0} \cos \gamma_{km}), \tag{B.7a}$$

$$E(Q_0 Q_m) = \frac{1}{2} \sum_i \sum_k |W_i W_k| E(|A_i(0)A_k(mT_s)|)$$

$$\times E(\sin \gamma_{i0} \sin \gamma_{km}), \tag{B.7b}$$

where γ_{im} is the phase shift for lag mT_s [in Eq. (4.3b)], and subtract Eq. (B.7b) from Eq. (B.7a) to get

$$E(I_0 I_m) - E(Q_0 Q_m) = \frac{1}{2} \sum_i \sum_k |W_i W_k| E[|A_i(0)A_k(mT_s)|]$$

$$\times E[\cos(\phi_i - \beta_i + \phi_k - \beta_k - \psi_i(0)$$

$$- \psi_k(mT_s) + 4\pi v_k mT_s/\lambda)]. \tag{B.8}$$

Because the ϕ_i are uniformly distributed, it follows that the argument of the cosine in Eq. (B.8) is likewise uniform, making the expected value zero, and Eq. (B.6a) results. A similar derivation establishes Eq. (B.6b).

Next we shall prove that

$$E(I_m Q_m) = 0 \qquad (B.9)$$

by considering

$$E(I_m Q_m) = \frac{1}{2} \sum_i \sum_k |W_i W_k| E[|A_i(0) A_k(mT_s)| \cos \gamma_{im} \sin \gamma_{km}]. \qquad (B.10)$$

Again, because we can assume that the A_i are independent of the γ_{im},

$$E[I_m Q_m] = \frac{1}{4} \sum_i \sum_k |W_i W_k| E[|A_i(0) A_k(mT_s)|]$$

$$\times E[\sin(\gamma_{im} + \gamma_{km}) + \sin(\gamma_{km} - \gamma_{im})], \qquad (B.11)$$

and because γ_{im} and γ_{km} are uniformly distributed across 2π, the expectation of the trigonometric function is zero; thus Eq. (B.9) is established.

Finally, we demonstrate that the correlation between two samples I_0, I_m is zero only if the distribution of $4\pi v_k m T_s$ is uniform over the interval of 2π. We start with Eqs. (B.7) and expand the product of cosine terms.

$$E(I_0 I_m) = \frac{1}{4} \sum_i \sum_k |W_i W_k| E[|A_i(0) A_k(mT_s)|]$$

$$\times \{ E[\cos(\phi_i - \beta_i + \phi_k - \beta_k - \psi_i(0) - \psi_k(mT_s) + 4\pi v_k m T_s / \lambda)]$$

$$+ E[\cos(\phi_k - \beta_k - \phi_i + \beta_i - \psi_k(mT_s) + \psi_i(0) + 4\pi v_k m T_s / \lambda)] \}. \qquad (B.12)$$

For $i \neq k$ expected values of both cosines are zero, but if $i = k$, Eq. (B.12) becomes

$$E(I_0 I_m) = \frac{1}{4} \sum_k |W_k|^2 C_k \sigma_{bk}(mT_s)$$

$$\times E[\cos(4\pi v_k m T_s / \lambda - \psi_k(mT_s) + \psi_k(0))], \qquad (B.13)$$

which in general is nonzero except when the cosine argument is uniform across the interval of 2π. The reader can verify that a result identical to Eq. (B.13) is obtained for $E(Q_0 Q_m)$.

C

Correlation of Echoes from Spaced Resolution Volumes

In this appendix we derive the correlation of the weather signal at the receiver input for samples from resolution volumes both (1) spaced in range by $c\,\delta\tau_s/2$, where $\delta\tau_s$ is the sample spacing, and (2) spaced in azimuth (or elevation) by $\Delta\phi$. The latter spacing is caused when the radar scans its beam and $\Delta\phi$ is the azimuthal displacement between successive signal samples taken at constant range delay τ_s but at sample times differing by T_s.

Assume the scatterers are randomly distributed in space but have uniform statistical properties (e.g., η, the cross section per unit volume, is constant). The signal increments dV from elemental volumes large compared to the wavelength but small compared to the resolution volume are then independent, but the expected elemental powers $E(dP)$ are equal. We assume the transmitted pulse to be rectangular and the angular pattern of radiation to be Gaussian. Scatterers in either case (1) or (2) are assumed to be fixed in place although they have random placement. The assumption of frozen scatterers is not strictly true; it is a good approximation when a correlation in range is sought because the time separation between echoes from overlapping range intervals is very small (microseconds). Nevertheless, there may be appreciable reshuffling of scatterers during the time T_s between two successive azimuth samples. To arrive at the decorrelation (and spectrum broadening) due solely to antenna motion, and to keep the derivation simple, we chose to treat frozen scatterers.

C.1 Signal Sample Correlation versus Range Difference $c\delta\tau_s/2$

For convenience, assume that signals from elemental volumes have been summed over θ, ϕ (r = const), producing a linear array of spherical shells of thickness dr along the range coordinate. One still obtains from each element of this radial array a voltage increment dV_m independent of all others. Thus the signal sampled at τ_s is

$$V_x(\tau_s) = \sum_{m=1}^{M} dV_m, \qquad (C.1)$$

where we have assumed M elemental shells contained in the range interval $c\tau/2$. Because the input signals are not filtered by the receiver, contributions to signal samples are confined to a range interval $c\tau/2$ (we use the subscript x to differentiate the input voltage from receiver output voltage). A signal sampled at $\tau_s + \delta\tau_s$ has a value

$$V_x(\tau_s + \delta\tau_s) = \sum_{m=l+1}^{M+l} dV_m, \qquad (C.2a)$$

where

$$l = \delta\tau_s M/\tau, \qquad (C.2b)$$

as can be seen in the following schematic for $M = 8$ and $l = 6$:

which depicts the elemental contributions dV_m to signals sampled at τ_s and $\tau_s + \delta\tau_s$; dV_m is an elemental signal contributed by scatterers located in a shell of thickness dr.

It is quickly realized that if $\delta\tau_s > \tau$, then the signal samples are independent because no shell has contributed a common dV_m to both samples. We now form the product

$$V_x^*(\tau_s)V_x(\tau_s + \delta\tau_s) = \sum_{m=1}^{M} dV_m^* \sum_{m=l+1}^{M+l} dV_m. \qquad (C.2c)$$

Taking the expectation of Eq. (C.2c), we have

$$E[V_x^*(\tau_s)V_x(\tau_s + \delta\tau_s)] = \sum_{m=1}^{M} \sum_{n=l+1}^{M+l} E(dV_m^* \, dV_n). \qquad (C.3)$$

Equation (C.3) gives a sum of elements in a matrix of products $E(dV_m^* dV_n)$. There are M^2 terms, but because the dV_m are independent and have zero mean, only the $M - l$ terms that have common elements differ from zero. Therefore,

$$E(dV_m^* dV_n) = \begin{cases} 0, & m \neq n, \\ 2\sigma_r^2 \, dr, & m = n, \end{cases} \qquad (C.4)$$

where $2\sigma_r^2$ is the power per unit length (see Section 4.2 for the factor of 2), which is multiplied by dr to give the power contributed by the elemental shell. Thus Eq. (C.3) reduces to

$$E[V_x^*(\tau_s)V_x(\tau_s + \delta\tau_s)] = 2(M - l)\sigma_r^2\, dr. \tag{C.5}$$

Recognizing that

$$\bar{P}(r) = 2\sigma_r^2 M\, dr \tag{C.6a}$$

is the expected (mean) power of the samples at τ_s (if $r_6 \ll r$), where

$$M \equiv c\tau/2\, dr, \tag{C.6b}$$

we can rewrite Eq. (C.5) as

$$E[V_x^*(\tau_s)V_x(\tau_s + \delta\tau_s)] \equiv R_{xx}(\delta\tau_s) = (1 - |l|/M)\bar{P}(r) \quad \text{for} \quad |l| \le M. \tag{C.7}$$

The absolute value $|l|$ is used because $R_{xx}(\delta\tau_s)$ is symmetric in l. Substituting Eqs. (C.6b) and (C.2b) in Eq. (C.7), we obtain the correlation of the input signal.

$$R_{xx}(\delta\tau_s) = (1 - |\delta\tau_s|/\tau)\bar{P}(r) \quad \text{for} \quad \delta\tau_s \le \tau. \tag{C.8}$$

C.2 Correlation of Signals from Azimuthally Spaced Resolution Volumes

The correlation of signals sampled as the antenna beam is pointed at different azimuths can be deduced by applying the principles derived for range correlation, but in a slightly different way. The input receiver voltage sampled at τ_s is

$$V_x(\tau_s, \theta_e, \phi_0) = \iint V(\theta, \phi)f^2(\theta - \theta_e, \phi - \phi_0)\sin\theta\, d\theta\, d\phi \tag{C.9}$$

when the antenna beam axis is pointed in the direction ϕ_0, θ_e, and where $V(\theta, \phi)\sin\theta\, d\theta\, d\phi$ is the elemental voltage contributed by the volume $(c\tau/2)r^2\sin\theta\, d\theta\, d\phi$ of scatterers. The voltage for a beam position at $\phi_0 - \Delta\phi$, θ_e is

$$V_x(\tau_s, \theta_e, \phi_0 - \Delta\phi) = \iint V(\theta, \phi)f^2(\theta - \theta_e, \phi - \phi_0 + \Delta\phi)\sin\theta\, d\theta\, d\phi. \tag{C.10}$$

Recognizing that we can set $\phi_0 = 0$ without loss of generality, we form the product

$$V_x^*(\theta_e)V_x(\theta_e, \Delta\phi) = \iiiint V^*(\theta, \phi)V(\theta', \phi')f^2(\theta - \theta_e, \phi)$$

$$\times f^2(\theta' - \theta_e, \phi' + \Delta\phi)\sin\theta\sin\theta'\, d\theta'\, d\theta\, d\phi\, d\phi', \tag{C.11}$$

where θ', ϕ' are dummy variables of integration. The expectation of the product is

$$E[V_x^*(\theta_e)V_x(\theta_e, \Delta\phi)] = 2\sigma_\Omega^2 \int\int f^2(\theta - \theta_e, \phi)$$

$$\times f^2(\theta - \theta_e, \phi + \Delta\phi) \sin\theta \, d\theta \, d\phi, \qquad \text{(C.12a)}$$

where

$$E[V^*(\theta, \phi)V(\theta', \phi')] = 2\sigma_\Omega^2 \delta(\theta - \theta', \phi - \phi') \qquad \text{(C.12b)}$$

and $2\sigma_\Omega^2$ is the power per unit solid angle. Equation (C.12b) signifies that signals received from different elemental volumes are independent. The Dirac delta function $\delta(\theta - \theta', \phi - \phi')$ has the property

$$g(\theta, \phi) \equiv \int\int g(\theta', \phi')\delta(\theta - \theta', \phi - \phi') \sin\theta' \, d\theta' \, d\phi', \qquad \text{(C.12c)}$$

which, when used with Eqs. (C.11) and (C.12b), produces Eq. (C.12a).

From Eq. (C.12a) one obtains the total power if the antenna is stationary $(\Delta\phi = 0)$.

$$\bar{P} = E[V_x^*(\theta_e)V_x(\theta_e)] = 2\sigma_\Omega^2 \int\int f^4(\theta - \theta_e, \phi) \sin\theta \, d\theta \, d\phi. \qquad \text{(C.13)}$$

Thus, solving for $2\sigma_\Omega^2$ and substituting in Eq. (C.12a), we have the correlation

$$R_{xx}[\theta_e, \Delta\phi] \equiv E[V_x^*(\theta_e)V_x(\theta_e, \Delta\phi]$$

$$= \bar{P} \int\int f^2(\theta - \theta_e, \phi)f^2(\theta - \theta_e, \phi + \Delta\phi) \sin\theta \, d\theta \, d\phi$$

$$\times \left[\int\int f^4(\theta - \theta_e, \phi) \sin\theta \, d\theta \, d\phi\right]^{-1}. \qquad \text{(C.14)}$$

On the assumption that the antenna pattern is product separable, Eq. (C.14) reduces to

$$R_{xx}[\theta_e, \Delta\phi] = \bar{P} \int f^2(\phi)f^2(\phi + \Delta\phi) \, d\phi \bigg/ \int f^4(\phi) \, d\phi. \qquad \text{(C.15)}$$

To proceed further we need to define an antenna pattern. Let us assume that the two-way *electric field* pattern can be approximated by the Gaussian function

$$f^2(\phi) = \exp(-\phi^2 \cos^2\theta_e/4\sigma_\theta^2), \qquad \text{(C.16)}$$

where σ_θ^2 is the second central moment of the two-way power pattern and the polar coordinate axis is perpendicular to the tangent plane at the radar location. The $\cos\theta_e$ term is required because the antenna pattern must be invariant to the direction of the beam axis. In terms of the one-way half-power width θ_1,

$$\sigma_\theta^2 = \theta_1^2/16\ln 2, \tag{C.17}$$

so

$$f^2(\phi) = \exp[-4(\ln 2)\phi^2 \cos^2 \theta_e/\theta_1^2]. \tag{C.18}$$

Substitution of Eq. (C.18) into Eq. (C.15) and integration leads to

$$R_{xx}(\theta_e, \Delta\phi) = \bar{P}\exp[-2(\ln 2)(\Delta\phi)^2 \cos^2 \theta_e/\theta_1^2], \tag{C.19}$$

where we have asssumed

$$4\sigma_\theta^2/\cos^2 \theta_e \ll 2\pi \tag{C.20}$$

in order to simplify the integration.

The change $\Delta\phi$ that occurs between samples spaced mT_s apart when the antenna rotates in azimuth at a rate α is $mT_s\alpha$, so

$$R_{xx}(\theta_e, mT_s) = \bar{P}\exp[-2(\ln 2)(mT_s)^2\alpha^2 \cos^2 \theta_e/\theta_1^2]. \tag{C.21}$$

Thus, even though scatterers are fixed in space, the successive signal samples are decorrelated if the antenna beam is moving. The width of the Gaussian correlation function (C.21) is

$$\sigma_\tau = \theta_1/2\alpha \cos \theta_e\sqrt{\ln 2}. \tag{C.22}$$

Because successive samples are less correlated, we have spectral broadening. (If the antenna beam were stationary, the spectrum would contain a single line at zero Doppler velocity.) The spectral width caused by this decorrelation is

$$\sigma_\alpha = \lambda/4\pi\sigma_\tau = \alpha\lambda \cos \theta_e\sqrt{\ln 2}/2\pi\theta_1, \tag{C.23}$$

in velocity units.

Geometric Optics Approximation to the Wave Equation

From Eq. (11.13b) we have, in absence of refractive index irregularities,

$$\nabla^2 \mathbf{E}_0 + k_0^2 n_a^2 \mathbf{E}_0 = -2\nabla[\mathbf{E}_0 \cdot \nabla \ln(n_a)] \tag{D.1}$$

as the equation satisfied by the electric field in a region of space in which n_a, the average refractive index, is a function only of the spatial coordinate $\mathbf{r}$. We shall determine the condition under which

$$\nabla(\mathbf{E}_0 \cdot \nabla \ln n_a) \ll k_0^2 n_a^2 \mathbf{E}_0 \tag{D.2}$$

everywhere, so that the term on the right of Eq. (D.1) can be neglected. Assume that

$$n_a = 1 + n_1(\mathbf{r}), \tag{D.3}$$

where $n_1(\mathbf{r}) \ll 1$, typical of the troposphere. Thus expanding $\ln[1 + n_1(\mathbf{r})]$ in a power series and retaining only first-order terms in $n_1(\mathbf{r})$, we obtain

$$\ln n_a \simeq n_1(\mathbf{r}),$$

and the inequality (D.2) reduces to

$$\nabla(\mathbf{E}_0 \cdot \nabla n_1) \ll k_0^2 n_a^2 \mathbf{E}_0. \tag{D.4}$$

The term on the left cannot be larger than $k_0 |\mathbf{E}_0| |\nabla n_1| + |\mathbf{E}_0| \nabla^2 n_1$, where

$$\mathbf{E}_0 = \mathbf{A}_0(\mathbf{r}) \exp(-j\mathbf{k}_0 \cdot \mathbf{r}) \tag{D.5}$$

has been assumed as a solution to Eq. (D.1) in which $\mathbf{A}_0(\mathbf{r})$ has variations changing spatially much more slowly than $\exp(-j\mathbf{k}_0 \cdot \mathbf{r})$. We now assume that the spatial Fourier spectrum of n_a contains structure wavelengths no smaller than l_0 and that $\lambda \ll l_0$. This value l_0 defines an effective cutoff of the spatial Fourier spectrum of time-invariant changes in the refractive index. Thus

$$\nabla^2 n_1 \le (2\pi)^2 \ (n_1/l_0^2) \ll k_0 |\nabla n_1|,$$

and if the inequality

$$k_0 |\mathbf{E}_0| |\nabla n_1| < k_0^2 n_a^2 |\mathbf{E}_0| \tag{D.6}$$

is satisfied, then (D.4) and therefore (D.2) must also be satisfied.

518

Condition (D.6) can be further reduced to the form

$$\lambda|\nabla n_a|/2\pi n_a^2 \ll 1. \tag{D.7}$$

Inequality (D.7) is satisfied if the change in n_a is small within a wavelength, and hence

$$\nabla^2 \mathbf{E}_0 + k_0^2 n_a^2 \mathbf{E}_0 = 0,$$

is a good approximation to the wave equation (D.1).

Derivation of Green's Function

In this appendix we derive Green's function for a general time-varying source $f(\mathbf{r}, t)$, as given in Eq. (11.20). The scattered field $\mathbf{E}_1(\mathbf{r}, t)$ and $f(\mathbf{r}, t)$ have the following Fourier transform pairs:

$$\mathbf{E}_1(\mathbf{r}, \omega) = \int_{-\infty}^{\infty} \mathbf{E}_1(\mathbf{r}, t)e^{-j\omega t}\, dt, \tag{E.1a}$$

$$\mathbf{F}(\mathbf{r}, \omega) = \int_{-\infty}^{\infty} \mathbf{f}(\mathbf{r}, t)e^{-j\omega t}\, dt, \tag{E.1b}$$

$$\mathbf{E}_1(\mathbf{r}, t) = \frac{1}{2\pi} \int_{-\infty}^{\infty} \mathbf{E}_1(\mathbf{r}, \omega)e^{j\omega t}\, d\omega, \tag{E.1c}$$

$$\mathbf{f}(\mathbf{r}, t) = \frac{1}{2\pi} \int_{-\infty}^{\infty} \mathbf{F}(\mathbf{r}, \omega)e^{j\omega t}\, d\omega. \tag{E.1d}$$

If Eqs. (E.1c) and (E.1d) are substituted into Eq. (11.19), it is seen that *each* frequency component obeys the differential equation

$$\nabla^2 \mathbf{E}_1(\mathbf{r}, \omega) + k^2 \mathbf{E}_1(\mathbf{r}, \omega) = -\mathbf{F}(\mathbf{r}, \omega), \tag{E.2}$$

where $k^2 \equiv \omega^2/c^2$. The solution for $\mathbf{E}_1(\mathbf{r}, \omega)$ is given in many textbooks on electromagnetism (e.g., Morse and Feshbach, 1953, Vol. II, pp. 1769–1778). Its solution for the assumption that all waves are outgoing from the scatter volume V is

$$\mathbf{E}_1(\mathbf{r}_{r0}, \omega) = \frac{1}{4\pi} \int_V \frac{\mathbf{F}(\mathbf{r}, \omega)\varepsilon e^{-jk|\mathbf{r}_{r0} - \mathbf{r}|}}{|\mathbf{r}_{r0} - \mathbf{r}|}\, dV, \tag{E.3}$$

where $\mathbf{r}_{r0}$ is the observation point (receiver location) and r is the distance from the origin to the elemental scatter volume dV (Fig. 11.2). The parameter ε is the unit dyadic or idemfactor. The solution (E.3) is an integral of the vector Green's function

$$\frac{(\mathbf{a}_x + \mathbf{a}_y + \mathbf{a}_z)e^{-jk(|\mathbf{r}_{r0} - \mathbf{r}|)}}{|\mathbf{r}_{r0} - \mathbf{r}|}, \tag{E.4}$$

weighted by the source function $\mathbf{F}(\mathbf{r}, \omega)$ where $\mathbf{a}_x$, $\mathbf{a}_y$, and $\mathbf{a}_z$ are unit vectors. The vector Green's function is a solution of Eq. (E.2) for a unit vector source $(\mathbf{a}_x + \mathbf{a}_y + \mathbf{a}_z)\delta(\mathbf{r}_{r0} + -\mathbf{r})$. The operator ∇ in $\mathbf{F}(\mathbf{r}, \omega)$ operates on the source location $\mathbf{r}$ dependence.

The scattered field in the time domain is found by applying Eq. (E.1c) to Eq. (E.3):

$$\mathbf{E}_1(\mathbf{r}_{r0}, t) = \frac{1}{8\pi^2} \int_{-\infty}^{+\infty} \int_V \mathbf{F}(\mathbf{r}, \omega) \frac{\exp[j(\omega t - k|\mathbf{r}_{r0} - \mathbf{r}|)]}{|\mathbf{r}_{r0} - \mathbf{r}|} dV \, d\omega, \tag{E.5}$$

in which we have made use of the identity $\mathbf{F}(\mathbf{r}, \omega)\varepsilon \equiv \mathbf{F}(\mathbf{r}, \omega)$. Now, $\mathbf{F}(\mathbf{r}, \omega)$ can be expressed in terms of the source function $\mathbf{f}(\mathbf{r}, t)$ through use of (E.1b), so

$$\mathbf{E}_1(\mathbf{r}_{r0}, t) = \frac{1}{8\pi^2} \int_{-\infty}^{+\infty} \int_{-\infty}^{+\infty} \int_V \mathbf{f}(\mathbf{r}, t')$$

$$\times \frac{\exp[j\omega(t - t') - j(\omega/c)|\mathbf{r}_{r0} - \mathbf{r}|]}{|\mathbf{r}_{r0} - \mathbf{r}|} dV \, d\omega \, dt', \tag{E.6}$$

where t' gives the time variation of the source $\mathbf{f}(\mathbf{r}, t')$. The integration over ω can be executed by using the identity

$$\int_{-\infty}^{+\infty} \exp\{j\omega[t - t' - |\mathbf{r}_{r0} - \mathbf{r}|/c]\} \, d\omega \equiv 2\pi\delta(t - t' - |\mathbf{r}_{r0} - \mathbf{r}|/c),$$

so that

$$\mathbf{E}_1(\mathbf{r}_{r0}, t) = \frac{1}{4\pi} \int_{-\infty}^{+\infty} \int_V \frac{\mathbf{f}(\mathbf{r}, t')\delta(t - t' - |\mathbf{r}_{r0} - \mathbf{r}|/c)}{|\mathbf{r}_{r0} - \mathbf{r}|} dV \, dt'. \tag{E.7}$$

Therefore the Green's function for the general time-varying source is

$$G(\mathbf{r}_{r0} - \mathbf{r}, t - t') = \frac{\delta(t - t' - |\mathbf{r}_{r0} - \mathbf{r}|/c)}{4\pi|\mathbf{r}_{r0} - \mathbf{r}|}. \tag{E.8}$$

References

Abramowitz, M., and Stegun, I. A. (1964)."Handbook of Mathematical Functions." Natl. Bur. Stand. Appl. Math. Ser. 55, 2nd printing, Supt. Doc. U.S. Govt. Printing Office, Washington, D.C.

Aden, A. L. (1951). Electromagnetic scattering from metal and water spheres. *J. Appl. Phys.* **22**, 601.

Al-Khatib, H. N., Seliga, T. A., and Bringi, V. N. (1979). "Differential Reflectivity and Its Use in the Radar Measurement of Rainfall." *Atmos. Sci. Prog. Rep.* **AS-S-106**. Ohio State University, Columbus.

Alberty, R. L., Burgess, D. W., Hane, C. E., and Weaver, J. F. (1979). "SESAME 1979 Operations Summary." NOAA, U.S. Dept of Commerce, U.S. Govt. Printing Office, Washington, D.C.

Anthes, R. A., Orville, H. D., and Raymond, D. J. (1982). Mathematical modeling of convection. *In* "A Social, Scientific, and Technological Documentary" (E. Kessler, ed.), Vol. 2, pp. 495–579.NOAA, U.S. Dept. of Commerce, U.S. Govt. Printing Office, Washington, D.C.

Appleton, E. V., and Barnett, M. A. F. (1926). On some direct evidence for downward atmospheric reflection of electric rays. *Proc. Roy. Soc., London Ser. A* **109**, 621–641.

Armijo, L. (1969). A theory for the determination of wind and precipitation velocities with Doppler radar. *J. Atmos. Sci.* **26**, 570–573.

Atlas, D. (1964). Advances in radar meteorology. *Adv. in Geophys.* **10**, 317–478.

Atlas, D., ed. (1990). "Radar in meteorology." American. Meterol. Society, Boston.

Atlas, D., and Chmela, A. C. (1957). Physical-synoptic variations of drop size parameters. *Proc. Weather Radar Conf., 65th, 1957* pp. 21–30.

Atlas, D., and Donaldson, R. J. (1954). The feasibility of the identification of hail and severe storms. *Air Force Surveys in Geophysics* **65**, A.F.C.R.C., p. 91.

Atlas, D., and Ludlum, F. H. (1961). Multi-wavelength radar reflectivity of hailstorms, *Quart. J. Roy. Meteor. Soc.* **87**, 523–534.

Atlas, D., and Ulbrich, C. W. (1974). The physical basis for attenuation–rainfall relationships and the measurement of rainfall parameters by combined attenuation and radar methods. *J. Rech. Atmos.* **8**(1-2), 275–298.

Atlas, D., and Ulbrich, C. W. (1977). Path- and area-integrated rainfall measurement by microwave attenuation in the 1–3 cm band. *J. Appl. Meteorol.* **16**, 1322–1331.

Atlas, D., Harper, W. G., Ludlam, F. H., and Macklin, W. C. (1960). Radar scatter by large hail. *Q. J. R. Meteorol. Soc.* **86**, 468–482.

Atlas, D., Hardy, K. R., Glover, K. M., Katz, I., and Konrad, T. G. (1966). Tropopause detected by radar. *Science* **153**, 1110–1112.

Atlas, D., Srivastava, R. C., and Sekhon, R. S. (1973). Doppler radar characteristics of precipitation at vertical incidence. *Rev. Geophys. Space Phys.* **2**, 1–35.

Atlas, D., Meneghini, R., and Moore, R. K. (1982). The outlook for precipitation measurements from space. *Atmos. Ocean* **20**(1), 50–60.

Atlas, D., Rosenfeld, D., and Short, D. A. (1990). The estimation of convective rainfall by area integrals: 1. The theoretical and empirical basis. *J. Geoph. Research* **95**, 2153–2160.

Austin, P. M., and Geotis, S. G. (1979). Raindrop sizes and related parameters for GATE. *J. Appl. Meteorol.* **18**, 569–575.

Aydin, K., Seliga, T. A., and Balaji, V. (1986). Remote sensing of hail with a dual linear polarization radar. *J. Clim. Appl. Meteorol.* **25**, 1475–1484.

Aydin, K., Zhao, Y., and Seliga, T. A. (1990). A differential reflectivity radar hail measurement technique: Observations during the Denver hailstorm of 13 June 1984. *J. Atmos. and Oceanic Tech.* **7**, 104–113.

Balakrishnan, N., and Zrnić, D. S. (1990a). Estimation of rain and hail rates in mixed phase precipitation. *J. Atmos. Sci.* **47**, 565–583.

Balakrishnan, N., and Zrnić, D. S. (1990b). Use of polarization to characterize precipitation and discriminate large hail. *J. Atmos. Sci.* **47**, 1525–1540.

Balakrishnan, N., Zrnić, D. S., Goldhirsh, J., and Rowland, J. (1989). Comparison of simulated rain rates from disdrometer data employing polarimetric radar algorithms. *J. Atmos. and Oceanic Tech.* **6**, 476–486.

Ball, F. K. (1961). Viscous dissipation in the atmosphere. *J. Meteorol.* **18**, 553–557..

Balsley, B., and Peterson, V. L. (1981). Doppler-radar measurements of clear air atmospheric turbulence at 1290 MHz. *J. Appl. Meteorol.* **20**, 266–274.

Banjanin, Z. B., and Zrnić, D. S. (1991). Clutter rejection for Doppler weather radars which use staggered pulses. *IEEE Trans. Geosci. Remote Sens.* **GRS-29**, 610–620.

Barnes, S. L. (1978). Oklahoma thunderstorms on 29–30 April 1970. Pt. I. Morphology of a tornadic storm. *Mon. Weather Rev.* **106**, 673–684.

Barratt, P., and Browne, I. C. (1953). A new method of measuring vertical air currents. *Q. J. R. Meteorol. Soc.* **79**, 550.

Batchelor, G. K. (1953). "The Theory of Homogeneous Turbulence." Cambridge Univ. Press, London and New York.

Battan, L. J. (1958). Use of chaff for wind measurement. *Bull. Amer. Soc.* **39**, 258–260.

Battan, L. J. (1973). "Radar Observation of the Atmosphere." Univ. of Chicago Press, Chicago, Illinois.

Bean, B. R., and Dutton, E. J. (1966). "Radio Meteorology." Natl. Bur. Stand., Monogr. 92, Supt. Doc. U.S. Govt. Printing Office, Washington, D.C.

Bebbington, D. H. O., McGuiness, R., and Holt, A. R. (1987). Correction of propagation effects in S-Band circular polarization-diversity radars. *IEEE Proc.* **134**, 431–437.

Bello, P. A. (1965). On the RMS bandwidth of nonlinearly envelope-detected narrow-band Gaussian noise. *IRE Trans. Inf. Theory* **IT-11**, 236–239.

Benjamin, T. B. (1968). Gravity currents and related phenomena. *J. Fluid Mech.* **31**(2), 209–248.

Bergen, W. R., and Albers, S. C. (1988). Two and three-dimensional dealiasing of Doppler radar velocities. *J. Atmos. Oceanic Tech.* **5**, 305–319.

Berger, M., and Doviak, R. J. (1979). "An Analysis of the Clear Air Planetary Boundary Layer Wind Synthesized from NSSL's Dual Doppler-radar Data." *NOAA Tech. Memo* **ERL NSSL-87**. Natl. Severe Storms Lab., Norman, Oklahoma.

Berger, T., and Groginsky, H. L. (1973). Estimation of the spectral moments of pulse trains. *Prepr., Int. Conf. on Information Theory, 1973*.

Bernella, D. (1991). Terminal Doppler Weather Radar Operational Test and Evaluation: Orlando 1990. Lincoln Laboratory Report No. DOT/FAA/NR-91/2, 101 pp.

Biron, P. J., and Sen Lee, T. (1990). Microburst Outflow Reflectivity Distributions, Internal Memo, Lincoln Laboratory.

Blake, L. B. (1970). Prediction of radar range. *In* "Radar Handbook" (M. I. Skolnik, ed), Chapter 2. McGraw-Hill, New York.

Bluestein, H. B. (1986). Fronts and Jet Streaks: A theorectical Perspective. *In* "Mesoscale Meteorology and Forecasting" (P. Ray, ed.). Chapter 9. American Meteorological Society, 793 pp.

Bluestein, H. B., and Hazen, D. S. (1989). Doppler-radar analysis of a tropical cyclone over land: Hurricane Alicia (1983) in Oklahoma. *Mon. Wea. Rev.* **117**, 2594–2611.

Bluestein, H. B., and Unruh, W. P. (1991). On the measurements of wind speeds in tornadoes with a portable CW/FM Doppler radar. *Prepr., Radar Meteorol. Conf.*, 25th, pp.848–851.

Bluestein, H. B., Hrebenach, S. D., and Brandes, E. A. (1991). A test of the synthetic dual-Doppler analysis technique: Two case studies. *Prepr., Internat. Conf. Radar Meteorol., 25th, 1991* pp. 634–637.

Bohne, A., and Srivastava, R. C. (1975). "Random Errors in Wind and Precipitation Fall Speed Measurement by a Triple Doppler Radar System," Rep. No. 37. Lab. Atmos. Probing, University of Chicago, Chicago, Illinois.

Bohne, A. R. (1981). Estimation of turbulence severity in precipitation environments by radar. *Prepr., Radar Meteorol. Conf., 20th, 1981* pp. 446–453.

Bohne, A. K. (1982). Radar detection of turbulence in precipitation environments. *J. Atmos.Sci.* **39**, 1819–1837.

Bohren, T. A., Cruz, J. R., and Zrnić, D. S. (1986). An artificial intelligence approach to Doppler weather radar velocity dealiasing. *Prepr., Conf. Radar Meteorol., 23rd, 1986* pp. 107–110.

Bonino, G., Lombardini, P. P., and Trivero, P. (1979). A metric wave radio-acoustic tropospheric sounder. *IEEE Trans. Geosci. Electron.* **GE-17**, 179–181.

Booker, H. G., and Gordon, W. E. (1950). A theory of radio scattering in the troposphere. *Proc. IRE* **38**, 401–412.

Borgeaud, M., Shin, R. T., and Kong, J. A. (1987). Theoretical models for polarimetric radar clutter. *J. Electron. Waves and Applic.* **1**, 73–89.

Born, M., and Wolf, E. (1964). "Principles of Optics," 2nd ed. Macmillan, New York.

Bowles, R. L., and Targ, R. (1988). Windshear detection and avoidance: Airborne systems perspective. *Congress of the ICAS, 16th, 1988.*

Boyenval, E. H. (1960). Echoes from precipitation using pulsed Doppler radar. *Prepr., Weather Radar Conf., 8th, 1960* pp. 57–64.

Brandes, E. A. (1975). Optimizing rainfall estimates with the aid of radar. *J. Appl. Meteorol.* **14**, 1339–1345.

Brandes, E. A. (1977). Severe thunderstorms observed by dual-Doppler radar. *Mon. Weather Rev.* **105**, 113–120.

Brandes, E. A. (1978). Mesocyclone evolution and tornadogenesis: Some observations. *Mon. Weather Rev.* **106**, 995–1011.

Brandes, E. A. (1990). Evolution and structure of the 6–7 May 1985 Mesoscale convective system and associated vortex. *Mon. Weather Rev.* **118**, 109–126.

Brandes, E. A., and Sirmans, D. (1976). Convective rainfall estimation by radar: Experimental results and proposed operational analysis technique. *Prepr., Conf. Hydro-Meteorol., 1976* pp. 54–59.

Breit, G., and Tuve, M. A. (1926). A test for the existence of the conducting layer. *Phys. Rev.* **28**, pp. 554–575.

Brekhovskikh, L. M. (1960). "Waves in Layered Media." Academic Press, New York.

Brewster, K. A., and Zrnić, D. S. (1986). Comparison of eddy dissipation rates from spatial spectra of Doppler velocities and Doppler spectrum widths. *J. Atmos, Oceanic Tech.* **3**, 440–452.

Bringi, V. N., and Hendry, A. (1990). Technology of polarization diversity radars of meteorology. *In* "Radar in Meteorology" American Meteorol. Society, Boston. 153–190.

Bringi, V. N., Seliga, T. A., and Aydin, K. (1984). Hail detection with a differential reflectivity radar. *Science* **225**, 1145–1147.

Bringi, V. N., Vivekanandan, J., and Tuttle, J. D. (1986). Multiparameter radar measurements in Colorado convective storms. Part II: Hail detection studies. *J. Atmos. Sci.* **43**, 2564–2577.

Bringi, V. N., Chandrasekar, V., Balakrishnan, N., and Zrnić, D. S. (1990). An examination of propagation effects in rainfall on radar measurements at microwave frequencies. *J. Atmos. Oceanic Tech.* **7**(6), 829–840.

Brown, R. A. (1970). A secondary flow model for the planetary boundary layer. *J. Atmos. Sci.* **27**, 742–757.

Brown, R. A. (1989). Initiation and propagation of thunderstorm mesocyclones. Ph.D. disseration, University of Oklahoma. (University microfilms, Ann Arbor, MI. order #89-19983), 321 pp.

Brown, R. A., and Crawford, K. C. (1972). Doppler radar evidence of severe storm high-reflectivity cores acting as obstacles to airflow. Preprints, *15th Radar Meteorology Conf.*, Champaign-Urbana, Illinois, American Meteor. Soc., pp. 16–21.

Brown, R. A., and Lemon, L. R. (1976). Single Doppler radar vortex recognition. Part II. Tornadic vortex signatures. *Prepr., Radar Meteorol. Conf., 17th, 1976* pp. 104–109.

Brown, R. A., and Wood, V. T. (1987). A guide for interpreting Doppler velocity patterns. *Report R400-DV-101*, NEXRAD Joint System Program Office, Silver Spring, Maryland.

Brown, R. A., and Wood. V. T. (1991). On the interpretation of single-Doppler velocity patterns within severe thunderstorms. *Weather and Forecast.* **6**, 32–48.

Brown, R. A., Lemon, L. R., and Burgess, D. W. (1978). Tornado detection by pulsed Doppler radar. *Mon. Weather Rev.* **106**, 29–38.

Browning, K. A., and Foot, G. B. (1976). Airflow and hail growth in supercell storms and some implications for hail suppression. *Q. J. R. Meteorol. Soc.* **102**, 499–522.

Browning, K. A. (1982). General circulation of middle-latitude thunderstorms. *In* "A Social, Scientific, and Technological Documentary" (E. Kessler, ed.), Vol. 2, pp. 211–247. NOAA, U.S. Dept. of Commerce, U.S. Govt. Printing Office, Washington, D.C.

Browning, K. A., and Wexler, R. (1968). A determination of kinematic properties of a wind field using Doppler-radar. *J. Appl. Meteorol.* **7**, 105–113.

Budden, K. G. (1961). "Radio Waves in the Ionosphere." Cambridge Univ. Press, London and New York.

Burgess, D. W. (1976). Single Doppler radar vortex recognition: Part I. Mesocyclone signatures. *Prepr., Conf. Radar Meteorol., 17th, 1976,* pp. 97–103.

Burgess, D. W., and Donaldson, R. J. (1979). Contrasting tornadic storm types. *Prepr., Conf. Severe Local Storms, 11th, 1979* 189–92.

Burk, S. D. (1978). "Use of a Second-moment Turbulence Closure Model for Computation of Refractive Index Structure Coefficients." Tech. Rep. TR-78-04. Naval Environmental Prediction Research Facility, Monterey, California.

Burrows, D. R., and Attwood, S. S. (1949). "Radio Wave Propagation." p. 219. Academic Press, New York.

Byers, H. R. (1944). "General Meteorology." McGraw-Hill, New York.

Byers, H. R. (1948). The use of radar in determining the amount of rain falling over a small area. *Trans. Amer. Geophys. Union* 187–196.

Byers, H. R., and Braham, R. R. (1949). "The thunderstorm." U.S. Government Printing Office, Washington, D.C., 287 pp.

Cadzow, J. W. (1982). Spectral estimation: An over-determined rational model equation approach. *Proc. IEEE* **70**(9), 907–939 (special issue on spectral estimation).

Cain, D. E., and Smith, P. L., Jr. (1976). Operational adjustment of radar estimated rainfall with rain gage data: A statistical evaluation. *Prepr., Conf. Radar Meteorol., 17th, 1976* pp. 533–538.

Campbell, W. C., and Strauch, R. C. (1976). Meteorological Doppler radar with double pulse transmission. *Prepr., Conf. Radar Meteorol., 17th, 1976* pp. 42–44.

Campbell, S. D., Merrit, M. W., and DiStefano, J. T. (1989). Microburst recognition performance of TDWR operational testbed. *Internat. Conf. Aviation Weather System, 3rd, 1989* no page numbers.

Carbone, R. E. (1982). A severe frontal rainband. Part I: Stormwide hydrodynamic structure. *J. Atmos. Sci.* **39**, 258–279.

Carbone, R. E. (1983). A severe frontal rainband. Part II: Tornado parent vortex circulation. *J. Atmos. Sci.* **40**, 2639–2654.

Carbone, R. E., and Nelson, L. D. (1978). The evolution of raindrop spectra in warm-based convective storms as observed and numerically modeled. *J. Atmos. Sci.* **35**, 2302–2314.

Carbone, R. E., Conway, J. W., Crook, N. A., and Moncreiff, M. W. (1990). The generation and propagation of a nocturnal squall line. Part I: Observations and implications for mesoscale predictability. *Mon. Wea. Rev.* **118**, 26–49.

Carlson, T. N., Benjamin, S. G., Forbes, G. S., and Li, Y. F. (1983). Elevated mixed layers in the regional severe strom environment: Conceptual model and case studies. *Mon. Wea. Rev.* **111**, 1453–1473.

Cataneo, R. (1969). A method for estimating rainfall rate–radar reflectivity relationships. *J. Meteorol.* **8**, 815–819.

Cataneo, R., and Stout, G. E. (1968). Raindrop-size distributions in humid continental climates, and associated rainfall rate–radar reflectivity relationships. *J. Appl. Meteorol.* **7**, 901–907.

Caton, P. A. F. (1963). Wind measurement by Doppler radar. *Meteorol. Mag.* **92**, 213–222.

Christie, D. R., and Muirhead, K. J. (1982). "Solitary Waves: A Hazard to Aircraft Operating at Low Altitudes," Rep. ACT 2600. Research School of Earth Sciences, Australian National University, Canberra.

Chadwick, R. B., and Moran, K. P. (1980). Long-term measurements of C_n^2 in the boundary layer. *Radio Sci.* **15**, 355–362.

Chandrasekar, S. (1955). A theory of turbulence. *Proc., R. Soc. London, Ser A* **229**, 1–9.

Chandrasekar, S. (1956). Theory of turbulence. *Phys Rev.* **102**, 941–953.

Chandrasekar, V., Bringi, V. N., Balakrishnan, N., and Zrnić, D. S. (1990). Error Structure of Multiparameter Radar and Surface Measurements of Rainfall. Part III: Specific Differential Phase. *J.Atmos. and Oceanic Tech.* **7**, 621–629.

Chen, W. Y. (1974). Energy dissipation rates of free atmosphere turbulence. *J. Atmos. Sci.* **31**, 2222–2225.

Cheng, L., and English, M. (1983). A relationship between hailstone concentration and size. *J. Atmos. Sci.* **40**, 204–213.

Chernikov, A. A., Mel'nichuk, Yu. V., Pinus, N. K., Shmeter, S. M., and Vinnechenko, N. K. (1969). Investigations of the turbulence in convective atmosphere using radar and aircraft. *Radio Sci.* **4**, 1257–1260.

Christie, D. R. (1989). Long nonlinear waves in the lower atmosphere. *J. Atmos, Sci.* **46**, 1462–1491.

Christie, D. R., and Muirhead, K. J. (1983). Solitary waves: A hazard to aircraft operating at low altitudes. *Aust. Meteorol. Mag.* **31**, 97–109.

Clarke, R. H. (1965). Horizontal mesoscale vortices in the atmosphere. *Aust. Meteorol. Mag.* **50**, 1–26.

Cobine, J. D. (1941). "Gaseous Conductors." McGraw Hill.

Collier, C. G. (1989). Applications of Weather Radar Systems. Ellis Horwood Ltd.Chichester, England, pp. 294.

Colwell, R. C., and Friend, A. W. (1936). The D region of the ionosphere. *Nature* **137**, p. 782.

Cooley, J. W., and Tukey, J. W. (1965). An algorithm for the machine calculation of complext Fourier series. *Math Comp.* **19**, 297–301.

Cooley, J. W., Lewis, P. A. W., and Welch, P. D. (1967). Historical notes on the fast Fourier transform. *Proc. IEEE* **55**, 1675–1677.

Crane, R. K. (1975). Comparison between reflectivity statistics at heights of 3 and 6 km and rain rate statistics at ground level. *Prepr. Conf. Radar Meteorol., 16th, 1975*, pp. 479–483.

Crane, R. K. (1976). "Low Elevation Angle Measurement Limitations Imposed by the Troposphere: An Analysis of Scintillation Observations Made at Haystack and Millstone." Tech. Rep. 518. Lincoln Lab., Massachusetts Institute of Technology, Lexington.

Crane, R. K. (1977). "Stratospheric Turbulence." Rep. No. AFGL TR 77-0207. Air Force Geophys. Lab., Hanscom AFB, Massachusetts.

Crane, R. K. (1979). Automatic cell detection and tracking. *IEEE Trans. Geosci. Electron* **GE-17** 250–261. Special Issue *Radio Meteorol.*

Crane, R. K. (1980). Radar measurements of wind at Kwajalein. *Radio Sci.* **15**, 383–394.

Cressman, G. P. (1959). An operational objective analysis system. *Mon. Weather Rev.* **87**, 367–374.

Crocker, S. C. (1988). TDWR PRF selection criteria. *Project Report*, ATC-147, MIT Lincoln Laboratory, MASS, 57.

Crook, N. A.(1986). The effect of ambient stratification and moisture on the motion of atmospheric undular bores. *J. Atmos. Sci.* **43**, 171–181.

Crook, N. A., and Miller, M. J. (1985). A numerical and analytical study of atmospheric undular bores. *Q. J. R. Meteorol. Soc.* **111**, 225–242.

Crozier, C. L., Joe, P. I., Scott, J. W., Herscovitch, H. N., and Nichols, T. R. (1989). First experiment with an operational Doppler radar, *Prepr., Radar Meteorol. Conf., 24th, 1989*, pp. 179–185.

Davenport, W. B., and Root, W. L. (1958). "An Introduction to the Theory of Random Signals and Noise." McGraw-Hill, New York.

Davies-Jones, R. J. (1979). Dual-Doppler radar coverage area as a function of measurement accuracy and spatial resolution. *J. Appl. Meteor.* **9**, 1229–1233.

Davis, J. I. (1966). Consideration of atmospheric turbulence in laser systems design. *Appl. Opt.* **5**, 139.

Deley, G. W. (1970). Waveform design. *In* "Radar Handbook: (M. I. Skolnik, ed.), pp. 3–1, 3–47. McGraw-Hill, New York.

Dicke, R. H., Beringer, R., Kyhl, R. L., and Vane, A. B. (1946). Atmospheric absorption measurements wtih a microwave radiometer. *Phys. Rev.* **70**, 340–348.

Donaldson, R. J., Jr. (1970). Vortex signature recognition by a Doppler radar. *J. Appl. Meteorol.* **9**, 661–670.

Doneaud, A. A., Ionescu-Niscov, S., Priegnitz, D. L., and Smith, P. L. (1984). The area–time integral as an indicator for convective rain volumes. *J. Climate Appl. Meteorol.* **23**(4), 555–561.

Doneaud, A. A., Miller, J. R. Jr., Johnson, L. R., Vonder Haar, T. H., and Laybe, P. (1987). The area–time integral technique to estimate convective rain volumes over areas applied to satellite data—preliminary investigation. *J. Climate Appl. Meteorol.* **26**, 156–169.

Douglas, R. H. (1964). Hail size distributions. *Proc. World Conf. on Radio Meteorol. and Prepr., Radar Meteorol. Conf., 11th, 1964*, pp. 146–149.

Doviak, R. J. (1972). Comparison of Bistatic and Monostatic Radar Detection of Clear Air Atmospheric Targets. AIAA Paper No. 72-175, Copies available from AIAA Library, 750 3rd Ave., New York NY 10017, 8 pp.

Doviak, R. J. (1983). A survey of radar rain measurement techniques. *J. Appl. Meteorol.* **22**, 832–849.

Doviak, R. J., and Berger, M. J. (1980). Turbulence and waves in the optically clear planetary boundary layer resolved by dual Doppler radars. *Radio Sci.* **15**, 297–317.

Doviak, R. J., and Christie, D. R. (1989). Thunderstorm-generated solitary waves: a wind shear hazard. *J. Aircraft* **26**. 423–431.

Doviak, R. J., and Christie, D. R. (1991). Buoyancy wave hazards to aviation. *Prepr., Internat. Conf. Aviation Weather Systems, 4th, 1991*, pp. 247–252.

Doviak, R. J., and Ge, R. S. (1984). An atmospheric solitary gust observed with a Doppler radar, a tall tower, and a surface network. *J. Atmos. Sci.* **41**, 2559–2573.

Doviak, R. J., Goldhirsh, J., and Miller, A. R. (1972). Bistatic-radar detection of high-altitude clear-air atmospheric targets. *Radio Sci.* **7**, 993–1003.

Doviak, R. J., and Lee, J. T. (1985). Radar for storm forecasting and weather hazard warning. *J. Aircraft* **22**, 1059–1064.

Doviak, R. J., and Sirmans, D. (1973). Doppler radar with polarization diversity. *J. Atmos. Sci.* **30**, 737–738.

Doviak, R. J., and Zrnić, D. S. (1984a). "Doppler radar and weather observations." Academic Press, San Diego, California. 458 pp.

Doviak, R. J., and Zrnić, D. S. (1984b). Reflection and scatter formula for anisotropically turbulent air. *Radio Sci.* **19**, 325–336.

Doviak, R. J., and Zrnić, D. S. (1985). Siting of Doppler weather radars to shield ground targets. *IEEE Tr. Anten. and Prop.* **AP-33**, 685–689.

Doviak, R. J., Goldhirsh, J., and Miller, A. R. (1971). Simultaneous bistatic and monostatic detection of tropospheric layers. *IEEE Trans. Antennas Propag.* **AP-19**, 714–716.

Doviak, R. J., and Zrnić, D. S. (1988). The Doppler Weather Radar, Chapter X. *In* Aspects of Modern radar (Eli Brookner, ed.) Artech House, Boston. 574 pp.

Doviak, R. J., Ray, P. S., Strauch, R. G., and Miller, L. J. (1976). Error estimation in wind fields derived from dual-Doppler radar measurement. *J. Appl. Meteorol.* **15**, 868–878.

Doviak, R. J., Sirmans, D., Zrnić, D. S., and Walker, G. B. (1978). Considerations for pulse Doppler radar observations of severe thunderstorms. *J. Appl. Meteorol.* **17**, 189–205.

Doviak, R. J., Rabin, R. M., and Koscielny, A. J. (1983). Doppler weather radar for profiling and mapping winds in the prestorm environment. *IEEE Trans. Geosci. Remote Sens.* **21**, 25–33.

Doviak, R. J., Thomas, K. W., and Christie, D. R. (1989). The wavefront shape, position, and evolution of a great solitary wave of translation. *IEEE Trans. Geo. Sci. and Remote Sens.* **27**, 658–665.

Doviak, R. J., Chen, S. S., and Christie, D. R. (1991). A thunderstorm-generated solitary wave observation compared with theory for nonlinear waves in a sheared atmosphere. *J. Atmos. Sci* **48**, 87–111.

Draper, W. R., and Smith, H. (1966). "Applied Regression Analysis." Wiley, New York.

Droegemeier, K. K., and Babcock, M. R. (1989). Numerical simulation of microburst downdrafts: Application to on-board and look-ahead sensor technology. *AIAA Aerospace Sciences Meeting.* 12 pp.

DuCastel, F., Misme, P., Spizzichino, A., and Voge, J. (1962). On the role of the process of reflection in radio wave propagation. *J. Res. Natl. Bur. Stand., Sect. D* **66**, 273–284.

Easterbrook, C. C. (1974). Estimating horizontal wind fields by two-dimensional curve fitting of single Doppler radar measurements. *Prepr. Radar Meteorol. Conf., 16th, 1974* pp. 214–219.

Eccles, P. J., and Atlas, D. (1973). A dual-wavelength radar hail detector, *J. Appl. Meteorol.* **12**, 847–854.

Eccles, P. J. (1979). Comparison of remote measurements by single- and dual-wavelength meteorological radars. *IEEE Trans. Geosci. Electron.* **GE-17**, 205–218.

Eccles, P. J., and Mueller, E. A. (1973). X-band attenuation and liquid water content estimation by a dual-wavelength radar. *J. Appl. Meteorol.* **10**, 1252–1259.

Ecklund, W. L., Carter, D. A., and Balsley, B. B. (1979). Continuous measurement of upper atmospheric winds and turbulence using a VHF radar: Preliminary results. *J. Atmos. Terr. Phys.* **41**, 983–984.

Ecklund, W. L., Carter, D. A., and Balsley, B. B. (1988). A UHF wind profiler for the boundary layer: Brief description and initial results. *J. Atmos. Oceanic Tech.* **5**, 432–441.

Eilts, M. D., and Doviak, R. J. (1987). Oklahoma Downbursts and their asymmetry. *J. Climate and Appl. Meteorol.* **26**, 69–78.

Eilts, D. M., and Smith, D. (1990). Efficient dealiasing of Doppler velocities using local environment constraints. *J. Atmos. Oceanic Tech.* **7**, 118–128.

Einaudi, F., and Lalas, D. P. (1974). On the correct use of the wet adiabatic lapse rate in stability criteria of a saturated atmosphere. *J. Appl. Meteorol.* **13**, 318–324.

Elmore, K. L., Politovich, M. K., and Sand, W. R. (1989). The 11 July 1988 Microburst at Stapleton International Airport, Denver, Colorado. *In* "Windshear Case Study: Denver, Colorado, July 11, 1988." (H. W. Schlickenmaier, ed.) Final Report DOT/FAA/DS-89/19, from the U.S. Department of Transportation, Federal Aviation Administrations, Advanced System Design Service, Washington, D.C. 20591, 552 pp.

Englund, C. R., Crawford, D. R., and Mumford, W. W. (1938). Ultra-short-wave transmission and atmospheric irregularities. *Bell Sys. Tech. J.* **17**, 489–519.

Federer, B., and Waldvogel, A. (1975). Hail and raindrop size distributions from a Swiss multicell storm. *J. Appl. Meteorol.* **14**, 91–97.

Fetter, R. W. (1961). Remote measurement of wind velocity by the electromagnetic-acoustic probe II: Experimental System. *Proc., Natl. Conv. Mil. Electron. 5th 1961* pp. 54–59.

Fitzgerald, J. W. (1972). "A study of the Initial Phase of Cloud Droplet Growth by Condensation: Comparison between Theory and Observation." Cloud Phys. Lab. Tech. Note, 44, p. 144. University of Chicago, Chicago, Illinois.

Foote, G. B., and duToit, P. S. (1969). Terminal velocity of raindrops aloft. *J. Appl. Meteorol.* **8**, 249–253.

Fornberg, B., and Whitham, G. B. (1978). A Numerical and theoretical study of certain nonlinear wave phenomena. *Phil. Trans. Roy. Soc.* **A289**, 373–404.

Frankel, M. S., and Peterson, A. M. (1976). Remote temperature profiling in the lower troposphere. *Radio Sci.* **11**, 157–166.

Friend, A. W., (1939). Continuous determination of air-mass boundaries by radio. *Bull. Amer. Meteorol. Soc.* **20**, 202–205.

Fritsch, J. M. (1975). Cumulus dynamics: Local compensating subsidence and its implications for cumulus parameterization. *Pure Appl. Geophys.* **113**, 851–867.

Frisch, A. S., and Clifford, S. F. (1974). A study of convection capped by a stable layer using Doppler radar and acoustic echo sounders. *J. Atmos. Sci.* **31**, 1622–1628.

Frisch, A. S., and Strauch, R. G. (1976). Doppler radar measurements of turbulent kinetic energy dissipation rates in a northeastern Colorado convective storm. *J. Appl. Meteorol.* **15**, 1012–1017.

Frisch, A. S., Weber, B. L., Straugh, R. G., Merritt, D. A., and Moran, K. P. (1986). The altitude coverage of the Colorado wind profilers at 50, 405, and 915 MHz. *J. Atmos. Oceanic Tech.* **3**, 680–692.

Frost, I. R., Illingworth, A. J., and Caylor, I. J. (1989). "Aircraft and Polarization radar measurements of a triggered lightning event." *Proc., Internatl. Lightning and Static Electr. Conf., 1989*, pp. 1A.1.1–1A.1.4.

Frost, I. R., Goddard, J. W. F., and Illingworth, A. J. (1991). Hydrometeor identification using cross polar radar measurements and aircraft verification. *Prepr., Radar Meteorol. Conf., 25th, 1991*, pp. 658–661.

Fujita, T. T. (1955). Results of detailed synoptic studies of squall lines. *Tellus.* **7**, 405–436.

Fujita, T. T. (1981). Tornadoes and downbursts in the context of generalized planetary scales. *J. Atmos. Sci.* **38**, 1511–1534.

Fujita, T. T. (1985). "The downburst, Microburst and Macroburst." Univ. Chicago Press, Chicago, Ill., 122 pp.

Fukao, S., Sato, T., Tsuda, T., Kato, S., Wakasugi, K., and Wakihara, T. (1985). The MU radar with an active phased array system. 2, In-house equipment *Radio Sci.* **20**, 1169–1176.

Fulton, R., Zrnić, D. S., and Doviak, R. J. (1990). Initiation of a solitary wave family in the demise of a nocturnal thunderstorm density current. *J. Atmos. Sci.* **47**, 319–337.

Gabel, R. A., and Roberts, R. A. (1973). "Signals and Linear Systems." Wiley, New York.

Gage, K. S., and Balsley, B. (1978). Doppler radar probing of the clear atmosphere. *Bull. Am. Meteorol. Soc.* **59**, 1074–1094.

Gage, K. S., Green, J. L., and VanZandt, T. E. (1980). Use of Doppler radar for the measurement of atmospheric turbulence parameters from the intensity of clear-air echoes. *Radio Sci.* **15**, 407–416.

Gage, K. S., Balsley, B. B., and Green, J. L. (1981). Fresnel scattering model for the specular echoes observed by VHF radar. *Radio Sci.* **16**, 1447–1453.

Gal-Chen, T. (1978). A method for the initialization of the anelastic equations: Implications for matching models with observations. *Mon. Weather Rev.* **106**, 587–606.

Gal-Chen, T. (1982). Errors in fixed and moving frames of references: Applications for conventional and Doppler rada· analysis. *J. Atmos. Sci.* **39**, 2279–2300.

Gal-Chen, T., and Kropfli, R. A. (1984). Buoyancy and pressure perturbations derived from dual-Doppler radar observations of the planetary boundary layer: Applications for matching models with observations. *J. Atmos Sci.* **41**, 3007–3020.

Gans, R. (1912). Ube die Form ultramikroskopischer Goldteilchen. *Ann. Phys. (Leipzig)* **37**, 881–900.

Girardin-Gondeau, J., Baudin, F., and Testud, J. (1991). Comparison of coded waveforms for an airborne meteorological Doppler radar. *J. Atmos. Oceanic Tech.* **8**, 234–246.

Goff, R. C. (1976). Vertical structure of thunderstorm outflows. *Mon. Wea. Rev.* **104**, 1429–1440.

Gold, B., and Rader, C. M. (1969). "Digital Processing of Signals." McGraw-Hill.

Golden, J. H. (1974). Scale-interaction implications for the waterspout life cycle. II. *J. Appl. Meteorol.* **13**, 693–709.

Goldhirsh, J. (1975). Improved error analysis of raindrop spectra, rain rate, and liquid water content using multiple wavelength radars. *IEEE Trans Antennas Propag.* **AP-23**, 718–720.

Goldhirsh, J., and Katz, I. (1974). Estimation of raindrop size distribution using multiple wavelength radar systems. *Radio Sci.* **9**, 439–446.

Golestani, Y., Chandrasekar, V., and Bringi, V. N. (1989). Intercomparison of multiparameter radar measurements. *Prepr., Radar Meteorol. Conf., 24th, 1989*, pp. 309–314.

Gossard, E. E. (1960). Power spectra of temperature, humidity, and refractive index from aircraft and tethered balloon measurements. *IRE Trans. Antennas Propag.* **AP-8**(2), 186–201.

Gossard, E. E. (1977). Refractive index variance and its height distribution in different air masses. *Radio Sci.* **12**, 89–105.

Gossard, E. E. (1981). Clear weather meteorological effects on propagation at frequencies above 1 GHz. *Radio Sci.* **6**, 589–608.

Gossard, E. E., and Hooke, W. H. (1975). "Waves in the Atmosphere: Atmospheric Infrasound and Gravity Waves—Their Generation and Propagation." Elsevier, Amsterdam, 456 pp.

Gossard, E. E., and Strauch, R. G. (1983). Radar observations of clear air and clouds. Elsevier, New York. 280 pp.

Gossard, E. E., Chadwick, R. B., Neff, W. D., and Moran, K. P. (1982). The use of ground-based Doppler radars to measure gradients, fluxes and structure parameters in elevated layers. *J. Appl. Meteorol.* **21**, 211–226.

Gossard, E. E., Strauch, R. G., and Rogers, R. R. (1990). Evolution of dropsize distribution in liquid precipitation observed by ground-based Doppler radar. *J. Atmos. and Oceanic Tech.* **7**, 815–818.

Gradshteyn, I. S., and Ryzhik, I. M. (1965). "Table of Integrals, Series, and Products." Academic Press, New York.

Gray, W. M. (1978). Hurricanes: Their formation, structure and likely role in the tropical circulation. *Meteorol. Trop. Oceans, R. Meteorol. Soc.* pp. 155–218.

Gray, G., Lewis, B., Vinson, J., and Pratte, F. (1989). A real-time implementation of staggered PRT velocity unfolding. *J. Atmos. Oceanic Tech.* **6**, 186–187.

Green J. L., and Gage, K. S. (1980). Observations of stable layers in the troposphere and stratosphere using VHF radar. *Radio Sci.* **15**, 395–406.

Green, J. L., Gage, K. S., and VanZandt, T. E. (1979). Atmospheric measurements by VHF pulsed Doppler radar. *IEEE Trans. Geosci. Electron., Spec. Issue Radio Meteorol.* **GE-17**, 262–280.

Green, A. W. (1975). An approximation for the shape of large raindrops. *J. Appl. Meteorol.* **14**, 1578–1583.

Groginsky, H. L., and Glover, K. M. (1980). Weather radar canceler design. *Prepr., Radar Meteorol. Conf., 19th, 1980* pp. 192–198.

Guerlac, H. E. (1987). Radar in World War II. Tomash Publishers, Amer., Inst. of Phys., 2 Volumes, 1171 pp.

Gunn, K. L. S., and East, T. W. R. (1954). The microwave properties of precipitation particles, *Q. J. R. Meteorol. Soc.* **80**, 522–545.

Gunn, K. L. S., and Marshall, R. S. (1958). The distribution of size of aggregate snowflakes. *J. Meteorol.* **15**, 452–466.

Gunn, R., and Kinzer, G. D. (1949). The terminal velocity of fall for water droplets in stagnant air. *J. Meteorol.* **6**, 243–248.

Gurvich, A. S., and Kon, A. I. (1992). The backscattering from anisotropic turbulent irregularities, *J. Electro. Waves and Appl.* **6**, 107–118.

Haase, S. P., and Smith, R. K. (1989). The numerical simulation of atmospheric gravity currents. Part II: Environments with stable layers. *Geophys. Astrophys. Fluid Dynamics* **46**, 35–51.

Habann, E. (1924). Eine Neue Generatorröhre, Z. Hochfrequenztech. **24**, 115–120; 135–141.

Hall, M. P. M., Cherry, S. M., Goddard, J. W. F., and Kennedy, G. R. (1980). Raindrop sizes and rainfall rate measured by dual-polarization radar. *Nature (London)* **285**, 195–198.

Hardy, K. R. (1963). The development of raindrop-size distributions and implications related to the physics of precipitation. *J. Atmos. Sci.* **20**, 299–312.

Hardy K. R., and Glover, K. M. (1966). Twenty-four-hour history of radar angel activity at three wavelengths. *Proc., Radar Meteorol. Conf., 12th, 1966* pp. 269–274.

Hardy, K. R., and Katz, I. (1969). Probing the clear atmosphere with high-power, high-resolution radars. *Proc. IEEE* **57**, 469–480.

Hardy, K. R., Atlas, D., and Glover, K. M. (1971). Multiwavelength backscatter from the clear atmosphere. *J. Geophys. Res.* **71**, 1537–1552.

Harju, A. J., and Puhakka, T. M. (1980). A method of correcting quantitative radar measurements for partial beam blocking. *Prepr., Radar Meteorol. Conf., 20th, 1980* pp. 234–239.

Harju, A. E., and Puhakka, T. M. (1980). A method of correcting quantitative radar measurements for partial beam blocking. *Prepr., Radar Meteorol. Conf., 19th, 1980* pp. 234–239.

Harris, C. M. (1966). Absorption of sound in air vs humidity and temperature. *J. of Acou. Soc. of America.* **40**, 148–159.

Harris, F. I. (1975). Motion field characteristics of the evaporative base region of a stratiform precipitation layer as determined by ANDASCE. *Prepr., Radar Meteorol. Conf., 16th, 1975* pp. 225–230.

Harrold, T. W., and Browning, K. A. (1971). Identification of preferred areas of shower development by means of high power radar. *Q. J. R. Meteorol. Soc.* **97**, 330–339.

Hartree, D. R., Michel, J. G. L., and Nicolson, P. (1946). Practical methods for the solution of the equations of tropospheric refraction. "Meteorological Factors in Radio Wave Propagation," pp. 127–168. Physical Society, London.

Hauser, D., and Amayenc, P. (1981). A new method for deducing hydrometeor-size distribution and vertical air motions from Doppler radar measurements at vertical incidence. *J. Appl. Meteorol.* **20**, 547–555.

Hendry, A, Antar, Y. M. M., and McCormick, G. C. (1987). On the relationship between the degree of preferred orientation in precipitation and dual-polarization radar echo characteristics. *Radio Sci.* **22**, 37–50.

Hennington, L., Doviak, R.J., Sirmans, D., Zrnić, D. S., and Strauch, R. G. (1976). Measurement of winds in the optically clear air with microwave pulse-Doppler radar. *Prepr., Radar Meteorol. Conf., 17th, 1976* pp. 342–348.

Hennington, L. (1981). Reducing the effects of Doppler radar ambiguities. *J. Appl. Meteorol.* **20**, 1543–1546.

Herman, B. M., and Battan, L. J. (1961). Calculations of Mie backscattering of microwaves from ice spheres. *Q. J. R. Meteorol. Soc.* **87**, 223–230.

Hess, S. L. (1959). "Introduction to Theoretical Meteorology." Holt, Reinhart and Winston, N.Y., 362 pp.

Hildebrand, P. H., and Moore, R. K. (1980). Meteorological radar observations from mobile platforms. "Radar in Meteorology" Atlas, ed. AMS, Boston. 287–314.

Hildebrand, P. H., and Sekhon, R. S. (1974). Objective determination of noise level in Doppler spectra. *J. Atmos. Sci.* **13**, 808–811.

Hill, F. F., Browning, K. A., and Bader, M. J. (1981). Radar and raingauge observations of orographic rain over South Wales. *Q. J. R. Meteorol. Soc.* **107**, 643–670.

Hill, R. D. (1990). Origins of Radar. *EOS. Trans. Amer. Geophys. Union* 3, 781–786.

Hill, R. J. (1978). Spectra of fluctuations in refractivity, temperature, humidity, and the tempera-ture-humidity cospectrum in the inertial and dissipation ranges. *Radio Sci.* **13**, 953–961.

Hitchfeld, W., and Stauder, M. (1965). The temperature of hailstones. Alberta Hail Studies, 1964, Sci. Rep. MW-42. Stormy Weather Group, McGill Univ. Montreal, Canada.

Hodara, H. (1966). Laser wave propagation through the atmosphere. *Proc. IEEE* **54**, 953–961d.

Holmes, C. R., Szymanski, E. W., Szymanski, S. J., and Moore, C. B. (1980). Radar and acoustic study of lightning. *J. Geophys. Res.* **85**, 7517–7532.

Hondl, K. D. (1990). Application and evaluation of wind retrieval algorithms for a single Doppler radar. A Master of Science Thesis in Meteorology, University of Oklahoma, Norman, Okla-homa.

Houze, R. A. (1989). Observed structure of mesoscale convective systems and implications for large-scale heating." *Q. J. R. Meteorol. Soc.* **115**, 425–461.

Houze, R. A., Jr., Rutledge, S. A., Biggerstaff, M. I., and Smull, B. F. (1989). Interpretation of Doppler weather radar displays of midlatitude mesoscale convective systems. *Bull. Amer. Meteorol. Soc.* **70**, 608–619.

Hubert, P., LaRoche, P., Eybert-Berard, A., and Barret, L. (1984). Triggered lightning in New Mexico. *J. Geophys. Res.* **89**, 2511–2521.

Huff, F. A. (1970). Sampling errors in measurement of mean precipitation, *J. Appl. Meteorol.* **9**, 35–44.

Hufnagel, R. E. (1978). "The Infrared Handbook" (W. L. Wolfe and G. J. Zissis, eds.), Libr. Congr. Cat. No. 77–90786, Chapter VI, U.S. Govt. Printing Office, Washington, D.C.

Hull, A. W. (1921). The Magnetron. *J. AIEE* **40**, pp. 715.

Humphries, R. G., and Barge, B. L. (1979). Polarization and dual-wavelength radar observations of the bright band. *IEEE Trans. Geosci. Electron.* **GE-17**, 190–195.

Husson, D., and Pointin, Y. (1989). Quantitative estimation of the hail fall intensity with a dual polarization radar and a hailpad network. *Prepr., Radar Meteorol. Conf., 24th, 1989* pp. 318–321.

Hynek, D. P. (1990). Use of clutter residue editing maps during the Denver 1988 terminal Doppler weather radar (TDWR) tests. *Project Report*, ATC-169, MIT Lincoln Laboratory, MASS, 65.

Illingworth, A. J., and Caylor, I. J. (1989). Cross polar observation of the bright band, *Prepr., Radar Meteorol. Conf., 24th, 1989* pp. 323–327.

Illingworth, A. J., Goddard, J. W. F., and Cherry, S. M. (1987). Polarization radar studies of precipitation development in convective storms. *Q. J. R. Meteorol. Soc.* **113**, 469–489.

Ioannidis, G. A., and Hammers, D. E. (1979). Optimum antenna polarizations for target discri-mination in clutter. *IEEE Trans. Antennas Propag.* **AP-27**, 357–363.

Istok, M. J. (1981). Analysis of Doppler spectrum broadening mechanisms in thunderstorms. *Prepr., Radar Meteorol. Conf., 20th, 1981* pp. 454–458.

Istok, M. J. (1983). "An analysis of the relation between Doppler spectrum width and thunderstorm turbulence." *MS. Thesis.*, University of Oklahoma, Norman, Oklahoma. pp. 81.

Istok, M. J., and Doviak, R. J. (1986). Analysis of the relation between Doppler spectral width and thunderstorm turbulence. *J. Atmos. Sci.* **43**, 2199–2214.

Jameson, A. R. (1985). Microphysical interpretation of multiparameter radar measurements in rain. Part III: Interpretation and measurement of propagation differential phase shift between orthogonal linear polarizations. *J. Atmos. Sci.* **42**, 607–614.

Jameson, A. R. (1987). Relations among linear and circular polarization parameters measured in canted hydrometeors. *J. Atmos. Oceanic Technol.* **4**, 634–645.

Jameson, A. R., and Davé, J. H. (1988). An interpretation of circular polarization measurements affected by propagation differential phase shift. *J. Atmos. Oceanic Technol.* **5**, 405–415.

Jameson, A. R., and Mueller, E. A. (1985). Estimation of differential phase shift from sequential orthogonal linear polarization radar measurements. *J. Atmos. Oceanic Technol.* **2**, 133–137.

Janssen, L. H., and Van der Spek, (1985). The shape of Doppler spectra from precipitation. *IEEE Trans. Aerosp. Electron. Syst.* **AES-21**, 208–219.

JDOP, (1979). "Final report on the Joint Doppler Operational Project (JDOP) 1976–1978." *NOAA Tech. Memo. ERL-NSSL-86*, National Severe Storms Laboratory, Norman, Oklahoma. 84 pp.

Jenkins, G. M., and Watts, D. G. (1968). "Spectral Analysis and Its Application." Holden Day, San Francisco, California.

Johnson, C. C. (1965). "Field and Wave Electrodynamics." McGraw-Hill, New York.

Jordan, E. C., and Balmain, K. G. (1968). "Electromagnetic Waves and Radiating Systems." Prentice-Hall, Englewood Cliffs, New Jersey.

Jorgensen, D. P., and Willis, P. T. (1982). A Z–R relationship for hurricanes. *J. Appl. Meteorol.* **21**, 356–366.

Joss, J., and Waldvogel, A. (1970). A method to improve the accuracy of radar-measured amounts of precipitation. *Prepr., Radar Meteorol. Conf., 14th, 1970* pp. 237–238.

Kedem, B., Chiu, L. S., and North, G. R. (1990). Estimating time mean areal average rainfall: A mixed distribution approach. *J. Geoph. Research* **95**, 1965–1972.

Keeler, R. J., and Carbone, R. E. (1986). A modern pulsing/processing technique for meteorological Doppler radars. *Prepr., Radar Meteorol. Conf., 23rd, 1986* JP357–JP360.

Kerr, D. W. (1951). "Propagation of Short Radio Waves." McGraw-Hill, New York.

Kessinger, C. J., Ray, P. S., and Hane, C. E. (1987). The 19 May 1977 Oklahoma squall line. Part I: A multiple Doppler analysis of convective and stratiform structure. *J. Atmos. Sci.* **44**, 2840–2864.

Kessler, E., ed. (1982). "Thunderstorms: A Social, Scientific, and Technological Documentary," Vols. I, II, and III. NOAA, U.S. Dept. of Commerce, U.S. Govt. Printing Office, Washington, D.C.

Kessler, E. (1985). Wind shear and aviation safety. *Nature* **315**, 179–180.

Kessler, E. (1987). Kinematic effect of vertical drafts on precipitation near earth's surface. *Mon. Weather Review* **115**, 2862–2864.

Kessler, E. (1991). *Aircraft Accident Report.* E. Kessler, Rt.2 Box 137, Purcell, Oklahoma. 92 pp.

Kingsmill, D. E., and Wakimoto, R. M. (1991). Kinematic, dynamic, and thermodynamic analysis of a weakly sheared thunderstorm over northern Alabama. *Mon. Wea. Rev.* **119**, 262–297.

Klemp, J. B., Wilhelmson, R. B., and Ray, P. S. (1981). Observed and numerically simulated structure of a mature supercell thunderstorm. *J. Atmos. Sci.* **38**, 1558–1580.

Klingle, D. L., Smith, D. R., and Wolfson, M. M. (1987). Gust front characteristics as detected by Doppler radar. *Mon. Weather Rev.* **115**, 905–918.

Knight, N. C. (1986). Hailstone shape factor and its relation to radar interpretation of hail. *J. Clim. Appl. Meteorol.* **25**, 1956–1958.

Knupp, K. R., and Cotton, W. R. (1982). An intense, quasi-steady thunderstorm over mountainous terrain, II, Doppler radar observations of the storm morphological structure. *J. Atmos. Sci.* **39**, 343–358.

Kolmogoroff, A. N. (1941). Dissipation of energy in the locally isotropic turbulence. *Dokl. Akad. Nauk SSSR* **32**, 16–18.

Komabayasi, M., Gonda, T., and Isono, K. (1964). Lifetime of water drops before breaking and size distribution of fragment droplets. *J. Meteorol. Soc. Jpn.* **42**, 330–340.

Koscielny, A. J., Doviak, R. J., and Rabin, R. (1982). Statistical considerations in the estimation of divergence from single-Doppler radar and application to prestorm boundary-layer observations. *J. Appl. Meteorol.* **21**, 197–210.

Koscielny, A. J., Doviak, R. J., and Zrnić, D. S. (1984). An evaluation of the accuracy of some radar wind profiling techniques. *J. Atmos. Oceanic Tech.* **1**, 309–320.

Kraus, J. D. (1966). "Radio Astronomy." McGraw Hill, New York.

Krehbiel, P. R., and Brook, M. (1979). A broad-band noise technique for fast-scanning radar observations of clouds and clutter targets. *IEEE Trans. Geosci. Electron., Spec. Issue Radio Meteorol.* **GE-17**, 196–204.

Krehbiel, P. R., Brook, M., and McCrory, R. (1979). Analysis of the charge structure of lightning discharges to ground. *J. Geophys. Res.* **84**, 2432–2456.

Krehbiel, P. R., Rison, W., McCrary, S., Blackman, T., and Brook, M. (1991). Dual-polarization radar observations of lightning echoes and precipitation alignment at 3 cm wavelength. *Prepr., Radar Meteorol. Conf., 25th, 1991* pp. 901–904.

Kropfli, R. A., Katz, I., Konrad, T. G., and Dobson, E. B. (1968). Simultaneous radar reflectivity measurements and refractive index spectra in the clear atmosphere. *Prepr., Radar Meteorol. Conf., 13th, 1968* pp. 270–271.

Kundu, P. K. (1990). "Fluid Mechanics." Academic Press Inc. San Diego, California. 638 pp.

Labitt, M. (1981). "Coordinated radar and aircraft Observations of turbulence," Proj. Rep. ATC 108. Massachusetts Institute of Technology, Lincoln Lab., Cambridge.

Lalas, D. P., and Einaudi, F. (1974). On the correct use of the wet-adiabatic lapse rate in stability criteria of a saturated atmosphere. *J. Appl. Meteorol.* **13**, 318–324.

Lamb, Sir Horace, (1932). "Hydrodynamics." Sixth Edition. Cambridge University Press, Dover Publications, Inc. 1945, 738 pp.

Larkin, R. P. (1991). Sensitivity of NEXRAD algorithms to echoes from birds and insects. *Prepr., Radar Meteorol. Conf., 25th, 1991* pp. 203–205.

Laws, J. O., and Parsons, D. A. (1943). The relationship of raindrop size to intensity. *Trans. Am. Geophys. Union* **24**, 452–460.

Leber, G. W., Merrit, C. J., and Robertson, J. P. (1961). WSR-57 analysis of heavy rains. *Proc., Weather Radar Conf., 9th, 1961* pp. 102–105.

Lee, A. C. L. (1988). The influence of vertical air velocity on the remote microwave measurement of rain. *J. Atmos. Oceanic Tech.* **5**, 727–735.

Lee, J. T. (1977). "Application of Doppler Radar to Turbulence Measurements Which Affect Aircraft," Final Rep. No. FAA-RD-77-145. FAA Syst. Res. Dev. Serv., Washington, D.C.

Lee, J. T., and Thomas, K. (1989). Turbulence Spectral widths view angle independence as observed by Doppler Radar. Technical Report, DOT/FAA/SA-89-2. 47 pp.

Leitao, M. J., and Watson, P. A. (1984). Application of dual linearly polarized radar data to prediction of microwave path attenuation at 10–30 GHz. *Radio Sci.* **19**, 209–221.

Lemon, L. R., and Doswell, C. A., III (1979). Severe thunderstorm evolution and mesocyclone structure as related to tornadogenesis. *Mon. Weather Rev.* **107**, 1184–1197.

Lenschow, D. H., and Wyngaard, J. C. (1980). Mean-field and second-moment budgets in a baroclinic, convective boundary. *J. Atmos. Sci.* **37**, 1313–1326..

Lhermitte, R. M. (1963). "Motions of Scatterers and the Variance of the Mean Intensity of Weather Radar Signals." SRRC-RR-63-57. Sperry-Rand Res. Cent., Sudbury, Massachusetts.

Lhermitte, R. M. (1968). Turbulent air motion as observed by Doppler radar. *Prepr., Radar Meteorol. Conf., 13th, 1968* pp. 14–17.

Lhermitte, R. M. (1969). Note on the observation of small-scale atmospheric turbulence by Doppler radar technique. *Radio Sci.* **4**, 1241–1246.

Lhermitte, R. M. (1970). Dual-Doppler radar observations of convective storm circulation. *Prepr., Radar Meteorol. Conf., 14th, 1970* pp. 153–156.

Lhermitte, R. M. (1982). Doppler radar observations of triggered lightning. *Geophys. Res. Lett.* **9**, 712–715, 1982.

Lhermitte, R. M., and Atlas, D. (1961). Precipitation motion by pulse Doppler radar. *Proc. Weather Radar Conf., 9th, 1961* pp. 218–223.

Ligda, M. G. H. (1950). Lightning detection by radar. Bull. Am. Meteorol. Soc. **31**, 279–283.

Ligda, M. G. H. (1956). The radar observation of lightning. *J. Atmos. Terr. Phys.* **9**, 329–346.

Lilly, D. K., Waco, D. E., and Adelfang, S. I. (1974). Stratospheric mixing estimated from high-altitude turbulence measurements. *J. Appl. Meteorol.* **13**, 488–493.

Linden, P. F., and Simpson, J. E. (1985). Microburst: Hazard for aircraft. *Nature*, **317**, 601–602.

Lindzen, R. S., and Tung, K. K. (1976). Banded convective activity and ducted gravity waves. *Mon. Wea. Rev.* **104**, 1602–1617.

List, R. (1986). Properties and growth of hailstones. *In* "Thunderstorm Morphology and Dynamics." (E. Kessler, ed.), University of Oklahoma Press, Norman, Oklahoma. 259–276.

Long, M. W. (1975). "Radar Reflectivity of Land and Sea." Heath, Indianapolis, Indiana.

MacCready, P. B., Jr. (1962). Turbulence measurements by sailplane, *J. Geophys. Res.* **67**, 1041–1050.

MacCready, P. B., Jr. (1964). Standardization of gustiness values from aircraft. *J. Appl. Meteorol.* **3**, 439–449.

Maddox, R. A. (1980). Mesoscale convective complexes. *Bull. Amer. Meteorol. Soc.* **61**, 1374–1387.

Maddox, R. A. (1983). Large-scale meteorological conditions associated with midlatitude, mesoscale convective complexes. *Mon. Weather Rev.* **111**, 1475–1493.

Mahapatra, P. R., Doviak, R. J., and Zrnić, D. S. (1991). Multi-sensor observations of an atmospheric undular bore. *Bull. Amer. Meteorol. Soc.* **72**, 1468–1480.

Marshall, J. S., and Gunn, K. L. S. (1952). Measurements of snow parameters by radar. *J. Meteorol.* **9**, 299–307.

Marshall, J. S., and Hitschfeld, W. (1953). Interpretation of the fluctuating echo from randomly distributed scatterers. *Can. J. Phys.* **31**, Pt. I, 962–995.

Marshall, J. S., and Palmer, W. (1948). The distribution of raindrops with size. *J. Meteorol.* **5**, 165–166.

Marshall, J. S., Hitschfeld, W., and Gunn, K. L. S. (1955). Advances in radar weather. *Adv. Geophys.* **2**, 1–56.

Marshall, J. M., Peterson, A. M., and Barnes, Jr., A. A. (1972). Combined radar-acoustic sounding system. *Appl. Opt.* **11**, 108–112.

Mason, B. J. (1971). "The Physics of Clouds." Oxford Univ. Press (Clarendon), London and New York, 672.

Masuda, Y. (1988). Influence of wind and temperature on the height limit of a radio acoustic sounding system. *Radio Sci.* **23**, 647–654.

Matson, R. J., and Huggins, A. W. (1980). The direct measurement of the sizes, shapes, and kinematics of falling hailstones. *J. Atmos. Sci.* **37**, 1107–1125.

Maxworthy, T. (1980). On the formation of nonlinear internal waves from gravitational collapse of mixed regions in two and three dimensions. *J. Fluid Mech.* **96**, 47–64.

May, P. T., and Strauch, R. G. (1989). An examination of wind profiler signal processing algorithms. *J. Atmos. Oceanic Tech.* **6**, 731–735.

May, P. T., Moran, K. P., and Strauch, R. G. (1989). The accuracy of RASS temperature measurements. *J. Appl. Meteor.* **28**, 1329–1335.

May, P. T., Strauch, R. G., Moran, K. P., and Ecklund, W. L. (1990). Temperature sounding by RASS with wind profiler radars: A preliminary study. *IEEE Trans. Geosci. Remote Sens.* **28**, 19–27.

Mazur, V. (1989). Triggered lightning strikes to aircraft and natural intracloud discharges. *J. Geophys. Res.* **94**, 311–3325.

Mazur, V., Fisher, B. D., and Gerlach, J. C. (1984). Lightning strikes to an airplane in thunderstorms. *J. Aircraft* **21**, 607–611.

Mazur, V., Zrnić, D. S., and Rust, W. D. (1985). Lightning channel properties determined with a vertically pointing Doppler radar. *J. Geophys. Res.* **90**, 6165–6174.

Mazur, V., Rust, W. D., and Gerlach, J. C. (1986). Evolution of lightning flash density and reflectivity structure in a multicell thunderstorm. *J. Geophys. Res.* **91**, 8690–8700.

McAnelly R. L., and Cotton, W. R. (1989). The Precipitation life cycle of mesoscale convective complexes over the central United States. *Mon. Wea. Rev.* **117**, 784–807.

McCarthy, J., and Blick, E. F. (1979). An airport wind shear detection and warning system. *Prepr., WMO Tech. Conf. Aviat. Meteorol. (TECAM), 1979* pp. 1–35.

McCarthy, J., and Serafin, R. (1984). The Microburst: A Hazard to Aviation. *Weatherwise* **37**, 120–127.

McCarthy, J., and Wilson, J. W. (1984). The Microburst as a Hazard to Aviation: Structure, Mechanisms, Climatology and Nowcasting. *Prepr., Nowcasting II Symposium, 1984* pp. 21–30.

McCarthy, J., Roberts, R., and Schreiber, W. (1984). JAWS data collection, analysis highlights, and Microburst Statistics. *Prepr., Radar Meteorol. Conf., 21st, 1984* pp. 596–601.

McCaul, E. W., Jr. (1991). Buoyancy and shear characteristics of hurricane–tornado environments. *Mon. Wea. Rev.* **119**, 1954–1978.

McCormick, C.G., and Hendry, A. (1975). Principles for the radar determination of the polarization properties of precipitation. *Radio Sci.,* **10**, 421–434.

McCormick, G. C., and Hendry, A. (1979). Radar measurement of precipitation-related depolarization in thunderstorms. *IEEE Trans. Geosci. Electron., Spec. Issue* **GE-17**, 142–150.

McCormick, G. C., and Hendry, A. (1985). Optimal polarizations for partially polarized backscatter. *IEEE Trans. Antennas Propag.* **AP-3**, 33–40.

Meneghini R., and Kozu, T. (1990). "Spaceborne Weather Radar." Artech House, Boston. pp. 199.

Merritt, M. W. (1984). Automatic velocity de-aliasing for real-time applications. *Prepr., Radar Meteorol. Conf., 22nd, 1984* pp. 528–533.

Mie, G. (1908). Beiträge zur Optik trüber Medien, speziell kolloidaler Metallösungen. [Contribution to the optics of suspended media, specifically colloidal metal suspensions.] *Ann. Phys.* 377–445.

Miller, K. S., and Rochwarger, M. C. (1972). A covariance approach to spectral moment estimation. *IEEE Trans. Inf. Theory* **IT-18**, 558–596.

Morgan, G. M., Jr., and Summers, P. W. (1986). Hailfall and hailstorm characteristics. *In* "Thunderstorm Morphology and Dynamics," (E. Kessler, ed.), University of Oklahoma Press, Norman, Oklahoma. 237–257.

Morse, P. M., and Feshbach, H. (1953). "Methods of Theoretical Physics," Vols. I and II. McGraw-Hill, New York.

Mueller, E. A. (1984). Calculation procedure for differential propagation phase shift. *Prepr., Radar Meteorol. Conf., 22nd 1984* pp. 397–399.

Nathanson, F. E. (1969). "Radar Design Principles." McGraw-Hill, New York.

Nathanson, F. E., and Smith, P. L. (1972). A modified coefficient for the weather radar equation. *Prepr., Radar Meteorol. Conf.,* 15th, 1972, pp. 228–230.

Neff, E. L. (1977). How much rain does a rain gage? *J. Hydrol.* **35**, 213–220.

Neiman, P. J., May, P. T., and Shapiro, M. A. (1992). Radio acoustic sounding system (RASS) and Wind Profiler observations of lower- and middle-tropospheric weather systems. *Monthly Weather Review*, **120**, 2298–2313.

Nelson, S. P. (1983). The influence of storm flow structure on hail growth. *J. Atmos. Sci.,* **40**, 1965–1983.

Nelson, S. P., and Brown, R. A. (1987). Error sources and accuracy of vertical velocities computed from multiple-Doppler radar measurements in deep convective storms. *J. Atmos. & Oceanic Tech.* **4**, 233–238.

Nelson, D. P., and Brown, R. A. (1982). Multiple Doppler radar derived vertical velocities in thunderstorms: Part I–Error analyses and solution techniques. *NOAA Tech. Memo ERL NSSL-94.* National Severe Storms Laboratory, Norman, Oklahoma.

Neter, J., and Wasserman, W. (1974). "Applied Linear Statistical Models." Richard D. Irwin, Inc., Homewood, Illinois.

Newton, C. W. (1950). Structures and mechanisms of the prefrontal squall line. *J. Meteorol.* **7**, 210–222.

Norbury, J. R., and White, W. J. K. (1972). Microwave attenuation at 35.8 GHz due to rainfall. *Electron. Lett.* **8**, 91–92.

Novlan, D. J., and Gray, W. M. (1974). Hurricane spawned tornadoes. *Mon. Weather Rev.* **102**, 476–488.

Nutten, B., Amayenc., P., Chong, M., Hauser, D., Rouse, F., and Testud, J. (1979). TThe Ronsard radars: A versatile C-band dual Doppler facility. *IEEE Trans. Geosci. Electron.* **GE-17**, 281–287.

Oguchi, T. (1983). Electromagnetic wave propagation and scattering in rain and other hydrometeors. *Proc. IEEE,* **71**, 1029–1078.

Ogura, Y. (1963). A review of numerical modeling research on small scale convection in the atmosphere in Severe Local Storms, (D. Atlas, ed.), *Meteorol. Mono.* **5**, 247 pp.

Ogura, Y., and Chen, Y. L. (1977). A life history of an intense mesoscale convective storm in Oklahoma. *J. Atmos. Sci.* **34**, 1458–1476.

Ogura, Y., and Phillips, N. A. (1962). Scale analysis of deep and shallow convection in the atmosphere. *J. Atmos. Sci.* **19**, 173–179.

Okabe, K. (1928). Production of intense extra-short electromagnetic waves by "split-anode magnetron" (The 3rd report). *J. Inst. Electr. Eng. JPN.* **48**, 284–290.

Ottersten, H. (1969a). Atmospheric structure and radar backscattering in clear air. *Radio Sci.* **4**, 1179–1193.

Ottersten, H. (1969b). Mean gradient of potential refractive index in turbulent mixing and radar detection of CAT. *Radio Sci.* **12**, 1247–1249.

Panchev, S. (1971). "Random Functions and Turbulence." Pergamon. Oxford.

Panofsky, H. A. (1969). Spectra of atmospheric variables in the boundary layer. *Radio Sci.* **4**, 1101–1109.

Papoulis, A. (1965). (2nd ed., 1984). "Probability, Random Variables, and Stochastic Processes." McGraw-Hill, New York.

Papoulis, A. (1977). 'Signal Analysis." McGraw-Hill, New York.

Parsons, D. B., Smull, B. F., Lilly, D. K. (1990). Mesoscale Organization and Processes: Panel Report. *In* "Radar in Meteorology" (D. Atlas, ed.), Amer. Meteorol. Soc., 461–476.

Passarelli, R. E., Jr., and Zrnić, D. S. (1989). An expression for phase noise. *Prepr., Radar Meteorol. Conf., 24th, 1989* pp. 433–435.

Pekeris, C. L. (1947). Note on the scattering of radiation in an inhomogeneous medium. *Phys. Rev.* **76**, p. 268.

Peters, G., Timmerman, H., and Hinzpeter, H. (1983). Temperature sounding in the planetary boundary layer by RASS-system analysis and results. *Intl. J. Remote Sensing* **20**, 49–63.

Plank, V. G. (1977). "Hydrometeor Data and Analytical–Theoretical Investigations Pertaining to the SAMS Rain Erosion Program of the 1972–73 Season at Wallops Island, Virginia," AFGL/SAMS Rep. No. 5, AFGL-TR-77-0149. Air Force Geophys. Lab., Hanscom AFB, Massachusetts.

Pointin, Y., Ramon, D., and Fournet-Fayard, J. (1988). Radar differential reflectivity Z_{DR}: a real-case evaluation of errors induced by antenna characteristics. *J. Atmos. Oceanic Tech.* **5**, 416–423.

Probert-Jones, J. R. (1984). Resonance component of backscattering by large dielectric spheres. *J. Opt. Soc. Am.* **A1**, 822–883.

Probert-Jones, J. R. (1990). A history of radar meteorology in the United Kingdom. In "Radar in meteorology." *American Met. Soc.*, Boston, pp. 54–60.

Pruppacher, H. R., and Beard, K. V. (1970). A wind tunnel investigation of the internal circulations and shape of water drops falling at terminal velocity in air. *Quart. J. Roy. Meteorol. Soc.*, **96**, 247–256.

Pruppacher, H. R., and Pitter, R. L. (1971). A semi-empirical determination of the shape of cloud and raindrops. *J. Atmos. Sci.* **28**, 86–94.

Purdom, J. F. W. (1982). Subjective interpretation of geostationary satellite data for nowcasting. *In* "Nowcasting," (K. A. Browning, ed), Academic, New York; chaps, 3,4; pp. 149–166.

Purdom, J. F. W., and Marcus, K. (1982). Thunderstorm trigger mechanisms over the southeast U.S. *Prepr., Severe Local Storms Conf., 12th, 1982* pp. 487–488.

Rabin, R. M., and Doviak, R. J. (1989). Meteorological and Astronomical Influences on Radar Reflectivity in the Convective Boundary Layer. *J. Appl. Meteorol.* **28**, 1226–1235.

Rabin, R., and Zrnić, D. S. (1980). Subsynoptic-scale vertical wind revealed by dual-Doppler radar and VAD analysis. *J. Atmos. Sci.* **37**, 644–654.

Rabin, R. M., Doviak, R. J., and Sundara-Rajan, A. (1982). Doppler radar observations of momentum flux in a cloudless convective layer with rolls. *J. Atmos. Sci.* **39**, 851–863.

Ray, P. S. (1972). Broadband complex refractive indices of ice and water. *Appl. Opt.* **11**, 1836–1844.

Ray, P. S., Ziegler, C. L., Serafin, R. J., and Bumgarner, W. (1980). Single- and multiple-Doppler radar observations of tornadic storms. *Mon Weather Rev.* **108**, 1607–1625.

Ray, P. S., Johnson, B. C., Johnson, K.W., Bradberry, J. S., Stephens, J. J., Wagner, K. K.,

Wilhelmson, R. B., and Klemp, J. B. (1981) The morphology of several tornadic storms on 20 May 1977. *J. Atmos. Sci.* **38**, 1643–1663.

Reed, I. S. (1962). On a moment theorem for complex Gaussian processes. *IRE Trans. Inf. Theory* **IT-8**, 194–195.

Reinking, R. F., Doviak, R. J., and Gilmer, R. O. (1981). Clear-air roll vortices and turbulent motions as detected with an airborne gust probe and dual-Doppler radar. *J. Appl. Meteorol.* **20**, 678–685.

Reiter, E. R. (1970). Recent advances in the study of clear-air turbulence (CAT). *Rev. Meteorol. Aeronaut.* **30**, 10–13.

Reiter, E. R., and Burns, A. (1960). The structure of clear-air turbulence derived from "TOPCAT" aircraft measurements. *J. Atmos. Sci.* **23**, 206–212.

Rhyne, R. H., and Steiner, R. (1964). Power spectral measurements of atmospheric turbulence in severe storms and cumulus clouds. *NASA Tech. Note* NASA TD D-2469, 1–48.

Rice, P. L., Langley, A. G., Norton, K. A., and Barses, A. P. (1966). "Transmission Loss Predictions for Tropospheric Communication Circuits." *Tech. Note 101.* Natl. Bur. Stand., Boulder, Colorado.

Richards, W. G., and Crozier, C. L. (1981). "Precipitation Measurement with C-Band Weather Radar in Southern Ontario," Int. Rep. No. APRB 112, p. 35. Cloud Phys Res. Div., Atmos. Environ. Serv. Downsview, Ontario, Canada.

Riley, J. R. (1989). Remote sensing in entomology. *Ann. Rev. Entomol.*, **34**, 247–271.

Rinehart, R. E. (1979). Internal storm motion from a single non-Doppler weather radar. Ph.D. Dissertation, Colorado State University, Ft. Collins.

Rinehart, R. E. (1991). "Radar for Meteorologists." Rinehart, P.O. Box 6124, Grandd Forks, ND, 58206–6124, p. 334.

Rinehart, R. E., and Isaminger, M. A. (1986). Radar Characteristics of Microbursts in the Mid-South, *Prepr., Radar Meteorol. Conf., and Cloud Physics Conf., 23rd, 1986,* pp. J116–J119.

Roberts, R. D., and Wilson, J. W. (1986). Nowcasting microburst events using single Doppler radar data. *Prepr. Radar Meteorol. Conf., 23rd, 1986,* pp. 14–17.

Rogers, R. R. (1964). An extension of the Z–R relation for Doppler radar. *Proc., Radar Meteorol. World Conf., 1964* pp. 158–161.

Rogers, R. R. (1971). The effect of variable reflectivity on weather radar measurements, *Q. J. R. Meteorol. Soc,* **97**, 154–167.

Rogers, R. R. (1976). "A Short Course in Cloud Physics." Pergamom Press, Oxford, 227.

Rogers, R. R. (1990). The Early Years of Doppler Radar in Meteorology. *In* "Radar in Meteorology" (D. Atlas, ed.), Chapter 16 Am. Met. Soc., Boston.

Rogers, R. R., and Tripp, B. R. (1964). Some radar measurements of turbulence in Snow. *J. Appl Meteorol.* **3**, 603–610.

Rosenfeld, D., Atlas, D., and Short, D. A. (1990). The estimation of convective rainfall by area integrals: 2. The height-area rainfall threshold (HART) method. *J. Geoph. Research* **95**, 2161–2176.

Röttger, J. (1981). Investigations of lower and middle atmosphere dynamics with spaced antenna drift radars. *J. Atmos. Terr. Phys.* **43**, 277–292.

Röttger, J., and Schmidt, G. (1979). High-resolution VHF radar sounding of the troposphere and stratosphere. *IEEE Trans. Geosci. Electron.* **GE-17**, 182–189.

Röttger, T., Czechowsky, P., and Schmidt, G. (1981). First low-power VHF radar observations of tropospheric, stratospheric and mesospheric winds and turbulence at the Arecibo observatory. *J. Atmos. Terr. Phys.* **43**, 789–800.

Rottman, J. W., and Simpson, J. E. (1989). The formation of internal bores in the atmosphere: A laboratory Model. *Q. J. Roy. Meteorol. Soc.* **115**, 941–963.

Rowland, J. R., and Arnold, A. (1975). Vertical velocity structure and the geometry of clear air convective elements. *Prepr. Radar Meteorol. Conf., 16th, 1975,* pp. 296–303.

Rust, W. D., Taylor, W. L., MacGorman, D. R., and Arnold, R. T. (1981). Research on electrical properties of severe thunderstorms in the great plains. *Bull. Am. Meteorol. Soc.* **62**, 1286–1293.

Rutkowski, W., and Fleisher, A. (1955). R-meter: An instrument for measuring gustiness. *MIT Weather Radar Research Rep.* **24,**

Ryde, J. W. (1946). The attenuation end radar echoes produced at centimeter wave-lengths by various meteorological phenomena. In "Meteorological Factors in Radio-Wave Propagation." Report of a conference held 8 April 1946, by the Phys. Soc. and the Roy. Meteorol. Soc., published by the Phys. Soc., London, S.W.7, England.

Sachidananda, M., and Zrnić, D. S. (1985). ZDR measurement considerations for a fast scan capability radar. *Radio Sci.* **20,** 907–922.

Sachidananda, M., and Zrnić, D. S. (1986). Differential propagation phase shift and rainfall rate estimation. *Radio Sci.* **21,** 235–247.

Sachidananda, M., and Zrnić, D. S. (1986). Recovery of spectral moments from overlaid echoes in a Doppler weather radar. *IEEE Trans. Geosci. Remote Sens.* **GE-24,** 751–764.

Sachidananda, M., and Zrnić, D. S. (1987). Rain rate estimates from differential polarization measurements. *J. Atmos. Oceanic Tech.* **4,** 588–598.

Sachidananda, M., and Zrnić, D. S. (1988). Efficient processing of alternately polarized radar signals. *J. Atmos. Oceanic Tech.* **4,** 1310–1318.

Sato, T., and Woodman, R. F. (1982). Spectral parameter estimation of CAT radar echoes in the presence of fading clutter. *Radio Sci.* **17,** 817–826.

Sauvageot, H. (1982). "Radarmétéorologie," Eyrolles, Paris, France. 296 pp.

Schaefer, J. T. (1986). The dryline. *In* "Mesoscale Analysis and Forecasting," (P. S. Ray, ed.), Amer. Meteorol. Soc., 549–572.

Schlesinger, R. E. (1982). Effects of mesoscale lifting precipitation and boundary-layer shear on severe storm dynamics in a three dimensional numerical modeling study. *Prepr., Severe Local Storms Conf., 12th, 1982* pp. 536–541.

Schneider, J. M. (1991). Dual Doppler Measurement of a Sheared, Convective Boundary Layer. Dissertation, University of Oklahoma, Norman, Oklahoma. 134 pp.

Scotti, R. S., and Corcos, G. M. (1969). Measurement on the growth of small disturbances in a stratified shear layer. *Radio Sci.* **4,** 1309–1313.

Segal, M., Pielke, R. A., and Mahrer, Y. (1984). Evaluation of surface sensible heat flux effects on the generation and modification of mesoscale circulations. *Proc., Nowcasting II Symp., 1984* pp. 263–270.

Seliga, T. A., Aydin, K., and Direskeneli H. (1986). Disdrometer measurements during an intense rainfall event in central Illinois: Implications for differential reflectivity radar observations. *J. Clim. Appl. Meteor.,* **25,** 69–76.

Sekhon, R. S., and Srivastava, R. C. (1970). Snow-size spectra and radar reflectivity. *J. Atmos. Sci.,* **27,** 299–307.

Sekhon, R. S., and Srivastava, R. C. (1971). Doppler radar observations of drop size distributions in a thunderstorm. *J. Atmos. Sci.* **28,** 983–994.

Seliga, T.A., and Bringi, V. N. (1976). Potential use of radar differential reflectivity measurements at orthogonal polarizations for measuring precipitation. *J. Appl. Meteorol.* **15,** 69–76.

Seliga, T. A., and Bringi, V. N. (1978). Differential reflectivity and differential phase shift: Applications in radar meteorology. *Radio Sci.* **13,** 271–275.

Setzer, D. (1970). Computed transmission through rain at microwave and visible frequencies. *Bell Syst. Tech. J.* **49,** 1873–1892.

Shapiro, L. J., and Willoughby, H. E. (1982). The response of balanced hurricanes to local sources of heat and momentum. *J. Atmos. Sci.* **39,** 378–394.

Shapiro, M. A., Hampel, T., Rotzoll, D., and Mosher, F. (1985). The frontal hydraulic head: A micro − α scale (~1 km) triggering mechanism for mesoconvective weather systems. *Mon. Wea. Rev.* **113,** 1166–1183.

Shea, L. V. (1965). "Handbook of Geophysics and Space Environments," pp. 2019, Air Force Cambridge Res. Lab., Office Aerosp. Res. USAF, Washington, D.C.

Sherman, J. W. (1970). Aperture-antenna analysis. *In* "Radar Handbook" (M. I. Skolnik, ed.), Chapter 9, McGraw-Hill, New York.

Shuman, F. G. (1957). Numerical methods and weather predictions, II. Smoothing and fitting. *Month. Weath. Rev.* **85**, 357–361.

Sikdar, D. N., Schlesinger, R. E., and Anderson, C. E. (1974). Severe storm latent heat release: Comparison of radar estimate versus a numerical experiment. *Mon. Weather Rev.* **102**, 455–465.

Sinclair, D. (1947). Light scattering by spherical particles. *J. Opt. Soc. Am.* **37**, 475–480.

Sinclair, P. C. (1973). Severe storm air velocity and temperature structure from penetrating aircraft, *Prepr., Conf. Severe Local Storms, 8th, 1973* pp. 25–32.

Sinclair, P. C. (1974). Severe storm turbulent energy structure. *Conf. Aerosp. Aeronaut. Meteorol., 6th, 1974, El Paso, Texas* (unpublished manuscript).

Sirmans, D., and Bumgarner, W. (1975). Numerical comparison of five mean frequency estimators. *J. Appl. Meteorol.* **14**, 991–1003.

Sirmans, D., and Doviak, R. J. (1973). Meteorological radar signal intensity estimation. NOAA Tech. Memo. ERL NSSL-64, National Severe Storms Laboratory, Norman, OK.

Sirmans, D., Zrnić, D.S., and Bumgarner, W. (1976). Extension of maximum unambiguous Doppler velocity by use of two sampling rates. *Prepr., Radar Meteorol. Conf., 17th, 1976* pp. 23–28.

Skolnik, M. I., ed. (1970). "Radar Handbook." McGraw-Hill, New York.

Smith, P. (1984). Equivalent radar reflectivity factor for snow and ice particles. *J. Clim. and Appl. Meteorol.* **23**, 1258–1260.

Smith, P. L. (1961). Remote measurement of wind velocity by the electromagnetic–acoustic probe I: System analysis. *Proc., Natl. Conv. Mil. Electron., 5th, 1961* pp. 48–53.

Smith, S. D. (1986). On the estimation of divergence in the prestorm boundary layer using single Doppler radar. Thesis, M. Sci. in Meteorol. Uni. of Oklahoma, Norman, Oklahoma. 132 pp.

Smith, S. D., and Doviak, R. J. (1984). Doppler velocity bias due to beam blockage by ground targets. *Prepr., Radar Meteorol. Conf., 22nd, 1984*, pp. 534–537.

Smythe, G. R., and Zrnić, D. S. (1983). Correlation analysis of Doppler radar data and retrieval of the horizontal wind. *J. Climate and Appl. Meteorol.* **22**, 297–311.

Sorbjan, J. (1989). "Structure of the Atmospheric Boundary Layer." Prentice Hall, Englewood Cliffs, New Jersey.

Spahn, J. F., and Smith, P. L. Jr. (1976). Some characteristics of hailstone size distributions inside hailstorms. *Prepr., Radar Meteorol. Conf., 17th, 1976*, pp. 187–191.

Srivastava, R. C. (1971). Size distribution of raindrops generated by their breakup and coalescence. *J. Atmos, Sci.* **28**, 410–415.

Srivastava, R. C., and Atlas, D. (1974). Effect of finite radar pulse volume on turbulence measurements. *J. Appl. Meteorol.* **13**, 472–480.

Srivastava, R. C., Jameson, A. R., and Hildebrand, P. H. (1979). Time-domain computation of mean and variance of Doppler spectra. *J. Appl. Meteorol.* **18**, 189–194.

Stackpole, J. D. (1961). The effectiveness of raindrops as turbulence sensors. *Proc., Weather Radar Conf., 9th, 1961* pp. 212–217.

Stapor, D. T., and Pratt, T. (1984). A generalized analysis of dual-polarization measurements of rain. *Radio Sci.* **19**, 90–98.

Steinhorn, I. and Zrnić, D. S. (1988). Potential uses of the differential propagation phase constant to estimate raindrop and hailstone size distributions. *IEEE Trans. Geosci. and Remote. Sensing.* **26**, 639–648.

Stoeffler, R. C. (1972). Additional Research on Instabilities in Atmospheric Flow, NASA Rep. CR-1985, April.

Stratton, J. A. (1941). "Electromagnetic Theory." McGraw-Hill, New York.

Strauch, R. G. (1988). A modulation waveform for short-dwell-time meteorologicl radars. *J. Atmos. Oceanic Tech.* **5**, 512–520.

Strauch, R. G., Merritt, D. A., Moran, K. P., Earnshaw, K. B., and van de Kamp, J. D. (1984). The Colorado wind-profiling network. *J. Atmos. Oceanic Tech.* **1**, 37–49.

Swingle, D. M., (1990). Weather radar in the United States Army's Fort Monmouth Laboratories. Ch. 2 *In* "Radar in Meteorology." (D. Atlas, ed.), Am. Met. Soc., Boston, pp. 7–15.

Swords, S. S., (1986). Technical history of the beginnings of RADAR. Peter Peregrinus Ltd. London, UK. 325 pp.

Szymanski, E. W., Szymanski, S. J., Holmes, C. R., and Moore, C. B. (1980). An observation of a precipitation echo intensification associated with lightning. *J. Geophys. Res.* **85**, 1951–1953.

Tatarskii, V. I. (1961). "Wave Propagation in a Turbulent Medium." McGraw-Hill, New York.

Tatarskii, V. I. (1971). "The Effects of the Turbulent Atmosphere on Wave Propagation" (translation from Russian edition by the Israel Program for Scientific Translations Ltd. IPST Cat. No. 5319). (Available from the U.S. Dept. of Commerce, UDC 551.510, ISBn 07065 0680 4, Natl. Tech. Inf. Serv., Springfield, Virginia.)

Taylor, J. W., and Mattern, J. (1970). Receivers. *In* "Radar Handbook" (M. I. Skolnik, ed.) Chapter 5, McGraw-Hill, New York.

Telford, J. W. (1955). A new aspect of coalescence theory. *J. Meteorol.* **12**, 436–444.

Tennekes, H., and Lumley, J. L. (1972). "A First Course in Turbulence." M.I.T. Press, Cambridge, Massachusetts.

Tesla, N. (1900). The problem of increasing human energy. *Century Magazine* June, **LX**. p. 175–211.

Testud, J., Breger, G., Amayenc, P., Chong, M., Nutten, B., and Sauvageot, A. (1980). A Doppler radar observation of a cold front: Three-dimensional air circulation, related precipitation system and associated wavelike motions. *J. Atmos. Sci.* **37**, 78–98.

Thrope, S. A. (1969). Experiments on the stability of stratified shear flows. *Radio Sci.* **12**, 1327–1331.

Thorpe, S. A. (1973). Turbulence in stably stratified fluid: A review of laboratory experiments. *Boundary Layer Meteorol.* **5**, 95–119.

Torlaschi, E., Humphries, R. G., and Barge, B. L. (1984). Circular polarization for precipitation measurement. *Radio Sci.* **19**, 193–200.

Torp, H. (1992). Signal processing in real-time, two-dimensional Doppler color flow mapping. Dissertation, University of Trondheim, Trondheim, Norway.

Tragl, K. (1990). Polarimetric radar backscattering from reciprocal random targets. *IEEE Trans. Geosci. and Remote Sensing* **28**, 856–864.

Tretter, S. A. (1976). "Introduction to Discrete-Time Signal processing." Wiley, New York.

Trout, D., and Panofsky, H. A. (1969). Energy dissipation near the tropopause. *Tellus* **21**. 355–358.

Tuttle, J. D., and Foote, G. B. (1990). Determination of the boundary layer airflow from a single doppler radar. *J. Atmos, and Oceanic Tech.* **7**, 218–232.

Tuttle, J. D., and Rinehart, R. E. (1983). Attenuation correction in dual-wavelength analysis. *J. Climate Appl. Meteorol.* **22**, 1914–1921.

Ulbrich, C. W. (1983). Natural variations in the analytical form of the raindrop-size distribution. *J. Climate Appl. Meteorol.* **22**, 1764–1775.

Ulbrich, C. W., and Atlas, D. (1984). Assessment of the contribution of differential polarization to improved rainfall measurements. *Radio Sci.* **19**, 49–57.

Uman, M. A. (1987). "The lighting discharge," Academic Press, Orlando FL., 377 pp.

Uyeda, H., Zrnić, D. S. (1986). Automatic detection of gust fronts. *J. Atmos. and Oceanic Tech.* **3**, 36–50.

Vasiloff, S. V., Brandes, E. A., Davies-Jones, R. P., and Ray, P. S. (1986). An investigation of the transition from multicell to supercell storms. *J. Clim. and Appl. Meteorol.* **25**, 1022–1036.

Vaughn, C. R. (1985). Birds and insects as radar targets: a review. *Proc. IEEE* **73**, 205–227.

Vinnichenko, N. K., and Dutton, J. A. (1969). Empirical studies of atmospheric structure and spectra in the free atmosphere. *Radio Sci.* **4**, 1115–1126.

Wait, J. R. (1962). "Electromagnetic Waves in Stratified Media." Macmillan, New York.

Wakasugi, K., Misutani, A., Matsuo, M., and Kato, S. (1986). A direct method for deriving dropsize distribution and vertical air velocities from VHF Doppler radar spectra. *J. Atmos. Oceanic Tech.* **3**, 623–629.

Wakasugi, K., Misutani, A., Matsuo, M., Fukao, S., and Kato, S. (1987). Further discussion on deriving dropsize distribution and vertical air velocities from VHF Doppler radar spectra. *J. Atmos. Oceanic. Tech.* **4**, 170–179.

Wakimoto, R. M., and Martner, B. E. (1992). Observations of a Colorado tornado. Part II: Combined photogrammetric and Doppler radar analysis. *Mon. Wea. Rev.* **120**, 522–543.

Wakimoto, R. M., and Wilson, J. W. (1989). Non-supercell tornadoes. *Mon. Wea. Rev.* **117**, 1113–1140.

Waldteufel, P. (1973). Attenuation des ondes hyperfrequence par la pluie: Une mise au point. *Ann. Tellecommun.* **28**, 255–272.

Waldteufel, P. (1976). An analysis of weather spectra variance in a tornadic storm. NOAA Technical Memorandum ERL-NSSL-76.

Waldteufel, P., and Corbin, H. (1979). On the analysis of single Doppler data. *J. Appl. Meteorol.* **18**, 532–542.

Waldvogel, A. (1974). The N_0 jump of raindrop spectra. *J. Atmos. Sci.* **31**, 1067–1078.

Waldvogel, A., Federer, B., and Grimm, P. (1979). Criteria for the detection of hail cells. *J. Appl. Meteorol.* **18**, 1521–1525.

Wang, K.-C., and Tseng, H.-Y. (1991). A case study of the circulation structure of typhoon Alex by using a single Doppler weather radar. *Prepr., Radar Meteorol. Conf., 24th, 1991*, pp. 537–544.

Watson, A. I., and Blanchard, D. O. (1984). The relationship between total area divergence and convective precipitation in south Florida. *Mon. Weather Rev.* **112**, 673–685.

Watson-Watt, Sir Robert (1957). "Three Steps to Victory." Odhams Press Ltd, Long Acre London, England. 480 pp.

Watson-Watt, R. A., Bainbridge-Bell, L. H., Wilkins, A. F., and Bowen, E. G. (1936). Return of radio waves from the middle atmosphere. *Nature* **137**, p. 866.

Weinstock, J. (1981). Using radar to estimate dissipation rates in thin layers of turbulence. *Radio Sci.* **16**, 1401–106.

Weisman, M. L., and Klemp, J. B. (1982). The effects of directional turning of the low level wind shear vector on modeled multicell and supercell storms. *Prepr. Severe Local Storms Conf., 12th 1982* pp. 528–531.

Westwater, R. R., and Grody, N. C. (1980). Combined surface- and satellite-based microwave temperature profile retrieval. *J. Appl. Meteorol.* **19**, 1438–1444.

Wexler, R., and Atlas, D. (1963). Radar reflectivity and attenuation of rain. *J. Appl. Meteorol.* **2**, 176–280.

Whalen, A. D. (1971). "Detection of Signals in Noise." Academic Press, New York.

Wheelon, A. D. (1959). Radio wave scattering by tropospheric irregularities. *J. Res. Natl. Bur. Stand., Sect. D.* **63**, 205–233.

Whitham, G. B. (1974). "Linear and Nonlinear Waves." John Wiley, New York, 636 pp.

Williams, E. R., Geotis, S. G., and Bhattacharya, A. B. (1988). A radar study of the plasma and geometry of lightning. *J. Atmos. Sci.* **46**, 1173–1185.

Wilson, J. W. (1970). Integration of radar and gage data for improved rainfall measurement. *J. Appl. Meteorol* **9**, 489–497.

Wilson, J. W. (1986). Tornadogenesis by nonprecipitation induced wind shear lines. *Mon. Wea. Rev.* **114**, 270–284.

Wilson, J. W., and Brandes, E. A. (1979). A radar measurement of rainfall—A summary. *Bull. Am. Meteorol. Soc.* **60**, 1048–1058.

Wilson, J. W., and Reum, D. (1988). The flare echo: Reflectivity and velocity signature. *J. Atmos, Oceanic Tech.* **5**, 197–205.

Wilson, J. W., and Schreiber, W. E. (1986). Initiation of convective storms at radar-observed boundary-layer convergence lines. *Mon. Weather Rev.* **114**, 2516–2536.

Wilson, J. W., Roberts, R. D., Kessinger, C., and McCarthy, J. (1984). Microburst structure and evaluation of Doppler Radar for airport wind shear detection. *J. Climate and Appl. Meteorol.* **23**, 898–915.

Wood, V. T., and Brown, R. A. (1983). Single Doppler velocity signatures: An atlas of patterns in clear air/widespread precipitation and convective storms. *NOAA Tech. Memo ERL NSSL-95*. NOAA Environmental Research Laboratories, Norman, Oklahoma.

Woodley, W. L., Olsen, A. R., Herndon, A., and Wiggert, V. (1975). Comparison of gage and radar methods of convective rain measurement. *J. Appl. Meteorol.* **14**, 909–928.

Woodman, R. F., ad Hagfors, T. (1969). Methods for the measurement of vertical ionospheric motions near the magnetic equator by incoherent scattering. *J. Geophys. Res. Space Phys.* **75**, 1205–1212.

Woods, J. D., Hogström, V., Misme, P., Ottersten, H., and Phillips, O. M. (1969). Fossil turbulence. *Radio Sci.* **4**, 1365–1367.

Wuertz, D. B., Weber, B. L., Strauch, R. G., Frisch, A. S., Little, C. G., Merritt, D. A., Moran, K. P., and Welsh, D. C. (1988). Effects of precipitation on UHF wind profiler measurements. *J. Atmos. Oceanic Tech.* **5**, 450–465.

Xu, M., and Gal Chen, T. (1993). Study of the convective boundary layer dynamics using single Doppler radar measurements. *J. Atmos. Sci.* [in press].

Yos, J. M. (1963). Transport properties of Nitrogen, Hydrogen, Oxygen, and Air to 30,000 K. AVCO Corp. Tech. Mem., RAD-TM-63-7.

Žáček, A. (1924). Nová metoda k vytvoření netlumených oscilací. *Casopis Piatovani. Math. Fys.* **53**, p. 378.

Zawadzki, I. (1981). The quantitative interpretation of weather radar measurements. *Prepr., Radar Meteorol. Conf., 20th, 1981* pp. 586–587.

Zeoli, G. W. (1971). IF versus video limiting for two-channel coherent signal processors. *IEEE Trans. Inf. Theory* **IT-17**, 579–587.

Zhang, D. L., and Fritsch, J. M. (1988). Numerical simulation of the Meso-β scale structure and evolution of the 1977 Johnstown flood. Part III: Internal gravity waves and the squall line. *J. Atmos, Sci.* **45**, 1252–1268.

Ziegler, C. L. (1978). "A Dual-Doppler Variational Objective Analysis as Applied to Studies of Convective Storms," NOAA Tech. Memo. ERL NSSL-85. Natl. Severe Storms Lab., Norman, Oklahoma. Available Natl. Tech. Inf. Serv., Operations Div., Springfield, Virginia.

Ziegler, C. L., Ray, P. S., and Knight, N. C. (1983). Hail growth in an Oklahoma multicell storm. *J. Atmos. Sci.* **40**, 1768–1791.

Ziegler, C. L., MacGorman, D. R., Dye, J. E., and Ray, P. S. (1991). A model evaluation of non-inductive graupel-ice charging in the early electrification of a mountain storm. *Jour. Geoph. Res.* **96**, 12833–12855.

Zrnić, D. S. (1975a). Moments of estimated input power for finite sample averages of radar receiver outputs. *IEEE Trans. Aerosp. Electron. Syst.* **AES-11**, 109–113.

Zrnić, D. S. (1975b). Simulation of weather-like Doppler spectra and signals. *J. Appl. Meteorol.* **14**, 619–620.

Zrnić, D. S. (1977a). Mean power estimation with a recursive filter. *IEEE Trans. Aerosp. Electron. Syst.* **AES-13**, 281–289.

Zrnić, D. S. (1977b). Spectral moment estimates from correlated pulse pairs. *IEEE Trans. Aerosp. Electron. Syst.* **AES-13**, 344–354.

Zrnić, D. S. (1975c). Signal to noise ratio at the output of nonlinear devices. *IEEE Trans. Inf. Theory.* **IT-21**, 662–663.

Zrnić, D. S. (1979a). Estimation of spectral moments for weather echoes. *IEEE Trans. Geosci. Electron.* **GE-17**, 113–128.

Zrnić, D. S. (1979b). Spectrum width estimates for weather echoes. *IEEE Trans. Aerosp. Electron. Syst.* **AES-15**, 613–619.

Zrnić, D. S. (1980). Spectral statistics for complex colored discrete-time sequence. *IEEE Trans. Acoust. Speech. Signal Process.* **ASSP-28**, 596–599.

Zrnić, D. S. (1987). Three-body scattering produces precipitation signature of special diagnostic value. *Radio Sci.* **22**, 76–86.

Zrnić, D. S. (1991). Complete polarimetric and Doppler measurements with a single receiver radar. *J. Atmos. and Oceanic Tech.* **8**, 159–165.

Zrnić, D. S., and Balakrishnan, N. (1990). Dependence of reflectivity factor–rainfall rate relationship on polarization. *J. Atmos. and Oceanic Tech.* **7**, 792–795.

Zrnić, D. S., and Doviak, R. J. (1975). Velocity spectra of vortices scanned with a pulse-Doppler radar. *J. Appl. Meteorol.* **14**, 1531–1539.

Zrnić, D. S., and Doviak, R. J. (1976). Effective antenna pattern of scanning radar. *IEEE Trans. Aerosp. Electron. Syst.* **AES-12**, 551–555.

Zrnić, D. S., and Doviak, R. J. (1978). Matched filter criteria and range weighting for weather radar. *IEEE Trans. Aerosp. Electron. Syst.* **AES-14**, 925–930.

Zrnić, D. S., and Doviak, R. J. (1989). Effect of drop oscillations on spectral moments and differential reflectivity measurements. *J. Atmos. and Oceanic Tech.* **6**, 532–536.

Zrnić, D. S., and Hamidi, S. (1981). Considerations for the design of ground clutter cancellers for weather radar. Interim Rep., Systems Research and Development Service, Rep. No. DOT/FAA/RD-81/72, 77 pp. [Available from NTIS, Springfield, VA 22151.]

Zrnić, D. S., and Istok, M. J. (1980). Wind speeds in two tornadic storms and a tornado, deduced from Doppler spectra. *J. Appl. Meteorol.* **19**, 0065–0075.

Zrnić, D. S., and Mahapatra, P. (1985). Two methods of ambiguity resolution in pulse Doppler weather radars. *IEEE Trans. Aerosp. Electron. Syst.* **AES-21**, 470–483.

Zrnić, D. S., Doviak, R. J., and Burgess, D. W. (1977). Probing tornadoes with a pulse Doppler radar. *Q. J. R. Soc.* **103**, 707–720.

Zrnić, D. S., Hamidi, S., and Zahrai, A. (1982). Considerations for the design of ground clutter cancelers for weather radar. Final Rep., Systems Research and Development Service, Rep. No. DOT/FAA/RD-82/68.

Zrnić, D. S., Rust, W. D., and Taylor, W. L. (1982). Doppler radar echoes of lightning and precipitation at vertical incidence. *J. Geophys. Res.* **87**, 7179–7191.

Zrnić, D. S., Doviack, R. J., and Mahapatra, P. R. (1984). The effect of charge and electric field on the shape of raindrops. *Radio Sci.* **19**, 75–80.

Zrnić, D. S., Burgess, D. W., and Hennington, L. (1985). Doppler spectra and estimated windspeed of a violent tornado. *J. Climate Appl. Meteorol.* **24**, 1068–1081.

Zrnić, D. S., Burgess, D. W., and Hennington, L. D. (1985). Automatic detection of mesocyclonic shear. *J. Atmos. Oceanic Tech.* **2**, 425–438.

Zrnić, D. S., Smith, S. D., Witt, A., Rabin, R. M., and Sachidananda, M. (1986). "Wind profiling of stormy and quiescent atmospheres with microwave radars," NOAA Tech. Memo. ERL NSSL-98, National Severe Storms Laboratory.

Index

Advection velocities, wind measurement, 289
Air density perturbations, wind field synthesis, 290–291
Aircraft, radar detection of, 3–4
Aliasing range/velocity ambiguities, 61
Alternating polarizations, Doppler spectrum
  coherent polarimetric radar, 147–149
  signal processing schematic, 157
Amplitude imbalance, Doppler radar artifacts, 190–192
Analog-to-digital (A/D) converters
  Doppler radar artifacts, quantization and saturation, 188–190
  Doppler spectrum, weather signal processing, 124–125
Anelastic approximation, 291
Anisotropic irregularities, common volume scatter, 463–464
Anomalous propagation, 23–28
Antenna
  angular velocity, velocity spectrum, 116–118
  beamwidth, 34
  effective area, 45–46
  effective beamwidth, 435–436
  far field, 435–436
  gain, 34–35
  patterns
    scanning radars, 193–197
    WSR-88D, 200
  sidelobe echoes, 197–199
Area of deeper convection (ADC), single Doppler wind analysis, 320–322
Area time integral (ATI) algorithm, 226–228
Artifacts
  Doppler radar, 187–193
    amplitude and phase imbalance, 190–192
    phase jitter, 192–193
    quantization and saturation noise, 188–190

receiver nonlinearities, 187–188
Attenuation
  in clouds, 43–44
  cross section, 38
  in gases, 44–45
  index, 35
  in rain, 39–43
  single-parameter precipitation measurement, 231–233
  in snow, 44
  specific, 39
  weather radar and, 38–45
Autocorrelation function
  coefficient, 125
  Doppler radar artifacts, 187–188
  pulse compression, 185–187
  refractive index irregularities, 444–446
  turbulence measurement, Chandrasekhar's theory, 396–398
  weather signal spectral analysis, 94–95
    periodogram variance, 106
    random sequences, power spectrum, 95–98
Autocovariance processing
  coherent polarimetric radar, 148, 150–153
  mean frequency estimators, 131–134
  width estimation, 136–139
Automatic gain control (AGC), 124–125

Backscattering
  covariance matrix and polarimetric measurands, 240–243
  cross section, 35–38
    common airborne objects, 48
    lightning, 382
  matrix, multiple-parameter precipitation measurement, 239–240
  per unit volume, 73

Backscattering (*continued*)
oblate spheroids
matrix coefficients, 248–250
matrix reflectivity, 250–252
reflectivity, fair weather observations, 452
Band
bright, 256, 258
cloud, 369
convergence, 317–320, 357
electromagnetic wave, 11
Bandwidth
filter, 6dB, 58
noise, 54
Batch processing, 176
Beamwidth
effective, 196
one-way, 31B, 34
Bessel function
antenna pattern, 33–34
correlation turbulence measurement, 391
Bias, weather signal spectral analysis, 98–100
Birds, radar clutter, 206–207
Bistatic radar, 430
Bores, buoyancy waves, 365–367
Born approximation
Doppler radar attenuation, 38
fair weather observations, 429
Boundary layer profiles (BLP), 492–495
Bragg scatter, fair weather observations, 437–439
frozen irregularities, 438
radio acoustic sounding system (RASS), 440–443
Buoyancy waves
large-amplitude waves, 365–371
bores, 365–367
buoyancy wave effects, 369–371
solitary waves, 367–369
single-Doppler data, 363–371
VAD observation, 364–365

Calibration, receiver, 74
Canting, 249
Central limit theorem, weather signal statistics, 69–72
Chaff, 35, 219
Chain Home radar network, 3–4
Chandrasekhar's theory, turbulence measurement, 395–398
Cheng-English hail size distribution, 218
Z, R_H relation, 262–264

Circular convolution and correlation, discrete Fourier transform (DFT), 92–95
Circular depolarization ratio (CDR), 255–261
Circularly polarized radar, polarization diversity, 244–245
Clear air observations
convective boundary layer kinematics, 496–502
reflectivity, 479–486
wind profiling, 487–495
Doppler radar, 492–495
profilers, 487–492
Cloud-to-ground (CG) lightning flashes, 380
Clouds
anvil, 283–284, 295–296
arcus, 284, 287–288
cirrus, 503
cumulus, 304
cumulus humilis, 479
Doppler radar attenuation, 43–44
drop size distributions, 210–212
thunderstorm structure, 281–288
wall, 284–285
Clutter
Doppler radar, 49–50, 199–207
ground clutter suppression, 200–206
map, 206
spectrum width, 204–205
Coalescence, 210
raindrop size distributions, 212–215
Coherency, signal, 165–167
Coherent polarimetric radar, Doppler spectrum, 145–157
correlation coefficient, 155–157
mean radial velocity and differential propagation phase, 147–153
alternate polarization signals, 147–149
autocovariance processing, 148, 150–153
reflectivity and differential reflectivity, 146–147
specific differential phase, 153–155
Coherent receivers, 50–53
Cold fronts, Doppler radar, 378–380
Common volume scatter
fair weather observations, 452–464
correlation length comparable to or longer than Fresnel length, 455–460
correlation length shorter than Fresnel length, 453–455
spectral representation, 460–462
scattering cross section, 455
Condensation nuclei, cloud drop size distributions, 211–212

Constant altitude plan position indicators (CAPPI), 376–378
Continuous wave (cw)
  weather signals, receiver calibration, 75
Continuity equation, 291
Convective boundary layer (CBL)
  clear-air reflectivity, 479–486
  fair weather observation, refractive index, 471–472
  kinematic structure, clear air observations, 496–502
Convergence
  band, 317–320, 357
  cold fronts and dry lines, 378–380
  downdrafts and outflows, 357–359
Convolution, circular, 93
Coplan technique, wind field synthesis, 289–291
Correlation coefficient
  applications, 264–267
  coherent polarimetric radar, 155–157
  weather signals, 125
Correlation functions
  receiver output, 127–128
  turbulence measurement, 390–391
    Chandrasekhar's theory, 396–398
    structure function, locally isotropic fields, 393–394
Correlation length, common volume scatter
  comparable to or greater than Fresnel length, 455–460
  shorter than Fresnel length, 453–455
Correlator, 186
Covariance analysis
  multiple-parameter precipitation measurement, 240–243
  turbulence measurement, single radar, 405–407
Cramer–Rao bounds, Doppler spectrum, 141–142
Cressman interpolation filter, 290–291
Cross-correlation functions, clear air observations, 500–502
Cross section
  absorption, 38
  attenuation, 38
  extinction, 38, 231
  spheroid, 249
  total scattering, 38
Curvature formulas, spherically stratified atmospheres, 19–23
Cyclones. *See also* Mesoscyclones
  supercell thunderstorms, 294–295

Debye theory, precipitation measurement, 250–252
Defense technology, radar development and, 3–4
Deformations, 313
Delay
  radar, 65, 78
  transmitter, 78
Density currents
  buoyancy waves, 365–371
  characteristics, 357
  front, 355–356, 369
  storm structure
    downbursts and microbursts, 359–363
    downdrafts and outflows, 352–357
Detector
  linear, 126
  logarithmic, 126–127
  square law, 126–128
  synchronous, 31–32, 50–52
  video, 48
Differential phase (K_{DP})
  coherent polarimetric radar, 155
  ice and liquid hydrometeors, 261–264
  single-parameter precipitation measurement, 234–235
Differential phase (ϕ_{DP}), 147–153
  autocovariance processing, 148, 150–153
  phase shift on scattering, 50, 150, 264
Differential reflectivity (Z_{DR})
  combined measurements, 268–271
  Doppler spectrum, coherent polarimetric radar, 146–147
  ice and liquid hydrometeors, 255–261
  multiple-parameter precipitation measurement, 253–255
  spheroids, 250–252
Discrete Fourier transform (DFT)
  weather signal spectral analysis, 87–92
    bias, variance, and the window effect, 98–100
    convolution and correlation, 92–95
    power spectrum of random sequences, 95–98
Dissipation, turbulence measurement, 394
Distributed scatterers
  antenna sidelobes, 197–199
  range measurements, 58
  signal-to-noise ratio (SNR), 83
  weather signal sample, 65–67
    statistical properties, 69–72
Doppler radar
  acquisition time reduction, 179–184
    frequency diversity, 180
    random signal transmission, 181–184

Doppler radar (*continued*)
  ambiguities, 60–61
  antenna sidelobes, 197–199
  artifacts, 187–193
    amplitude and phase imbalance, 190–192
    phase jitter, 192–193
    quantization and saturation noise, 188–190
  attenuation, 38–45
    in clouds, 43–44
    in gases, 44–45
    in rain, 39–43
    in snow, 44
  bandwidth, 57–58
  clutter, 199–207
    ground clutter, 200–206
  coherent receivers, 50–53
  effective antenna patterns, 193–197
  filtered waveform, 58–60
  incoherent receivers, 48–50
  mitigation of ambiguities, 167–179
    aliased velocity correction, 177–179
    interlaced sampling, 175–177
    phase diversity, 167–170
    PRT staggering to increase unambiguous
      velocity, 171–175
    spaced pairs with polarization coding,
      170–171
  pulse compression, 184–187
  range ambiguities, 160–164
  receiving aspects, 45–53
    coherent receiver, 50–53
    incoherent receiver, 48–50
    radar equation, 46–48
  scattering cross section, 35–38
  schematic of, 30–31
  signal coherency, 165–167
  signal-to-noise ratio, matched filters, 60
  storm structure, 8
  system noise temperature, 54–57
  transmitting aspects, 30–35
    antenna gain, 34–35
    electromagnetic beam, 32–34
  velocity ambiguities, 164–165
  wind profiling, 492–495
Doppler processing
  coherent polarimetric radar, 145–157
    alternating polarization echoes, 157
    correlation coefficient, 155–157
    mean radial velocity and differential
      propagation phase, 147–153
    reflectivity and differential reflectivity,
      146–147

    specific differential phase, 153–156
    width estimation, 155
  mean frequency estimators, 130–135
    autocovariance processing, 131–134
    spectral processing, 135
  performance on data, 143–145
  signal power estimation, 125–130
    range time averaging, 129–130
    sample time averaging, 125–129
  spectral moment estimates, 122–123
  weather signal processing, 123–125
  width estimation, 136–141
    autocovariance processing, 136–139
    minimum variance bounds, 141–143
    spectral processing, 140–141
Doppler shift
  Doppler spectrum, coherent polarimetric radar,
    150–153
  fair weather observations, 440–443
  weather radars, 67
Downbursts, storm structure, 359–363
Downdrafts
  storm structure, 351–363
    convergence bands, 357–359
    density currents, 352–357
    downbursts and microbursts, 359–363
    thunderstorm structure, 284–288
  turbulence measurement, 414–422
Drizzle, 271
Drop diameter, equivalent volume, 248
Drop size distribution
  cloud drop sizes, 212
  Doppler radar, attenuation, 39
  hailstone sizes, 215–218
  multiple-parameter precipitation measurement
    dual polarization, 255
    dual wavelength, 238–239
  raindrop sizes, 212–215
  single-parameter precipitation measurement,
    224–226
    attenuation and, 232–233
    R, Z reflectivity factors, 224–226
Dry lines, 378–380
Dual polarization, 252–271
  combined measurements, 268–271
  correlation coefficient, 264–267
  ice and liquid hydrometeors
    reflectivity factors, 255–261
    reflectivity and specific differential phase,
      261–264
  linear depolarization ratio, 267–268
  rainfall rates, 252–255

Dual Doppler radar
  turbulence measurement, 407–408
  wind measurement, 288–304
    ordinary thunderstorms, 301–304
    supercell thunderstorms, 293–301
    wind field synthesis, 289–293
Dual radars
  clear air observations
    convective boundary layer (CBL), 496–502
  storm observations, 288–293
Dual wavelength
  multiple-parameter precipitation measurement,
    236–239
Dwell time, sample time averaging, 127–128
Dynamic range
  radar, 47, 124
  weather signals, 123

Earth, effective radius, 21
Echo power
  Doppler radar receiving, 45–53
    coherent receivers, 51–53
    incoherent receivers, 48–50
    radar equations, 46–48
  weather radar equation, 72–82
  weather signal statistics, 69–71
Eddy
  diffusion momentum, 467
  diffusion refractive index, 465, 475–477
  turbulence measurement
    Doppler spectrum width, 408–409
    single radar
      large-scale eddies, 404–405
      smaller than resolution volume, 405–407
Electromagnetic waves
  defined, 10–14
  phasor diagrams, 12–13
  polarization, 11–12
  power density, 13
  propagation, 10–28
  propagation paths, 14–28
    refractive index N, 16–18
    refractive index of air, 14–16
    schematic of, 10–11
    spatial dependence, 11–12
    spherically stratified atmospheres, 18–28
  spectrum, 11
Energy dissipation rate, 393
Ensemble average, 93
Equivalent earth models, spherically stratified
  atmospheres, 19–23

Equivalent number of independent samples, 127
Equivalent reflectivity factor, weather signals, 82
Errors
  Doppler radar, filter waveforms, 58–60
  wind field synthesis, two-Doppler radar data,
    291–293
Estimation theory, weather signal spectral analysis,
  98–100
Extinction cross section, Doppler radar attenuation,
  40–41

Fair weather observations
  clear air observations, 487–502
    reflectivity factor, 479–486
  common volume scatter, 452–464
    anisotropic irregularities, 463–464
    correlation length comparable to or larger than
      Fresnel length, 455–460
    correlation length shorter than Fresnel length,
      453–455
    spectral representation, 460–462
  convective boundary layer (CBL), 496–502
    dual Doppler radar, 496–502
  reflection, refraction and scatter, coherent radar,
    424–426
  reflectivity factor Z, 503–505
  refractive index
    detection criteria, 477–478
    inertial subrange, 475–477
    structure parameter, 464–478
    structure parameter, height dependence of,
      469–475
  wave equation for inhomogeneous and turbulent
    media, 426–430
    scattered fields, 430–435
  wind profiling, 487–495
    Doppler radar, 492–295
    profilers, 487–492
Fall speed, 275
False alarm radar, tornado warnings, 351
Fast Fourier transform (FFT). *See also* Discrete
  Fourier transform (DFT); Fourier transform
  Doppler spectrum
    mean frequency estimates, spectral processing,
      135
    weather signal spectral analysis, 92
      random sequences, 97–98
Filter
  bandwidth, 58
  Gaussian, 59, 76–77
  impulse response, 76

Filter (*continued*)
  matched, 60, 77
  propagation delay, 59
  response, 58–59
  WSR-88D, 47
Finite bandwidth power loss, 79–80
Flow models, single Doppler wind analysis,
    327–328
Fog particles, Doppler radar attenuation, 43–44
Forward-flank downdraft (FFD), 285–286
Fossil turbulence, 476
Fourier transform
  Doppler radar artifacts, 190–192
  Doppler spectrum
    minimum variance bounds, 142
    spectral processing, 140–141
    tornadoes, 344–351
    von Hann window, 143–145
  fair weather observations
    Bragg scatter, 437–439
    radio acoustic sounding system (RASS),
        440–441
  linear wind analysis, velocity azimuth display
      (VAD), 312–317
  spectral representation theory, 9
  turbulence measurement, 388–390
    spatial spectra, 398–403
  weather signal spectral analysis
    discrete Fourier transform (DFT), 88–92
    true spectrum, 100–105
  weather signals, 8
  wind profiling, 489–492
Frequency estimation, averaging time reductions,
    180
Fresnel
  lengths, 453
  term, 458
Frontal boundaries, clear air observations, 481–486

Gage/Radar (GIR) ratios, rainfall measurements,
    272–274
Gain
  directional, maximum, 34
  power, 58, 75
  receiver, 74
  system, 75
Gamma function, precipitation measurements,
    220–221
Gans' theory, 249
Gaseous absorption loss, Doppler radar attenuation,
    44–45

Gating circuits, weather radar, 64–67
Gaussian distribution, weather signal spectral
    analysis, 105–106
Gaussian filters
  matched, 60, 118
  sample correlation along range time, 84–85
Gaussian function
  Doppler radar, filtered waveforms, 58–60
  weather signal
    density function, 69–72
    finite bandwidth power loss, 79–80
    resolution volume, 80–81
    velocity spectrum, 117–118
Geometric optics approximation, 518–519
Graupel
  combined measurements, 268–271
  hailstone size distributions, 216–218
Green's function
  derivation, 520–521
  solution to the wave equation, 432–435
Ground-based ducts, spherically stratified
    atmospheres, 23–28
Ground clutter
  antenna sidelobes, 197–199
  suppression, 200–206
Gust front, 354

Hail
  dual polarization
    correlation coefficient, 265–267
    ice and liquid hydrometeors, 258–261
  ice and liquid hydrometeors
    Cheng–English distribution, 262–264
    dual polarization, 265–267
  multiple-parameter precipitation measurement,
      238–239
  single-parameter precipitation measurement
    R, Z relationships, 228–229
    signatures in reflectivity field, 229–231
  size distributions, 215–218
  terminal velocity, 219
Hamming window, weather signal spectral analysis,
    104–105
Histograms, ground clutter suppression, 205–206
Homodyne radar, 30
Hook, reflectivity, 328
Horizontal divergence, single Doppler wind
    analysis, 320–322
Horizontal winds
  large-scale wind, single Doppler wind analysis,
      323–325

Hurricanes, Doppler radar, 376–378
Hydrometeors
    ice and liquid, reflectivity factors, 255–261
    polarization diversity
        backscattering covariance matrix and
            polarimetric measurands, 240–243
        backscattering matrix, 239–240
        scattering cross section, 35–38
        signal statistics, 69–72
        size distribution from Doppler radar, 274–278
        weather radar equation, 72–82
Hygroscopic liquids, 211–212

Ice
    backscattering cross section, 36–38
    pellets, hailstone size distributions, 216–218
Image suppression, Doppler radar artifacts,
    190–192
Incoherent receivers, 48–50
Inertial subrange
    height dependence, 475–477
    turbulence measurement, 409
        structure function, locally isotropic field, 394
Inhomogeneous media, 426–430
In-phase and quadrature phase components
    sample time averaging, signal sample correlation,
        510–513
    weather signal
        samples, 113
        statistics, 71–72
Insects, radar clutter, 206–207
Instantaneous power, weather signals, 68–69
Interlaced sampling, range and velocity ambiguities,
    175–177
Intermediate frequency (if) circuit, 124–125
Intracloud (IC) flashes, 380
Inverse discrete Fourier transform (IDFT), 87–92
Inversion layers
    electromagnetic propagation paths, 17–18
    temperature, 469
Isodop surfaces
    mesocyclones and tornadoes, 335–338
    weather signal spectrum, 108–109
Isotropic fields, turbulence measurement,
    387–388
    spatial spectra, 399–403
    structure function, 393–394

Jitter probability density function, 193
Joint probability, 71

Kelvin-Helmholtz (K-H) waves
    growth rate, 468
    generation of reflectivity, 469
    radar observations, 479–486
Klystron amplifiers, Doppler radar operation, 30–31
Kolmogorov-Obukhov spectrum, turbulence
    measurement, 400–403
Kolmogorov's hypothesis, turbulence measurement
    eddy dissipation rate, 408–409
    structure function, locally isotropic fields,
        393–394

Lag window
    Doppler spectrum, 140–141
    weather signal spectral analysis, 100–105
Lagrangian operator, turbulence measurement,
    395–398
Laplace transform, Doppler radar artifacts, 187–188
Large-scale winds, single-Doppler radar
    horizontal wind, 323–325
    vertical wind, 325–327
Large weather systems, Doppler radar
    and, 371–380
    cold fronts and dry lines, 378–380
    hurricanes and typhoons, 376–378
    mesoscale convective systems (MCSs), 371–375
Lightning, Doppler radar, 380–383
    buoyancy, 382
    electron density, 382
    influences on hydrometeors, 382–383
    physical characteristics, 380–382
    reflectivity, 380
    spectra, 383
    storm structure and, 382–384
Linear depolarization ratio
    dual polarization, 267–268
Linear wind analysis, velocity azimuth display
    (VAD), 311–317
Liquid water density
    Doppler radar attenuation, clouds and fog, 43–44
    precipitation measurements, 220–221
Locally homogeneous fields, turbulence
    measurement, 392–393
Logarithmic amplifier-detector (LOG), 124–125
Longitudinal correlation, turbulence measurement,
    388–391
Lorentz–Lorenz formula, 15–16
Loss
    factor, 54
    transmission, 39
Low-noise amplifier (LNA) detector, 55–57

Magnetrons, radar technology, 4–7
Marshall–Palmer (M-P)
   raindrop size distribution, 213
   rainfall rate
      integrand, 222–223
      polarization effect, 253
      Z relation, 226
Master oscillator and power amplifier (MOPA),
   30–32
Matched filters, Doppler radar, 60
Maximum-likelihood estimation, Doppler spectrum,
   141–142
Maxwell's equations, 426–430
Mean frequency estimators, 130–135
   autocovariance processing, 131–134
   spectral processing, 135
Mean velocity
   estimate variance
      pulse pair, 133–134
      spectral processing, 135
   mesocyclones and tornadoes, vortex signature
      smoothing, 338–340
   radial velocity, coherent polarimetric radar,
      147–153
      alternate polarization signals, 147–149
      autocovariance processing, 148, 150–153
   weighting function, 109–110
   wind measurement
      two Doppler radars, 288–304
      wind field synthesis, 291–293
Median volume diameter, 220–221
Mesocale convective systems (MCSs), 371–375
Mesocyclones
   Doppler radar, 335–351
      vortex signature smoothing, 338–340
   signature track, 329–331
   supercell thunderstorms, 294–297
Microbursts
   reflectivity, 354
   storm structure, 361–363
Microwaves
   defined, 10–11
   rain/cloud penetration of, 13–14
Mie scattering
   attenuation, 40–41
   multiple-parameter precipitation measurement,
      236–239
Minimum detectable signal (MDS), 60
Minimum variance bounds
   mean velocity, 141–142
   spectrum width, 142
MTI (moving target indication) radars, 6

Near-ground convergence, single Doppler wind
   analysis, 371–322
Noise
   bandwidth, 54–55, 58
   Doppler radar. See Signal-to-noise ratio (SNR)
   power, system, 47, 55
   sky, 56–57
   temperature, 54–57
Nyquist frequency, 89–92
Nyquist limits, 131–134
   severe storm structure, 334–335
Nyquist wave number, 312–317

Oblate spheroids, backscattering matrix,
   248–252
Optimum spectral estimation, 106, 142
Orthogonal channels, polarization diversity,
   244–246
Outflows, storm structure, 351–363
   convergence bands, 357–359
   density currents, 352–357
   downbursts and microbursts, 359–363

Parseval's theorem, averaging time reductions,
   183–184
Periodogram
   weather signal spectral analysis
      bias and variance, 99–100
      random sequences, 98
      variance, 105–106
Perturbation analysis
   pulse pair estimator, 133–134
   width estimation, 137–139
Phase diversity, range and velocity ambiguities,
   167–170
Phase imbalance, Doppler radar artifacts, 190–192
Phase jitter, Doppler radar artifacts, 192–193
Phase shift, differential, 243–247
Phased array antennas, wind profiling, 487–488
Phasor, 11–13, 50–53
Phugoid frequency, 361
Plan position indicators (PPI)
   large-scale horizontal wind, 323–325
   mesocyclones and tornadoes, 336–337
   spherically stratified atmospheres, temperature
      inversion, 27–28
   storm structure
      downdrafts and outflows, 351–352
      severe storm structure, 328–329
      thunderstorm profile, 331–335

Planetary boundary layer (PBL), 407–408,
    496–502
Point scatterer, 49
Point velocities, turbulence measurement, 403–404
Polar molecules, electromagnetic propagation paths,
    14–16
Polarization diversity
    antenna sidelobes, 198–199
    multiple-parameter precipitation measurement,
        239–252
        backscattering matrix, 239–240
            coefficients and reflectivities for oblate
                spheroids, 248–252
            covariance matrix and polarimetric
                measurands, 240–243
            propagation effects, 243–247
        plane of polarization, 250
Polarized samples, range and velocity ambiguities,
    170–171
Power
    average transmitter, 34
    density, 13, 34
    density pattern, 34
    estimation, 127–128
    instantaneous, 68
    mean, 68
    peak, 184
    spectrum, 96
    weather signals, 67–69
        probability distribution, 72
Power law approximation, attenuation, 231–233
Power spectra
    tornadoes, 344–351
    weather signal spectral analysis
        independent meteorological processes,
            112–116
        probability distribution, 115–116
        random sequences, 95–98
        true spectrum, 100–105
        uniform shear and reflectivity, 110–112
Poynting vector
    expected scattered power density, 444–447
    scattering cross sections, 449
    waves, 13
Prandtl number, 476–477
Precipitation measurement
    drop size distribution, 210–218
        cloud drop size 210–212
        hailstone size, 215–218
        raindrop size, 212–215
    hydrometeor distribution from Doppler spectra,
        274–278

liquid water content, 220–221
multiple-parameter measurements, 235–274
    dual polarization, 252–271
        combined measurements, 268–271
        correlation coefficient, 264–267
        ice and liquid hydrometeors, 255–264
        linear depolarization ratio, 267–268
        rainfall rate estimates, 252–255
    dual wavelength, 236–239
    polarization diversity, 239–252
        backscattering coefficients and reflectivities,
            248–252
        backscattering matrices, 239–240
        covariance matrix and polarimetric
            measurands, 240–243
        propagation effects, 243–247
    rain gauge and radar, 271–274
    rainfall rate, 222–223
    reflectivity factor Z, 221–222
    single-parameter technique, 223–235
        attenuation method, 231–233
        differential phase method, 234–235
        reflectivity factor, 223–231
    terminal velocities, 218–220
Pressure dependence, refractivity N, 17–18
Prestorm data, single Doppler radar, 317–322
Probability density function, turbulent velocities,
    115–116
Probability of detection (POD), tornadoes, 351
Profilers
    boundary layer, 493
    temperature, 440–443, 489–491
    wind, 487–495
Propagation
    buoyancy waves, 365
    effects on polarization diversity, 243–247
    electromagnetic, 10–12
Pulse
    duration, 32
    modulator, 31
Pulse compression, Doppler radars, 184–187
Pulsed-Doppler radar, historical background,
    5–7
Pulse-modulated single-frequency radar, averaging
    time reductions, 181–184
Pulse pair processor
    coherent polarimetric radar, 152–153
    mean frequency estimators, 131–134
    spectrum width estimators, 136–139
Pulse repetition frequencies (PRFs), 61
Pulse repetition time (PRT), 32
    coherent receivers, 52–53

Pulse repetition time (PRT) (*continued*)
  range and velocity ambiguities
    interlaced sampling, 175–177
    phase diversity, 167–170
    spaced pairs, 170–171
    staggering, 171–175
Pulse techniques, radar technology, 3–4

Quadrature components
  correlation, 511–513
  weather signal spectral analysis
    periodogram variance, 105–106
    signal statistics, 71–72
Quadrature phase signal, weather signal spectrum,
    113–115
Quantization noise
  Doppler radar artifacts, 188–190
  weather signal spectral analysis, 97–98

Radar
  amplitude and phase fluctuations, 8–9
  delay, 65, 78–79
  detection capability, historical background, 1–2
  environment, defined, 6–7
  international research on, 4
  microwave capability
    historical background, 1–7
    overview, 1
  ranging capabilities, 2
  research methodology, 7–9
  theory, 7–8
Radar equation
  Doppler radar receiving aspects, 46–48
  weather signals, 72–82
    finite bandwidth power loss, 79–80
    range-weighting function, 75–79
    receiver calibration, 74–75
    reflectivity factors, 82
    resolution volume, 80–81
Radar-estimated rainfall (RER), 224–226
Radial velocities
  single Doppler wind analysis, 325–327
  storm structure, downdrafts and outflows,
    351–352
  weather signal spectral analysis, 106–116
    velocity spectrum, 117–118
  wind field synthesis, 290–291
Radial velocity shear, turbulence measurement,
    410–422

Radio Acoustic Sounding System (RASS)
  fair weather observations, 424, 440–443
    continuous acoustic pulse, 441–443
    short acoustic pulse, 440–441
  wind profiling, 490–492
Radio frequency (rf) filter, 124–125
Radio waves. *See* Electromagnetic waves
Radiosonde, radar detection of, 3–4
Radome losses, 56–57
Raindrops, size distributions, 212–215
Rainfall estimation
  area time integral, 226–228
  attenuation methods, 231
  differential phase methods, 234
  dual polarization methods, 239
  dual wavelength, 236
  errors, 224–226
  gauge, 271–274
  multiple parameter, 235
  single parameter, 223
Random phase techniques, 168–170
Random signal transmission, reflectivity and
    velocity estimations, 181–184
Random variables
  structure functions, locally homogeneous fields,
    391–393
  turbulence measurement, 386–391
Range ambiguities
  Doppler radar, 60–61
  thunderstorm cells, 160–164
Range height indicator (RHI)
  clear-air reflectivity, 481–486
  fair weather observation, refractive index,
    471–473
  severe storm structure, 334–335
Range time averaging
  incoherent receivers, 48–49
  weather signal power estimation, 129–130
  weather signals, sample correlation, 84–85
Range-weighting function, weather signals
  finite bandwidth power loss, 79–80
  radar equation, 75–79
Rankine vortex
  mesocyclones and tornadoes
    Doppler radar, 335–336
    smoothing of signature, 338–340
  single Doppler wind analysis, 327–328
  tornadoes, Doppler spectra, 344–351
Ray paths
  integral solutions for, 507–508
  spherically stratified atmospheres
    circular paths, 19–23

ground-based ducts, reflection height, 23–28
Rayleigh approximation
    attenuation, 40–42
    dual polarization correlation coefficient, 264–267
    scattering cross section, 36–38
Rayleigh–Gans theory, 243
Rayleigh probability density, weather signal
    statistics, 71–72
Rayleigh scattering
    multiple-parameter precipitation measurement,
        236–239
        coefficients and reflectivities, oblate spheroids,
            249–252
    single-parameter precipitation measurement,
        234–235
Rear-flank downdraft (RFD), 285–288
Receiver calibration, weather signals
    inputs and outputs, 77–79
    radar equation, 74–75
Receiver-detector combinations, Doppler spectrum
    signal power estimation
    range time averaging, 129–130
    sample time averaging, 126–128
Rectangular windows, weather signal spectral
    analysis, 101, 103–104
Recursive filters, ground clutter suppression,
    202–206
Reflectivity, 73, 250
    averaging time reductions, 179–184
        frequency diversity, 180
        random signal transmission, 181–184
    clear-air observations, 479–486
    differential, 250
    Doppler spectrum
        coherent polarimetric radar, 146–147
        mean frequency estimators, 130–135
        spectral moments, 122–123
        von Hann windows, 144–145
    downdrafts and outflows
        convergence bands, 358–359
        density currents, 355–356
        storm structure, 363
    factor Z
        equivalent, 82
        liquid water content, 221–222
    fair weather observations, 424–426
        refractive index, 469–475
        scattering cross sections, 449–452
        Z factor, 503–505
    ground clutter suppression, 206
    ice and liquid hydrometeors, 261–264
        dual polarization, 255–261

reflectivity and specific differential phase,
    261–264
lightning and, 380–382
mesocyclones and tornadoes, 335–337
mesoscale convective systems (MCSs), 372–375
multiple-parameter precipitation measurement,
    237–239
    backscattering coefficients, oblate spheroids,
        250–252
polarization diversity, 245–247
radar clutter, 206–207
severe storm structure, 328–335
single-parameter precipitation measurement,
    223–231
    area time integral, 226–228
    hail signatures, 229–231
    R, Z relations, 223–226
    snow and hail R, Z relationships, 228–229
    supercell thunderstorms, 293–294
    wind fild synthesis, 299–301
tornadoes, 340–344
    Doppler spectra, 347–351
turbulence measurement, 396–398
weather signal spectrum, 106–112
    power spectrum, 110–112
    radar equation, 82
wind field synthesis, 291–293
Refraction, fair weather observations, 424–426
Refractive index
    backscattering cross section, 35–38
    electromagnetic propagation paths, 14–16
    fair weather observation
        expected scattered power density, 443–452
        inertial subrange, 475–477
        inhomogeneous and turbulent media, 426–430
        irregularities, 430–435, 477–478
        radio acoustic sounding system (RASS),
            441–443
        structure parameter, 464–478
            height, 469–475
    spherically stratified atmospheres, 21–23
        ground-based ducts, reflection height, 23–28
Refractivity N
    electromagnetic propagation paths, 16–18
    spherically stratified atmospheres, 23–28
Relative permittivity, electromagnetic propagation
    paths, 16–18
Resolution volume
    antenna patterns 195–197
    common volume scatter, 456–460
    Doppler spectrum signal power estimation,
        128–129

Resolution volume (*continued*)
  echo correlation, 513–517
    azimuthally space volumes, 515–517
    range weighting, 75–79
    signal sample correlation vs. range difference,
      513–515
  tornadoes, Doppler spectra, 344–351
  turbulence measurement
    severe thunderstorms, 417–422
    single radar, eddies smaller than, 405–407
  weather signals, radar equation, 80–81
Reynolds numbers, 219–220
Richardson number, 467–470
R, *Z* relations, 223–226

Sample time averaging
  signal power estimation, 125–129, 125–130
    equivalent independent samples, 127–128
    resolution volume location, 128–129
    statistical properties from linear and nonlinear
      receiver detectors, 126–127
  signal sample correlation, 510–513
Sample time mean power, weather signals, 68–69
Saturation noise, Doppler radar artifacts, 188–190
Scalar field, turbulence measurement, 387–390
Scanning radar, antenna patterns, 193–197
Scatterers
  antenna sidelobes, 197–199
  point, 49
  range resolution, 58
  statistical properties, 94–95
  velocity, 113
Scattering
  common volume scatter, anisotropic
    irregularities, 463–464
  fair weather observations, 424–426
    Bragg scatter, 437–439
Scattering cross section
  backscattering, 35
  blobs, 454, 455
Scattering geometry
  common volume scatter, 454–455
  fair weather observations, 430–435
  multiple-parameter precipitation measurement,
    249–250
Severe storms, turbulence measurement,
  410–422
Shear. *See* Wind shear
Signal coherency, Doppler radar, 165–167
Signal power estimation, 125–130
  range time averaging, 129–130

sample time averaging, 125–129
  equivalent number of independent samples,
    127–128
  linear and nonlinear receiver-detectors,
    126–127
  resolution volume location shifts,
    128–129
Signal statistics, weather signals, 69–72
Signal-to-noise ratio (SNR)
  Doppler radar artifacts
    matched filters, 60
    quantization and saturation, 188–190
    system noise, 55–57
  Doppler spectrum
    coherent polarimetric radar, 147–149
    mean frequency estimators, 130–134
    minimum variance bounds, 141–142
    spectral processing, 140–141
    width estimation, 136–139
  fair weather observation, 477–478
  pulse compression, 185–187, 186–188
  weather signals, distributed scatterers, 83
  wind profiling, 488–492
    Doppler radars, 492–495
Single Doppler radar
  turbulence measurement
    eddies smaller than resolution volume,
      405–407
    large-scale eddies, 404–405
  wind measurement, 304–328
    flow models, 328
    large-scale horizontal winds, 323–325
    large-scale vertical winds, 325–327
    linear wind models, 306–310
    linear wind over a circle (VAD), 311–317
    prestorm applications, 317–322
    uniform wind, circular arc, 310–311
Sky noise temperature, 56–57
Small volume scatter, 435–452
  Bragg scatter, 437–440
  expected scattered power density, 443–452
  radio acoustic sounding system (RASS),
    440–443
Smoke plumes, 504
Snell's Law, 507
Snow
  attenuation, 44
  single-parameter precipitation measurement,
    228–229
  terminal velocity, 219–220
Spaced pairs technique, range and velocity
  ambiguities, 170–171

Spatial spectra
    turbulence measurement, 398–404
    weighting function filtering, 398–403
Specific differential phase (K_{DP})
    Doppler spectrum, coherent polarimetric radar,
        153–155
    ice and liquid hydrometeors, 261–264
Spectral broadening
    mechanisms, 112–115
    turbulence, 409
Spectral density factor, turbulence measurement,
    388–390
Spectral moment estimates, weather signals,
    122–123
Spectral processing
    coherent polarimetric radar, 148–149
    mean frequency estimates, 135
    spectrum width estimates, 140–141
Spectral representation, common volume scatter,
    460–462
Spherically stratified atmospheres, 18–28
    equivalent earth models, 19–23
    ground-based ducts, reflection height, 23–28
Squall lines, mesoscale convective systems (MCSs),
    373–375
Stabilized local oscillator (STALO)
    coherent receivers, 50–53
    Doppler radar operation, 30–32
Standard deviation (SD)
    differential phase, 151
    differential reflectivity, 146
    mean frequency, 131–135
    power, 127–130
    sample time averaging, 128
    specific differential phase, 155
    spectrum width, 136–141
Stationary signals
    weather signal spectral analysis, discrete Fourier
        transform (DFT), 94–95
    wind measurement, 289
Statistical theory, turbulence measurement,
    386–398
    Chandrasekhar's theory, 395–398
    spatial spectra, point and average velocities,
        398–408
        dual radars, 407–408
        single-radar parameters, 404–407
        velocity variances, 403–404
        weighting function filtering, 398–403
    spectra and correlation function, 386–391
    structure functions
        locally homogeneous fields, 391–393

    locally isotropic fields, 393–394
Storm data, Doppler radar observation, 280
    severe storms, 328–335
Storm structure
    downbursts and microbursts, 359–363
    downdrafts and outflows, 351–357
    hook reflectivity, 351–359
    hurricanes and typhoons, 376–378
    mesocyclones and tornadoes, 335–351
    mesoscale convective systems, 371–375
    ordinary cells, 301–304
    RHI, 334
    super cell, 282, 286, 293–301
Structure functions, turbulence measurement
    locally homogeneous fields, 391–393
    locally isotropic fields, 393–394
Supercell thunderstorms, two-Doppler radar data,
    293–301
Synchronous detectors, coherent receivers, 50–53
System noise temperature, 54–57

Tangential velocities, tornadoes, 340–344
Temperature. See Sky noise temperature
Temperature dependence, electromagnetic
    propagation paths, 17–18
Temperature inversion
    fair weather observation, 469
    spherically stratified atmospheres
        equivalent earth paths, 23
        spatial resolution, 25–28
Terminal Doppler weather radars (TDWR)
    fair weather observation, 473
    hurricane tracking, 378
    storm structure, 361
Terminal velocities, precipitation measurements,
    218–220
Thin line, reflectivity, 379
Three-body scattering signature, 229–231
Three-dimensional spectra, turbulence
    measurement, 388–391
Thunderstorms. See also Storm structure
    air mass storms, wind field synthesis, 301–304
    benefits of, 8
    Doppler radar observation, 281–288
    hailstone size distributions, 215–218
    mesoscale convective systems (MCSs), 371–375
    ordinary thunderstorms, 301–304
    range ambiguities, 160–164
    severe storm structure, 331–335
    supercell storms, 293–301
    turbulence measurement, 410–422

Time-averaged power density, 13
Time dependence
    electromagnetic fields, 12–13
    in phase, quadrature phase, 53
Tornadoes
    cyclones, 164
    Doppler radar, 335–351
        high-resolution images, 340–344
        power spectrum, 344–351
        vortex signature smoothing, 338–340
        warning performance accuracy, 328, 330
    spectral signature, 345–351
        window effects, 103, 105
    supercell thunderstorms
        in mesocyclones, 296–297
    vortex signature (TVS)
        Doppler vs. non-Doppler radar, 350–351
        mesocyclones and tornadoes, 336–337
Transmission matrix, 245
Transmit/receive (T/R) switch, 32
Transverse correlation function, 388–391
Tropopause, thunderstorm structure, 284–285
Troposphere
    electromagnetic propagation paths, 15–16
TSS, 345
Turbulence
    Doppler spectrum width
        contributions to, 408–409
        eddy dissipation rate, 408–409
        in severe thunderstorms, 410–422
    fair weather observation
        refractive index, 465–469
        wave equations, 426–430
    parameters
        from dual radar, 407–408
        from single radar, 404–407
    spatial spectra, point and average velocities,
        398–408
        velocity variances, 403–404
        weighting function, 398–403
    statistical measurements
        Chandrasekhar's theory, 395–398
        spectra and correlation function, 386–391
        structure functions
            locally homogeneous fields, 391–393
            locally isotropic fields, 393–394
    statistical theory, 386–398
    velocities, 115–116
Turbulent kinetic-energy dissipation rate, 417–422
TVS, 336
Typhoons, 376–378

UHF bands, fair weather observation, 474
Unambiguous
    range, 60
    velocity, 61
Undular bores
    buoyancy waves, 365–367
    solitary waves, 367–368, 370–371
Uniform block averaging, 127
Uniform shear, power spectrum, 110–112
Uniform wind analysis, single Doppler radar
    circular arc, 310–311
    VVP analysis, 318–320
Updrafts
    air mass thunderstorms, 304
    supercell thunderstorms, 297–301
    thunderstorm structure, 281–288
    turbulence measurement, severe thunderstorms,
        414–422

Variance, weather signal spectral analysis, 98–100
Vector field, turbulence measurement, 387–391
Vector wind estimation, dual-Doppler radar,
    292–293
Velocity
    ambiguities, 60–61
        aliases, 177–179
        PRT staggering, 171–175
        thunderstorms, 164–165
    distributions in tornadic storms, 165
    estimation
        averaging time reductions, 179–184
        frequency diversity, 180
        random signal transmissions, 181–184
        PRT staggering, unambiguous velocities,
            172–175
    mesocyclones and tornadoes, 335–338
    storm structure, downdrafts and outflows,
        355–357
    tornadoes, high-resolution single-Doppler radar,
        340–344
    turbulence measurement
        Chandrasekhar's theory, 396–397
        point and average variances, 403–404
        spatial spectra, 402–403
    up- and downdrafts, 275–278
    vertical velocity, 317–322
    weather signal spectrum, 108–109
        radial velocity, 109–116
Velocity azimuth display (VAD)
    buoyancy waves, 364–365

linear wind analysis over a circle, 311–317
range and velocity ambiguities, 179
single Doppler radar, wind measurement, 305–306
turbulence measurement, 404–407
Vertical large-scale winds, 325–327
Vertical wind error, 292–293
VHF bands, fair weather observation, 474
Voltage output, weather signals, 77–79
Volume velocity processing (VVP)
    prestorm applications, 318–322
    wind measurement, 305–306
von Hann (raised cosine window), 104
Vortex signature. *See also* Rankine vortex:
    Tornadoes vortex signature (TVS)
    mesocyclones and tornadoes, smoothing, 338–340

Wall clouds, thunderstorm structure, 284–285
Water spheres, backscattering cross section, 36–38
Wave equation
    fair weather observations, 426–430
    geometric optics approximation, 518–519
Weather radar. *See also specific weather topics*
    antenna-reflector, structure of, 32–34
    data performance, 143–145
    historical development of, 5–7
    mean frequency estimators, 130–135
        autocovariance processing, 131–134
        spectral processing, 135
    minimum variance bonds, 141–143
    polarimetric radar signal processing, 145–157
        correlation coefficient, 155–157
        mean radial velocity and differential propagation phase, 147–153
        reflectivity and differential reflectivity, 146–147
        specific differential phase, 153–155
        spectrum width, 155
    signal power estimation, 125–130
        range time averaging, 129–130
        sample time averaging, 125–129
    signal processing
        block diagram, 123–125
        spectral moments, 122–123
    spectrum width estimators, 136–141
        autocovariance processing, 136–139
        spectral processing, 140–141

Weather signals
    Doppler spectra
        bias, variance and window effect, 98–100
        contributions by independent meteorological processes, 112–115
        convolution and correlation, 92–95
        discrete Fourier transform, 87–92
        periodogram variance, 105–106
        power spectrum of random sequences, 95–98
        probability distribution, turbulent velocities, 115–116
        reflectivity and radial velocity fields, 106–116
        spectral estimates, 100–105
        velocity spectrum width, shear, turbulence, 116–118
        weather signal spectral analysis, uniform shear and reflectivity, 110–112
    power sample, 67–69
    radar equation, 72–82
        finite bandwidth power loss, 79–80
        range-weighting function, 75–79
        receiver calibration, 74–75
        reflectivity factors, 82
        resolution volume, 80–81
            range time correlation of samples, 84–85
    samples, 64–67
    signal coherency, 165–167
    signal statistics, 69–72
    signal-to-noise ratio, distributed scatterers, 83
Weighting function
    angular, 74, 80–81
    composite, 73, 107
    normalized, 109
    range-weighting function, 75–79
    turbulence measurement, spatial spectra, 398–403
    vortex signature smoothing, 338–340
    weather signal spectrum, 107–108
White noise, weather signal spectral analysis, 95–98
Width estimation
    Doppler spectrum, 136–141
        autocovariance processing, 136–139
        coherent polarimetric radar, 155
        spectral moments, 122–123
        spectral processing, 140–141
    turbulence measurement
        eddy dissipation rate, 408–409
        severe thunderstorms, 411–422

Wind field synthesis
  supercell thunderstorms, 297–301
  two-Doppler radar data, 289–293
    coplan, 289–291
    errors, 291–293
Wind measurement
  single Doppler radar, 304–328
    linear wind fields, 306–310
    linear wind over a circle (VAD), 311–317
    uniform wind along a circular arc, 310–311
    vertical wind, 335–336
  two Doppler radars, 288–304
    wind field synthesis, 289–293
      Coplan technique, 289–291
      errors, 291–293
Wind profiles
  clear air observations, 487–495
    profilers, 487–492
    weather radars, 492–495

coherent sum of weather signals, 130–131
hydrometeor distribution, 276–278
Wind shear
  storm structure, 361–363
  turbulence measurement, 411–422
Window effect, weather signal spectral analysis
  bias and variance, 98–100
  true spectrum, 100–105
  velocity spectrum, 116–118
WSR-57 radar, plan position indicator (PPI)
    display, 160–164
WSR-88D radar
  antenna pattern, 200
  antenna sidelobes, 198–199
  parameters, 47–48
  transmitted pulse sequences, 176–177

Zeroth moment, weather signals, 122–123